FINE STRUCTURE OF PROTEINS AND NUCLEIC ACIDS

BIOLOGICAL MACROMOLECULES

A Series of Monographs

SERIES EDITORS

SERGE N. TIMASHEFF and GERALD D. FASMAN

Graduate Department of Biochemistry
Brandeis University
Waltham, Massachusetts

VOLUME 1

Poly-α-Amino Acids: Protein Models for Conformational Studies
edited by Gerald D. Fasman

VOLUME 2

Structure and Stability of Biological Macromolecules
edited by Serge N. Timasheff and Gerald D. Fasman

VOLUME 3

Biological Polyelectrolytes
edited by Arthur Veis

VOLUME 4

Fine Structure of Proteins and Nucleic Acids
edited by Gerald D. Fasman and Serge N. Timasheff

In Preparation

Subunits in Biological Systems
edited by Serge N. Timasheff and Gerald D. Fasman

FINE STRUCTURE OF PROTEINS AND NUCLEIC ACIDS

Edited by
Gerald D. Fasman and Serge N. Timasheff

Graduate Department of Biochemistry
Brandeis University
Waltham, Massachusetts

1970

MARCEL DEKKER, INC., New York

MARCEL DEKKER, INC.
95 Madison Avenue, New York, New York 10016

LIBRARY OF CONGRESS CATALOG CARD NUMBER 77-134272

ISBN 0–8247–1669–8

PRINTED IN THE UNITED STATES OF AMERICA

PREFACE

In a continuation of the philosophy of the *Biological Macromolecules* series to bring together principles and approaches that cross various specific areas of biological research, the fourth volume presents four chapters dealing with both proteins and nucleic acids. Two chapters utilizing nuclear magnetic resonance discuss its application to studies of conformation. The chapter by C. C. McDonald and W. D. Phillips, entitled "Protein Magnetic Resonance Spectroscopy of Proteins," discusses the elucidation of structure and the interactions of small molecules in solution and provides insight into the binding of enzymes, cofactors, substrates and inhibitors by this technique. The chapter by P. O. P. Ts'o, entitled "Monomeric Units of Nucleic Acids–Bases, Nucleosides, and Nucleotides," is an exhaustive review of the physical-chemical properties of these monomeric units and their relation to nucleic acids, and utilizes data assembled from X-ray diffraction, polarizability, nuclear magnetic resonance, infrared and ultraviolet spectroscopy, fluorescence, optical rotatory dispersion, etc. Their individual properties and the interactions of these monomers build a solid foundation to the understanding of the conformation of the nucleic acids.

The nucleic acids are further explored in the chapter by J. Brahms and S. Brahms entitled, "Circular Dichroism of Nucleic Acids." The review presents both the theoretical and experimental results which allow the interpretation of the spectra and facilitate the correlation with macromolecular conformation. This brings together for the first time an evaluation of the theories and use of circular dichroism for nucleic acid studies.

The fourth chapter, entitled "Buried and Exposed Groups in Proteins" by M. J. Kronman and F. M. Robbins, discusses the use of the techniques of differential ultraviolet spectroscopy, in particular, solvent perturbation, fluorescence, nuclear magnetic resonance, infrared, spin labeling, chemical reactivity, hydrogen-ion binding, etc., as probes of the topography and molecular environment of proteins. Thus, the comprehensive literature review contained in this chapter, surveys the present state of our knowledge regarding the location of various residues in the three-dimensional structure of proteins.

The object of this volume as specified in the title, "Fine Structure of Proteins and Nucleic Acids," has been handled with great finesse by the individual authors and it is hoped that this volume will assist those interested in the conformational aspects of biological macromolecules.

Gerald D. Fasman
Serge N. Timasheff

CONTRIBUTORS TO THIS VOLUME

J. Brahms,* *Centre of Molecular Biophysics* (*C.N.R.S.*), *Orleans, France*

S. Brahms,* *Centre of Molecular Biophysics* (*C.N.R.S.*), *Orleans, France*

Martin J. Kronman,† *Pioneering Research Laboratory, U.S. Army Natick Laboratories, Natick, Massachusetts*

C. C. McDonald, *Central Research Department, Experimental Station, E. I. Dupont, de Nemours and Company, Wilmington, Delaware*

W. D. Phillips, *Central Research Department, Experimental Station, E. I. Dupont, de Nemours and Company, Wilmington, Delaware*

Frederick M. Robbins, *Pioneering Research Laboratory, U.S. Army Natick Laboratories, Natick, Massachusetts*

Paul O. P. Ts'o, *Department of Radiological Sciences, The Johns Hopkins University, Baltimore, Maryland*

* Present Address: Institut de Biologie, Faculté des Sciences de Paris, Paris, France.

† Present Address: Department of Biochemistry, Upstate Medical Center, State University of New York, Syracuse, New York.

CONTENTS

FINE STRUCTURE OF PROTEINS AND NUCLEIC ACIDS

CHAPTER I

PROTON MAGNETIC RESONANCE SPECTROSCOPY OF PROTEINS

C. C. McDonald and W. D. Phillips

CENTRAL RESEARCH DEPARTMENT
EXPERIMENTAL STATION
E. I. DUPONT DE NEMOURS AND COMPANY
WILMINGTON, DELAWARE

I. INTRODUCTION

A major current problem in biological research is the elucidation of the catalytic chemistry of enzymatic proteins in terms of their molecular structure. It is now clear that most enzymatic and electron transport proteins derive their highly specific chemical properties from the folding of the polypeptide chain and any pendant prosthetic groups into a unique conformation that brings specific functional groups together into a particular reactive geometry—the native, biologically active state. Thus, the problem, from the physical point of view, is to specify the spacial arrangement of all atoms in the native conformation of a protein and to assess any changes that occur in this spatial arrangement when the protein is perturbed by substrates, inhibitors, changes in environment, etc. Some such information has been obtained both by chemical and by physical techniques—perhaps most successfully by X-ray spectroscopy. The power of these techniques has steadily increased as they have become more sophisticated. Still, it is generally agreed that new complementary approaches could provide significant advances in the field of protein structure, and such tools are being actively sought and tested in many laboratories.

Nuclear magnetic resonance spectroscopy has been developed into a very powerful approach for elucidation of structures and interactions of small molecules in the liquid state. This branch of spectroscopy encompasses a number of different, specialized techniques. Many of these have been evaluated for their ability to provide information on protein structure, and some have proved quite useful in this respect. For example, studies of the relaxation rate of the protons of water in enzymatic systems involving a paramagnetic cofactor such as Mn^{2+} have provided a detailed description of the binding of enzyme, cofactor, and substrate and have, as well, contributed to the understanding of the kinetics of these processes (*1*). Another approach that is finding increasing application is the study of the high-resolution proton magnetic resonance (PMR) spectra of small molecule substrates and inhibitors as they are modified by binding to enzymes (*2*). Investigations involving spectroscopy of magnetic resonances of other nuclei, particularly fluorine (*3*), chlorine (*4*), and phosphorus (*5*), have also provided information on proteins.

The most general and potentially the most powerful NMR approach to structures and interactions of proteins is the direct examination of the resonances of the hydrogen atoms (protons) of the protein itself. Until recently, this technique was severely limited by insufficient sensitivity and resolution but pioneering studies (*6–9*) suggested its potential once these limitations were overcome.

The problem of sensitivity has been alleviated, though certainly not eliminated, by development of more sensitive PMR spectrometers and by improving the signal-to-noise output of spectrometers by computer averaging techniques. Currently, a practical lower concentration limit for adequate detection of a single proton in a native protein under favorable circumstances is about 1×10^{-3} *M*. However, it appears that this concentration limit may be reduced to about 1×10^{-4} *M* by the application of Fourier transform spectroscopy (*10*) that is now under development.

The problem of resolving individual proton resonances in the PMR spectrum of a native protein has been and remains a severe one. Considering the 20 different amino acids that commonly comprise proteins, a very wide variety of shielding environments is inevitable. Proton resonances in proteins occur at many different field positions, but these resonance positions are confined to a relatively narrow field region. Consequently, resonances frequently overlap. The degree of overlap is worsened by the rather broad resonance widths of proton resonances of native proteins (as compared to small molecules). This broadening arises because anisotropic dipole–dipole interactions between magnetic nuclei are not as completely averaged out for rigid, slowly rotating large molecules as they largely are by the more rapid tumbling of small molecules. Both the multiplicity of resonances and resonance linewidths generally increase so that resonances overlap more severely as the size of the molecule under consideration increases. One approach to improving resolution of overlapped resonances is to employ higher resonance fields since chemical shifts are linear functions of this parameter. There thus has been considerable incentive to develop PMR spectrometers that operate at higher fields and frequencies to improve effective resolution. Furthermore, gains in sensitivity are to be expected as spectrometer frequency is increased. Thus, over the years the fixed frequency commonly used in commercial PMR spectrometers was advanced from 40 to 60 MHz, then to 100 MHz, and most recently to 220 MHz. Ability to resolve PMR spectra of proteins in aqueous solution and to extract useful information from these spectra has improved concomitantly with these technological developments (*11–26*). The great potential power of the technique where resolution of individual resonances is possible has been demonstrated in studies at 100 MHz wherein it has been possible to completely resolve and assign C-2 resonances of histidine in small proteins such as ribonuclease and lysozyme and to detect changes in the spectra of these proteins with pH and by addition of inhibitors or certain metal ions (*23,24*). At 220 MHz, extended regions of the PMR spectra of low molecular weight proteins (MW $< 25{,}000$) become sufficiently resolved to provide access to detailed information on protein structure and the critical dependence of the characteristics of the PMR spectra on protein conformation becomes more manifest (*20,21,26*). Thus, it appears that high-resolution PMR spectroscopy has reached a stage in its development where it can be used productively to provide new information on structures and interactions of proteins in aqueous solution. In this chapter, we discuss some of the characteristics of PMR spectra of proteins (particularly as observed at 220 MHz), information that can be obtained from these spectra, and some probable future developments in this field.

II. PMR SPECTRA OF RANDOM-COIL PROTEINS

If PMR spectra of native proteins in neutral aqueous solution are to provide information on the folded native structure, a prerequisite is that the characteristics of the spectra are responsive to elements of secondary and tertiary structure. The compelling promise of the PMR technique arises from observations that, in general, this requirement is amply met—characteristics of spectra of native proteins are usually markedly different from those of the corresponding random-coil proteins where the elements of secondary and tertiary structures are absent. A typical example is shown in

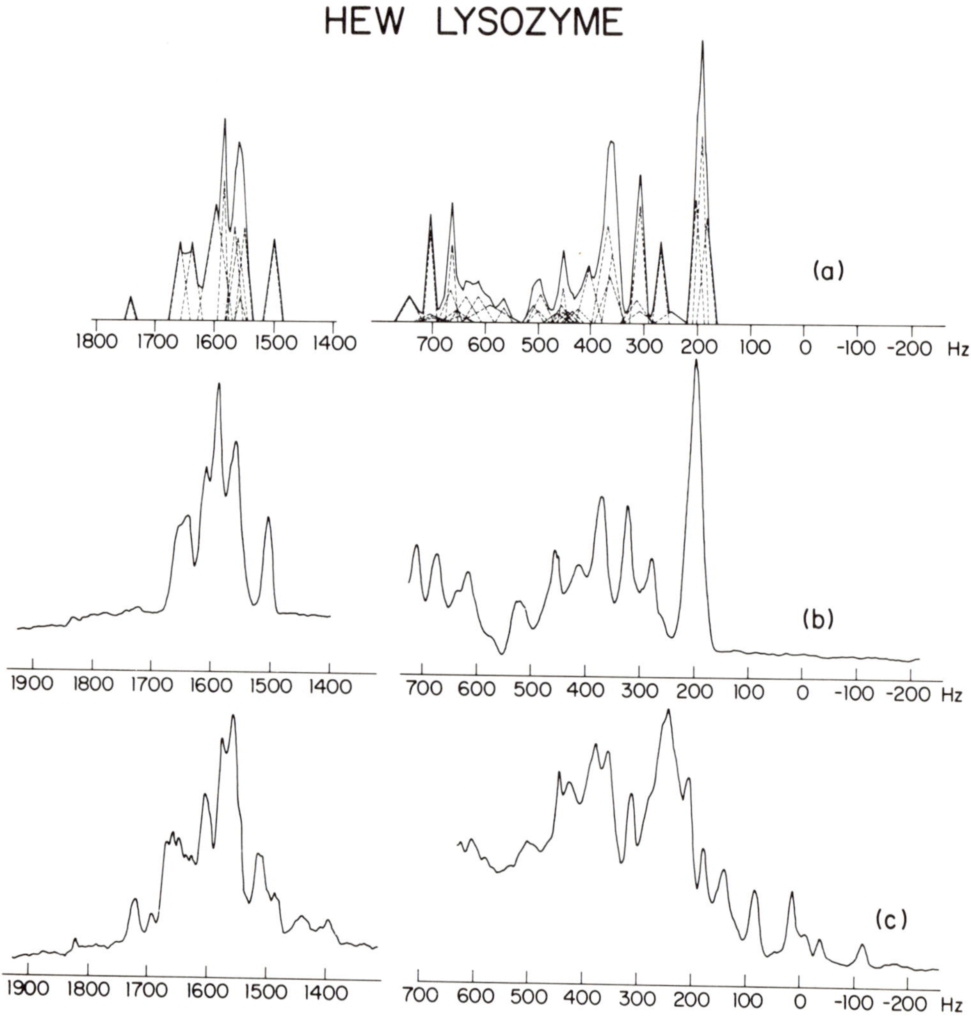

Fig. 1. 220-MHz PMR spectra of hen egg white lysozyme. (a) Schematic representation pD 7.0 40°C. Low-field intensity ×4; (b) and (c) 10% lysozyme in D_2O, pD 5.0; (b) random-coil spectrum 80°C, low field amplified ×4·1 relative to high field; (c) native spectrum 65°C, low field amplified ×2·3 relative to high field, pre-heated to remove N*H* resonances.

Fig. 1 where regions of the PMR spectra of native (1c) and random-coil (1b) hen egg white (HEW) lysozyme are presented.* Elucidation of elements of the native structure of a protein then depends on one's ability to explain changes in the PMR spectrum that accompany the conformational transition from the random-coil to the native structure. Consequently, one must first examine the origin of the characteristics of the PMR spectra of random-coil proteins. While these spectra provide no direct information on native structure, they do constitute a necessary base for comparison with the spectra of the native conformations.

The side chains of the amino acid residues of a random-coil protein reside in an environment characterized largely by the solvent. Consequently, we expect protons of a given chemical type to experience similar environments and to exhibit resonance at the same field position unless the proton environments are significantly influenced by neighbor residues. Experimentally, it has been shown (*27,28*) that only the resonance positions of protons bonded to α-carbon atoms are significantly affected by the nature of neighbor residues.† For other chemical types of protons on the various side chains (for example, the methyl-group protons of alanine residues), the respective resonances for a given chemical type very nearly coincide for all such protons in the random-coil protein or indeed for the same proton type in any other random-coil protein in the same solvent environment. While experience so far indicates that the above statement is generally true, we do not exclude the possibility that some exceptions may arise. For example, the degree of perturbation of resonance positions by ring-current magnetic fields from a nearest-neighbor aromatic residue has not been assessed for a random-coil configurational relationship between various residue types.

A set of resonance positions for most of the chemical types of hydrogen atoms of proteins, which is believed to be generally applicable to all random-coil proteins in neutral D_2O, has been derived from actual PMR spectra of proteins in extended configurations in neutral D_2O (*28*). Some of these resonance positions are listed in Table I. In general, they are similar to the corresponding resonance positions of the same proton types in amino acids

* Except where designated otherwise, spectra presented here were obtained with a Varian Associates 220-MHz PMR spectrometer and C-1024 computer of average transients using experimental techniques we have described previously (*21*). Chemical shifts are referred to the proton resonance position of the methyl groups of sodium 2,2-dimethyl-2-silapentane-5-sulfonate (DSS) employed as an internal reference. Positive numbers denote shifts to low field. Shift units are given either in hertz (cps) or parts per million (ppm) of the polarizing magnetic field (shift in Hz $\times 10^6$ divided by spectrometer frequency).

† This discussion refers to proteins in neutral D_2O. In this solvent environment and for random-coil proteins, protons bonded to nitrogen or oxygen exchange rapidly with deuterons of the solvent and are not observable in the PMR spectra. These remarks, therefore, refer only to hydrogen atoms bonded to carbon.

or dipeptides; they are not usually identical because of the effects of ionized amino and carboxyl groups in the small molecules (*27*).

The rotations permitted by the flexible polypeptide chain of a random-coil protein tend to reduce broadening of proton resonances by anisotropic dipolar interactions. Proteins in random-coil configurations consequently exhibit relatively narrow resonances, typically about 10-Hz half-width for

TABLE I

Expected Resonance Positions of Proton Resonances of Random-Coil Proteins in Neutral D_2O at 40°C

Proton type	Resonance position (Hz from internal DSS at 220 MHz)	Estimated half-width (Hz)	Relative peak height per residue
Leucine CH_3	195	15	40.0
Isoleucine CH_3	183	20	30.0
Valine CH_3	205	17	35.2
Alanine CH_3	310	18	16.7
Threonine CH_3	270	16	18.7
Lysine $\delta CH_2 + \beta CH_2$	370	30	13.3
Methionine CH_3	454	10	30.0
Cystine βCH_2	665	12	16.6
Histidine C-4	1555	10	10.0
C-2	1740	10	10.0
Tyrosine ortho to OH	1500	17	11.8
meta to OH	1560	17	11.8
Phenylalanine aromatic	1598	30	16.7
Tryptophan indole C-2	1584	10	10.0
indole C-5, C-6	1549	15	6.7
	1566	15	6.7
indole C-4, C-7	1638	18	5.6
	1658	18	5.6

proteins in the 10,000 to 20,000 molecular weight range. However, actual overall resonance profiles of many chemical types of hydrogens in random-coil proteins appear somewhat broader because of separate but unresolved components arising from spin–spin interactions. This broadening (and concomitant reduction of resonance peak heights) can be evaluated by inspection of appropriate PMR spectra of amino acids. In Table I, estimates of expected apparent resonance half-widths are listed for the proton types shown. The expected peak height per residue (on a relative scale) for a resonance type can be estimated from consideration of the resonance half-width and the number of equivalent protons of that particular chemical type in a residue. These amplitudes are also listed in Table I. With such values for resonance positions, half-widths, and peak heights per residue and a knowledge of the amino acid composition of a protein, the computation of the expected appearance of the PMR spectrum of a protein in a random-coil configuration in neutral D_2O is straightforward. The only remaining requirement is to select a resonance line shape. A simple but satisfactory first approximation is to represent resonances by triangles. Such computed PMR spectra for random-coil proteins serve as analyses of the components of the resonance contours that are obtained experimentally for random-coil proteins in neutral D_2O. They also provide random-coil spectra for comparison with PMR spectra of native proteins for proteins that cannot be readily examined in a random-coil configuration in neutral D_2O by PMR spectroscopy. These computational procedures and their applicability have been described in detail elsewhere (*28*). The same procedures can be used to compute PMR spectra of random-coil proteins in other solvent environments if appropriate values of resonance positions, half-widths, and peak heights are used. In general, many of these values will be quite different from those displayed in Table I. In this connection, PMR spectra of random-coil insulin in trifluoroacetic acid have been analyzed in detail by Bak et al. (*29,30*) through comparison of the protein spectra with spectra of amino acids in this solvent.

Turning to a more detailed examination of a PMR spectrum with characteristics typical of relatively low molecular weight random-coil proteins, we refer again to the spectrum of hen egg white (HEW) lysozyme in Fig. 1(b). This protein undergoes reversible thermal denaturation from 68 to 76°C and at 79°C is believed to be in an extended random-coil configuration except for four disulfide bridges which remain intact. [Recent studies of lysozyme by 100-MHz spectroscopy (*23*) show that the resonances of random-coil lysozyme are somewhat narrower if these disulfide linkages are broken.] The PMR spectrum computed for random-coil lysozyme in neutral D_2O using the procedures described above is shown in Fig. 1(a). This computed spectrum does not include protons bonded to nitrogen or oxygen since they exchange in D_2O, nor does it include the spectral region from 750 to 1400 Hz. Thus, we have excluded all α hydrogens, the β hydrogen of threonine, and the two β hydrogens of serine. These resonances occur between 750 and 1100 Hz, a spectral region that is partially obscured by the resonance of HDO in the D_2O solvent and by spinning side bands of this strong resonance. In Fig. 1(a) the individual resonance triangles for each proton type are shown as dashed lines; the predicted spectrum is obtained by summation of these triangles. The resonance region to high field from 325 Hz arises mainly from protons of methyl groups of alanine, threonine, valine, leucine, and isoleucine at 310, 270, 205, 195, and 183 Hz, respectively, and is therefore termed the "methyl" resonance region. This region contains minor contributions from lysine γ protons and isoleucine methylene protons. The only random-coil protein methyl resonances not in the "methyl" region are those of methionine at 454 Hz. The spectral region from 1400 to 1800 Hz comprises resonances of protons bonded to the aromatic rings of histidine, tyrosine, tryptophan, and phenylalanine and is termed the "aromatic" region. Because of their relative simplicity, the methyl and aromatic regions are among the most attractive for detailed analysis in PMR spectra of random-coil or native proteins.

Comparing Figs. 1(a) and 1(b) we see good agreement between the computed and actual random-coil spectra, particularly in the methyl and aromatic regions. The histidine C-2 proton resonance at 1740 Hz is not evident in Fig. 1(b). In neutral D_2O at 80°C, this proton is completely exchanged in a time short compared to the interval of about 1 hr required to obtain the PMR spectrum of Fig. 1(b) (*24,28*). One can, of course, obtain PMR spectra of proteins in random-coil configurations denatured by urea or guanidine, changes in pH, or changes in solvent (e.g., trifluoroacetic acid). Resonances in such spectra are generally shifted sufficiently so that they are less useful for comparison with spectra of native proteins in neutral D_2O than are random-coil protein spectra measured or computed for the latter solvent environment. With this brief discussion of the general characteristics of random-coil protein spectra, we proceed to examine the spectral changes that occur when a protein assumes its native conformation.

III. PMR SPECTRA OF NATIVE PROTEINS

When a protein in neutral aqueous solution undergoes the random-coil → folded conformational transition, many of the side chains of the component amino acids no longer are in a solvent environment but in one defined by their locations in the protein interior. Resonances of protons of residues in altered environments may be shifted from their random-coil position, and we may no longer expect protons of a given chemical type to be equivalent. Thus, the PMR spectra will become much more complex and difficult to resolve, but to the extent that such shifts do occur and can be resolved and rationalized, they can provide information on differences between the native conformation and the random-coil configuration. The characteristics of the PMR spectra of the considerable number of native proteins we have examined so far are strongly modified by secondary and tertiary structure so that differences between the spectra of the random-coil and native proteins are readily observable. For native proteins, it appears that the averaging of anisotropic dipolar interactions is governed largely by the overall rotation of the protein molecules so that resonance linewidths increase progressively with protein molecular weight. However, for small proteins with molecular weights in the 10,000 to 15,000 range (e.g., lysozyme, ribonuclease, and cytochrome c) resonance half-widths are only around 15 Hz. One finds, using 220-MHz PMR spectroscopy, that it is possible to resolve resonances in extensive regions of the spectra of such proteins. For smaller proteins, resolution is even better, but as protein size increases the resolution of resonances becomes progressively more difficult so that when proteins of molecular weights around 25,000 are examined, resolution has diminished to about the degree obtainable for lysozyme (MW 14,400) with 100-MHz spectroscopy. The following examples of PMR spectra of native proteins illustrate characteristic features of these spectra and means for deducing structural information from them. Many of the illustrations are taken from studies of HEW lysozyme since a great deal is known about the structure (*31*) and the biochemistry (*32*) of this protein.

A. Native to Random-Coil Transition

HEW lysozyme in aqueous solution at pD 3.0 to 5.0 and 65°C is folded in its native, active conformation. On heating from about 65 to 75°C, the protein is denatured to an extended random-coil configuration (retaining four disulfide bridges). Thermal denaturation extends over only about 6°C, but the median temperature is a few degrees lower at pD 3.0 than at pD 5.0. Regions of the PMR spectra of the native and random-coil protein are compared in Figs. 1 and 2. The denaturation process is completely reversible on cooling. It is clear that changes occur throughout the PMR spectrum as lysozyme folds to its native conformation. These changes are particularly evident and well-resolved in the methyl region, the aromatic region, and the region of resonance absorption of the indole N*H* protons of the six tryptophan residues of lysozyme (*33*).

Figure 3 shows the temperature dependence of PMR spectra of lysozyme in D_2O in more detail. In the thermal denaturation temperature range, the random-coil and native states of the protein exist in equilibrium, the proportion of native to random-coil decreasing regularly as temperature is increased. Through this temperature range, the PMR spectrum is a composite of the spectra of the native and random-coil states with relative intensities of the component spectra proportional to the populations of the individual states. This may be seen by examination of the spectrum at 74°C in Fig. 3 where the ratio of the random-coil state to native state is about 2. Since

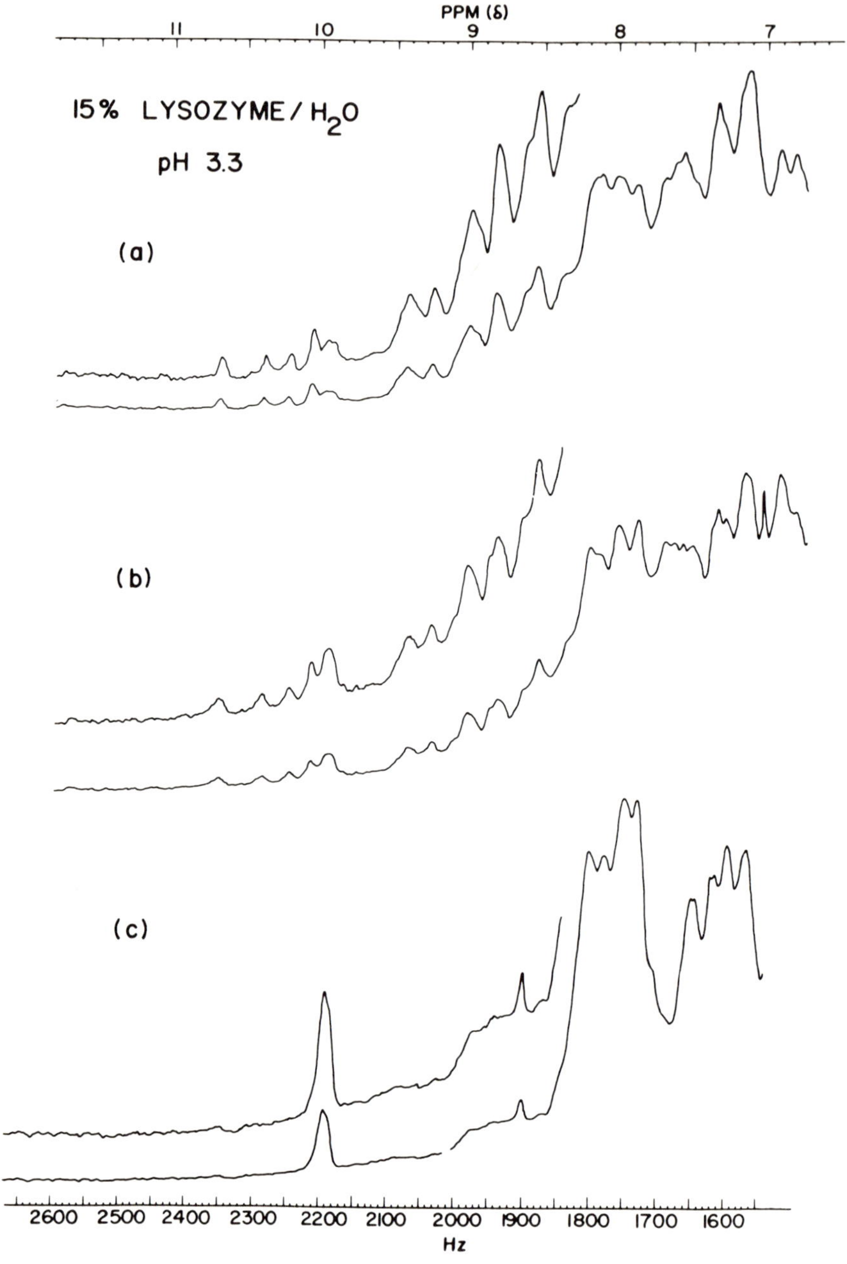

Fig. 2. Low-field 220-MHz PMR spectra of hen egg white lysozyme showing resonances of N*H* protons and aromatic C*H* protons: (a) native at 65°C, (b) partially denatured at 67°C, (c) denatured at 71°C. Tryptophan indole N*H* protons from 2080 to 2350 Hz (Ref. *34*).

individual spectra for the random-coil and native states are exhibited by these states throughout the thermal denaturation temperature range rather than a single exchange-averaged spectrum, and since there is not even evidence of exchange broadening of the resonances, we can say that the average lifetimes

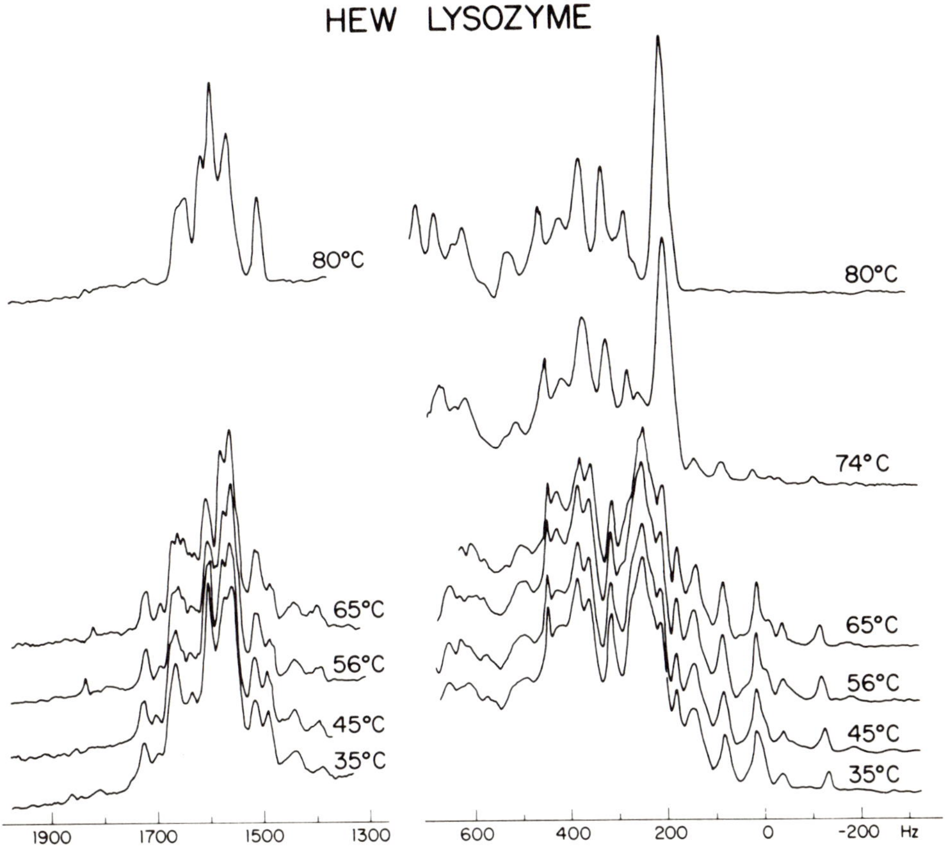

Fig. 3. 220-MHz PMR spectra of HEW lysozyme (10% in D_2O, pD 5.0) at several temperatures. Low-field spectra amplified more than high-field spectra (Ref. *34*).

of these states are longer than 10 msec.* However, these lifetimes must not be much greater than this lower limit because if a solution of lysozyme at 80°C is cooled in a few seconds to room temperature and examined within

* A discussion of effects on PMR spectra of intramolecular and intermolecular rate processes that exchange protons between nonequivalent environments may be found in *High-Resolution Nuclear Magnetic Resonance*, by J. A. Pople, W. G. Schneider, and H. J. Bernstein, McGraw, New York, 1959, p. 218.

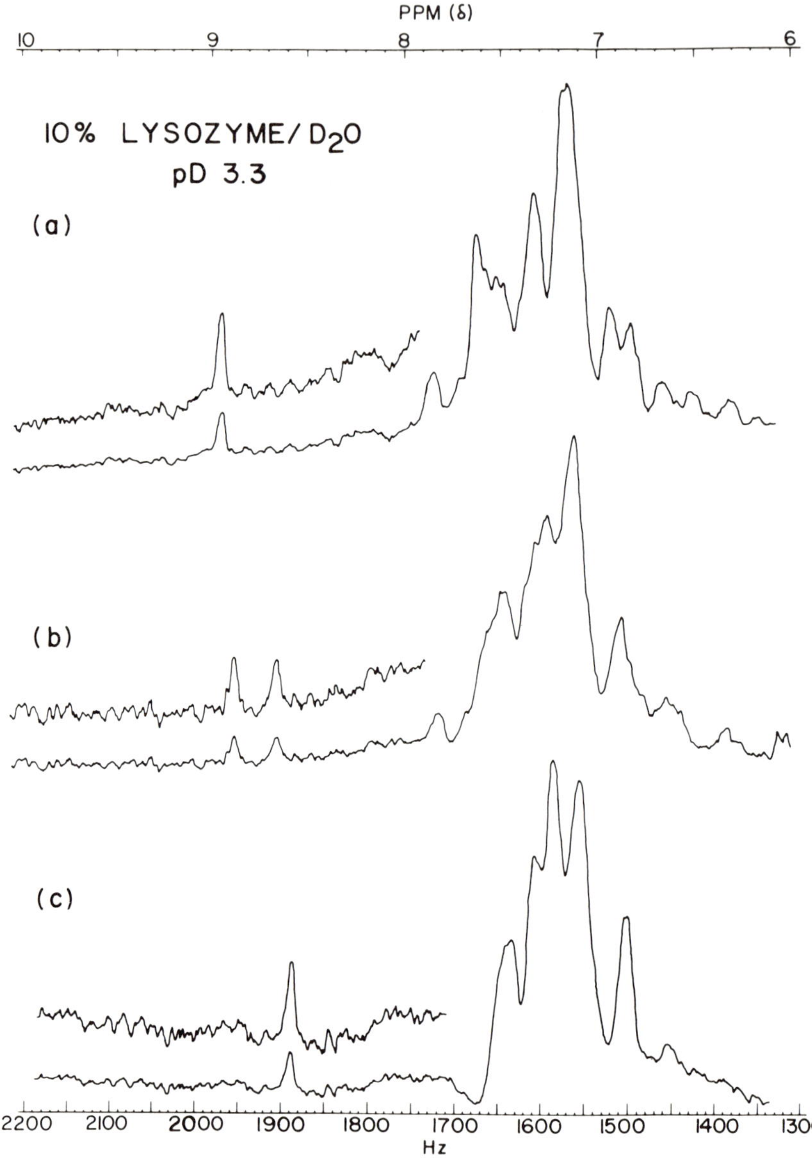

Fig. 4. 220-MHz PMR spectra of aromatic *CH* resonances of HEW lysozyme: (a) 60°C, (b) 69°C, (c) 80°C. C-2 resonances of his-15 in 1880–1950 Hz range (Ref. *34*).

a few minutes by PMR spectroscopy, one finds the random-coil form completely converted to the native conformation.

There is no evidence in the PMR spectrum of lysozyme of spectra of intermediate forms between the random-coil and native states. Either the denaturation and renaturation processes are sufficiently rapid so that detectable concentrations of intermediate forms are not present in the equilibrated solutions or such intermediates may undergo sufficiently fast exchange (average lifetimes $< 5 \times 10^{-4}$ sec) with the random-coil or native states (but not both) so that their PMR spectra are exchange-averaged components of the spectra of the latter states. However, it seems improbable that there are substantial amounts ($> 10\%$ of the total protein) of intermediate forms undergoing "fast-exchange" with the native form because one would expect the contribution of these forms to diminish as temperature is reduced from 70 to 35°C and thereby produce readily detectable changes throughout the "native" PMR spectrum. Figure 3 shows that small changes do occur in the positions of some resonances through this temperature range, but we feel that these reflect minor conformational changes in a predominant native state conformation rather than a redistribution of concentrations of several components jointly contributing to an "exchange-averaged" native state spectrum.

We have discussed the thermal denaturation of lysozyme in terms of a two-state cooperative process and based this conclusion on a general superposition of resonances characteristic of the native and denatured states in the region of thermal transition (*34*). Specifically, in this transition region separate resonances for the C-2 proton of the single histidine residue of lysozyme (his-15) are observed, one characteristic of native lysozyme and the other of denatured lysozyme (Fig. 4). A similar manifestation of two-state denaturation of lysozyme is derived from the separately resolvable indole N*H* protons of the six tryptophan residues of lysozyme in the thermal denaturation range. Finally, the same behavior is shown by methyl resonances in the high-field region assigned to leu-8, leu-17, ileu-98, and met-105.* Thus, for peaks assignable to several amino acid residues, rather widely distributed with respect both to primary and tertiary structure, we have evidence for two-state thermal denaturation in lysozyme.

Clearly then, PMR spectroscopy can be used to follow the rather drastic structural change of lysozyme that occurs on thermal denaturation from the native to the random-coil state. Similarly, one can follow native to random-coil transitions induced by other means: guanidine, urea, change in pH, change in solvent environment, etc. For example, changes in the high-field region of the PMR spectrum of lysozyme that accompany addition of

* Assignments of these resonances and others are discussed in Section IV.

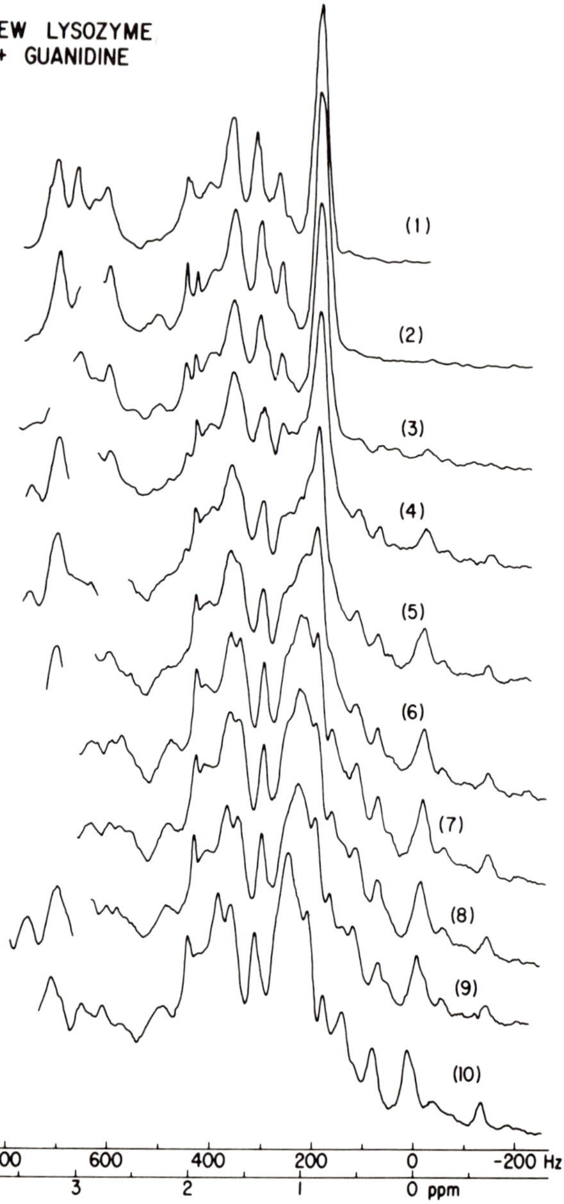

Fig. 5. 220-MHz PMR spectra of HEW lysozyme in solutions containing guanidine. Lysozyme 10% in D_2O. Spectrum (1) pD 5.9, 79°C all others pD 5.5, 40°C. Guanidine concentrations (1) 0 (2) 6 *M* (3) 5 *M* (4) 4.5 *M* (5) 4.25 *M* (6) 4 *M* (7) 3 *M* (8) 2 *M* (9) 1 *M* (10) 0. Amplification not constant between spectra (Ref. *34*).

guanidine to the protein solution are shown in Fig. 5. Conversion to the random-coil form is detectable in 4 *M* guanidine solutions and is virtually complete when the guanidine concentration reaches 5 *M*.

B. Perturbation of Native Conformations

Next, it is of interest to inquire whether more subtle conformational changes of a protein may be detected by PMR spectroscopy—in particular, whether there are manifestations in the PMR spectra of perturbations of the native conformation that accompany changes of temperature, of pH or ionic environment, in the oxidation state of a protein, or in reaction of a protein with substrates, inhibitors, metal ions, or cofactors.

1. Temperature

If one dissolves lysozyme in D_2O (pD 5 to 6) at room temperature and examines the PMR spectrum in the region from 1300 to 2300 Hz, a number of discrete resonances are observed that do not appear in the spectra of native lysozyme shown in Figs. 1 or 3. These are resonances of protons bonded to nitrogen atoms in the interior of the protein that do not have free access to the solvent. They do exchange slowly and at different rates suggesting that the native protein unfolds occasionally to expose the protein interior to D_2O. It is probably necessary for a given protein molecule to expose a given N*H* proton only once to D_2O to accomplish exchange, so slow exchange implies infrequent "breathing" of a protein molecule. The exchange rates increase, of course, at higher temperatures, and evidence of these resonances is usually absent for lysozyme in PMR spectra obtained at 50°C or higher. Other proteins (e.g., reduced cytochrome c) exhibit much more slowly exchanging N*H* protons and may, therefore, be considered to unfold very infrequently to an extent which exposes the protein interior to solvent. Specific N*H* protons are discussed further in later sections.

The effect of temperature on C*H* proton resonances of native lysozyme from 35 to 65°C is shown in Fig. 3. The solutions used for these experiments were preheated to exchange all N*H* protons. As the temperature is increased from 35 to 65°C, a resonance at −132 Hz shifts to −116 Hz, a resonance at about 0 Hz shifts to −11 Hz (these two resonances are tentatively assigned to leu-17), a resonance at about 33 Hz shifts to 48 Hz, and some small changes are observable in the aromatic region at 1600 Hz and between 1630 and 1680 Hz. Otherwise, the resonance positions appear unaffected by temperature. One may conclude either that the characteristics of the PMR spectra are insensitive to small, temperature-induced modifications of the native conformation of lysozyme or that this conformation changes very little from 35 to 65°. We favor the latter interpretation as the PMR spectra do reflect conformation changes caused by other means. Furthermore, we believe, for reasons discussed in Section IV, that positions of resonances to high field from 170 Hz are particularly sensitive to very small conformation changes. A minor local conformation change may account for the small temperature-induced shifts of three resonances in this field region. For example, the resonance positions of the leu-17 methyl groups appear to depend to a small extent on their geometric relationship to the aromatic ring of tyr-20 (see IV, B). This ring on the exterior of the protein may well shift somewhat as temperature is changed without any change in the spacial arrangement of the lysozyme polypeptide chain.

Below 35°C, the resonances of lysozyme broaden progressively, undoubtedly because an increasing degree of aggregation of lysozyme molecules provides species with higher molecular weights and hence, longer rotational correlation times and broader resonances (*35*). Such aggregation becomes quite evident even at elevated temperatures if pD is increased above about 6.

2. pD

Regions of the PMR spectrum of HEW lysozyme at 55°C are shown at pD values from 1.9 to 5.5 in Fig. 6. In the pD range from 2.8 to 5.5, there are a number of small but readily observable changes. These may arise for various reasons: shifts of proton resonances caused by a change in protonation of a group on the same residue; shifts caused by a change in charge of a neighbor residue; and shifts caused by conformation changes resulting from modification of protein ionization. The spectral changes are mostly rather small and indicate that the protein conformation changes little until the pD is reduced below 2.8. At pD 1.9, the spectrum is quite different and indicates that the protein is largely converted to an extended configuration. As pD is increased from 2.8 to 5.5, the spectra become somewhat less resolved because of line broadening. This again results from an increasing amount of aggregation (*35*) and becomes more pronounced as the pD is further increased so that it has not been possible to obtain useful spectra of this protein at pD values above about 8.5.

One very apparent change in the spectra of Fig. 6 is the shift of a resonance at 1795 Hz at pD 5.5 to 1990 Hz at pD 2.8. This is the resonance of the C-2 proton of the single histidine residue of HEW lysozyme (his-15). The change in the chemical shift of this resonance is a consequence of protonation of the imidazole ring and has been used to measure the p*K* of this process (*17*). This resonance is not evident in all of the spectra in Fig. 6 because of exchange of this proton in some samples during preheating to exchange N*H* protons whose resonances also occur in this field region.

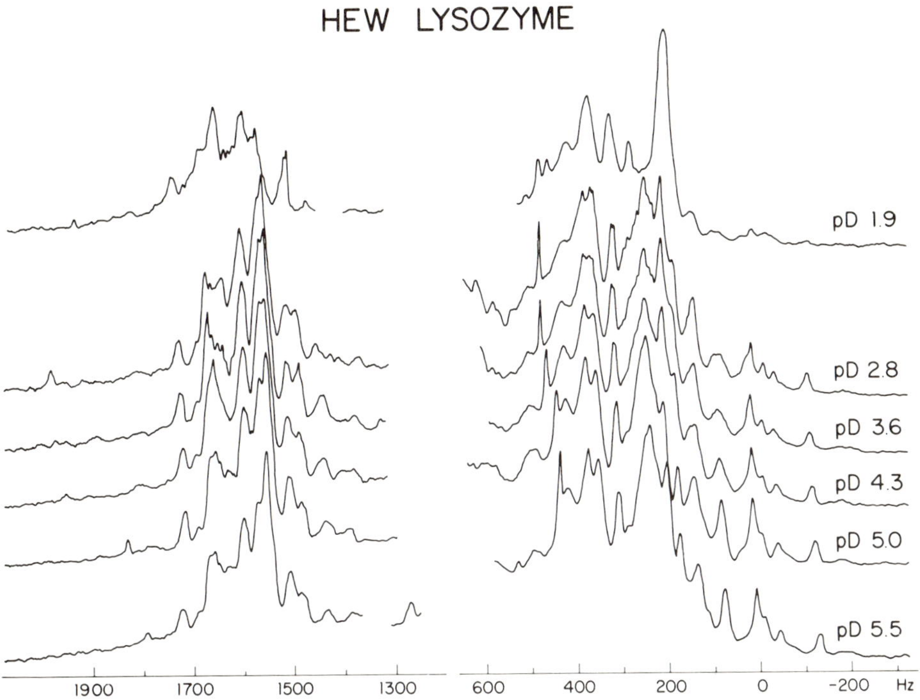

Fig. 6. 220-MHz PMR spectra of HEW lysozyme at several pD values. Lysozyme 10% in D_2O, 55°C. Preheated to exchange all N*H* resonances (Ref. *34*).

3. Oxidation State

A number of small proteins have prosthetic groups that can exist in two stable oxidation states. Cytochrome c, for example, has a heme group in which the central iron atom can be in either a ferric (paramagnetic, $S = \frac{1}{2}$) or a ferrous (diamagnetic, $S = 0$) state. Regions of the PMR spectra of horse ferri- and ferrocytochrome c are shown in Fig. 7. The spectra are quite different; these differences support the position that a considerable conformational change accompanies the change in oxidation state of the heme iron. However, as discussed below (IV, A), a part of the change on oxidation of ferrocytochrome c, independent of any conformational change, may arise from contact shifts of resonances caused by the paramagnetism of the ferric iron atom. Large changes in the PMR spectrum have also been observed on changing the oxidation state of myoglobin (*25,26*), leghemoglobin (*36*), and rubredoxin (*37*). However, reduction of flavodoxin, which has a flavin mononucleotide prosthetic group (*37*), did not affect the PMR spectrum significantly, suggesting that reduction of this protein is not accompanied by a substantial conformational change (*39*), nor were the PMR spectra of plant (*40*) or bacterial (*37*) ferredoxin greatly changed on reduction of these proteins.

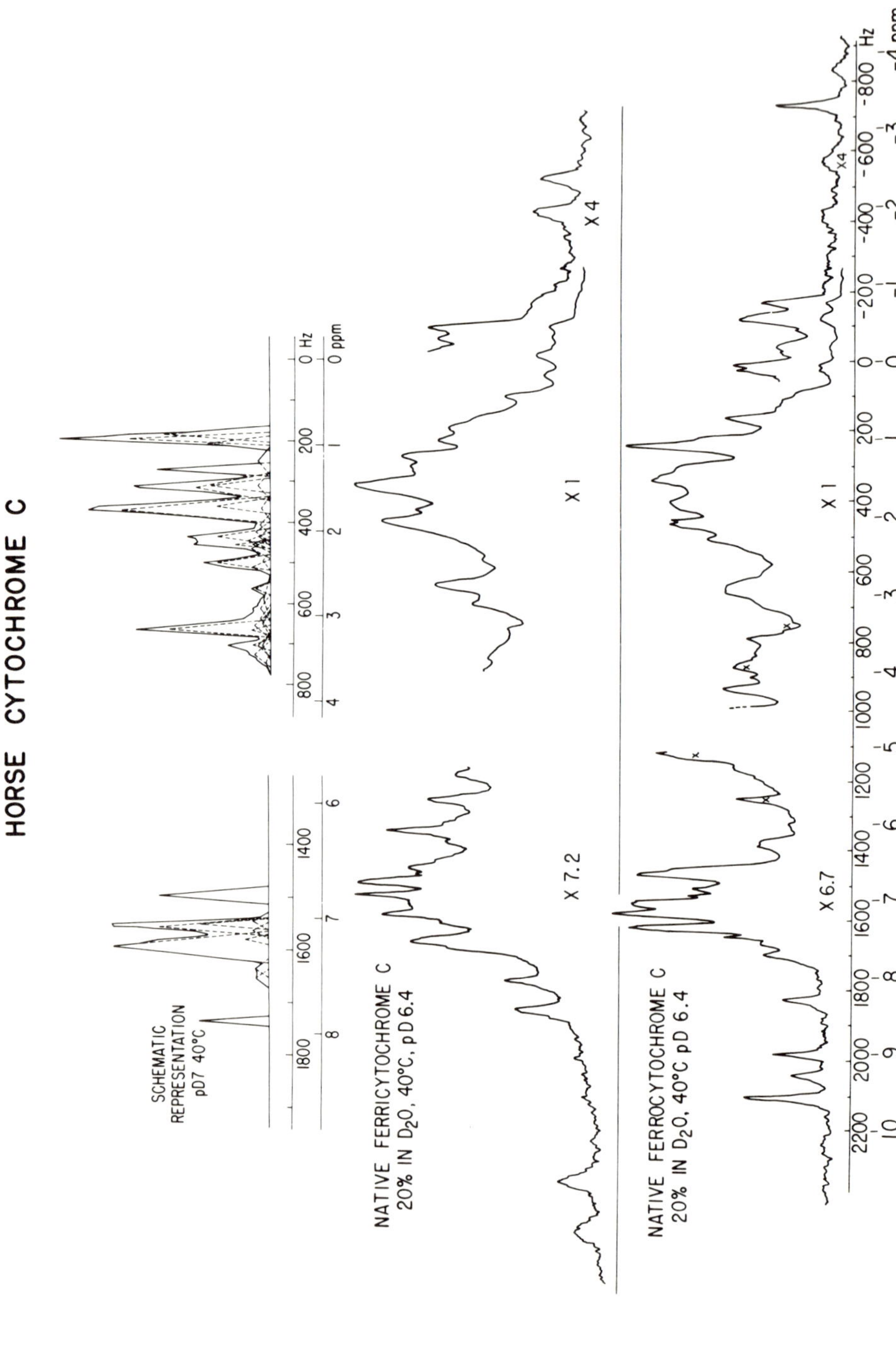

Fig. 7. 220-MHz PMR spectra of horse cytochrome c (Ref. *48*).

4. Interactions with Other Molecules

It is of interest to inquire whether changes in protein conformation that may occur upon the interaction of a protein with substrates, inhibitors, or cofactors can be detected through changes in the protein PMR spectrum. An extensive study of effects of binding by the inhibitor 3′-cytidine monophosphate (3′-CMP) on the PMR of ribonuclease has been carried out by Jardetzky and co-workers (*17,41*). Bovine pancreatic ribonuclease contains four histidine residues located at positions 12, 48, 105, and 119 along the polypeptide chain. For RNase in which all N*H* protons have been replaced

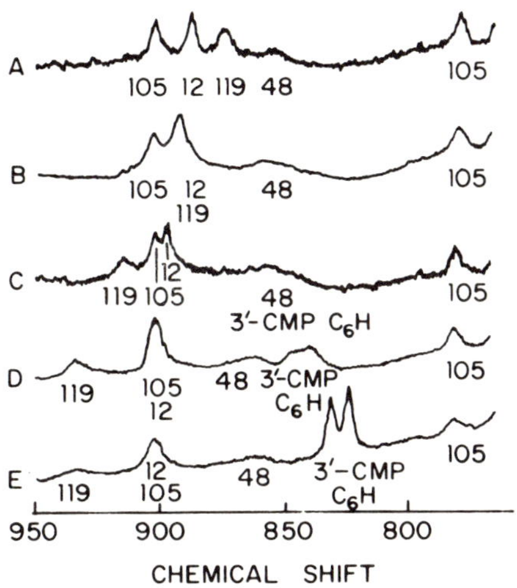

Fig. 8. 100-MHz PMR spectra of histidine resonance absorption region of ribonuclease (6.5×10^{-3} *M*, pH 5.5, 32°C, O.2 *M* NaCl). Chemical shifts in hertz from resonance of hexamethyldisiloxane used as an external reference. Concentration of 3′-CMP (A) 0, (B) 2×10^{-3} *M*, (C) 5×10^{-3} *M*, (D) 1×10^{-2} *M*, (E) 3×10^{-2} *M*. Peak 105 at 780 Hz is C-4 proton of his-105. Other labeled peaks are C-2 protons of respective histidine residues (Ref. *41*).

by deuterium through exchange, the C-2 protons from the four histidine residues (as well as the C-4 proton of his-105) are well resolved in the extreme low-field region of the spectrum (Fig. 8). Identification of each of these resonances with specific histidine residues is discussed below (IV, C). Chemical (*42*) and physical (*43*) studies had indicated his-12 and his-119 to be involved in the active site of the enzyme.

The effect of 3′-CMP on the PMR absorption of the C-2 protons of RNase is shown in Fig. 8. With increasing concentrations of 3′-CMP, the C-2 proton resonances of his-119 and his-12 are shifted to low field and that of his-105 is unaffected. The resonance of his-48 exhibits a small high-field shift with added 3′-CMP. The large shifts of the C-2 resonances of his-119 and his-12 upon addition of 3′-CMP to RNase are compatible with their involvement in binding the inhibitor at the active site. In addition to being slightly perturbed by the addition of 3′-CMP to RNase, the C-2 resonance of his-48 also appears to be anomalously broad over the entire concentration range. It was suggested that his-48, which is somewhat buried in the interior of the folded enzyme, may have available to it two or more shielding environ-

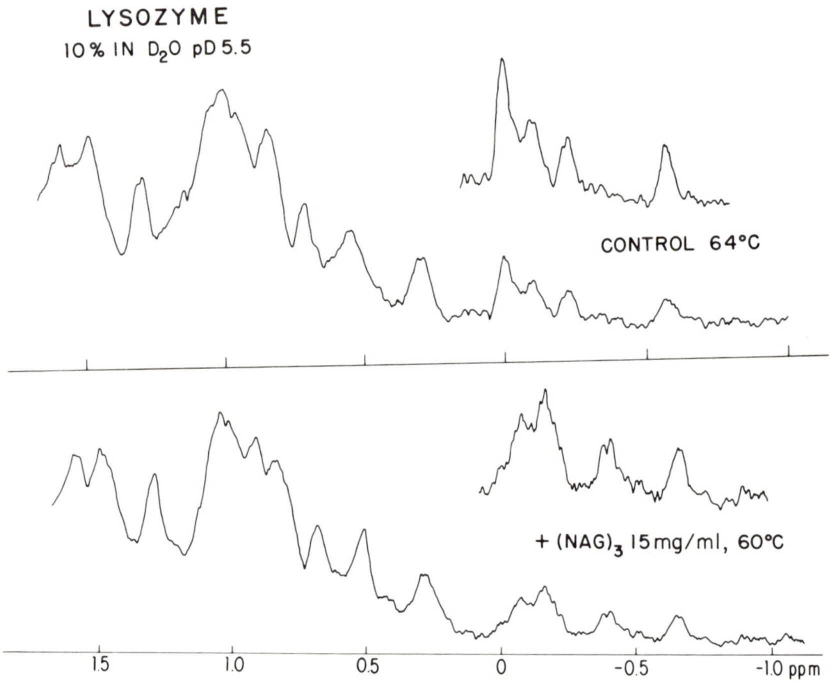

Fig. 9. 220-MHz PMR spectra of HEW lysozyme in the absence and presence of N-acetylglucosamine trimer (Ref. *47*).

ments between which it is exchanging at a rate comparable to the chemical shift between C-2 protons in the multiple environments (*41*).

HEW lysozyme contains six tryptophan residues located at positions 28, 62, 63, 108, 111, and 123 along the polypeptide chain. Employing H_2O rather than D_2O as solvent, indole N*H* proton resonances from five of the six tryptophan residues are clearly resolved as separate resonances of unit intensity in the extreme low-field region of the spectrum (Fig. 2). X-ray studies (*31,44,45*) have clearly placed tryptophan residues 62, 63, and 108 at or near the active site of the enzyme. The indole N*H* protons of try-62 and try-63 are believed from the X-ray studies to be directly involved in binding inhibitors such as the dimer and trimer of N-acetylglucosamine (NAG), $(NAG)_2$ and $(NAG)_3$, to the cleft containing the active site. And indeed, two and only two of the resolved resonances of the indole N*H* protons of the tryptophan residues of lysozyme are displaced upon addition of $(NAG)_2$. The N-acetylglucosamine trimer has been shown through biochemical studies to bind to HEW lysozyme in a manner similar to that of the natural substrate of this protein (*46*). Further, the X-ray structural studies indicated a small conformational change of the protein on binding of this inhibitor (*31*). The C*H* PMR spectrum of lysozyme is modified on addition of $(NAG)_3$ to a solution of the protein (*21,23*). Readily observable resonance shifts occur throughout the C*H* spectrum reflecting the conformational change. These differences in the high-field methyl region are illustrated in Fig. 9 (*47*). Pronounced changes in the field region from 0 to −0.5 ppm are attributed, at least in part, to shifts of the methyl resonances of ileu-98 (see IV, B) that are caused by a change in the relationship of this residue to try-63.

The shifts of tryptophan N*H* resonances and of methyl resonances that accompany binding of $(NAG)_2$ and $(NAG)_3$ to lysozyme were employed in assignment of these residues. Similar effects were used in assigning the resonances of the histidine C-2 protons of RNase (Section IV, B, C, D). These studies provide further examples of the close working relationship that is developing between X-ray and NMR spectroscopies in the elucidation of protein structure.

We have reported (*21*) that the PMR spectra of ferricytochrome c and metmyoglobin are greatly altered when these proteins react with cyanide ion. These variations are attributed both to conformational changes and to changes in magnetic properties of these proteins that accompany reaction of cyanide with the heme iron. More recently (*48*) we have examined effects of carboxymethylation of methionine residues of horse ferricytochrome c. Reaction of cytochrome c with iodoacetic acid at pH 5.6 is thought (*49*) to cause preferential S-carboxymethylation of met-65 on the exterior of cytochrome c and this derivative exhibits PMR spectra similar to unreacted cytochrome c in both the ferri and ferro forms. Reaction at pH 3 is thought (*50*) to lead to S-carboxymethylation of both met-65 and met-80, the latter residue being implicated as a coordinating group to the heme iron (*51*). The cytochrome c derivative prepared by this reaction gave spectra quite different from those of native ferri- or ferrocytochrome c and consistent with a major conformational change.

PMR spectra have been obtained for a HEW lysozyme derivative in which the try-62 residue was converted to the oxindole form (*47*). Only minor spectral changes were observed from the spectrum of the native protein. Examination of PMR spectra of the derivative in which try-108 is specifically modified to the oxindole form is in progress. Initial results indicate a considerably larger conformational change on modification of try-108 which is intimately involved in the tertiary structure of lysozyme than on modification of try-62 which is at the surface of the protein.

We conclude that PMR spectroscopy provides a very sensitive tool to assess whether chemical modification of a protein causes a conformational change from the native conformation.

IV. IDENTIFICATION OF RESONANCES AND INTERPRETATION OF RESONANCE POSITIONS OF NATIVE PROTEINS

From the foregoing discussion it is clear that characteristics of PMR spectra can be used to follow changes in conformation of a protein. However, to take full advantage of the PMR technique, one must be able to identify in the native protein PMR spectrum the specific protons that give rise to each resolved resonance and to relate changes in resonance positions from the random-coil positions to structural features of the protein. Assignment is not straightforward because, during renaturation or denaturation, at least for the proteins we have examined, the resonances do not shift gradually from the random-coil positions to the native positions (or vice-versa) but "grow in" at their new positions as one form converts to the other. In correlating changes in resonance positions with protein structure, it is necessary initially to develop relationships from proteins of known structure, such as lysozyme, ribonuclease, and myoglobin, which can be expected to apply generally to proteins of unknown structure. Some approaches to resonance assignment and to chemical shift interpretation have been developed which, when used together, take one along the road to these desired goals. Some examples of these approaches are discussed below but first some of the types of interactions likely to cause shifts in resonance positions are reviewed.

The largest shifts of protein proton resonances from their normal random-coil positions occur when the protons are in the vicinity of a paramagnetic center in the protein. Such shifts, with magnitudes ranging up to 90 ppm have been detected in PMR spectra of several proteins having heme-iron prosthetic groups and are discussed in IV, A. Paramagnetic centers in certain non-heme iron proteins and flavoproteins may also be expected to influence proton resonance positions of these proteins.

Aside from the very profound effects of a paramagnetic center on proton resonance positions, the major cause of large changes in resonance positions when a protein folds into a close-packed conformation is likely to be the effects of ring-current fields of cyclic conjugated structures on nearby protons (*52*). Such ring-current fields in proteins arise not only from the aromatic rings of phenylalanine, tyrosine, tryptophan, and histidine residues but also from prosthetic groups such as flavin and porphyrin. Ring-current fields shift resonances of nearby protons to low field if the protons are in or nearly in the plane of the aromatic ring (but external to the ring) and shift resonances to high field for protons located within the aromatic ring or for nearby protons on or near the sixfold axis normal to the ring plane. Neither magnitudes nor geometrical distributions of ring-current fields of aromatic structures have been well established by theoretical calculations or experimental measurements (*53*). The ring-current field of benzene calculated by Johnson and Bovey (*54*) is generally used for semiquantitative description of ring-current fields for phenyl groups and with appropriate modifications for other aromatic structures. The data of the Johnson–Bovey tables (Ref. *53*, Appendix B) indicate that ring-current fields from aromatic structures may persist at significant magnitudes for a considerable distance through a protein molecule. Assuming that a phenylalanine or tyrosine ring is equivalent to benzene, the low-field shift of the resonance of an "in-plane" proton does not decrease to 20 Hz (at 220 MHz) until the proton is about 6.5 Å from the ring center; the high-field shift of the resonance of a proton on the sixfold axis has only decreased to 68 Hz at 5.55 Å. One might expect the ring-current field of a histidine residue to be comparable in magnitude and that of a tryptophan residue to be considerably larger. The large heterocyclic aromatic structures of flavin or porphyrin prosthetic groups are expected to provide even larger and more complex ring-current fields (*55*).

Electric fields from charged groups may also cause large shifts of resonances of protons in the vicinity. However, the majority of charge-bearing residues (aspartic acid, glutamic acid, lysine, argine, and histidine) are likely to be on the exterior of the protein molecule in contact with the aqueous phase. In such locations, they are not likely to influence protons of many other residues, but where charged groups are internal they may cause considerable perturbation of resonance positions. Other factors that may have appreciable influence on resonance positions of native proteins are electric fields from permanent dipoles, magnetic anisotropy of peptide linkages, hydrogen bonding, and location of residues in internal hydrophobic regions as opposed to their solvated condition in random-coil proteins. The potential importance of several of these factors has been discussed in considerable detail by Sternlicht and Wilson (*18*). However, the relative contributions of the various factors that may contribute to the many differences between PMR spectra of native and random-coil spectra have not yet been worked out in detail for any protein and remain an important research area for the future.

We now return to the discussion of assignment of resonances and to the interpretation of chemical shifts in terms of protein structure.

A. Contact Shifts

The profound effects of paramagnetic centers on the characteristics of nuclear resonances are well established (*56*) and in favorable circumstances can be employed most advantageously in studying biological systems by NMR. The saturation and linewidth characteristics of nuclear resonances are dominated by dipolar and spin-exchange phenomena. Therefore, introduction of a paramagnetic center with a magnetic moment some 658 times as large as that of the proton into an otherwise diamagnetic species can be expected to produce effects on nuclear longitudinal (T_1) and transverse (T_2) relaxation times (*57,58*). Such effects have been employed to study the interactions of paramagnetic ions with nucleic acids (*58*) and proteins and in particular have led to new insight into the role of the manganous ion in activating the enzymes enolase and pyruvate kinase (*1*).

Such studies are however beyond the scope of this review since we are principally interested here in factors that affect the positions of nuclear resonances of proteins. As implied above, spin-exchange effects can affect the positions of nuclear resonances. The first of these effects to be considered is the pseudocontact shift, ΔH_c, which in solution for a complex of tetragonal symmetry is given by

$$\Delta H_c = -(3\cos^2\zeta - 1)(g_{\parallel} - g_{\perp})(g_{\parallel} + 2g_{\perp})\frac{\beta^2 H_0 S(S+1)}{27kTd^3} \qquad (1)$$

Here, ζ is the angle between the distance vector and the tetragonal axis of the complex; the other symbols have their usual significance. It is to be noted that for a complex of octahedral or tetrahedral symmetry where $g_{\parallel} = g_{\perp}$ the pseudocontact shift vanishes. However, in molecules of biological significance, such high symmetry about the paramagnetic center seldom is the case, and one can expect pseudocontact displacements of nuclear resonances in paramagnetic systems with striking angular and distance dependences.

The second effect of paramagnetic centers on the positions of nuclear resonances is the Fermi or isotropic hyperfine contact interaction. Such shifts, ΔH_j, are given by the expression

$$\frac{\Delta H_j}{H_0} = -A_j \frac{\gamma_e}{\gamma_N} g \frac{\beta S(S+1)}{3kT} \qquad (2)$$

where A_j is a hyperfine coupling constant and the other symbols have their usual significance. A_j customarily is defined as

$$A_j = \frac{Q\rho_j}{2S} \qquad (3)$$

where ρ_j is a spin density centered at the jth atom and Q is a proportionality constant that is characteristic of the particular nucleus–spin density interaction. The isotropic hyperfine or Fermi contact interaction has been employed extensively to elucidate the geometrical and electronic structures of complexes and coordination compounds of transition metals (*56*).

The isotropic hyperfine contact interaction is of particular interest because it is capable in principle of providing information regarding the distribution of spin density over a ligand and the extent to which spin density is delocalized from a coordinated paramagnetic transition metal ion to the ligand. However, isotropic hyperfine contact interactions are only manifested as contact shifts in NMR when one or both the conditions $T_{1e}^{-1} \gg A$ and $T_e^{-1} \gg A$ obtain. Here A is the relevant hyperfine coupling constant of Eq. (2), T_{1e} is the electronic relaxation time, and T_e is the characteristic exchange time of the system.

If circumstances are such that observation of contact shifts is possible, two situations frequently are encountered when the ligand is conjugated, as

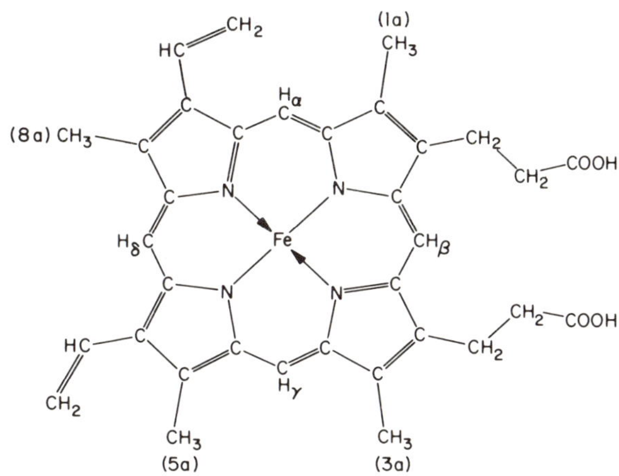

Fig. 10. The heme group of sperm whale metmyoglobin, Fe^{3+}–protoporphyrin IX.

with a heme-containing protein. For a $H—C\cdot$ fragment (as for the H_α, H_β, H_γ, H_δ positions of the heme of Fig. 10), spin density localized at the $p\pi$ orbital of the sp^2 carbon of the fragment causes a shift in the resonance of the proton bound to the carbon that is given by Eqs. (2) and (3) with $Q \cong -6.3 \times 10^7$ Hz. For a $CH_3—C\cdot$ fragment (as for the 1a, 3a, 5a, and 8a positions of the heme of Fig. 10), the methyl protons are subjected to an isotropic hyperfine contact interaction shift given by Eqs. (2) and (3) with $Q \cong +7 \times 10^7$ Hz. Note that for a given spin density centered on carbon, isotropic

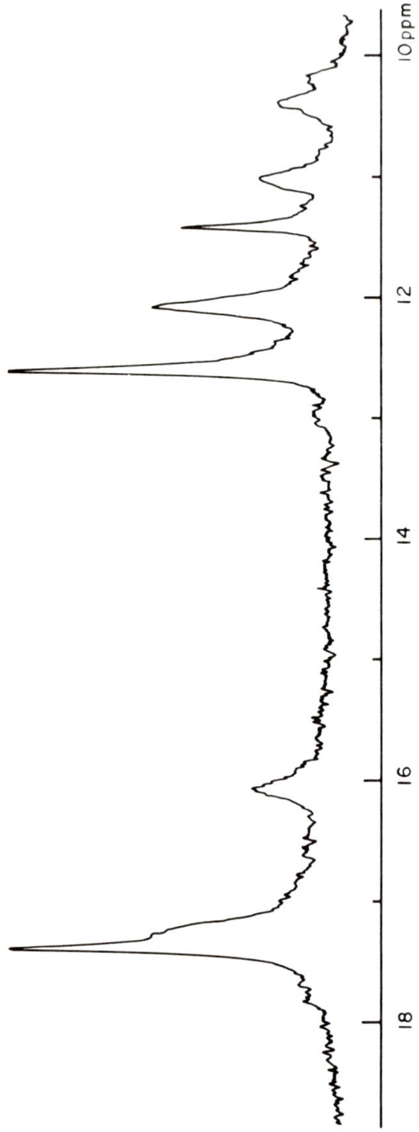

Fig. 11. 220-MHz PMR spectrum of cyanometmyoglobin (5×10^{-3} *M* in deuterated phosphate buffer, pD 6.6, 35°C). Low-field resonances -10 to -18 ppm from internal DSS (Ref. *26*).

hyperfine contact interaction shifts for the two situations cited above differ not only in magnitude but in sign as well. Note also that not only magnitudes but signs of spin densities are derivable from these Fermi contact shifts.

Contact shifts were first observed in the paramagnetic forms of cytochrome c and myoglobin by Kowalsky (*14*) in 1965. Shifts of up to 35 ppm were observed and these were tentatively assigned to spin densities on the porphyrin ring resulting from delocalization from iron through $d\pi$–$p\pi$ bonding. Both high- and low-field contact shifts were observed and these were attributed to the existence of both negative and positive spin densities on the porphyrin ring. Both positive and negative spin densities are expected for the porphyrin ring since it is a nonalternant π-electron system by virtue of possessing the five-membered pyrrole rings.

Recently contact shifts in cytochrome c, myoglobin, and hemoglobin have been reexamined by Kurland et al. (*25*) and Wüthrich et al. (*26*) employing improved instrumentation. Kurland et al. observed displacements of 30–90 ppm downfield for metmyoglobin and methemoglobin which they attributed to protons subjected to Fermi contact interactions resulting from spin density distributed over the porphyrin ring.

Wüthrich et al. have studied in some detail the contact shifted resonances of cyanometmyoglobin. The cyano derivative was chosen for study because the introduction of the cyanide ligand alters the ligand field about Fe^{3+} from weak-field to strong-field. The number of unpaired electrons centered on iron is correspondingly reduced from five to one, and electronic dipole relaxation effects on nuclear resonances are reduced by a factor of 35/3. Resonances of the cyanide complex therefore are somewhat sharper and easier to detect than for the weak-ligand-field case. As a matter of fact, only in the cyanide derivative were the expected high-field contact shifted resonances of metmyoglobin detected. In the weak ligand field (cyanide absent), these resonances presumably have eluded detection because of excessive linewidths (*21*).

A portion of the PMR spectrum of cyanometmyoglobin in the region of low-field contact shift resonances is shown in Fig. 11. In order to achieve this excellent signal-to-noise ratio, computer averaging over a period of 12 hr was required. In both high- and low-field regions of the spectrum, 15 of 16 expected contact shifted resonances of cyanometmyoglobin have been detected. Resonances exhibiting contact shifts were distinguished from the rest of the spectrum by the Curie Law ($1/T$) dependence of the magnitudes of contact shifts on temperature (Fig. 12). In the absence of structural or environmental perturbation, chemical shifts uninfluenced by contact interactions are temperature independent. The temperature dependence of contact shifts is obviously of great importance in distinguishing them from the more commonly encountered chemical shifts.

It was concluded that the contact shifts for cyanometmyoglobin are primarily Fermi contact in nature with small if not negligible pseudocontact contributions. It was also concluded that the mechanism for the transfer of spin density from iron to the porphyrin ring was principally via the π-electron

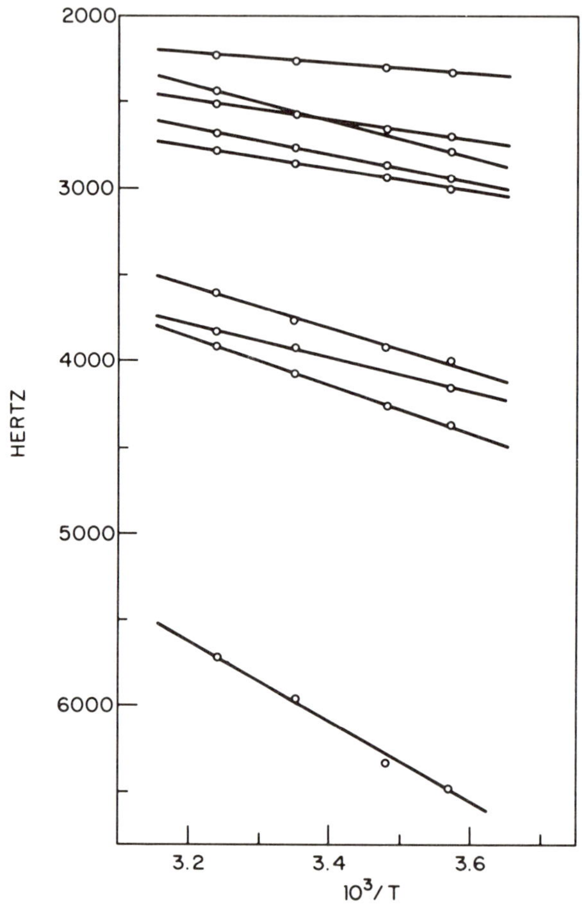

Fig. 12. Dependence of the proton resonances of cyanometmyoglobin in the field range from −10 to −27 ppm on the reciprocal of temperature. Referred to internal DSS, 220 MHz (Ref. *26*).

system rather than the σ-bond system of the porphyrin. Estimates of the extent of transfer of spin density from Fe^{3+} to the porphyrin moiety of cyanometmyoglobin are of the order of 1% of an electron (*26*).

Ho and co-workers (*59*) have examined the low-field contact shifted region of resonance absorption of the cyanomethemoglobins, human adult hemoglobin A ($\alpha_2\beta_2$), human fetal hemoglobin F ($\alpha_2\gamma_2$), and horse hemoglobin ($\alpha_2\beta_2$). The observed differences in patterns of heme methyl resonance absorption for the three hemoglobins were attributed tentatively to differences in interactions between the heme groups and the surrounding amino acid residues among the three sequentially different proteins.

It is expected that studies of contact shifts in hemoglobin in progress will yield definitive answers regarding the presence or absence of conformational changes of the protein associated with reversible combination with oxygen. In light of the above most important results, it can confidently be expected that NMR studies based on contact shifted resonances will be of great future value.

Contact shifts of protein resonances result in some cases as a consequence of the binding of paramagnetic transition metal ions to proteins. A particular use of this phenomenon is described in IV, E.

B. The Methyl Resonance Region

We turn next to a consideration of ring-current shifts and in particular to the effect of these shifts on the high-field methyl resonance region. Qualitative analyses of this spectral region of native HEW lysozyme have been made (*18,21*). These analyses were based on consideration of the environments of the residues of native lysozyme as defined by the lysozyme structure deduced by X-ray spectroscopy by Phillips and his collaborators (*31,44,45*). The general conclusions drawn from these analyses and from a more detailed study now in progress in our laboratory (*60*) are similar; they differ somewhat in detail. Only a brief summary is given here.

Lysozyme contains 40 methyl groups as components of 8 leucine, 6 isoleucine, and 6 valine residues. In the random-coil lysozyme spectrum these are nearly equivalent (as regards PMR positions) and produce a strong resonance with a maximum at 195 Hz.

This is the resonance to highest field for the random-coil protein. In the spectrum of native lysozyme, resonances of about 72 protons occur to high field from 195 Hz. Nine of these proton resonances are attributed to unperturbed isoleucine methyl groups at 183 Hz. The other resonances are at new resonance positions which are attributed to ring-current shifts. Estimation of the ring-current fields of aromatic residues at proton positions for protons lying close to the faces of aromatic rings in native lysozyme (*18,21,60*) indicate that 45 proton resonances of methyl groups of leu-8, -17, -56, -75, ileu-55, -88, -98, val-92, thr-51, and met-105 are shifted into the high-field region above 195 Hz. In addition, 20 single proton resonances of leu-17, ileu-88, -98, met-105, arg-61, and lys-33, -96, and -116 are expected to be shifted into this high-field region. Thus, these calculations predict 65 proton resonances shifted into the field region from 194 to −450 Hz which, taken with nine unperturbed proton resonances of isoleucine methyl groups, agrees well with the approximately 72 resonances observed in this field region. From these calculations and from examination of perturbation of the native lysozyme spectrum by temperature, $(NAG)_3$ binding, and Co^{2+} (IV, E), it has been possible to dissect some of the partially resolved resonance regions in the field region above 195 Hz and to make some tentative assignments of specific methyl resonances in this region (*60*). Results of these studies are summarized in Table II. Less certain assignments are in parentheses. Resonances designated by number in column 2 are either clearly resolved in the native spectrum or separately resolved as discrete resonances in the spectra of lysozyme in the presence of Co^{2+}. The met-105 methyl resonance at 15 Hz is considered to be most firmly established. This resonance is sharper than any of the other methyl resonances which exhibit somewhat rounded peaks because of unresolved spin–spin structure. The methionine methyl resonance does not have observable spin–spin fine structure.

Compared to the random-coil spectrum of lysozyme, the spectrum of the native protein shows a large decrease around 195 Hz and a large increase to lower field in the region from 223 to 300 Hz. Part of the decrease at or near 195 Hz is accounted for by the shift to higher fields of approximately 42 of the 120 methyl resonances that were originally in this field region. Intensity is also lost from the 195-Hz methyl region because some leucine, isoleucine, and valine methyl groups in the native protein are in the low-field directing ring-current fields of aromatic residues. These ring-current fields are not expected to produce shifts greater than about 60 Hz for the closest approach of a methyl group to the edge of an aromatic ring but appear to produce shifts of 10 to 50 Hz for a considerable number of methyl resonances. The resonance region from 223 to 300 Hz, corresponding to about 86 protons, arises from methyl groups of threonine residues, from leucine, valine, and isoleucine methyl groups that are subjected to a low-field ring-current shift, and from resonances to lower field in the random-coil protein that are shifted into this region by high-field ring-current shifts. Sternlicht and Wilson (*18*) have postulated that a considerable amount of the absorption in this region arises from resonances of methyl groups of leucine, isoleucine, and valine that are shifted into this region because the residues are in an internal, hydrophobic environment in the protein.

The resonances from 150 to −200 Hz represent high-field ring-current shifts of at least 40 to 390 Hz. The positions of these resonances, particularly those exhibiting the larger shifts, are expected to change rapidly in magnitude as the proton relationship to the aromatic ring face is changed. For example, a proton along the sixfold axis of a phenylalanine ring exhibiting a high-field shift of 325 Hz might well shift 10 Hz if the relative position of the proton

TABLE II

HEW Lysozyme Resonances Observed to High-Field from 330 Hz pD 5.0, 55°

Resonance position or region [Hz (220 MHz) from DSS]	Resonance designation	Estimated number of component protons	Assignment	Major source of ring-current shift
−435	11	1	(Ileu-98 γ)	Try-63
−183	10	1		
−124	1	3	Leu-17 CH_3	Try-28 (Tyr-20 minor)
−64	9	1		
−38	2	3	Ileu-98 δCH_3	Try-63
−8	3	3	Leu-17 CH_3	Try-28 (Tyr-20 minor)
0	4	3	Ileu-98 γCH_3	Try-63, Try-108
15	5	3	(Leu-8 CH_3)	Phe-3
15	6	3	Met-105 CH_3	Try-28, Try-108, Try-111, Tyr-23
≈42	15, 16	2		
57–104	7, 8 plus others	10	2 or 3 methyl groups (Thr-51 CH_3, Ileu-88 δCH_3) plus single protons	Thr-51—Tyr-53 Ileu-88—Phe-3, His-15
104–163		23	Includes several methyl groups (Leu-8, Leu-56, Ileu-55)	
163–194		16	Includes several methyl groups [Val-92, Leu-56, Ileu-58, Ileu-78(2)]	
194–223		26		
223–300		86		
300–330		26		

to the aromatic ring changed by only 0.05 Å. Thus, identification of the resonances from 150 to −200 Hz is important since they are very sensitive indicators of conformational changes in the protein regions where their residues are situated. If the assignments listed in Table II are correct, these resonances correspond to residues in several different regions of lysozyme and provide probes for conformational changes at these locations. As was indicated earlier, if all the high-field resonances (150 to −200 Hz) remain virtually constant when a parameter such as temperature is varied, one can

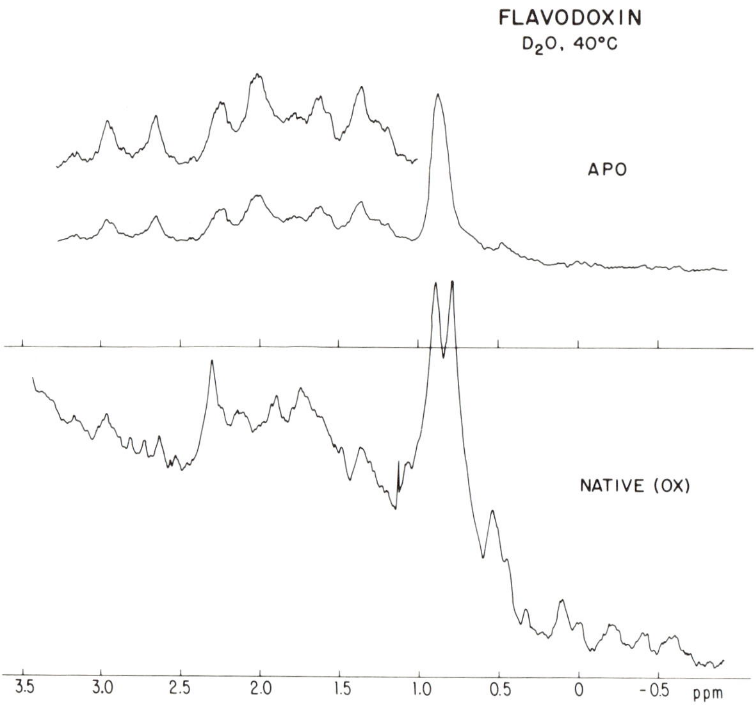

Fig. 13. 220-MHz PMR spectra of flavodoxin and apo-flavodoxin.

be assured that the protein conformation is essentially unchanged. Shifts of some of these resonances when the environment of lysozyme is changed or the protein is reacted with other species were discussed in III, B.

The 220-MHz PMR spectrum of native ribonuclease does not exhibit resonances substantially shifted to high field from the random-coil positions of methyl resonances, indicating that for this protein there are no close interactions of methyl groups with the faces of the conjugated rings of aromatic residues. The PMR spectrum of native flavodoxin (*37*) (Fig. 13) exhibits a rather sharp division of methyl proton resonances into two categories at 175 and 195 Hz as well as a number of resonances shifted to higher field (*38*). Here ring-current shifts may arise from interactions with the flavin mononucleotide prosthetic group as well as with aromatic residues. The largest high-field shifts that we have observed for a diamagnetic protein are in the spectrum of ferrocytochrome c where resonances of single protons are observed at -420, -570, and -830 Hz and a resonance of three equivalent protons is at -730 Hz. In this protein, very substantial ring currents are provided by the heme ring, and these resonances are tentatively assigned to the three methyl group protons, two γ protons, and one β proton of methionine-80 (horse cytochrome c) (*48*). If the sulfur atom of this methionine residue is bound to the heme iron (*51*), these protons are located in the strong heme ring-current field so that their resonances should exhibit large high-field shifts.

C. Aromatic Resonances

The aromatic region of resonance absorption of proteins is of particular interest because only four residue types contribute to this region (histidine, phenylalanine, tyrosine, and tryptophan), and the aromatic residues (particularly histidine, tyrosine, and tryptophan) often are implicated in the active site of an enzyme or in crucial structural interactions. This spectral region, therefore, is highly attractive for detailed analysis, but has not yet been elucidated in detail for any protein. However, in D_2O solution where N*H* protons are replaced by deuterons, the C-2 protons of component histidine residues customarily are the lowest field resonances encountered in the PMR spectra of proteins. (An exception is encountered in heme-containing proteins where ring-current and contact interaction effects can displace resonances to low field of the histidine C-2 protons.) Jardetzky, Scheraga, and their co-workers have effectively used the C-2 protons of the four histidine residues of ribonuclease to elucidate structure and interaction in this enzyme (*15,17,24,41*).

Bovine pancreatic ribonuclease A (RNase A) contains four histidine residues whose C-2 protons are separately resolved at 100 MHz. The pH dependences of these four resonances along with the C-4 proton of one of these residues are shown in Fig. 14. As can be seen, the four histidine residues are differentiated by their p*K* values which occur at 6.7, 6.2, 5.8, and 6.4. The single C-4 peak observed is associated with the C-2 resonance of curve 1 of Fig. 14 since both exhibit a p*K* of 6.7.

The four histidine residues of RNase A are located at positions 12, 48, 105, and 119 in the primary sequence, and his-12 and his-119 are known from other studies to be at the active site of the enzyme. Meadows et al.

succeeded in associating each of the C-2 proton resonances of RNase A with a specific histidine residue in the following manner.

Peaks 2 and 3 of Fig. 14 were displaced, and peaks 1 and 4 were unaffected by the presence of mononucleotide inhibitors. Peaks 2 and 3 were therefore to be associated with residues his-12 and his-119, although on this information alone specific assignment was not possible. Peak 4 was assigned to his-48, the buried histidine of RNase A, because of its anomalous chemical shift in the acid pH region and its greater linewidth which was taken to reflect a lower mobility than the three other relatively exterior histidine residues. Peak 1 therefore remained to be assigned to his-105.

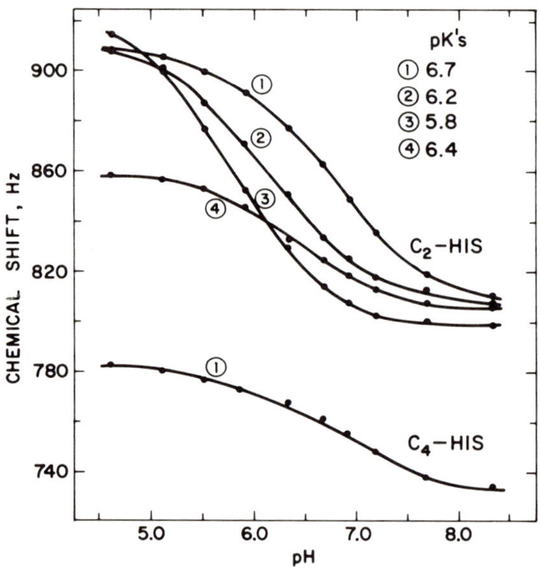

Fig. 14. pH dependences of the chemical shifts of the four C-2 and one of the C-4 resonances of the histidine residues of RNase. Referred to external HMS, 100 MHz (Ref. *24*).

To assign peaks 2 and 3 to the appropriate his-12 and his-119 residues, use was made of the fact that RNase could be cleaved by the enzyme subtilisin at the 20–21 peptide bond. Earlier work had shown that RNase S, formed by bringing together without covalent linkage the two subtilisin-produced fragments of RNase A, has virtually the same enzymatic activity and specificity as the parent RNase A. The C-2 proton of his-12 was replaced by a deuteron by exposing the S-peptide (residues 1–20) to D_2O at 40°C for five days. Then RNase S was reconstituted from the S-peptide with deuterated C-2 of his-12 and S-protein (residues 21–124) with nondeuterated his-119. From this, it was relatively straightforward to associate peak 2 with his-12 and peak 3 with his-119.

We have examined the C-2 histidine resonance region of horse ferrocytochrome c as a function of pD. Only one proton resonance is observed to shift to low field as pD is decreased from 7 to 3 although the protein contains three histidine residues. (The resonances at 1985, 2042, and 2108 Hz in Fig. 7 are assigned to heme ring C*H* protons.) We conclude that the C-2 resonance of his-18, which is bound to the heme iron (*51*), is shifted to high field out of the aromatic region; that his-26 is internally bound with the imidazole ring in the basic form and is inaccessible to protonation; and that the observed resonance can be assigned to the C-2 proton of his-18 situated on the outside of the protein molecule.

Similar analyses involving histidine residues and other proteins should be possible. The C-2 protons of histidine residues thus provide very useful access to information on protein structure and interactions by NMR. In passing it might be mentioned that the resonance of the C-2 proton of the single histidine residue of lysozyme is readily observable, but is of somewhat secondary interest since it is not involved in the active site of the enzyme.

D. N*H* Proton Resonances

The proton magnetic resonance spectroscopy of proteins normally is carried out in D_2O solution in order to suppress the strong H_2O resonance that can overwhelm the central and aromatic regions of absorption of a protein. As a matter of fact, the exchangeable N*H* and O*H* protons of proteins usually are replaced by exchange preliminary to running in D_2O to further decrease the intensity of the residual HDO resonance and attendant side bands.

However, it is apparent that while replacement of N*H* protons by deuterium results in great simplifications of the PMR spectra of proteins, much potentially valuable information is lost in our resultant inability to observe such resonances. Amide N*H* protons are intimately involved in the secondary and tertiary structures of proteins, and the indole N*H* protons of tryptophan residues often are major participants, as in lysozyme, in substrate and inhibitor binding. Recently N*H* proton resonances have been observed to good advantage in synthetic polypeptides (*61*) and in natural cyclic peptides (*62,63*), and we feel that such studies will become increasingly attractive for protein chemists and NMR spectroscopists in their attempts to unravel the subtleties of protein structure and interactions.

Optical rotary dispersion (ORD) and circular dichroism (CD) studies have established that whereas poly-L-leucine is random-coil in trifluoroacetic acid, it assumes a right-handed helix in chloroform. Ferretti (*61*) examined the region of resonance absorption of N*H* protons of a solution of poly-L-leucine in a solvent consisting of 42% trifluoroacetic acid and 58% $CDCl_3$, a composition in which ORD measurements suggested that comparable concentrations of the random-coil and helical forms could coexist. And indeed Ferretti was able to partially resolve two N*H* proton resonances in this system, one attributable to the polypeptide in the random-coil form and the other, shifted about 0.2 ppm to low field, assignable to the helical form. The observation of separate resonances led to the important conclusion that the rate of interconversion between the helical and random-coil forms must be less than about 10^2 sec^{-1}, a surprisingly slow rate.

The question then arises as to how profitably one can expect to study the N*H* proton region of resonance absorption of proteins in the molecular weight range of, say, 10,000 and higher and to carry this study out in H_2O rather than nonbiological solvents such as trifluoroacetic acid/$CDCl_3$ mixtures. These questions can best be answered through a specific example. The PMR spectrum of the region of resonance absorption low field to that of the solvent H_2O is shown for hen egg white lysozyme in Fig. 15(a) under conditions of thermal denaturation (71°C). Resonances in the range 1550 to 1700 Hz arise from aromatic C*H* protons of component aromatic residues of lysozyme (1 his, 3 phe, 3 tyr, 6 try). The intense, incompletely resolved resonances from 1700 to 1850 Hz arise from amide N*H* protons. The peak to extreme low field at 2200 Hz is attributable to the indole N*H* protons of the six tryptophan residues of lysozyme.

It is immediately apparent from the above that even for the denatured form of the protein, the rate of exchange of amide and tryptophan N*H* protons is less than about 10^3 sec^{-1}, since otherwise resonances from these protons would be severely broadened; for fast exchange on the NMR time scale, the tryptophan N*H* resonances would of course actually merge with the proton resonance of the solvent water. As might be expected, however, the resonance characteristics of the N*H* region are highly pH dependent since the exchange is base catalyzed; at pH's > 7 the N*H* resonances of lysozyme at 81°C are severely broadened and in part merged with that of H_2O.

As the temperature of the solution is reduced and the protein reassumes the folded, biologically active conformation, the peaks attributed to amide N*H* are reduced in intensity and replaced by a complex distribution of peaks [Fig. 15(b)] largely displaced to low field. The complexity of the N*H* region below the denaturation temperature undoubtedly is attributable to the multiplicity of shielding environments and hydrogen-bonding situations experienced by the amide N*H* protons in the folded conformation. Differential exchange rates against D_2O have been observed for amide N*H* protons in the native conformation of the protein. We are attempting to analyze this complex but potentially highly rewarding region of resonance absorption of proteins, to study exchange rates quantitatively, and to correlate these rates with structural parameters such as the α-helical content and the accessibility of the proton to the solvent.

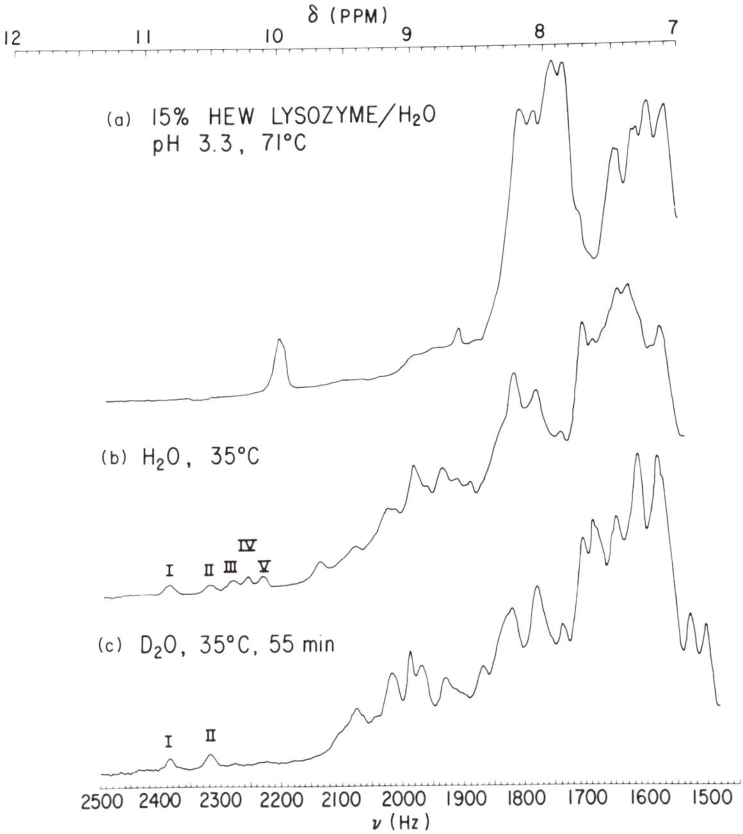

Fig. 15. 220-MHz PMR spectra of low-field resonances of HEW lysozyme: (a) thermally denatured in H_2O, (b) native in H_2O, (c) native after 55-min exchange in D_2O at 5°C (Ref. *33*).

A considerably simpler situation is encountered with respect to the 2200-Hz resonance that arises from the six tryptophan residues of hen egg white lysozyme. It is worth noting that even though all six tryptophan residues are structurally nonequivalent with respect to the primary sequence of amino acids, the resonances from all six indole N*H* protons are coincident in the denatured state of the protein. When the protein is refolded by reducing the temperature, this coincidence is removed and five of the six indole N*H* protons exhibit separate and resolved resonances [Fig. 15(b)]. The resonance of the indole N*H* proton of the sixth tryptophan residue presumably has been displaced into the complex amide N*H* region. Similarly, human lysozyme which contains five rather than six tryptophan residues exhibits five resolved indole N*H* resonances in the folded conformation.

For the HEW lysozyme, a probable assignment of the five resolved indole N*H* resonances to the six tryptophan residues has been made on the basis of exchange and inhibitor perturbation studies (*33*) and the X-ray structure. For example, two of the six indole N*H* protons are replaced by deuterium at a slow rate [Fig. 15(c)] indicating that they do not have free access to the solvent and are assigned to try-28 and try-108. Chemical shifts of two N*H* protons are displaced by interaction with the dimer of N-acetylglucosamine, an inhibitor, and are assigned to tryptophan residues 62 and 63. X-ray studies have shown these residues to be involved in substrate and inhibitor binding and try-28 and try-108 to be involved in intramolecular hydrogen bonding (*31*). The sixth indole N*H* resonance which has not yet been positively identified in the spectrum of HEW lysozyme presumably arises from try-111. However, there still remains some ambiguity in the assignments of try-28 and try-111; neither participates in inhibitor binding and both appear from X-ray studies to be involved in intramolecular hydrogen bonding and to be located in relatively interior positions of the enzyme. It is expected that this ambiguity will be removed once the sixth indole N*H* resonance is identified. Studies in progress on human lysozyme may be helpful in this regard.

E. Perturbation of Protein PMR Spectra by Co^{2+}

The two regions of the C*H* PMR spectra of diamagnetic proteins that are most accessible to detailed analysis because of their relative simplicity are the low-field aromatic region and the methyl region to high field of 0.8 ppm. Resonances in the latter region have been attributed to protons (largely methyl protons) that in the native protein are in the ring-current field of aromatic residues and have their resonance positions shifted to higher field. Even in these regions with the resolution available with a 220-MHz spectrometer, considerable overlap of resonances is usually encountered. Techniques to dissect out the components of such groups of resonances are required for complete analysis of these spectral regions. Changes in spectra induced by changing temperature or pD, by adding reactive molecules, etc., can provide this kind of information, but in most cases studied so far, such changes are either small or discontinuous. What is needed is a parameter that will cause substantial and continuous perturbation of the relationships between resonance positions. We have found that, for some proteins, addition of paramagnetic Co^{2+} ion to the protein solution provides such a parameter; we have described the use of this technique to elucidate the high-field spectrum of HEW lysozyme (*64*).

When Co^{2+} is added to a solution of HEW lysozyme in neutral D_2O, changes occur throughout the PMR spectrum. Some resonances are shifted to high field, some to low field, and some remain virtually unchanged. The magnitudes of the shifts are different for different resonances but all increase in a continuous manner with increasing Co^{2+} over a considerable concentration range. Thus, resonances move with respect to each other and the

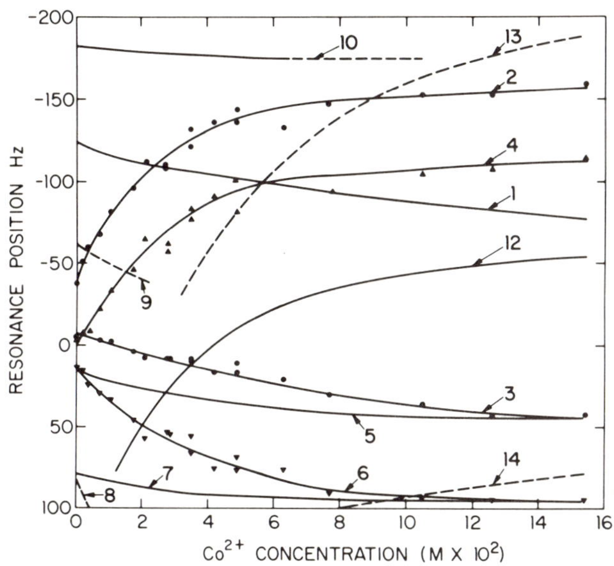

Fig. 16. Co^{2+}-induced shifts of HEW lysozyme resonances vs. Co^{2+} concentration. Lysozyme 7×10^{-3} *M* in D_2O, pD 5.5, at 55°C. Referred to internal DSS, 220 MHz (Ref. *64*).

components of overlapped resonance groups can be discerned. The changes in resonance positions of resonances in the field region from 100 to −200 Hz are plotted as a function of Co^{2+} concentration in Fig. 16. Representative spectra and the numbering of the resonances are shown in Fig. 17. Examination of these spectra indicates that at least 10 individual resonances contribute to this spectral region in the absence of Co^{2+}. In particular, the resonance group initially between 30 and −30 Hz can be seen in spectra (b) to (e) to separate into four separate resonances (3, 4, 5, and 6). Resonances 1 through

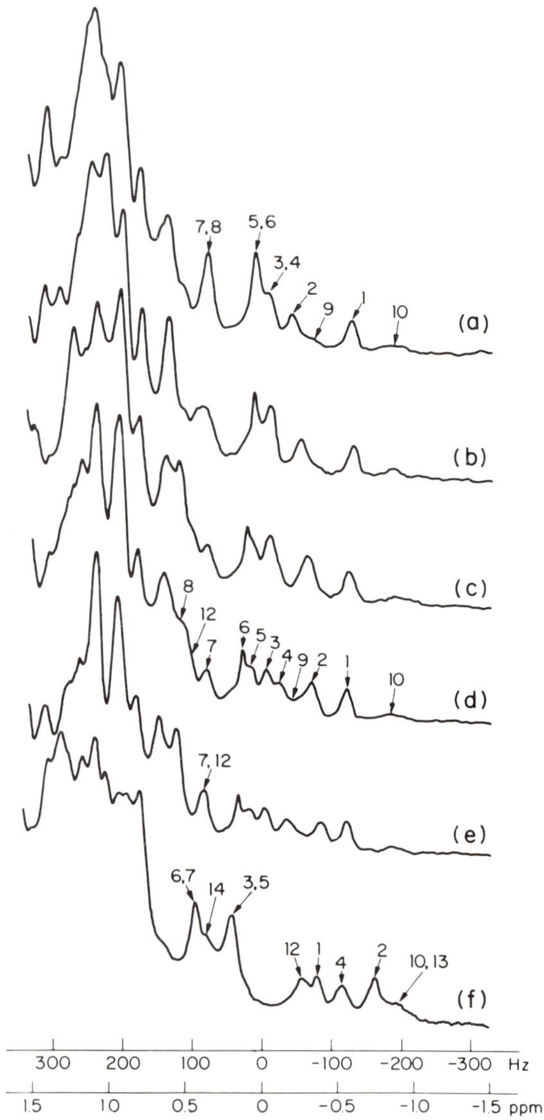

Fig. 17. Perturbation of the 300 to -300 Hz PMR spectrum of HEW lysozyme by Co^{2+}. Lysozyme 7×10^{-3} *M* in D_2O, pD 5.5, 55°C. Co^{2+} concentrations: (a) 0, (b) 1.75×10^{-3} *M*, (c) 3.50×10^{-3} *M*, (d) 7.00×10^{-3} *M*, (e) 1.05×10^{-2} *M*, and (f) 1.54×10^{-1} *M* (Ref. *64*).

7 appear to have intensity corresponding to methyl groups. Resonances 12, 13, and 14 move into this field region only after addition of Co^{2+} to the protein solution.

The Co^{2+}-induced shifts are presumed to arise from binding of Co^{2+} to lysozyme and to be caused by contact shifts arising from the paramagnetism

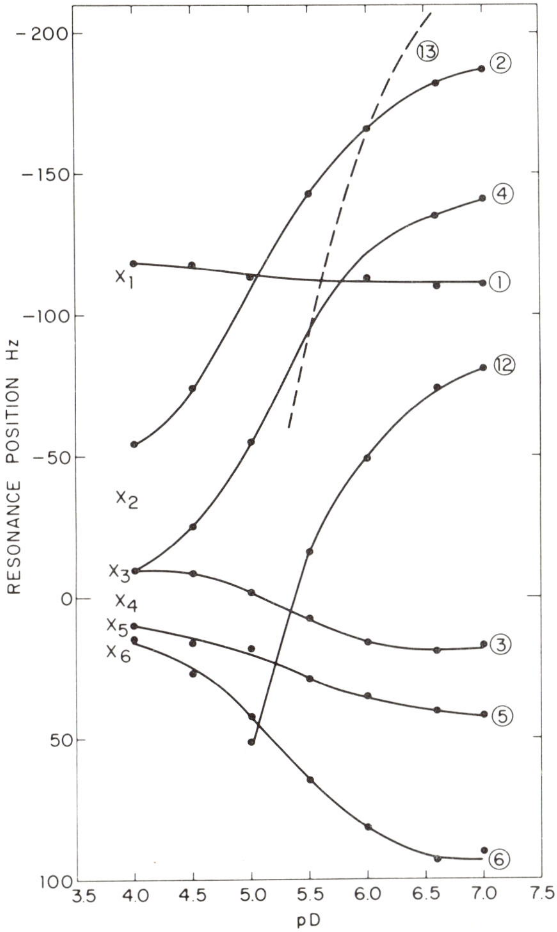

Fig. 18. Co^{2+}-induced shifts of HEW lysozyme resonances vs pD. Lysozyme 7×10^{-3} *M* in D_2O, 55°C. Co^{2+} 4.2×10^{-2} *M* (Ref. *64*).

of the Co^{2+}. The effects appear to be rather specifically related to Co^{2+} as a similar perturbation is not produced by Fe^{2+}, Ni^{2+}, or Zn^{2+}. Exchange between free and complexed protein is fast on a millisecond time scale since an exchange-averaged spectrum is observed. The curves for resonances 2, 4, and 6 in Fig. 16 indicate that the Co^{2+} binding site causing the shifts of these resonances is almost saturated at 0.15 *M* Co^{2+}. In fact, these resonances appear to be associated with the same Co^{2+} binding site, and analysis of the dependences of the Co^{2+}-induced shifts on Co^{2+} concentration indicates a 1 : 1 association of protein and Co^{2+} (at this binding site) with an association constant of 57 liters mole^{-1} at pD 5.5 and 55°C. Shifts of other resonances (5, 12, and 13) also appear to be associated with the Co^{2+} binding at this site. However, the Co^{2+}-induced shifts of resonances 1 and 3 are still increasing at 0.15 *M* Co^{2+} and may therefore be influenced by Co^{2+} at a second, weaker binding site.

For a given Co^{2+} concentration, many of the Co^{2+}-induced resonance shifts are strongly dependent on the pD of the protein/Co^{2+} solution. Representative data for one Co^{2+} concentration are shown in Fig. 18. At pD 4, the pD-dependent resonance positions which are also those associated with binding site 1 (e.g., resonances 2, 4, and 6) are similar to those of a Co^{2+}-free solution, but shift with increasing pD following an apparent titration curve with a p*K* of about 5.2. At higher Co^{2+} concentrations, this apparent p*K* is shifted to lower pD values. We have concluded that binding site 1 is probably the negatively charged carboxyl groups of glu-35 and asp-52. The dependence of this binding on pD is a consequence of the competition of Co^{2+} and hydrogen ions at the glu-35 carboxyl group which is thought (*65*) to have a p*K* value (in H_2O and in the absence of Co^{2+}) of 6.3.

The use of Co^{2+} to assist in the analysis of protein PMR spectra is not limited to HEW lysozyme. Similar perturbations (some greater and some smaller) have been observed for Co^{2+} with other proteins such as human lysozyme, bovine pancreatic ribonuclease, and horse ferrocytochrome c.

F. Specific Residue Substitutions

Another approach to elucidation of PMR spectra of native proteins is through comparison of spectra of proteins that have slightly different (but known) amino acid sequences. At the present time, cytochrome c is best for this approach since sequence information is available for a considerable number of cytochromes c from different organisms (*66*). These cytochromes c have a further advantage in having a relatively small number of amino acid substitutions for species that are closely related phylogenetically.

In this connection, we have examined PMR spectra of cytochrome c in both the ferri and ferro forms from horse, cow, hog, chicken, turkey, duck, pigeon, tuna, and yeast (*Candida krusei*) (*48,67,68*). The methyl and high-field regions of the spectra of ferrocytochromes c from horse, hog, turkey, and pigeon are shown as typical examples in Fig. 19. As perhaps might be anticipated, these spectra have a number of constant features but also a

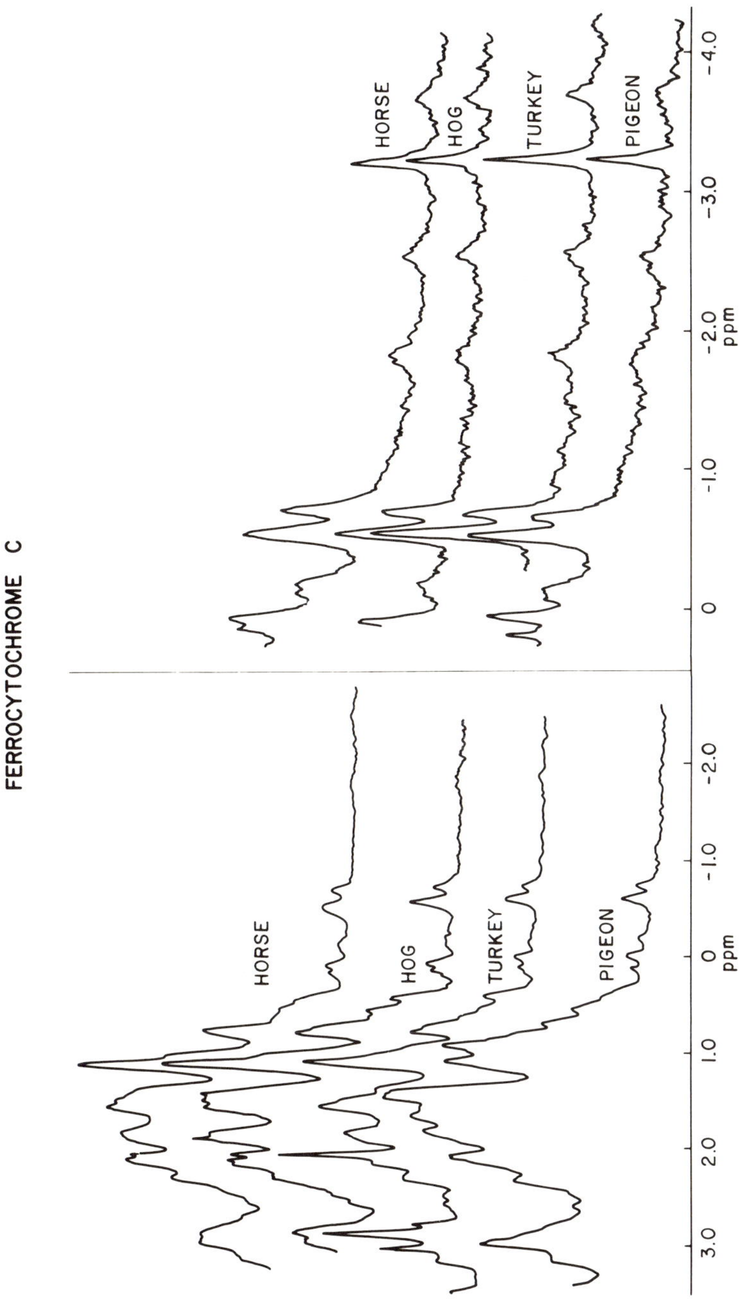

Fig. 19. 220-MHz PMR spectra of ferrocytochromes c from horse, hog, turkey, and pigeon.

considerable number of readily observed differences. The four resonances at −420, −570, −730, and −830 Hz, already tentatively assigned to met-80 bound at the sixth ligand position to the heme iron, are present for all the species shown here and for the other cytochromes c we have examined. The two resonances at −120 and −165 Hz appear to arise from three methyl groups and again are a common feature of all the cytochromes c studied except *Candida krusei* which has only two methyl group resonances in this spectral region. Good possibilities for assignment to these resonances are methyl protons of thr-19, thr-78, and ileu-81 (ala-81 in yeast). In the normal methyl region around 200 Hz, differences between the species (e.g., between turkey and pigeon) cannot be accounted for by addition or subtraction of protons corresponding to the residue differences. Indeed, similar characteristic differences can be observed in the aromatic region for species which have no differences in *aromatic* residues. These differences then must arise from small conformational differences induced by the residue substitutions that have occurred.

G. Selective Deuteration of Proteins

A technique, quite different from those that have been discussed thus far, for simplifying analysis of C*H* PMR spectra of proteins has recently been introduced and appears to have considerable future potential. The technique is based on preparation of a protein with known specific replacements of residue C*H* protons by deuterons. This approach has, in fact, been described above in Section IV, C in the discussion of the work of Meadows et al. (*24*). Here, rather unique circumstances permitted these investigators to deuterate one specific proton in pancreatic ribonuclease—the C-2 proton of his-12—and then to assign the resonance of this proton in the PMR spectrum of the protonated protein by its absence from the spectrum of the deuterated molecule. In general, such specific deuteration is not feasible. However, there are general approaches to the deuteration of all protons of a given chemical type (e.g., all alanine methyl protons) in a protein molecule or to deuteration of many chemical proton types simultaneously. Then the PMR spectrum of the protein comprises only resonances of the remaining proton types and is greatly simplified. Illustrations of the use of this technique have been provided by Markley, Putter, and Jardetzky (*69*) and by Crespi, Rosenberg, and Katz (*70*).

In the work of Markley et al., deuterated amino acids were obtained by hydrolysis of protein isolated from algae grown in D_2O. These amino acids were completely deuterated except for a few proton types which back-exchanged to various extents to the protonated form during hydrolysis of the completely deuterated protein. These proton types were at the 2- and 6-

position of the tyrosine ring, the 2-position of the histidine imidazole ring, the β-positions of aspartic acid, and the γ-positions of glutamic acid. The bacterium *Staphylococcus aureus* was grown on a chemically defined nutrient medium containing the deuterated protein hydrolysate with an excess of the fully protonated forms of tryptophan and methionine. The protein, designated staphylococcal nuclease, was isolated, and PMR spectra were obtained at 100 MHz both for this protein (deuterated protein) and for the protein from bacteria grown in a fully protonated medium. The spectrum of the deuterated protein was indeed greatly simplified, comprising only resonances of tryptophan, methionine, and the back-exchanged protons of tyrosine, histidine, aspartic acid, and glutamic acid mentioned above.

The histidine C-2 proton resonances of the deuterated and fully protonated proteins appeared identical providing evidence that the conformation of the protein was not changed by the extensive deuteration. In the low-field aromatic region, resonances of the one tryptophan and seven tyrosine residues were incompletely resolved, An analysis distinguishing between tyrosine and tryptophan resonances was made, but it was not possible to assign specific resonances to particular residue protons. Several of the tyrosine resonances were shifted when the inhibitor deoxythymidine-3′,5′-diphosphate was added to the solution of deuterated nuclease. Further shifts were observed when Ca^{2+} was added to the protein/inhibitor solution.

Comparison of the PMR spectra of the protonated and deuterated proteins in the field region from 1 to 2.5 ppm (referred to external hexamethyldisiloxane) was particularly striking. For the protonated protein, many different aliphatic protons exhibit resonances in this spectral region so that only an envelope of resonances is observed. The spectrum of the deuterated protein exhibited only quite well-resolved methionine resonances.

Crespi et al. have described rather similar experiments in which they grew blue-green algae and the bacterium *Rhodospirillum* in D_2O with fully protonated L-leucine added to the growth medium. (Deuterated algal hydrolysate also was added in the latter case.) The proteins C-phycocyanin and C-phycoerythrin were extracted from the algae. PMR spectra of hydrolysates of these proteins showed that the leucine content was largely in the protonated form, but all other amino acids appeared to be fully deuterated. PMR spectra of phycocyanin and phycoerythrin from the algae and of cytochrome c from *Rhodospirillum* all indicated multiple resonances for leucine, reflecting multiple environments for leucine side chains in these native proteins.

It is apparent there are two major ways in which specific deuteration of amino acid side chains can be used to elucidate and simplify protein PMR spectra: (1) preparation of a protein with all but one or two amino acid types deuterated by growing microorganisms on a deuterated medium as has been described or (2) preparation of protein with one amino acid type deuterated and the rest protonated. This latter type of protein would be prepared by growing an appropriate microorganism on a protonated medium except for the one amino acid type which would be fed in the deuterated form. This procedure could, of course, also be used for higher organisms for specific amino acids that they require as nutrients. In comparing the PMR spectra of a lightly deuterated protein prepared by (2) with that of its fully protonated homolog, information is obtained by noting resonances that are missing in the former. This procedure will be generally less useful than (1) but may be helpful in some instances. It requires a supply of a purified deuterated amino acid whereas procedure (1) requires growth on D_2O with addition of deuterated protein hydrolysate in some instances.

Specific deuterated amino acids can be obtained by fractionation of a deuterated protein hydrolysate. There are also now available several chemical procedures for specific deuteration of amino acids. The exchange of the C-2 imidazole proton of histidine in D_2O has been described. The experiments of Markley et al. (*69*) indicate that, for the acidic conditions used in protein hydrolysis, exchange of the following protons in D_2O is feasible: the 2- and 6-protons of the tyrosine ring, the β protons of aspartic acid, and the γ protons of glutamic acid. A procedure for deuteration of amino acids at the α- and β-positions is indicated by the work of Abbott and Martell (*71*) and of Gansow and Holm (*72*). In this technique, a protonated amino acid is treated in alkaline D_2O with pyridoxyl. Exchange occurs because of formation of a Schiff's base and the process is catalyzed by metal ions such as Al^{3+}, Cu^{2+}, Zn^{2+}. Specific deuteration of valine and alanine by this method has been demonstrated. Bak et al. (*73*) have reported exchange of tryptophan indole C*H* protons when tryptophan is dissolved in mixtures of CF_3COOD and D_2O.

REFERENCES

1. For example, see A. S. Mildvan and M. Cohn, *J. Biol. Chem.*, **240,** 238 (1965); **241,** 1178 (1966); W. J. O'Sullivan and M. Cohn, *J. Biol. Chem.*, **241,** 3104, 3116 (1966); G. Navon, R. G. Shulman, B. J. Wyluda, and T. Yamane, *Proc. Natl. Acad. Sci. U.S.*, **60,** 86 (1968).

2. For example, see J. J. Fischer and O. Jardetzky, *J. Am. Chem. Soc.*, **87,** 3237 (1965): O. Jardetzky and N. G. Wade-Jardetzky, *Mol. Pharmacol.*, **1,** 214 (1965); J. T. Gerig, *J. Am. Chem. Soc.*, **90,** 2681 (1968).

3. T. McL. Spotswood, J. M. Evans, and J. H. Richards, *J. Am. Chem. Soc.*, **89,** 5052 (1967).

4. T. R. Stengle and J. D. Baldeschwieler, *Proc. Natl. Acad. Sci. U.S.*, **55,** 1020 (1966).

5. C. Ho and R. J. Kurland, *J. Biol. Chem.*, **241,** 3002 (1966).

6. M. Saunders, A. Wishnia, and J. G. Kirkwood, *J. Am. Chem. Soc.*, **79,** 3289 (1957).

7. F. A. Bovey, G. V. D. Tius, and G. Filipovich, *J. Polymer Sci.*, **38,** 73 (1959).

8. A. Wishnia and M. Saunders, *J. Am. Chem. Soc.*, **84,** 4235 (1962).

9. A. Kowalsky, *J. Biol. Chem.*, **237,** 1807 (1962).

10. R. R. Ernst and W. A. Anderson, *Rev. Sci. Instr.*, **37,** 93 (1966).
11. A. Kowalsky, *J. Biol. Chem.*, **239,** 3299 (1964).
12. M. Mandel, *Proc. Natl. Acad. Sci. U.S.*, **52,** 736 (1964).
13. M. Mandel, *J. Biol. Chem.*, **240,** 1586 (1965).
14. A. Kowalsky, *Biochemistry*, **4,** 2382 (1965).
15. J. H. Bradbury and H. A. Scheraga, *J. Am. Chem. Soc.*, **88,** 4240 (1966).
16. D. P. Hollis, G. McDonald, and R. L. Biltonen, *Proc. Natl. Acad. Sci. U.S.*, **58,** 758 (1967).
17. D. H. Meadows, J. L. Markley, J. S. Cohen, and O. Jardetzky, *Proc. Natl. Acad. Sci. U.S.*, **58,** 1307 (1967).
18. H. Sternlicht and D. Wilson, *Biochemistry*, **6,** 2881 (1967).
19. J. H. Bradbury and P. Wilairat, *Biochem. Biophys. Res. Commun.*, **29,** 84 (1967).
20. C. C. McDonald and W. D. Phillips, *Proceedings of the Second International Conference on Magnetic Resonance in Biology*, Pergamon, New York, New York, 1967.
21. C. C. McDonald and W. D. Phillips, *J. Am. Chem. Soc.*, **89,** 6332 (1967).
22. E. M. Bradbury, C. Crane-Robinson, H. Goldman, H. W. E. Rattle, and R. M. Stephens, *J. Mol. Biol.*, **29,** 507 (1967).
23. J. S. Cohen and O. Jardetzky, *Proc. Natl. Acad. Sci. U.S.*, **60,** 92 (1968).
24. D. H. Meadows, O. Jardetzky, R. M. Epand, H. H. Ruterjans, and H. S. Scheraga, *Proc. Natl. Acad. Sci. U.S.*, **60,** 766 (1968).
25. R. J. Kurland, D. G. Davis, and C. Ho, *J. Am. Chem. Soc.*, **90,** 2200 (1968).
26. K. Wüthrich, R. G. Shulman, and J. Peisach, *Proc. Natl. Acad. Sci. U.S.*, **60,** 373 (1968).
27. A. Nakamura and O. Jardetzky, *Proc. Natl. Acad. Sci. U.S.*, **58,** 2212 (1967).
28. C. C. McDonald and W. D. Phillips, *J. Am. Chem. Soc.*, **91,** 1513 (1969).
29. B. Bak and E. J. Pedersen, *J. Biol. Chem.*, **242,** 2637 (1967).
30. B. Bak, C. Dambmann, F. Nicolaisen, and E. J. Pedersen, *J. Mol. Spectr.*, **26,** 78 (1968).
31. D. C. Phillips, *Proc. Natl. Acad. Sci. U.S.*, **57,** 484 (1967).
32. J. A. Rupley, L. Butler, M. Geering, F. J. Hartdegen, and R. Pecoraro, *Proc. Natl. Acad. Sci. U.S.*, **57,** 1088 (1967).
33. J. D. Glickson, C. C. McDonald, and W. D. Phillips, *Biochem. Biophys. Res. Comm.*, **35,** 492 (1969).
34. J. D. Glickson, C. C. McDonald, and W. D. Phillips (to be published).
35. A. J. Sophianopoulous and K. E. Van Holde, *J. Biol. Chem.*, **239,** 2516 (1964).
36. From collaborative studies with L. L. Anderson, R. W. F. Hardy, and E. K. Jackson (to be published).
37. From collaborative studies with W. Lovenberg, National Institutes of Health (to be published).
38. E. Knight, Jr., and R. W. F. Hardy, *J. Biol. Chem.*, **242,** 1370 (1967).
39. From collaborative studies with E. Knight, Jr., and R. W. F. Hardy (to be published).
40. From collaborative studies with G. Palmer, University of Michigan (to be published).
41. D. H. Meadows and O. Jardetzky, *Proc. Natl. Acad. Sci. U.S.*, **61,** 406 (1968).
42. A. M. Crestfield, W. H. Stern, and S. Moore, *J. Biol. Chem.*, **238,** 2413 (1963).
43. G. Kartha, J. Bello, and D. Harker, *Nature*, **213,** 862 (1967).
44. C. C. F. Blake, G. A. Mair, A. C. T. North, D. C. Phillips, and V. R. Sarma, *Proc. Roy. Soc.*, Ser. B, **167** (1009), 365 (1967).
45. C. C. F. Blake, L. N. Johnson, G. A. Mair, A. C. T. North, D. C. Phillips, and V. R. Sarma, *Proc. Roy. Soc.*, Ser. B, **167** (1009), 378 (1967).
46. J. A. Rupley and V. Gates, *Proc. Natl. Acad. Sci. U.S.*, **57,** 496 (1967).
47. From collaborative studies with J. A. Rupley, Univeristy of Arizona.

48. C. C. McDonald, W. D. Phillips, and S. N. Vinogradov, *Biochem. Biophys. Res. Comm.*, **36,** 442 (1969).
49. K. Ando, H. Matsubara, and K. Okunuki, *Biochem. Biophys. Acta*, **118,** 240 (1966); *Proc. Japan. Acad.*, **41,** 79 (1965).
50. H. J. Tsui and G. R. Williams, *Can. J. Biochem.*, **43,** 1409 (1965).
51. S. N. Vinogradov and H. A. Harbury, *Biochemistry*, **6,** 709 (1967); M. W. Fanger, T. P. Hettinger, and H. A. Harbury, *ibid.*, **6,** 713 (1967).
52. J. A. Pople, *J. Chem. Phys.*, **24,** 1111 (1956).
53. J. W. Emsley, J. Feeney, and L. H. Sutcliffe, *High Resolution Nuclear Magnetic Resonance Spectroscopy*, Vol. 1, Pergamon, New York, New York, 1965, pp. 140–149
54. C. E. Johnson and F. A. Bovey, *J. Chem. Phys.*, **29,** 1012 (1958).
55. E. D. Becker and R. B. Bradley, *J. Chem. Phys.*, **31,** 1413 (1959).
56. For a review of the effects of paramagnetic centers on the nuclear resonance characteristics of molecules, see D. R. Eaton and W. D. Phillips, *Advan. Magnetic Resonance*, **1,** 103 (1965).
57. N. Bloembergen and L. O. Morgan, *J. Chem. Phys.*, **34,** 842 (1961).
58. J. Eisinger, R. G. Shulman, and W. E. Blumberg, *Nature*, **192,** 963 (1961).
59. D. G. Davis, N. L. Mock, V. R. Loman, and C. Ho, *J. Mol. Biol.*, **40,** 311 (1969).
60. C. C. McDonald and W. D. Phillips (in preparation).
61. J. A. Ferretti, *Chem. Commun.*, **20,** 1030 (1967).
62. A. Stern, W. A. Gibbons, and L. C. Craig, *Proc. Natl. Acad. Sci. U.S.*, **61,** 734 (1968).
63. A. M. Liquori and F. Conti, *Nature*, **217,** 635 (1968).
64. C. C. McDonald and W. D. Phillips, *Biochem. Biophys. Res. Comm.*, **35,** 43 (1969).
65. J. A. Rupley, L. Butler, M. Gerring, F. J. Hartdegen, and R. Pecoraro, *Proc. Natl. Acad. Sci. U.S.*, **57,** 1088 (1967).
66. E. Margoliash and A. Schejter, *Advan. Protein Chem.*, **21,** 113 (1966).
67. C. C. McDonald, W. D. Phillips, S. N. Vinogradov, and R. Zand (in preparation).
68. C. C. McDonald, W. D. Phillips, R. Estabrook, and A. Kowalsky (in preparation).
69. J. L. Markley, I, Putter, and O. Jardetzky, *Science*, **161,** 1249 (1968).
70. H. L. Crespi, R. M. Rosenberg, and J. J. Katz, *Science*, **161,** 795 (1968).
71. E. H. Abbott and A. E. Martell, Abstracts of the 156th Natl. Meeting Am. Chem. Soc., Atlantic City, N.J., Sept., 1968, Biol. Abstr. 98.
72. O. A. Gansow and R. H. Holm, *J. Am. Chem. Soc.*, **90,** 5629 (1968).
73. B. Bak, C. Dambmann, and F. Nicolaisen, *Acta Chem. Scand.*, **21,** 1674 (1967).

CHAPTER 2

MONOMERIC UNITS OF NUCLEIC ACIDS—BASES, NUCLEOSIDES, AND NUCLEOTIDES

Paul O. P. Ts'o

DEPARTMENT OF RADIOLOGICAL SCIENCES
THE JOHNS HOPKINS UNIVERSITY
BALTIMORE, MARYLAND

I. INTRODUCTION

The importance of nucleic acid in biology is so well acknowledged that it requires no further emphasis. Therefore, it is of great value to know the basic principles governing the structure, the chemical reactivity, the conformation, and the interaction of nucleic acid. The primary structure of nucleic acid is a single-stranded, linear (or circular) copolymer of nucleotides that contain predominantly five heterocyclic bases (adenine, guanine, cytosine, uracil, and thymine), a D-ribose or D-deoxyribose attached to the N(9) of the purines or the N(1) of the pyrimidines, and a phosphate group attached to the 3′-position or the 5′-position of the pentose. The chemical structures and the numerical designation of the atoms of the nucleotides are shown in Fig. 1. RNA contains, in the main, the four ribosyl nucleotides i.e., AMP, GMP, CMP, and UMP, while DNA contains, in the main, the four deoxyribosyl nucleotides i.e., dAMP, dGMP, dCMP, and dTMP. Therefore, from the viewpoint of the primary structure, nucleic acid is already much less com-

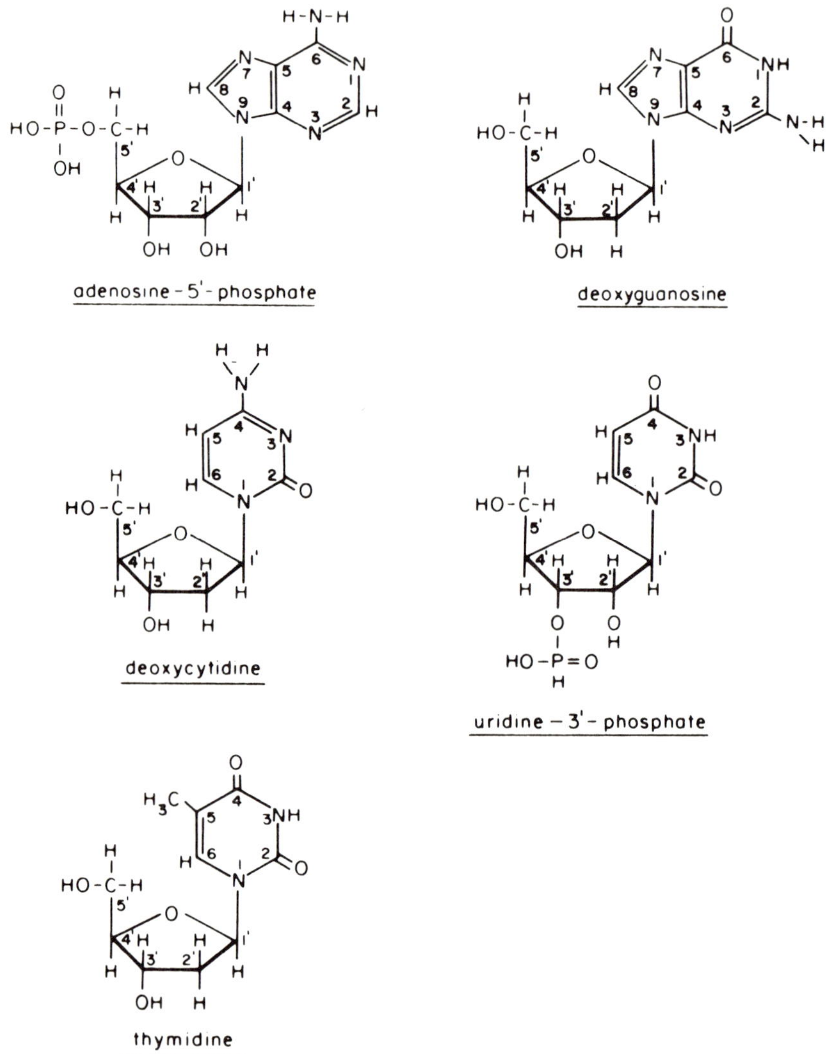

Fig. 1. Structural formula and numbering system for the atoms of the five commonly occurring nucleosides and nucleotides.

plicated than protein, which usually has about 20 kinds of amino acids as monomeric units. As is shown in this review, a systematic study of the physicochemical properties of the monomeric units of nucleic acid, i.e., bases, nucleosides, and nucleotides, provides a valuable understanding about the basic principles governing the properties of nucleic acids. The monomeric units, being much smaller in size and more homogeneous in preparation, can be studied much more thoroughly and vigorously. In fact, it appears this this is the most logical and rewarding first step in studying the basic principles of nucleic acid chemistry.

The scope of this review encompasses mainly the physicochemical properties of the monomeric units. Special emphasis is given to two areas (1) nuclear magnetic resonances (NMR) studies—the reason for this special emphasis is partially the keen interest of our laboratory in this field and also the absence of a comprehensive review in the area of NMR studies on monomeric units of nucleic acids. (2) Interaction studies—the interaction and association in crystalline complexes, in organic solvents, and in aqueous solution, the interaction between the polymers and the monomers, etc. The reason for the special emphasis in this area is that through these intensive studies on the interaction of the monomeric units, we hope to learn about the basic principles governing the conformation and interaction of these polymers—the nucleic acids. Other selected topics such as various physicochemical studies, conformational states, and intramolecular interaction are also discussed.

Many important areas and subject matter concerned with the monomeric units of nucleic acid, such as organic synthesis, chemical reactivities, structure proof, analogs, polarographic studies, etc., are not included in this review. A list of recent general references (reviews, comprehensive articles, etc.) was selected for the purpose of providing the reader with a source of information about areas and subjects not included or not discussed sufficiently in this review.

II. Intrinsic Properties of the Bases, Nucleosides, and Nucleotides

The first six subsections of Section II are concerned with certain recent findings about the monomers by various physicochemical methods. The last two subsections in Section II discuss the problem of the hindered rotation of the glycosyl linkage between the N atom of the base and the C(1′) atom of the pentose, and the problem of intramolecular hydrogen bonding between the 2′-OH group and the base.

A. X-Ray Diffraction Studies

Through the application of the principles of regularity and symmetry, X-ray diffraction investigation of crystals is perhaps the most informative tool for studies on bases, nucleosides, and nucleotides. The precise data on bond distances and bond angles derived from this technique not only provide unambiguous proof of structure and configuration but also provide valuable information on the electronic properties of these molecules. In addition, together with other data on the packing arrangement and the intermolecular distances, etc., this technique yields useful knowledge about the conformation, the hydrogen-bonding scheme, the site of protonation, and so on. The crystallographic data on these monomers are used extensively as guidelines for solving the problems of structure of the nucleic acids. While all the X-ray diffraction studies on nucleic acids previously were done on oriented fibers, crystals of tRNA were obtained very recently (*1–3*). Even so, the minute features required for the building of a most plausible structure for the polymers still must be defined by the crystallographic data on the monomers.

Concerning the problems of subtle conformation and intermolecular interaction, it is often asked, however, "How relevant are the data collected in the solid crystal state to the actual situation of the molecule in aqueous solution?" While this is a legitimate and simple question, there is no short and common answer for all the individual inquiries on this general problem. All the stereochemical and geometrical relationships ultimately depend on the resultant energy levels from the interaction of all participating forces, and each individual question concerning a particular relationship (hydrogen bonding, restricted rotation angle, site of protonation, etc.) has to be defined and examined separately. Several experimental approaches to this problem are briefly discussed here as examples.

It is most gratifying that there have been three penetrating, comprehensive reviews written recently about crystallographic studies on the monomers and the polymers (*4–6*). These papers were very helpful in the preparation of this section. The important question concerning the rotation of the glycosyl bond (C—N bond)† and the study of the paired bases and paired nucleosides is considered in Section II, G and Section III, A.

1. Influence of Crystal Structure on Intermolecular Interaction and Conformation

Many bases, nucleosides, and nucleotides form hydrated crystals. Among these are guanine·HCl·H_2O (*7*), cytosine·H_2O (*8*), thymine·H_2O (*9*),

† Frequently, the term glycosidic bond was used in the literature in the X-ray diffraction field to designate the C—N linkage between the C(1′) of the pentose and the N(9) (purine) or N(1) (pyrimidine) in the nucleoside. It was pointed out by Cohn, however (IEG #7, Scientific Memo #378, May 20, 1966), that since this is a C—N linkage and not a C—O linkage, the linkage should be designated as a glycosyl bond and not as a glycosidic bond in accordance with the rules of carbohydrate nomenclature [*J. Org. Chem.*, **28**, 281 (1963)].

deoxyadenosine·H_2O (*10*), Ca·thymidylate·$6H_2O$ (*11*), adenosine 3′-phosphate·$2H_2O$ (*12*), Ba·uridine 5′-phosphate hydrate (*13*), Na·inosine 5′-monophosphate hydrate (*14*), etc. Studies of these hydrated crystals provide valuable information about the interaction of water molecules with these monomers through hydrogen bonding and packing arrangement. Many bases, nucleosides, and even nucleotides form anhydrous crystals from aqueous solutions. Among these are 1-methylthymine (*15*), purine (from methanol–H_2O mixture, Ref. *16*), guanosine and inosine (*17*), adenosine 5′-phosphate (*18*), cytidine 3′-phosphate (from ethanol–H_2O mixture, Ref. *19*), etc. Even more interestingly, as in the case of cytosine, for example, both anhydrous and hydrated crystals are formed (*8*,*20*). Comparison (*20*) of the data obtained from the anhydrous orthorhombic crystal (*20*) with that from the monohydrated, monoclinic plate (*8*) has been made. As expected, bond distances and bond angles are very similar in both cases (*20*). As for the hydrogen-bonding arrangement, N(1) is bonded to N(3) in both crystals; the 4-amino N is bonded to two 2-oxo O from two neighboring molecules in the anhydrous crystals, but in the hydrated crystal, the 4-amino N is bonded to one 2-oxo O and to one water —O, and the 2-oxo O in turn is bonded to one water —O and one 4-amino N. This comparison clearly indicates the competition of water molecules for the hydrogen-bond donor and acceptor groups of the base. The bases are packed mainly parallel to each other in forming ribbons of molecules.

Another relevant situation is that of cytidine-3′-phosphate which crystallizes simultaneously from aqueous ethanol solutions into an orthorhombic form (*21*) and a monoclinic form (*19*). These two anhydrous forms are very different in density; the orthorhombic form has a density of 1.64 g/cm^3 and the monoclinic form has a density of 1.72 g/cm^3. Comparison of the data on 3′-CMP in these two crystal forms, however, reveals that the molecular structures and conformation in these two forms differ only trivially, and the difference in density seems to be due entirely to differences in hydrogen bonding and packing efficiency (*19*). The torsion angle of the O(3′)—P bond (semifolded or *gauche–trans* conformation as defined by Sundaralingam in Ref. *12*) and the torsion angle of the C(1′)—N(1) glycosyl bond (*anti* conformation as defined in Section G) of 3′-CMP are very similar in the two crystal forms. The two protons are located at the N(3) of the cytosine ring and at one phosphate oxygen atom in both forms; the 2-oxo O fails to participate in hydrogen bonding in both forms. In the orthorhombic form, there is a hydrogen bond between phosphate groups and there are hydrogen bonds between the 4-amino N and the two O of a single phosphate group, while in the monoclinic form, there is no hydrogen bonding between the phosphate groups and the 4-amino N is hydrogen bonded to two different phosphate groups. Therefore, it is interesting to note that the difference in the packing and the hydrogen-bonding arrangement in these two different forms is not sufficient to change significantly the torsion angles of the O(3')—P bond and the C(1')—N(1) bond in 3'-CMP.

Another very interesting situation is the existence of two crystallographically independent molecules in an asymmetric unit as found recently in the crystal of uridine 3':5'-cyclic phosphate (*22*) and in the crystal of adenosine 3':5'-cyclic phosphate (*23*). In the case of the uridine compound, the two molecular structures are rather similar; the ϕ_{cn} angles for the C(1')—N(1) bond (see definition in Section G) are $-72°$ and $-56°$, both in the *anti* range; the C(3') atoms are both in the *endo* position (see further discussion for definition), 0.64 Å and 0.58 Å out of the plane; the dihedral angles for C(5')—C(4') are both very similar in *trans–gauche* conformation. In the hydrated adenosine 3':5'-cyclic phosphate, however, the ϕ_{cn} angle of one molecule is $-50°$ (in the *anti* range, mean at $-30°$) and the ϕ_{cn} angle for the other molecule in the asymmetric unit is $+102°$ (in the *syn* range, mean at $+150°$). In this case also, the ribose ring in each molecule is puckered with C(4') displaced by at least 0.6 Å from the best four-atom plane of the furanose ring. Very recently, such a situation was also found in the crystal of inosine and the crystal of guanosine (*17*; Buggs, private communication). These crystals have two crystallographically independent molecules which have two different furanose conformations. The furanose in one of the nucleosides has a common C(2') *endo* conformation and a ϕ_{cn} angle of $-120°$ while the furanose in the other nucleoside has an unusual conformation with none of the four atoms defining a satisfactory plane and a ϕ_{cn} angle of $-45°$. These findings are further discussed in Section G.

One of the most interesting findings on the solvent–solute interaction was made in the study on the adenosine 3'-phosphate dihydrate (*12*). The structure exhibits a C(8)—H···O (water) hydrogen bond with a distance of H(8) to O(water) of 2.24 Å or C(8) to O(water) of 3.262 Å. As shown in Section C on the NMR studies, there is ample evidence for the hydrogen bonding of H(8) of purine derivatives with the solvent molecules. It will be interesting to see whether such a C—H···O bonding can be found in the highly hydrated crystal (6–8 water molecules per nucleotide) of inosine 5'-monophosphate (*14*). Hydrogen bonding with a C—H system has also been detected by Shefter and Mautner (*24*) for the C(6)—H of pyrimidine in their study on 2,4-dithiouracil. They found a 2.73 Å contact between the S(4) and the C(6), while the sum of the van der Waals radii for H and S is 3.05 Å. They proposed, therefore, a possible structure of C—H···S. They also stated that "it is worthy to note that in the crystal structures of barium uridine phosphate ..., adenosine uridine phosphate ..., and calcium thymidylate ..., a short intermolecular contact was found between the pyrimidine residue's

C(6) hydrogen and a sugar oxygen" (*24*). The C(6) to O(1′) distance is 2.77 Å for the calcium thymidylate (*11*). The information on barium uridine phosphate (*13*) and adenosine uridine phosphate (*25*) presumably was from the unpublished work done or participated in by Shefter. As shown in Section C on NMR studies, the temperature effect on the chemical shifts of H(6) suggests a hydrogen-bonding interaction of H(6) [not H(5)] with water molecules in an aqueous solution.

Nonlinear intermolecular hydrogen bonding has been cited, for instance, by Kraut and Jensen (*18*) in their study on adenosine 5′-phosphate. They found that the hydrogen on N(1) is 27° away from the line joining N(1) and one of the phosphate-O from the neighboring molecule.

2. Planarity and Rotation

The bases themselves or the base moiety in the nucleosides and nucleotides are mostly planar with displacement usually much less than 0.1 Å (*6*). There are expected exceptions, such as 5-ethyl-6-methyluracil (*26*) and dihydrouracil (*27*). In the case of 5-ethyl-6-methyluracil, the pyrimidine ring is significantly nonplanar, being buckled about the C(2)—C(5) line apparently to relieve strain between the substituent groups. For dihydrouracil, many of the ring bonds show marked lengthening in comparison with uracil. As expected, the greatest difference is in the C(5)—C(6) bond, which is 1.50 Å in dihydrouracil but only 1.34 Å in uracil. The six-membered ring has a half-chair conformation with atoms C(5) and C(6) displaced by about 0.3 Å. For the common nucleosides or nucleotides, the C(1′) atoms are usually displaced 0.1–0.2 Å from the base plane.

For the conformation of the furanose ring, in general four atoms of the ring will be approximately coplanar, with the fifth, either C(2′) or C(3′), displaced from this plane by 0.5 to 0.6 Å. (For an exception, see previous discussion on one of the conformations of guanosine and inosine.) The out-of-plane atom may be displaced either on the same side of the C(5′) atom, designated as the *endo* conformation, or the opposite side of the C(5′) atom, designated as the *exo* conformation. As mentioned above, in the case of adenosine 3′:5′-cyclic phosphate (*23*), it was found that C(4′) instead is displaced. Interestingly, this is not found in uridine 3′:5′-cyclic phosphate (*22*). For uridine cyclic phosphate, the furanose has a C(3′) *endo* conformation. In all cases but deoxyadenosine (*28*), the displaced atoms have been displaced in the *endo* direction (*6*). All the monomeric adenosine compounds such as 5′-AMP (*18*), 3′-AMP (*12*), and adenosine (in a cocrystal form with 5-bromouridine, Ref. *29*) have a C(3′) *endo* conformation, while the adenosine moiety linked at the 2′-position in adenosine (2′-5′) uridine phosphate (*25*) has a C(2′) *endo* conformation. There is only one recent study on a guanosine compound (*17*; Buggs, private communication) and as mentioned above this molecule can assume a C(2′) *endo* conformation or an unusual furanose conformation in the same crystal. This is the same situation for inosine (*17*; Buggs, private communication) since these two compounds, inosine and guanosine, have isomorphous crystal structure. Inosine 5′-monophosphate also assumes a C(2′) *endo* conformation (*14*). In the case of cytidine (*30*), it has a C(3′) *endo* conformation, while 3′-cytidylic acid has a C(2′) *endo* conformation instead (*19,21,42*). As for uridine 5′-phosphate (*13*), it has a C(2′) *endo* conformation. Interestingly for 5-bromouridine, it assumes a C(2′) *endo* conformation when it crystallizes by itself (*31*), but it assumes a C(3′) *endo* conformation when it *co*crystallizes with adenosine (*32*). There is also no general rule for the deoxyribosyl compounds: 5-bromodeoxyuridine (*31*) and 5-fluorodeoxyuridine (*33*) both are C(2′) *endo*; Ca thymidylate, C(3′) *endo* (*11*); 5′-bromodeoxycytidine, C(2′) *endo* (in cocrystal with deoxyguanosine, Refs. *34*, *35*); deoxyadenosine, C(3′) *exo* as mentioned above; deoxyguanosine, C(2′) *endo* (in cocrystal with 5-bromodeoxycytidine, Refs. *34*, *35*). Since there are clear examples that the furanose ring of a given nucleoside can assume either C(2′) *endo* or C(3′) *endo*, or even other conformations in different crystals, it will be reasonable to conclude that the furanose conformation is subjected to the influence of the environment and that there is no large energy barrier between these conformational states (*6*).

Sasisekharan et al. (*5*) have performed a very valuable task in analyzing the rotational freedom of various bonds in a monomeric unit which are relevant to the conformational states of the polynucleotides. Their original paper is highly recommended for those who are interested in this topic. A very brief account of their findings will be described here. As far as rotation is concerned, they noted that the conformation of the polynucleotides is governed by the rotation of five bonds in the monomeric units, discounting the glycosyl bond of C(1′) to N(1 or 9). (They have also studied the ϕ_{cn} angle, a subject to be treated in Section G.) These bonds are: P(3)—O(3′) as ϕ, O(3′)—C(3′) as $\theta 3$, C(4′)—C(5′) as $\theta 2$, C(5′)—O(5′) as $\theta 1$, and the O(5′)—P(5) as ψ. These bonds and their bond distances are shown in Fig. 2. They considered that these three internal rotations, $\theta 1$, $\theta 2$, and $\theta 3$, completely specify the conformation of the sugar phosphate unit, since the five-membered ring of the ribose itself may be considered to be a rigid unit. They defined these angles in the following statement: "The rotations are taken to be positive when they are anti-clockwise, looking along the bond, and to increase from 0° to 360°. The initial configuration (viz., the definition of $\theta 1$, $\theta 2$, $\theta 3 = 0°$) is defined as follows: for $\theta 1 = 0°$, P5 is taken to be in the plane of C4′, C5′, and O5′ and *trans* to C4′; for $\theta 2 = 0°$, O5′ is taken to be in the plane of C3′, C4′, and C5′ and *trans* to C3′; for $\theta 3 = 0°$, P3 is taken to be in the plane of C4′, C3′, and O3′ and *trans* to C4′."

After a systematic computer analysis on the stereochemical data based on a set of allowable, minimum contact distances, they concluded that the rotation of these three angles are severely restricted. The allowable regions for these angles are the following: $\theta 1$, $-30°$ to $30°$; $\theta 2$, $-20°$ to $10°$, $110°$ to $140°$, and $220°$ to $280°$; and $\theta 3$, $270°$ to $280°$. They went on to state "Thus the position of the P5 atom is nearly *trans* to C4′; the O5′ atom assumes the threefold staggered positions; and the P3 atom is restricted to the 270–280° region

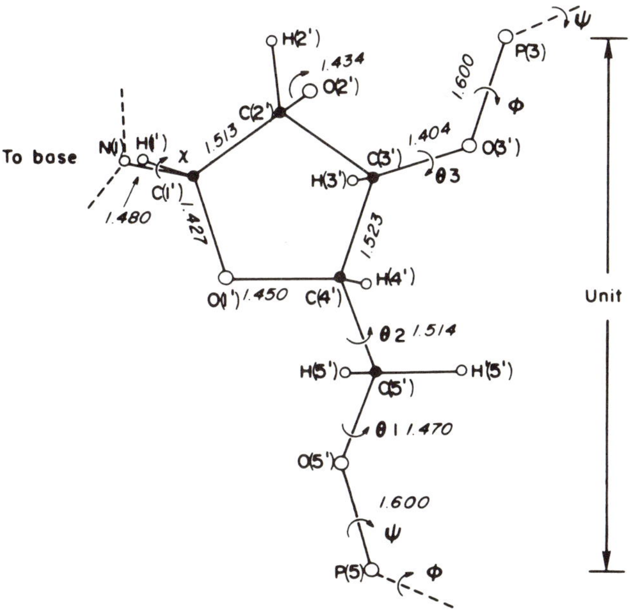

Fig. 2. The rotational angles and the bond distances in the geometrical D-ribose phosphate unit (after Sasisekharan et al., Ref. 5).

between the C4′ and H3′ atoms . . . . Thus the conformation of the ribose phosphate unit comes down to manageable proportions and there seems to be only three possible conformations, which may be characterized by (0°, 0°, 270°), (0°, 120°, 270°) and (0°, 240°, 270°). This simplifies the calculation for a nucleotide when the bases are included." They compared this theoretical evaluation to the actual situation for 13 compounds in crystals. They concluded that "for the rotation $\theta 1$, the agreement is fairly good for the three cases reported. For the rotation $\theta 2$, all the observed conformations lie within the predicted ranges. Interestingly, nine out of the thirteen cases observed so far have $\theta 2$ in the range 110–140°. Of the remaining, three occur near 0° and one near 240°. Thus the agreement for $\theta 2$ is remarkable. However, for $\theta 3$, i.e., the rotation about the C3′—O3′ bond, of the two reported structures, one falls slightly outside the allowed range of 270–280°, but both the values of $\theta 3$ lie within the range of 270–335°." The two structures referred to in the quotation are those of 3′-AMP and 3′-CMP. This study is indeed very valuable for conformational analysis of mononucleotides, dinucleotides, and perhaps of polynucleotides.

Study of the CPK models of dinucleoside monophosphate (*36*) indicates that there is considerable amount of freedom in the rotation of ϕ (P(3)—O(3′)) and ψ (P(5)—O(5′)) bonds (Fig. 2). As for the O—P bond in the crystal of nucleotides, Sundaralingam (*12*) has made the following study on 3′-AMP and 3′-CMP concerning the conformations of the chain —C(3′)—O(3′)—P —O—H. The conformational angle was defined as the angle made by the O(3′)—P bond relative to the O(7)—H bond when viewed in the direction of the P → O(7) bond; or by the C(3′)—O(3′) bond relative to the P—O(7) bond when viewed in the direction of the O(3′) → P bond. The conformation for the first angle, O(3′)—P → O(7)—H is *gauche* for both compounds (ca., 70°, see original paper for sign designation); and the conformation for the second angle, C(3′)—O(3′) → P—O(7), is *gauche* (ca., 77°) for 3′-AMP and *trans* (ca., 171°) for 3′-CMP. Therefore, the 3′-AMP may be designated as having the folded (*gauche–gauche*) conformation and the 3′-CMP as having the semifolded (*gauche–trans*) conformation. Sundaralingam (*12*) further stated at that time "A literature survey of the known phosphates reveals the fact that they fall into one of the two conformations cited above, irrespective of whether the proton is substituted by a carbon, as in the phosphodiester anion, or not . . . . "

3. Tautomerism and Site of Protonation

The hypoxanthine, guanine, thymine and uracil moieties in the monomers were shown to have the oxo form (instead of the enol form) (*11,13–15,17,22,* etc.). The cytosine and adenine moieties in the monomers were shown to have the amino form (instead of the imino form) (*8,12,18,19,21,37*). For purine, the hydrogen was found to be located at N(7) in the crystal (*38*). According to the quantum-mechanical calculation of Pullman et al. (*39*), the dipole moment is higher for the N(7)H tautomer (5.5–6.3 D) than for the N(9)H tautomer (3.6–4.1 D). They proposed that the occurrence of N(7)H in the crystal may be due to the relatively higher dipole moment of this tautomer which may lead to greater electrostatic interaction energy, since these two tautomers were calculated to have the same energy level in the free state. The predicted difference in dipole moment between the N(7) substituted compound versus the N(9) substituted compounds have been experimentally verified as described in Section B.

The site of protonation for adenine (*40*) and the adenine moiety in 3′-AMP (*12*), and 5′-AMP (*18*) was found to be on N(1) of the ring. For guanine (*7*), there is strong suggestive evidence from the hydrogen-bond distribution that N(7) is the site of protonation. The site of protonation for the cytosine in 1-methyl cytosine (*41*), cytidine-3′-phosphate (*21,42*), and cytosine-5-acetic acid (*43*) was found to be at the N(3) of the ring. The hydrogen-bonding scheme and the protonation scheme of cytosine-5-acetic acid is particularly interesting. A symmetrical relationship is found to exist between two molecules: a proton is shared commonly by both N(3) atoms, and the 2-oxo group of the first molecule is hydrogen bonded to the 4-amino group of the second molecule, while the 2-oxo group of the second molecule is hydrogen bonded to the 4-amino group of the first molecule. This hydrogen bond-protonation scheme was adopted by our laboratory as a basis to interpret our observation on properties of poly C in acidic solution, and as a model for the scheme of hydrogen bonding and of the protonation for the poly C helix (*44*). This proposal was confirmed soon after by the X-ray diffraction study on the fiber prepared from the helical poly C and from other physicochemical investigations (*45,46*).

The above conclusions on the tautomeric forms and the site of protonation derived from X-ray diffraction on crystals are in full agreement with those obtained from studies on these monomers in solution. Some of these studies are described in later sections.

B. Ionization Potential, Polarizability, and Dipole Moment of the Bases

Several electronic properties of the bases in the ground state are important for the theoretical evaluation of the forces involved in the interaction. Among these are the ionization potential (I), dipole moment (μ), and polarizability (α). For instance, for the calculation of the well-known van der Waals–London dispersion interaction energy between compound A and B based on "dipole" approximation, the ionization potentials (I_A and I_B) and the polarizabilities (α_A and α_B) appear in the Eq. (1).

$$E_L = -\frac{3}{2}\frac{I_A I_B}{I_A + I_B}\frac{\alpha_A \alpha_B}{R_{AB}^6} \tag{1}$$

where R_{AB} is the distance between points of location of these two dipoles. The average group polarizability, α_i, is usually estimated from the atomic refractions (*47*) or by the use of the usual additivity rules (*48*). The range of the values for α so obtained for uracil, cytosine, and thymine is from 10–12 Å^3 (*47,48*) and about 14 Å^3 for adenine and guanine (*47,48*). For calculations based on "monopole" approximation, anisotropic polarizability values have to be used instead of the isotropic value (*49*). In the dipole approximation (*48,49*), the values of dipole moment are used for the calculation of the dipole–dipole interaction energy [$E_{\mu\mu}$, Eq. (2)] and the dipole-induced dipole interaction energy [$E_{\mu\alpha}$, Eq. (3)]:

$$E_{\mu\mu} = \frac{1}{R_{AB}^3}\left[\mu_A\mu_B - \frac{3}{R_{AB}^2}(\mu_A \mathbf{R}_{AB})(\mu_B \mathbf{R}_{AB})\right] \tag{2}$$

$$E_{\mu\alpha} = -\frac{1}{2}\frac{1}{R_{AB}^6}\left\{\alpha_A\left[3\left(\frac{\mu_B \mathbf{R}_{AB}}{R_{AB}}\right)+1\right]+\alpha\beta\left[3\left(\frac{\mu_A \mathbf{R}_{AB}}{R_{AB}}\right)+1\right]\right\} \tag{3}$$

During the last three years, attempts were made to study theoretically these interactions by a more elaborate approach, the "monopole" approximation. For a more extensive discussion on this subject, two original papers and two recent reviews are recommended (*49–52*). Nevertheless, the values of dipole moment of the bases determined accurately by experimental methods are of

TABLE I

Ionization Potential of Bases[a]

Compounds	I, ev
Pyrimidine[b]	9.9 ± 0.5
Uracil	9.8 ± 0.1
Thymine	9.4 ± 0.1
6-Methyluracil[c]	9.7 ± 0.1
5,6-Dimethyluracil[c]	9.4 ± 0.1
6-Azauracil	10.2 ± 0.1
Cytosine	8.9 ± 0.2
Purine	9.7 ± 0.1
6,9-Dimethylpurine[c]	9.3 ± 0.1
Adenine	8.9 ± 0.1
Xanthine	9.3 ± 0.2
Hypoxanthine	9.2 ± 0.1
Tetramethyluric Acid	7.9 ± 0.1
Theophylline[c]	9.3 ± 0.1

[a] Most of the data in the table were from Lifschitz, Bergmann, and Pullman (1967), Ref. *53*.

[b] From Omura, Baba, and Higasi (1957), Ref. *54*.

[c] From Bergmann and Lifschitz (1969), Ref. *55*.

importance for evaluation of MO calculation and other physical properties. Because of the substantial solubility problem, values of dipole moment were only obtained very recently.

In Table I, the ionization potentials of various bases, their derivatives, and analogs are cited. These values were determined by the mass spectroscopy method in comparison with the known standard of the rare gases (*53,55*). From these values, Bergmann, Lifshitz, and Pullman (*53,55*) have made the following conclusions. (a) Since the range of the ionization potentials of all these compounds is sufficiently small (9–10 eV) and yet the structural variation within this group is sufficiently large, one can conclude that this representative ionization potential is due to the loss of one of the π-electrons in the first ionization step. (b) The sequence of order of the ionization potentials of these compounds is the same as that of the energies of the highest occupied molecular orbital (HOMO) and as that of the calculated values of I based in MO procedure, though the calculated values are lower than the observed values by 0.5–0.9 eV. Purines and pyrimidines may form charge-transfer complexes with various organic compounds, and these complexes are formed by a partial transfer of an electron from the highest occupied MO of the "donor" to the lowest empty MO of the acceptor. As predicted from theoretical considerations and shown from the above experimental results, the energies of the HOMO of these bases are related to the ionization potentials.

In Table II, the experimental values of dipole moment of various bases, derivatives, and analogs are cited. The observed values of μ for adenine (3.0 ± 0.2 D), purine (4.3 ± 0.2 D), and 1,3-dimethyluracil (3.9 ± 0.1 D) are in reasonable agreement with those values obtained from MO calculation by DeVoe and Tinoco (*47*) and by Pullman (*48*). The calculated values are 2.8–3.2 D for adenine, 3.6–4.2 D for purine, and 3.7 for uracil. The calculated values based on vector addition of appropriate molecule and bond moment values are 2.3 D for adenine, 4.2 D for 1,3-dimethyluracil, 3.5 D for purine, and 4.5 D for 5-bromouracil (*47*). Owing to the solubility problem, no experimental values for cytosine and guanine are available. The calculated value for cytosine based on MO methods was reported to be 8.0 D by DeVoe and Tinoco (*47*) and 7.2 D by Pullman (*48*), and the value for cytosine based on bond moments addition is 6.1 D by DeVoe and Tinoco (*47*). The calculated values for guanosine based on MO was reported to be 6.8–6.9 D (*47,48*) and the value based on bond moments addition was 7.2 D (*47*).

The next important question is the direction of the moment. The directions of the moment of the bases calculated by DeVoe and Tinoco are shown in Fig. 3. These theoretical results are supported by the above observation that the calculated values are in substantial agreement with the experimental values, at least for certain bases.

TABLE II

Dipole Moment (μ) of Bases

Compounds	Solvent	μ, D
3-Methyl-5-bromouracil[a]	Dioxane	4.5 $\pm$0.3
5-Bromouracil[b]	Dioxane	4.6 $\pm$0.1
1,3-Dimethyluracil[a]	Dioxane	3.9 $\pm$0.1
5,6-Dimethyluracil[b]	Dioane	4.7 $\pm$0.1
Pyrimidine	Dioxane[c]	2.42
	Benzene[d]	2.27
2-Chloropyrimidine[d]	Benzene	3.58
4,6-Dichloropyrimidine[b]	Benzene	1.09$\pm$0.03
2,4,6-Trichloropyrimidine[b]	Benzene	2.34$\pm$0.03
2,4,5,6-Tetrachloropyrimidine[b]	Benzene	1.57$\pm$0.05
9-*n*-Butylpurine[a]	Dioxane	4.3 $\pm$0.2
9-*n*-Butyladenine[a]	CCl_4	3.0 $\pm$0.2
6-Chloropurine[b]	Dioxane	5.3 $\pm$0.15
2,6-Dichloropurine[b]	Dioxane	5.6 $\pm$0.1
2,6,8-Trichloropurine[b]	Dioxane	3.4 $\pm$0.1
6, 9-Dimethylpurine[b]	Benzene	3.69$\pm$0.05
6,9-Dimethyl-2-methylaminopurine[b]	Benzene	4.70$\pm$0.05
6-Methylthio-7-methylpurine[e]	Dioxane	5.6 $\pm$0.2
6-Methylthio-9-methylpurine[e]	Dioxane	3.0 $\pm$0.1
1,3,7-Trimethyl-8-decylthioxanthine[e]	Dioxane and benzene	4.2 $\pm$0.1
1,3,9-Trimethyl-8-decylthioxanthine[e]	Dioxane	6.54$\pm$0.2
	Benzene	5.90$\pm$0.2
3,7-Dimethyl-8-decylthioxanthine[e]	Dioxane	4.90$\pm$0.1
3,9-Dimethyl-8-decylthioxanthine[e]	Dioxane	8.2 $\pm$0.1
Theophylline[b]	Dioxane	3.9 $\pm$0.10
Caffeine[b]	Benzene	3.70$\pm$0.05
Tetramethyluric Acid[b]	Benzene	3.21$\pm$0.05

[a] From DeVoe and Tinoco (1962), Ref. *47*.
[b] From Weiler-Feilchenfeld and Bergmann (1968), Ref. *56*.
[c] From Smyth (1955), Ref. *57*.
[d] From Lumbroso et al. (1967), Ref. *58*.
[e] From Pullman et al. (1969), Ref. *59*.

One experimental approach for the determination of the direction of moment is to measure the values of μ of a variety of derivatives substituted with a halogen at different positions. If the bond moment of the halogen substitution is known, then the direction of μ of the parent molecule can be ascertained by vector addition. This method was employed successfully by Weiler-Feilchenfeld and Bergmann (*56*) in the assignment of the direction of μ for pyrimidine [from C(5) → C(2)] by comparing the μ values of pyrimidine (2.3 D), 2-chloropyrimidine (3.6 D), 4,6-dichloropyrimidine (1.1 D), and 2,4,6- trichloropyrimidine (2.3 D same as pyrimidine). This study, however, suggests that the μ [$C_{aromatic}$—C(1)] is 1.3 D for the pyrimidine instead of 1.6 D for μ [$C_{aromatic}$—C(1)] for chlorobenzene. Weiler-Feilchenfeld and Bergmann (*56*) showed concern about the substantial difference of μ values observed between 1,3-dimethyluracil (3.9 D) and 5,6-dimethyluracil (4.7 D), and the large μ value observed for 5-bromouracil (4.6 D). They suggested that the moment of uracil should be determined in relation to these problems.

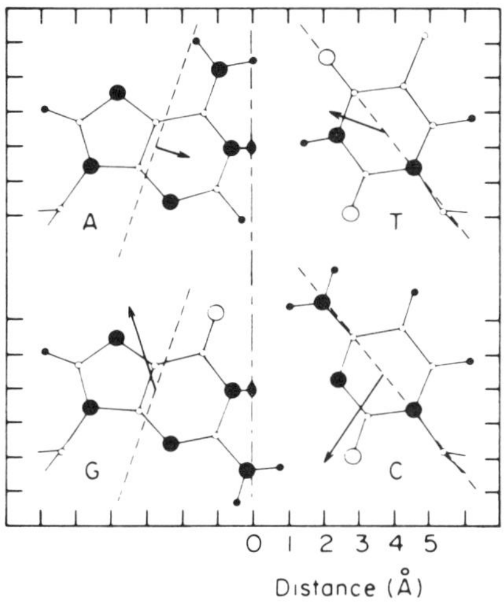

Fig. 3. The orientation and the magnitude of the calculated permanent dipole moments in the A·T and G·C base pairs of DNA (after DeVoe and Tinoco, Ref. *47*). The calculated dipole moment of each base is drawn to scale, with the positive end located at the position of the point dipole. Atom symbols: ●, nitrogen; ○, oxygen; •, hydrogen; ○, carbon.

A theoretical and experimental study by Pullman, Bergmann, and their associates (*59*) on the dipole moments of purine and xanthine was able to provide some interesting answers about the tautomerism of these molecules. Their calculations based on quantum chemistry indicates that the μ values for the N(7)—H and the N(9)—H tautomers of purine are 5.5 D and 3.6 D, respectively, and the μ value for the N(7)—H and the N(9)—H tautomers of xanthine are 4.3 D and 7.2 D, respectively. These calculations are supported by the experimental results. As shown in Table II, the 7-methyl and 9-methyl purine derivatives do have corresponding μ values of 5.6 D and 3.0–3.7 D; the 7-methyl and 9-methylxanthine derivatives do have corresponding μ values of 4–5 D and 6–8 D as predicted. The computed angles of μ with respect to the C(4)—C(5) bond (counterclockwise) are 150° and 49° for the N(7)—H and N(9)—H tautomers of purine, respectively, and are 99° and 24° for the N(7)—H and N(9)—H tautomers of xanthine, respectively. Weiler-Feilchenfeld and Bergmann (*56*) concluded that μ values of 6-chloropurine, 2,6-dichloropurine, and 2,6,8-trichloropurine indicate that these compounds are derived mainly from the N(9)—H tautomer of purine.

It should be noted, however, all the theoretical and experimental studies were done on bases or on bases substituted by a methyl group (or other hydrocarbon side chain) at the N(3) position of the pyrimidines or at the N(9) position of the purines. In the nucleotides, the substitution is that of the pentosyl group. The electron-withdrawing property of the pentose might make a small difference in the dipole moment.

C. Nuclear Magnetic Resonance (NMR)

Recent advances in theory and in instrumentation (in both sensitivity and resolution) of NMR have made this technique most promising for studies of bio-organic molecules, even the biopolymers. Since the pioneering work of the Jardetzky's published in 1960 (*60–62*), much progress has been made in the field of NMR studies on the bases, nucleosides, and nucleotides. Most of the work, however, is in the area of proton magnetic resonance (PMR). Some of the studies on other nuclei will be briefly mentioned.

1. Assignment and Calculation of the Chemical Shifts of The Base Protons in Bases and in Nucleosides

Surely, the first task of a PMR study is the assignment of the various spectral lines to the appropriate protons in the molecule. Based on existing theories, the following factors which are known to influence the positions of the proton resonances were enumerated in one of our early papers (*63*): (1) electron densities at various carbon atoms bearing the proton; (2) the ring-current effect associated with the mobile π-electron cloud; (3) the magnetic anisotropy of the hetero atoms in the aromatic rings; (4) solvent effects; (5) solute–solute interaction. The problem of solute–solute interaction of the bases, nucleosides, and nucleotides is discussed extensively in a later section. The question of solvent effects has not been explored adequately, except as one of the variables in relation to the study of solute–solute interaction. The main reason is the problem of solubility; the bases, nucleosides, and nucleotides are not even sparingly soluble in most solvents, and usually they are soluble only in water (or D_2O for PMR studies). Owing to the special interest of our laboratory, we have made an attempt (*63–66*) to relate the observed chemical shifts of the base protons to the electronic structure of these heterocyclic compounds in a semiquantitative fashion.

A linear correlation between the observed chemical shifts and the calculated π-electron charge densities in the aromatic homocyclic and heterocyclic systems can be established with a certain degree of satisfaction, notably as shown by the work of Schaefer and Schneider (*67*), Veillard (*68*), and more recently by the work of Lynch and Dou (*69*), on protonated nitrogen heterocyclic compounds. These papers pointed out as well the significant contributions of ring current and nitrogen atom magnetic anisotropy, which are considered in the following paragraphs. In following the method developed in our previous paper (*63*), the calculated excess π-charge densities at each atom of the base ring are handled in terms of a field effect upon the proton chemical shifts by the following equation:

$$\Delta\sigma_{\Delta\rho} = 12.5\times 10^{-6}\sum_i \frac{\Delta\rho_i}{R_i^2}\cos\theta - 17.0\times 10^{6}\left(\frac{\Sigma_i \Delta\rho_i}{R_i^2}\right)^2 \tag{4}$$

where σ is the screening constant (ppm), $\Delta\rho_i$ is the excess π-electron density on the *i*th atom of the base, R_i is the distance (Å) from the *i*th atom to the proton in question, and θ_i is the angle in degrees of the field vector with respect to the C—H bond axis for the appropriate proton. R_i and θ_i were obtained from scale drawings of the bases according to appropriate X-ray diffraction results. Equation (4) takes into account the calculated π-electron density at each atom of the base, rather than just that at the carbon atom to which the proton is bonded. This expression was obtained from the familiar equation for field effect calculations:

$$\Delta\sigma = -aE_z - bE^2$$

For our case, a has been chosen to fit the linear correlation found experimentally by Schaefer and Schneider (*67*). This value of a is 2.60×10^{-12} which is close to the value of 2.0×10^{-12} obtained by Buckingham (*70*) and 2.9×10^{-12} by Musher (*71*). Musher's value for b, which was obtained from Marshall and Pople's calculation, was used. The adoption of several sets of π-charge density calculations for the bases is discussed later in greater detail.

The contribution to the screening constants due to ring currents was calculated with the following equation from Schaefer and Schneider (*67*):

$$\Delta\sigma_\pi\ (\text{ppm}) = -12.0\sum_i\left(\frac{A_i^2}{R_i^2}\times \text{ring-current intensity}\right) \tag{5}$$

where A and R in angstrom units are, respectively, the radius of each ring and distance of the proton being considered from the center of each ring. Fortunately, ring-current intensities of various bases relative to benzene have been calculated by Giessner-Prettre and Pullman (*72*).

The final problem is the effect of nitrogen atom magnetic anisotropy. For pyrimidine, this effect was also evaluated in our earlier papers (*63,64*). It was noticed that only slightly more than half of the shift difference between H(2) and H(6) (0.50 ppm) (*73*) could be accounted for by the difference in charge densities calculated by two different groups (*74,75*). If the remaining 0.21 ppm (based on charge densities calculated from Ref. *74*) or 0.24 ppm (based on charge densities calculated from Ref. *75*) can be attributed to the difference in the effects of the nitrogen magnetic anisotropy at the two protons, one can compute a magnetic anisotropy for the nitrogen atoms in the pyrimidine ring by the use of the following general equation based upon the dipole approximation (*76*):

$$\Delta\sigma_{\chi N} = \sum_i \frac{\Delta\chi_i}{3R_i^3}(1 - 3\cos^2\gamma_i) \tag{6}$$

where R_i is the distance between N_i and the proton in question, $\Delta\chi_i$ is the anisotropy of the susceptibility tensor, and γ_i is the angle between the axis of symmetry in N_i and the N_i—H line. $\Delta\chi$ was computed to be $+2.72 \times 10^{-6}$ and $\Delta\sigma$ for H(2) due to N(1) in purine or adenine was computed to be -0.183 ppm. Considering the uncertainties and assumptions involved in both calculations, this value is in reasonable agreement with the -0.25 ppm calculated by Gil and Murrell (*77*) as the effect of the magnetic anisotropy of the N atom on the α protons in pyridine. As in the treatment of these authors, our reference for calculation was sp^2 carbon. But the actual comparison between the bases of the nucleosides (see later section) is the difference in the magnetic anisotropy between pyridine-type nitrogen [N(1) in adenine and N(3) in cytosine] and cyclic amide nitrogen [N(1) in hypoxanthine, guanine as well as N(3) in uracil]. At present, we have no reliable evaluation of the magnetic anisotropy of the cyclic amide nitrogen, so the value of -0.183 ppm may be either an over- or underestimate of the difference in the anisotropic effect exerted on the adjacent hydrogen by these two different types of nitrogen atoms.

Now we attempt to provide a theoretical basis to account for the observed chemical shifts of the ring protons of these heterocyclic compounds based on the summation of the effects calculated from these three equations, i.e., (4), (5), and (6).

a. *Purine and Purine Nucleosides*

Though the unsubstituted purine is not a base component of nucleic acid, because of its relatively simple structure and its high solubility it is an excellent model for the properties of 6-aminopurine (adenine), 6-oxopurine (hypoxanthine), and 2-amino-6-oxopurine (guanine), which are the bases present in nucleic acid (Fig. 1). There are three resonance lines for the protons H(2), H(6), and H(8), and the correct assignment of these lines to the appropriate protons was achieved independently by Matsuura and Goto (*78*), by ourselves (*63*), and by Bullock and Jardetzky (*79*). This achievement was made possible by selective deuteration of the molecule: 2-deutero purine and 6-deutero purine were synthesized by Matsuura and Goto; and 2-deutero purine and 8-deutero purine were synthesized by ourselves and by Bullock and Jardetzky. This is an important experimental finding since the result is

TABLE III

Calculations of the Proton Chemical Shifts for Purine (ppm)[a]

		$\Delta\sigma_{\Delta\rho}$	$\Delta\sigma_{\pi}$[b]	$\Delta\sigma_{\chi N}$	Summation
H(8)					
	I.[c]	−1.202	−1.532, −1.306	−0.206	−2.940, −2.714
	II.[d]	−1.464	−1.532, −1.306	−0.206	−3.202, −2.976
	III.[e]	−0.994	−1.532, −1.306	−0.206	−2.732, −2.506
	IV.[f]	−2.328	−1.532, −1.306	−0.206	−4.066, −3.840
H(2)					
	I.[c]	−1.194	−1.952, −1.766	−0.383	−3.529, −3.343
	II.[d]	−0.326	−1.952, −1.766	−0.383	−2.771, −2.475
	III.[e]	−1.427	−1.952, −1.766	−0.383	−3.762, −3.576
	IV.[f]	−0.685	−1.952, −1.766	−0.383	−3.020, −2.834
H(6)					
	I.[c]	−1.469	−2.062, −1.850	−0.210	−3.741, −3.529
	II.[d]	−0.509	−2.062, −1.950	−0.210	−2.781, −2.569
	III.[e]	−1.922	−2.062, −1.850	−0.210	−4.194, −3.982
	IV.[f]	−1.002	−2.062, −1.850	−0.210	−3.274, −3.062

[a] Calculations are based on Eqs. (*4*), (*5*), (*6*).
[b] Ring-current intensity values from Ref. *72* (See text).
[c] Excess π-charge density values from Refs. *75* and *81*.
[d] Excess π-charge density values from Ref. *82*.
[e] Excess π-charge density values from Ref. *74*.
[f] Excess π-charge density values from Ref. *83*.

different from the two previous attempts based on more speculative grounds (*60,80*).

The calculated values for the proton chemical shifts of purine based on Eqs. (4), (5), and (6) are listed in Table III. These values were calculated using four sets of excessive π-charge density calculations (*72,74,81–83*). From four sets of ring-current values reported by Giessner-Prettre and Pullman (*72*), the two which were results of most recent calculations were adopted for comparison. It can be seen that the major contributions to the calculation of the chemical shifts comes from the excess π-charge density and the ring-current effect, while the contribution of the nitrogen magnetic anisotropy is relatively small. Table IV provides a comparison between the observed differences among the chemical shifts of the three protons in purine and the calculated values. The observed values in D_2O, dimethyl sulfoxide (DMSO), and dimethylformamide were from our work (*84*), and the observed values in chloroform were from the work of Reddy et al. (*80*). The large solvent effect on the difference between the chemical shift of the H(8) proton and that of the

TABLE IV

Comparison of the Observed Differences among the Chemical Shifts of the Three Protons in Purine with the Calculated Values (ppm)[a]

	H(8) − H(2)		H(2) − H(6)	
Observed values				
D_2O, infinite dilution[b]	0.36		0.18	
DMSO, 0.2 *m*[b]	0.29		0.22	
DMF 0.2 *m*[b]	0.24		0.21	
$CHCl_3$, uncertain[c] conc.	0.55		0.17	
Calculated values				
I.[d]	0.589	0.629	0.212	0.186
II.[d]	−0.541	−0.501	0.120	0.094
III.[d]	1.029	1.070	0.432	0.406
IV.[d]	−1.046	−1.006	0.254	0.228

[a] The calculated values are computed from values in Table III.
[b] From Ref. *84*.
[c] From Ref. *80*.
[d] See Table III for the references.

H(2) proton is most likely due to the interaction of the H(8) proton with the solvent. Recently Hruska et al. (*85*) have studied the chemical shifts of purine in a series of solvents and binary solvent mixtures with varying proton acceptor strengths and dielectric constants. They concluded that the H(8) proton forms a hydrogen bond with the proton acceptor group of the solvent molecule and the formation of such a bond will shift the H(8) to a lower field. Therefore, this bonding will also reduce the difference between the chemical shifts of H(8) and that of H(2) as noted in Table IV. Interaction of H(8) proton in the nucleosides and nucleotides has also been observed independently by our group (*86*). As shown in the above section on X-ray diffraction studies, there is evidence to support hydrogen bonding between water and the H(8) proton of 3′-AMP in the crystal (*12*). Therefore, the value of 0.55 ppm obtained for the difference between the H(8) and the H(2) protons in $CHCl_3$ is more suitable for comparison with the calculated value. It can be seen that the calculated values of Set I agree semiquantitatively with the experimental values. These results encourage us to continue our effort further in comparing the calculated values of the base protons among various nucleosides as shown in later paragraphs. Recently, two sets of X-ray data have been obtained for purine (*38*). These dimensions are slightly different from those adopted in our calculation based on the model of adenine from Cochran (*40*) and the C—H bond length of 1.08 for aromatic systems. One reason why these values are not recalculated is that these two sets of data (*38*) differ significantly in the C—H bond length, for instance, varying from 0.95 to 1.06 Å in C(8)—H. It is felt that the values previously adopted for our calculations are sufficiently close for the purpose of a preliminary attempt.

The carbon-13 resonance of purine has been studied by Pugmire et al. (*87*), with the aid of deuterated compounds and the decoupling technique. The ^{13}C chemical shift values for the 2, 6, and 8 carbons in purine were assigned unequivocally. The two remaining peaks found at the lowest- and highest-field positions are assigned to carbons 4 and 5, respectively, on the strength of the theoretical prediction that the highest π-electron density is at C(5). They also observed that a gross correlation of ^{13}C chemical shift data with theoretical estimates of charge distribution in purine does exist. They pointed out that in an increasing magnetic field, the resonance pattern for ^{13}C (2, 8, 6) differs from that for the corresponding directly bonded hydrogens (6, 2, 8). This is perhaps expected now in view of the large contribution from the ring-current magnetic anisotropy (Table III). The H(6) proton is subjected to the deshielding effect from both the six-membered ring and the five-membered ring and thus is moved substantially downfield.

In the course of preparing the deuteropurine, it was found that the H(8) proton of the purine and the purine nucleosides is exchangeable at elevated temperatures (*63,79,87*). By this method, the resonance lines of H(2) and H(8) of many purine nucleosides were assigned (*79*). For the adenine nucleoside and nucleotides, this procedure was checked independently (*88*). Table V summarizes the pertinent information on the chemical shifts of 18 nucleosides which have been studied in our laboratory.

Up to the present time, no serious attempt has been made to account quantitatively for the effect of ring substituents upon base proton chemical shifts in the purine nucleosides. Coburn et al. (*89*) have reported a correlation of proton chemical shifts of 6-substituted purines with the reactivity parameters (Brown's electrophilic substituent constants) of the substituents.

TABLE V

Proton Chemical Shifts for Nucleosides in D_2O at 33–35°C (ppm from TMS capillary)

	Molal conc.	H(2)	H(8) or H(5)	H(6)	H(1′)	CH_3
Ribosylpurine	0.0	9.425	9.16	9.63	6.70	
	0.1	9.30	9.09	9.48	6.60	
Ribosyl-6-chloropurine	0.03	9.26	9.26		—	
	0.05	9.20	9.20		6.66	
1-Methylinosine[a]	0.0	8.87	8.80		6.56	4.15
	0.1	8.77	8.77		6.48	4.09
6-Thioinosine	0.03	8.81	8.93		6.59	
	0.05	8.79	8.92		6.57	
Adenosine[b]	0.0	8.73	8.82		6.50	
	0.1	8.48	8.67		6.47	
N-6-Methyladenosine	0.0	8.73	8.73		6.49	3.55
	0.1	8.31	8.50		6.32	3.31
Inosine[c]	0.0	8.72	8.83		6.62	
	0.1	8.64	8.76		6.44	
N-6-Methyl-2′-deoxyadenosine	0.0	8.71	8.73		6.90	3.58
	0.1	8.34	8.50		6.72	3.36
2′-O-Methyladenosine	0.0	8.60	8.76		6.58	3.86
	0.1	8.53	8.68		6.50	3.845
2′-Deoxyadenosine	0.0	8.69	8.76		6.93	
	0.1	8.435	8.585		6.765	
3′-Deoxyadenosine	0.0	8.68	8.76		6.51	
	0.1	8.42	8.61		6.35	
N-6-Dimethyladenosine	0.0	8.63	8.69		6.51	3.89
	0.1	8.25	8.48		6.31	3.54
Ribosyl-2,6-diaminopurine	0.03		8.42		6.35	
	0.05		8.40		6.35	
1-Methylguanosine	0.025		8.42		6.375	3.92
Cytidine	0.0		6.51[H(5)]	8.28	6.355	
	0.1		6.50	8.275	6.35	
Uridine	0.0		6.33[H(5)]	8.28	6.35	
	0.1		6.325	8.275	6.34	
Thymidine	0.0			8.075	6.72	2.31
	0.1			8.07	6.70	2.31
7-Methylinosine[d]	0.0	8.71	9.68(H_2O)		6.60	4.61
	0.1	8.60	9.67(H_2O)		6.51	4.58

[a] At 28°, and at extrapolation to infinite dilution, the chemical shift of H(2) was found to be 8.77, H(8) to be 8.72, H(1′) to be 6.49, and CH_3 to be 4.09 ppm.

[b] At 28°, and at extrapolation to infinite dilution, the chemical shift of H(2) was found to be 8.67, H(8) to be 8.75, and H(1′) to be 6.495 ppm.

[c] At 28°, and at extrapolation to infinite dilution, the chemical shift of H(2) was found to be 8.65, H(8) to be 8.75, and H(1′) to be 6.52 ppm.

[d] Measured at 28°.

Diminution of ring currents by polar substituents has been proposed by Jardetzky (*90*), to explain the reduction of $\Delta\delta$ and intrinsic δ of anions of purine and substituted purines. This explanation may be invoked in the case of ribosylpurine vs. adenosine, but fails to account for the practically identical shifts observed for adenosine and inosine (Table V). Inosine, existing in the lactam form, should have less ring current in the pyrimidine ring than does adenosine which exists in the benzenoid form as shown in the calculation (*72*). Thus according to this ring-current argument one would expect H(2) in inosine to be shifted upfield significantly, contrary to what has been observed.

Additional uncertainty is encountered in applying Eqs. (4) and (6) for the quantitative estimation of the chemical shifts of the base protons in the nucleosides, since the theoretical values of the excess π-charge densities and the ring current were calculated on the free bases and not on the pentosyl bases, i.e., the nucleosides. The comparison of experimental δ's of purine and ribosylpurine at infinite dilution in water indicates that H(6) and H(2) are deshielded by about 0.16 ppm and the H(8) by about 0.25 ppm owing to the N(9) ribosyl substitution. This observation is expected from the electron-withdrawing effect of this group. It is reasonable as a first approximation to consider that the effect of pentosyl substitution at N(9) on all the atoms in the base is the same for each of the several nucleosides in question. Thus, we assume that a comparison of the calculated π-charge densities of the bases can represent the comparison of the π-charge densities of the bases in the nucleosides. At present we shall *not* attempt to compare the observed difference in δ values between the H(2) and the H(8) protons of the nucleoside to the calculated difference based on the computed values for the free bases, owing to the uncertainty about the effect of the pentosyl substitution. As for the problem of ring current, the validity of employing the figures for the five-membered ring is again somewhat questionable since they were calculated for the bases and not for the nucleosides. Fortunately, the calculated ring-current differences for various five-membered rings among these bases is rather small (*72*). As for the problem of nitrogen magnetic anisotropy, the contribution in this category to H(8) is likely to be the same for all the nucleosides and, therefore, not important in the following comparison. However, in comparing the H(2) of adenosine and ribosylpurine with the H(2) of inosine, the N(1) of the latter is changed from the pyridine type to that of the amide type. As mentioned above, we have estimated that this change of bonding to the nitrogen atom in N(1) position may bring about a positive contribution of about 0.183 ppm to the H(2) of inosine in comparison with the H(2) of adenosine or ribosylpurine because of the reduction in the magnetic anisotropy. The anisotropic effect of N(1) upon H(8) is sufficiently small to be neglected. The results of the calculation of the chemical shifts of the base protons are listed in Table VI. In Table VII the observed differences of the

TABLE VI

Calculations of the Base Protons' Chemical Shifts for the Purine Nucleosides (ppm)[a]

	$\Delta\sigma_{\Delta\rho}$	$\Delta\sigma_{\pi}$[b]	$\Delta\sigma_{\chi N}$
A. Adenosine			
H(8)			
I.[c]	−1.073	−1.470, −1.254	Same[d]
II.[e]	−0.976	−1.470, −1.254	Same[d]
IV.[f]	−1.792	−1.470, −1.254	Same[d]
H(2)			
I.[c]	−1.155	−1.715, −1.540	Same[d]
II.[e]	−0.063	−1.715, −1.540	Same[d]
IV.[f]	−0.460	−1.715, −1.540	Same[d]
B. Inosine			
H(8)			
I.[c]	−0.984	−1.159, −1.032	Same[d]
II.[e]	−0.489	−1.159, −1.032	Same[d]
IV.[f]	−1.158	−1.159, −1.032	Same[d]
H(2)			
I.[c]	−1.721	−0.557, −0.575	−0.200[g]
II.[e]	−2.264	−0.557, −0.575	−0.200[g]
IV.[f]	−2.742	−0.557, −0.575	−0.200[g]
C. Guanosine			
H(8)			
I.[c]	−0.817	−1.140, −1.012	Same[d]
II.[e]	−0.168	−1.140, −1.012	Same[d]
IV.[f]	−0.904	−1.140, −1.012	Same[d]

[a] Calculations are based on Eqs. (4), (5), (6).

[b] Ring-current intensity values from Ref. *72* (See text).

[c] Excess π-charge density values from Refs. *75* and *81*.

[d] $\Delta\sigma_{\chi N}$ value is the same as that listed for the appropriate protons of purine listed in Table III.

[e] Excess π-charged density values from Ref. *82*.

[f] Excess π-charged density values from Ref. *83*.

[g] The N(1) atom in hypoxanthine is changed from the pyridine type to cyclic amide type. A contribution of +0.183 ppm was calculated to compensate this change in comparison to purine (see text).

H(2) protons as well as the H(8) protons among several nucleosides are compared to the calculated differences based on values enumerated in Table III (for ribosylpurine) and in Table VI (for adenosine, inosine, and guanosine). The experimental results are those taken in D_2O at infinite dilution (Table V). Since guanosine is too insoluble for measurement, the δ value for H(8) of 1-methylguanosine is used instead. The justification for

TABLE VII

Comparison of the Observed Differences of the Chemical Shifts of the Base Protons among Several Purine Nucleosides with the Calculated Differences Based on the Values in Table VI (ppm)

	P−A		P−I			
A. H(2)						
Observed values[a] (D_2O, infinite dilution)	−0.695		−0.697			
Calculated values						
I.[b]	−0.276	−0.265	−1.051	−0.847		
II.[b]	−0.500	−0.489	+0.250	+0.564		
IV.[b]	−0.462	−0.451	+0.479	+0.683		
	P−A		**P−I**		**P−G (1=MG)[c]**	
B. H(8)						
Observed values[a] (D_2O, infinite dilution)	−0.342		−0.332		−0.744	
Calculated values						
I.[b]	−0.191	−0.181	−0.591	−0.492	−0.777	−0.679
II.[b]	−0.550	−0.540	−1.348	−1.249	−1.688	−1.590
IV.[b]	−0.598	−0.588	−1.543	−1.443	−1.816	−1.718

[a] From Ref. *64* and Table V.
[b] See Table VI for the references.
[c] 1=MG is 1-methylguanosine.

this substitution is that δ of H(8) is the same for both inosine and 1-methylinosine (Table V). It should be mentioned also that the value for H(8) for 1-methylguanosine is not extrapolated to infinite dilution because of solubility limitations, and therefore is likely to be about 0.05 ppm closer to the purine than listed (Table VII). In general, it appears that the calculated values from Set I are in better agreement with the observed values.

b. *Pyrimidines and Pyrimidine Nucleosides*

The assignment of the high-field doublet to H(5) and the low-field doublet to H(6) of uracil (or uridine) and of the cytosine (or cytidine) was first made by the Jardetzkys (*60*). This assignment was substantiated by studies on a large number of 5- and 6-substituted uracils (*91*) as well as by the studies on methyl derivatives of cytosine involving the use of ^{15}N and spin-coupling data (*92*). Kokko, Mandel, and Goldstein (*91*) reported that in DMSO the

TABLE VIII

Calculations of the Base Protons' Chemical Shifts of the Pyrimidine Nucleosides[a]

	Calculation I[b]		Calculation II[c]	
	H(5)	H(6)	H(5)	H(6)
1. Uracil[d]				
$\Delta\sigma_{\Delta\rho}$	+0.105	−0.990	+0.292	−0.824
$\Delta\sigma_{\pi}$	−0.098, −0.140	−0.107, −0.153	−0.098, −0.140	−0.107, −0.153
	+0.007, −0.035	−1.097, −1.143	+0.194, +0.152	−0.931, −0.977
2. Cytosine[e]				
$\Delta\sigma_{\Delta\rho}$	+0.323	−1.141	+0.930	−1.006
$\Delta\sigma_{\pi}$	−0.402, −0.439	−0.398, −0.435	−0.402, −0.439	−0.398, −0.435
$\Delta\sigma_{\chi N}$	−0.050	−0.036	−0.050	−0.036
	−0.129, −0.166	−1.575, −1.612	+0.478, +0.441	−1.440, −1.477

[a] Calculations were based on Eqs. (4), (5), and (6). Ring-current intensity values were from Ref. *72*. See text.
[b] Excess π-charge density values from Ref. *93*.
[c] Excess π-charge density values from Ref. *49*.
[d] Dimensions from calcium thymidylate, Ref. *11*.
[e] Dimensions from 3′ cytidylic acid, Ref. *42*.

chemical shifts of H(6), N(3)—H, and N(1)—H of uracil are related approximately linearly to the group dipole moment of the 5-substituents ranging from nitro to methyl groups. Similarly, the data on the average chemical shifts of N(3)—H and N(1)—H can be correlated linearly with the Hammett σ-constants of the 5-substituents as well; groups with high positive values (electron-withdrawing), such as nitro, cause the shift to go downfield, and groups with low or negative values (electron-releasing), such as methyl, cause the shift to go upfield.

We have employed Eqs. (4), (5), and (6) for computation of the chemical shifts of H(5) and H(6) of uracil (U) and cytosine (C) with two sets of excess π-charge density calculations (*49,93*). Again, two sets of ring-current values from Giessner-Prettre and Pullman (*72*) were adopted. The results of these computations are given in Table VIII. Because of the long distance, the contribution of the magnetic anisotropy of N(3) to H(5) or H(6) of cytosine is very small. Comparison between the observed and the calculated differences in the chemical shifts of the H(5) and H(6) of cytosine (or cytidine) and those

TABLE IX

Comparison of the Differences among the Calculated Chemical Shifts of the Pyrimidine Nucleoside Base Protons with the Observed Values (ppm)[a]

	$C_{H(5)} - U_{H(5)}$		$C_{H(6)} - U_{H(6)}$		$C_{H(6)} - C_{H(5)}$		$U_{H(6)} - U_{H(5)}$	
Observed[b] values								
DMSO	−0.172		+0.03		−1.760		−1.668	
D_2O	−0.175		0.00		−1.772		−1.945	
Calculated[c] values								
I	−0.136	−0.131	−0.478	−0.469	−1.446[d]	−1.446[d]	−1.104[d]	−1.108[d]
II	+0.284	+0.289	−0.509	−0.500	−1.918[d]	−1.918[d]	−1.125[d]	−1.129[d]

[a] The calculated values were obtained from Table VIII.

[b] Values in DMSO were obtained from Ref. (*91*) and values in D_2O were obtained from Table V.

[c] See Table VIII for the corresponding references.

[d] A correction factor of about -0.15 ± 0.05 ppm should be added to all the calculated differences. When compared to the observed difference of Set I (bases), this factor is due to the omission of the differential contribution of N(1) magnetic anisotropy to H(6) versus that to H(5). When compared to the observed differences of Set II (nucleosides) now, this factor is due to the omission of the differential deshielding effect of the N(1) pentosyl substitution on H(6) versus that on H(5).

of uracil (or uridine), i.e., H(6) of cytosine, H(6) of uridine, etc., is given in Table IX. The observed values are those of the bases in DMSO (I) (*91*) and those of the nucleosides in water (Table V). These two sets of experimental data are very close to each other ($\sim$ 0.05 ppm) except for the difference between H(6) and H(5) of uridine ($\sim$ 0.3 ppm). Both sets of calculations agree with the experimental results that the H(6) proton should be downfield from the H(5) proton; the agreement is better with cytosine than with uridine. One set of the calculated results (I) is also very close to the observed difference between the δ of H(5) of cytidine and the δ of H(5) of uridine. It is somewhat surprising that experimentally the H(6) of cytidine is very close to the H(6) of uridine, while in both sets of calculations, the difference is about 0.50 ppm. In terms of the diamagnetic contribution from the excess π-charge density to the calculated shift, H(5) of uracil is less than H(5) of cytidine, but this difference is minimized to about 0.13 ppm by the larger paramagnetic contribution of the ring current in cytosine. On the other hand, the greater paramagnetic contribution from the deficient π-charge density to H(6) of cytosine compared with H(6) of uracil, is amplified by the large paramagnetic ring-current effect of cytosine, causing H(6) of cytosine to be computed significantly downfield from H(6) of uracil, a result contrary to the experimental observation. Clearly, more work is needed in this area, since much can be learned from these relatively simple molecules.

In conclusion, chemical shifts of the base protons of the bases and the nucleosides all have been properly assigned. A preliminary attempt to account for these assignments on a theoretical and quantitative basis has been presented here. Undoubtedly, NMR spectra contain pertinent and much needed information about the electronic structure of these important heterocyclic bases in the ground state. This information is most valuable, not only for the quantum mechanical studies of these types of compounds, but also for the theoretical prediction of other properties of these molecules such as hydrogen-bonding and vertical stacking. Obviously, better theoretical treatments of the observed NMR spectra of these compounds are urgently needed. We hope that our preliminary effort (*63–66*) will bring about more interest and more significant advances in these directions.

2. Tautomerization, Protonation, Methylation, and Other Structure Assignments

Effective application of NMR has provided decisive information about various important but subtle structural features such as "*cis-trans*" configuration, 2′- or 3′-isomers, tautomerization, site of protonation and methylation, etc. Some of the outstanding examples in relation to the monomers of nucleic acids are briefly described here.

One of the most difficult problems for structural proof of the products in a nucleoside synthesis is the problem of the anomeric configuration. Two general procedures have been developed to solve this problem through the use of PMR technique. The first procedure developed by Robins and Robins (*94*) is for the 2′-deoxyribosyl-D-purines and possibly for the deoxyribosyl pyrimidine nucleosides as well. A "pseudo-triplet" where $J_{H(1')} = 7$ cps and the peak width equals 14 cps was observed for the anomeric proton of the β-anomer. The corresponding peak for the α-anomer appeared as a multiplet of four with $J_{H(1')} = 3$ and 7 cps and the peak width equal to

10 cps. This method of anomeric assignment for 2′-deoxy-D-ribofuranosyl nucleosides appears to be general, irrespective of the heterocyclic base. The second procedure was developed by Cushley, Watanabe, and Fox (*95*) for the pentofuranosyl and the pyranosyl pyrimidine nucleosides. These authors have observed the anisotropic effect of the double bond in pyrimidine aglycon on the chemical shifts of the acetyls substituted at the C(2′) position of the sugars. This anisotropic field effect can be clearly illustrated by the comparison between the chemical shifts of the acetyl groups before and after the catalytic hydrogenation of the 5,6-double bond of the pyrimidine base. With the pyranosyl pyrimidine nucleosides, when the C(2′)-acetoxy group and the base are in a *cis* relationship, the acetyl resonance is shifted upfield (about 0.2 ppm) after hydrogenation of the double bond. When a *trans*-diequatorial relationship exists at C(1′)–C(2′), then the hydrogenation of the double bond causes a small downfield shift instead. With 2′-acetylated pentofuranosyl nucleosides, hydrogenation of the double bond causes a downfield shift of the acetyl resonance when a *cis*-C(1′)–C(2′) relationship exists between the base and the acetyl group (for example, α-anomers of ribosyl and xylosyl, and β-anomers of arabinosyl and lyxosyl compounds). The same hydrogenation treatment causes a small upfield shift of the acetyl resonance for nucleosides having a *trans* relationship (such as the β-anomers of ribosyl and xylosyl, or α-anomers of arabinosyl and lyxosyl compounds). These observations have a direct bearing on the question of the nucleoside conformation with respect to the rotation of the glycosyl C—N bond, a subject which is discussed in a later section.

Another useful PMR application is the solution of difficult structural assignment problems concerned with the immediate environment of the H(1′) anomeric proton of a nucleoside as indicated by the chemical shifts and the coupling constant $J_{H(1')-H(2')}$. Fromageot et al. (*96*) have found that for a large variety of substituents at the OH(2′) or the OH(3′) of the ribonucleosides (such as phosphoryl, acetyl, benzoyl, benzyl, trityl, etc.) and for a pair of 2′- and 3′-isomers, the H(1′) resonance is at lower field (0.05 to 0.2 ppm); usually $J_{H(1')-H(2')}$ is greater for the 3′- than for the 2′-isomer. Our work (*97*) done independently on 2′-CMP, 3′-CMP, 2′-AMP, and 3′-AMP supported the conclusions of Fromageot et al. (*96*). In addition, we found that the H(1′) resonance of the ribosyl nucleoside is substantially upfield (over 0.3 ppm) from the corresponding 2′-deoxyribosyl nucleoside and arabinose nucleoside (*97*). This phenomenon was attributed to the magnetic shielding of H(1′) by the *cis*-oxygen atom in the 2′-position and it is very useful for identification purposes, for instance, between a 2′-deoxyadenosine and a 3′-deoxyadenosine (cordycepin).

In comparing the PMR spectra of 1,9-dimethylguanine and guanosine in DMSO, Miles, Howard, and Frazier (*98*) have properly assigned the resonance for the protons of the 2-amino group, therefore eliminating the possibility of a 2-imino structure. The assignment of N(1)—H in the PMR spectra together with other pertinent infrared data allow them to conclude the correct tautomeric structure of guanine or guanosine (Fig. 1). In comparing the ^{14}N and ^{15}N spectra of 1-methylcytosine in DMSO, Becker, Miles, and Bradley (*92*) have established the chemical shift of the 4-amino protons and therefore proved the amino form to be the correct tautomeric structure (Fig. 1). The work of Kokko et al. (*91*) has also contributed to the establishment of the keto form as the correct form for the 4-oxo group in uracil or thymine.

An elegant quantitative study has been made by Read and Goldstein (*99*) on the rather complex question concerning the site of protonation of purine. They found that the δ's of H(6), H(2), and H(8) are affected by the change of pD to about the same extent (deshielding at lower pD), indicating that N(1), N(3), and N(7) are all affected. However, the change of the ^{13}C—H coupling constants versus the change of pD was found to be in the following order: H(2) > H(6) > H(8). Analyses of these data led them to conclude that the per cent of protonation at each site is: N(1), $54 \pm 12\%$, N(3), $17 \pm 12\%$, and N(7) [or N(9)], $29 \pm 7\%$. In a study of the site of protonation for cytosine derivatives, Becker, Miles, and Bradley (*92,100*) compared the spectrum of 1-methylcytosine hydrochloride (and hydroiodide) with its ^{15}N derivatives in DMSO at 23° and SO_2 at −60°. They observed a splitting of H(5) (by 2.5 cps) due to the long-range coupling of N(3)—H which they have located at 11.43 ppm (downfield from TMS internal standard) in SO_2 at −60°. These data unequivocally established the site of protonation for cytosine. In the studies of the δ values of the pyrimidines in anhydrous trifluoroacetic acid, Jardetzky, Pappas, and Wade (*101*) proposed that the order of basicity in pyrimidines is N ring > N amino > oxygen.

Confirmation of the site of methylation in 7-methylinosine versus 1-methylinosine was made possible by the observation that the methyl resonance of 7-methylinosine occurs at about 0.55 ppm downfield from that of 1-methylinosine in D_2O indicating the location of the methyl group at a positively charged nitrogen atom (*86*). In addition, the fact that the H(8) of the 7-methylinosine is located about 1 ppm downfield from the H(8) of 1-methylinosine in DMSO, and that this proton cannot be found in D_2O due to rapid exchange (*86*), further supports the assignment of the site of methylation for the 7-methylinosine. Comparison of the data on 1-methylinosine with inosine and other nucleosides reveals some interesting features. 1-methylinosine is the only 6-substituted nucleoside to have a δ of H(2) at a lower field than the δ of H(8). H(2) of 1-methylinosine is shifted downfield by 0.15 ppm as compared to H(2) of inosine, while the H(8) proton of 1-methylinosine is unchanged as compared to inosine (Table V).

In view of the large downfield shift (1 ppm) of the H(8) in 7-methylinosine, in comparison with 1-methylinosine due to quaternization of the imidazole ring, we proposed (*64*) that the substantial and selective deshielding of H(2) in 1-methylinosine is due to a more extensive degree of quaternization occurring at N(1) in the 1-methylinosine as compared with inosine itself. The electron-releasing inductive effect of the methyl group will tend to stabilize the positive charge on the N(1) nitrogen.

Similar methyl-stabilized resonance forms can be adopted as an explanation for the difference in base proton chemical shifts of adenosine, N-6-methyladenosine, and N-6-dimethyladenosine (Table V). At infinite dilution, H(2) and H(8) are shifted progressively to higher fields as methyl groups are added to the N(6) amino position. H(2) and H(8) of adenosine are 8.731 and 8.821 ppm, those of the monomethyl compound 8.731 and 8.73, and those of the dimethyl derivatives 8.63 and 8.69. On the other hand, the methyl proton resonance shifts downfield from monomethyl to dimethyl compound, 3.550–3.886 ppm. These observations suggest that a progressively greater degree of resonance donation of electrons from the amino group to the ring takes place as hydrogen is being replaced by CH_3, due to methyl inductive stabilization of positive charge on N(6).

A comprehensive survey of chemical shifts of N-methyl derivatives has been published by Ma and Warnhoff (*102*). Comparison of data on monomethyl and dimethyl *N*-substituted aromatic compounds, such as *N*-methylaniline vs *N*,*N*-dimethylaniline, shows a similar deshielding of the methyl protons in the dimethylated compounds. In saturated systems, methyl protons are more shielded in N,N-dimethyl compounds as compared to monomethyl derivatives.

3. Conformation

During the early sixties, C. D. Jardetzky made a substantial effort to study the pentose conformation of the nucleosides and nucleotides (*61,62, 103*) by measuring the coupling constants, especially the $J_{H(1')-H(2')}$, and by the application of the Karplus analysis (*104*) of the H—C—C—H system. Though the application of this Karplus relationship between the vicinal coupling constants and the dihedral angle of a H—C—C—H system to the analyses of sugar conformation is subject to the influence of other effects such as substituent perturbations, C(1')—C(2') bond lengths, etc. (*105,106*), it is generally recognized that this provides a valuable guide. The coupling constants of various nucleosides and nucleotides are shown in Table X. These values differ slightly from those previously published by the Jardetzky's (*60,61,103*) probably because of the improved quality of the spectra obtained from 60- and 100-Mc instruments. It can be seen that while all the average $J_{H(1')-H(2'),H(2'')}$ values for the deoxyribose compounds are similar to each

TABLE X

The Coupling Constant, $J_{H(1')-H(2')}$, for Various Nucleosides and Mononucleotides[a]

	$J_{H(1')-H(2')}$, cps
A. Ribose derivatives	
Adenosine	6.1
3′-Deoxyadenosine	2.2
2′-O-Methyladenosine	5.3
5′-AMP	5.6
5′-AMP monomethyl ester	5.1
3′-AMP	5.8
2′-AMP	5.9
2′-AMP monomethyl ester	5.1
1-Methylguanosine	5.7
5′-GMP	5.8
3′(2′)-GMP	5.4
Inosine	5.6
5′-IMP	5.5
Cytidine	3.7
5′-CMP	3.2
3′-CMP	3.4
2′-CMP	4.6
Uridine	4.2
5′-UMP	4.5
5-Bromouridine-5′-phosphate	4.4
3′(2′)-UMP	3.5
B. Deoxyribose derivatives[b]	
2′-Deoxyadenosine	6.5
Thymidine	7.0
5′-dAMP	7.0
5′-dCMP	6.8
5′-dUMP	7.0
5′-TMP	7.0

[a] Solution conditions: Neutral D_2O, 0.02–0.1 *M*. The accuracy of the values is about ± 0.2 cps.

[b] $J_{H(1')-H(2'),\,H(2'')} = \frac{1}{2}(J_{H(1')-H(2')} + J_{H(1')-H(2'')})$

other, the $J_{H(1')-H(2')}$ of the ribosyl purine compounds (5.5 cps) are significantly higher than that of the ribosylpyrimidine compounds (4.0 cps). Based on these observations, C. D. Jardetzky has proposed that in the ribose nucleosides, the C(2′) is *endo* for the purine compounds while the C(3′) is *endo* for the pyrimidine compounds. Though she could not exclude the possibility of a C(3′) *exo* conformation for the purine compounds, her arguments appear to be self-consistent, especially in the study of cyclic 3′:5′ AMP (*103*). However, her conclusion received very little support from the X-ray diffraction data of these compounds in the crystalline state (*4,6*). As discussed in Section II, A, adenosine (in a complex with 5-bromouridine) and 5′-adenylic, and 3′-adenylic acid all have C(3′) instead of C(2′) in *endo* conformation. As for the pyrimidine compounds, the X-ray results show that cytidine has C(3′) in *endo* form (*30*) as concluded by C. D. Jardetzky, but 3′-cytidylic acid has C(2′) in *endo* form instead (*19,21,42*), yet both compounds have about the same $J_{H(1')-H(2')}$ values (3.4–3.7 cps). It should be noted, however, that we are comparing conformational data on the same compound but for different states, i.e., solid-crystal versus the solution state. Whether this comparison is meaningful is debatable as discussed in Section II, A. Nevertheless, it appears that it is still an open question at present whether this difference of $J_{H(1')-H(2')}$ values for ribosyl purine compounds and for ribosyl pyrimidine compounds is due to the difference in spatial arrangement (differences in dihedral angle) or to the electronic effect of having different substituents at the C(1′) position, because it is not entirely clear why the ribose conformation will be changed by having different substituents at the C(1′) position. However, if this difference is due to the electronic effect of different substituents at the C(1′) position, then this effect is not observed in the average values of $J_{H(1')-H(2'),H(2'')}$ of deoxyribose compounds. Nevertheless, we feel that under certain conditions, the influence of the dihedral angle must be decisive as in the case of comparing the value of $J_{H(1')-H(2')}$ of 3′-deoxyadenosine (2.2 cps) with that of adenosine (6.1 cps) (*64*). In this case, the substituent at C(1′) is the same (both adenine), and there is no change of the groups at C(2′) either. The only difference between these two compounds is a change of OH to H at C(3′), two carbon atoms removed from the C(1′). Therefore, we feel that this large difference in the $J_{H(1')-H(2')}$ value must reflect a difference in the dihedral angle between H(1′) and H(2′) if the Karplus relationship can be applied at all to this type of system.

Attempts have been made by Tsuboi et al. (*107,108*) to ascertain the conformation of the H—C—O—P group in nucleotides or nucleic acid by analyzing the J_{HP}, the proton phosphorus spin coupling constant, in a manner similar to the H—C—C—H system just discussed. In studying phosphoesters, it is believed that the value of J_{HP} is sensitive to its dihedral angle, and it was further assumed that the internal rotation around the C—O single bond in the CH_3—O—P group or C—CH_2—O—P group has mainly two kinds of potential minima, a *trans* or a *gauche* relation between the proton and the phosphorus (or between the C in C—CH_2—O—P and the phosphorus). The value of J_{HP} in the *gauche* relation was assigned to be 1.5 cps and the value of J_{HP} in the *trans* relation was assigned to be 28 cps from an analysis of the data of the J_{HP}'s of a series of model compounds, including some cyclic phosphates. Experimentally, the J_{HP} values in the H—C(5')—O—P group were measured to be about 4.5 cps for both 5'-IMP and 5'-GMP (*108*). These authors conclude that in these two nucleotides the phosphorus atom is estimated to be ~80% *trans* to C(4'). This approach and the results are of great interest and should be confirmed by measurements with other model compounds and be extended to di- or polynucleotides. We have observed a splitting of the H(5') and H(5") resonance of the nucleosides into a doublet of about 3 cps that is independent of field strength, most likely indicating that this splitting is due to the coupling with H(4') (Schweizer and Ts'o, unpublished data). Substitution of a phosphate group at the 5'-position further splits this doublet into a quartet with a $J_{H(5')}$ C—O—P value of 4.0–5.0 cps for the following compounds in agreement with the observation of Tsuboi (*108*) on 5'-IMP and 5'-GMP: 5'-dTMP, 5'-AMP methyl ester, 5'-dAMP, 5'-dCMP, 5'-dUMP, as well as 5'-IMP.

Based on the available X-ray crystallographic data and a hard-sphere model, Ramachandran's group (*5*) has analyzed the allowable conformation of the nucleosides and nucleotides as discussed in Section II, A. Because of the restriction due to the contact between the P(5') atom and the C(4') atom, they also reached the conclusion that the P atom must rotate away from the C(4'), and therefore must be in a *trans* position to the C(4').

Hindered rotation of the 4-amino group ($H_2N(7)$—C(4)) in protonated cytosine derivatives has been observed by Becker, Miles, and Bradley (*92*) and by Shoup, Miles, and Becker (*109*), especially in SO_2 at −40 to −60°. The existence of geometric isomers can be demonstrated by the splitting of the H(7) or $CH_3(7)$ resonance by being *cis* or *trans* to the $N(3)^+$—H. Again, the use of ^{15}N compounds and the application of the decoupling techniques have been very helpful in the assignment of the resonance lines.

The question about the conformation of the nucleosides and nucleotides in respect to rotation of the glycosyl C—N bond is discussed extensively in Section II, G.

4. Binding of Metal Ions

Owing to the charge and the unpaired electron (for the paramagnetic ion) of the metal ion, NMR is a very useful tool for the study of binding of metal ions.

Eichhorn, Clark, and Becker (*110*) have found that in D_2O, copper(II) broadens the H(2) and H(8) peaks in dAMP, H(5) and H(6) of dCMP, and the H(8) peak of dGMP; dTMP is unaffected. In DMSO the protons on the amino group and the heterocyclic nitrogen atoms are not broadened by copper ions. It is concluded the copper binds to the N(7) position of adenosine and guanosine and to N(3) of cytidine but does not bind thymidine. The ^{31}P resonance in dAMP and dTMP is broadened by copper ions, showing that copper also binds phosphate. Therefore, the Cu(II) apparently can be bound simultaneously to N(7) (A and G) or N(3) (cytosine) and to the phosphate group of the mononucleotides. Wang and Li (*111*) have also found that the N(7) of purine and the N(3) of cytosine are the binding sites for Zn in DMSO as indicated by the downfield shift of the H(2), H(6), and especially H(8) resonance of purine and the H(6) and H(5) resonance of cytosine in increasing concentrations of Zn ions. For the nucleosides, Wang and Li (*112*) have investigated the formation of 1 : 1 Zn-nucleoside binary complexes and the formation of ternary complexes of Zn-adenosine–guanosine and Zn–adenosine–cytidine in DMSO. On the other hand, the base-pairing complexes of adenosine–uridine and guanosine–cytidine are broken up by the addition of Zn to give the Zn-nucleoside binary complexes. The results were based on the paramagnetic effect of Zn ion on the chemical shifts of the proton resonance.

Cohn and Hughes (*113*) have investigated the binding of Mg^{2+}, Ca^{2+}, and Zn^{2+} to ADP and ATP by studying the proton and phosphorus resonance. The chemical shifts of ^{31}P in the presence of equimolar concentrations of these metal ions indicate that these metals form complexes with β- and γ-phosphate of ATP and α- and β-phosphate of ADP. The chemical shifts of the protons in ATP complexes showed that only Zn^{2+} causes a change; namely, the H(8) resonance is shifted to a lower field indicating the binding of Zn to N(7) of ATP. For the binding studies of the paramagnetic ions as shown by the line broadening of the ^{31}P resonance, Cu^{2+} interacts solely with the β and γ peaks, but Mn^{2+} and Co^{2+} interact with all α-, β-, and γ-phosphate groups. With ADP, Cu^{2+} and Mn^{2+} both interact with the α- and β-phosphate groups. In addition all these paramagnetic ions specifically broaden the H(8) proton of ATP. Ward and Happe (*114*) have studied the linewidth of the ^{35}Cl resonance, which is influenced by the concentration and the physical state of Zn^{2+}, as an indicator of the binding of Zn to ADP. They drew conclusions about the presence of Zn_2ADP, ZnADP, and $Zn(ADP)_2$ in solution from the study of linewidth of ^{35}Cl in relationship to the ADP/Zn ratio in solution. In a series of papers from the Bell Laboratories, Shulman, Sternlicht, and their associates (*115–117*) have studied the binding of various paramagnetic ions to AMP and ATP, etc., by the magnetic resonance technique. Their extensive studies quantitatively confirm the basic conclusions of the early findings by Cohn and Hughes (*113*). In a recent paper, Sternlicht, Jones, and Kustin (*118*) reported about the existence of the 1:2 Mn^{2+}–ATP complex at high total ATP concentration and the existence of the 1:1 complex at low concentration. In the Mn^{2+}–$(ATP)_2$ complex, the metal ion simultaneously binds to the phosphate moiety of one nucleotide and to the N(7) of the second nucleotide.

5. Temperature Studies

Though there are rather few studies done on the effect of temperature on the NMR properties of the monomers, some significant information was obtained (*119*). The values of the coupling constant of the H(1′)–H(2′) protons, $J_{H(1')-H(2')}$ of the following nucleotides were found to be the same ($\pm$ 0.2 cps) at 30° to 60°, 3′-AMP, 5′-AMP, 3′-GMP, 5′-GMP, 5′-CMP, 3′-UMP, 2′-AMP, and probably also 5′-dTMP and 5′-UMP ($\pm$0.4 cps) (*119*). Essentially the same result was observed by Hruska and Danyluk (*120*) for 3′- and 5′-AMP at the temperature range of about 20° to 80°. Those results suggest that the conformation of the furanose ring of these nucleotides is unlikely to change within this range of temperature variation.

Another interesting observation is that with the present external reference system (TMS capillary), the differentials between the δ value at 30° and the δ value at 60° (downfield) are not identical for different protons from the same compound or from different compounds. For the H(2) protons, the H(5) protons, and many H(1′) protons the differential is about 0.1 ppm at infinite dilution, a value similar to that observed for the CH_2- and CH_3-protons of ethanol (0.02 *M*) in D_2O (*119*). This value of the differential (0.1 ppm) is considered to reflect mainly the change of the bulk magnetic susceptibility of the medium at these two temperature levels. The temperature differential, however, is considerably less than 0.1 ppm for the H(6) protons of the pyrimidine compounds (only around 0.04 ppm) and for some of the H(8) of the purine compounds as well as for some H(1′) protons. As a general rule, the temperature differential for the H(6) proton is about half of that for the H(5) proton of the same pyrimidine compound; the temperature differential for the H(8) is about 70–80% of that for the H(2) proton of the same adenine compound at infinite dilution. This phenomenon is most likely due to a greater solvent–solute interaction of water molecules with the H(6) and H(8) protons than with the H(5) proton or H(2) proton at room temperature. As mentioned in the above section on X-ray studies, there are crystallographic evidences for the hydrogen bonding between water molecules and the H(8) and the H(6) of purine and pyrimidine monomer, respectively. This explanation is also supported by the studies on the solvent effect on chemical shifts as discussed above in this Section II, C, 1. Therefore, our interpretation is that the small temperature differentials of the H(6) (pyrimidine) and H(8) (purine) protons are due to a weak hydrogen bonding between these protons and the water molecules at low temperatures. This weak hydrogen bonding diminishes at 60°, thus moving the chemical shifts of these protons upfield. This contribution of the shielding effect counteracts the deshielding effect from the change of bulk magnetic susceptibility at higher temperatures. The net result is that the chemical shifts of these protons are less sensitive to temperature elevation because of these counteracting factors.

D. Infrared (IR) and Raman Spectroscopy

Infrared spectroscopic studies on the nucleotides (*121–124*) and phosphate group (*125*) began in the mid- and late fifties, concurrent with the beginning of the IR studies on nucleic acids (*126–131*, etc.). Since that time, most of the IR studies on nucleic acids and their monomers have been made by Miles, Frazier, Howard, and their associates at NIH, and by Tsuboi, Shimanouchi, Kyogoku, and their associates in Tokyo. In addition, the group in Prague, Pitha and others, has also made substantial contributions, especially in the area of pyrimidine nucleosides and azapyrimidine nucleosides (for example, see Refs. *132–135*).

IR spectroscopy has a unique asset unrivaled by any other physicochemical technique—the sample can be studied by IR in any physical state: gas, liquid solution, mull, disk, crystal, oriented film, etc. Dichroic studies with polarized irradiation on oriented solid samples [such as oriented films of nucleic acid (*126,127,129*, etc.)] can provide valuable information about the geometry of the chromophore. Sensitivity of this technique for compounds in solution generally is not as high as that of UV spectroscopy or fluorescent spectroscopy and is at about the level of PMR technique, i.e., 10^{-3} *M* and higher in D_2O solution. In the early stage, most of the IR studies on the monomers were concerned with the problem of band assignment, tautomeric forms, and site of protonation. Recently, most of the studies are related to "interaction," a subject considered in Section III.

Shimanouchi, Tsuboi, and Kyogoku have written a comprehensive review on the IR studies of nucleic acid and related compounds in 1964 (*136*). By studying the neutral and the protonated form of the nucleosides in H_2O and D_2O (*136,137*), they have assigned various IR bands to the nucleosides and have made conclusions about the tautomeric forms and the sites of protonation. These conclusions from IR studies are in essential agreement with those quoted in the X-ray diffraction studies. They have also paid special attention to the IR bands of the phosphate group in the low-frequency regions (500–1200 cm^{-1} wave number region). This background information is most useful for the dichroic study of the oriented nucleic acid sample concerning the geometry of the phosphate group (*137,138*). The original papers are recommended for the details.

In their studies on the monomers, the NIH group has investigated the problem of band assignment, tautomerism, and site of protonation by comparative studies of model compounds and by ^{18}O substitution. In a comparative study on guanosine, 1,9-dimethylguanine, 2-amino-6-methoxy-9-β-D-ribofuranosyl purine (the enol model), 1,7,9-trimethylguanine and 7,9-bis-(2′-hydroxylethyl)-guanine (*139*), they concluded that guanine has a keto-amino structure in neutral aqueous solution and undergoes protonation at N(7) in acid solution. In this study, they also assigned the 1665 cm^{-1} band in D_2O to be a "C_6-carbonyl stretching vibration (probably strongly coupled to the $C_4{=}C_5$ bond but not to $C_2{=}N_3$ bond)." Subsequently, Howard and Miles (*140*) were able to confirm this assignment unambiguously by the enzymic synthesis of [^{18}O] guanosine and [^{18}O] inosine derivatives with adenosine deaminase and adenylic acid deaminase in [^{18}O]D_2O. In the [^{16}O]D_2O and [^{18}O]D_2O solution, the enzyme converts the 5′-AMP to 5′-IMP with the disappearance of a 1625–1627 cm^{-1} band [assigned to be a "ring vibration in which the C=N stretching motions predominate" by Shimanouchi et al. (*136*)] and the concomitant appearance of a 1672 cm^{-1} band in [^{16}O]D_2O and a 1659 cm^{-1} band in [^{18}O]D_2O. Similarly, the 2,6-diamino-9-β-D-ribofuranosyl purine is converted enzymatically to guanosine to give a band at 1665 cm^{-1} in [^{16}O]D_2O and a band at 1652 in [^{18}O]D_2O. This shift to a lower frequency by 13 cm^{-1} from [^{16}O] keto to [^{18}O] keto unambiguously assigns these new bands which appear after the enzymic reactions. The work of the NIH group on the interaction of the monomers is discussed in Section III.

Recently, a comprehensive study on the Raman spectra of the ribosyl monomers and related model compounds has been made by Lord and Thomas (*141*). In comparison to the IR study, Raman spectroscopy has the advantage that the spectrum can be obtained in aqueous solution from 3200–200 cm^{-1} frequency with little interference, and the 2000–200 cm^{-1} region of the vibrational spectrum is accessible for spectra taken in either H_2O or D_2O solution, a condition favorable for studying the effect of isotopic exchange. Raman spectroscopy has the disadvantage of requiring relatively large sample specimens (100 mg or more usually) and of giving poor results with samples which are optically inhomogeneous or which fluoresce at the excitation frequency employed. By the employment of lasers as exciting sources, these limitations may soon be significantly reduced. Raman spectroscopic results on both solid and aqueous solution samples are in agreement with those from IR studies that the derivatives of uracil, adenine, cytosine, and guanosine exist in the keto and amino forms. The conjugated carbonyl groups of uracil, cytosine, and guanine derivatives in aqueous media and in the absence of vibrational coupling have their stretching frequencies in the narrow region of 1670–1655 cm^{-1}. The nonconjugated carbonyl group of uracil derivatives has a higher frequency, ca. 1695 cm^{-1}, under identical conditions. The ring C=C and C=N stretching frequencies are found generally in the 1650–1550 cm^{-1} range. The carbonyl stretching vibrations are considerably broader than the other ring double-bond stretching modes; the carbonyl lines are more strongly polarized than the lines of the ring-stretching vibrations. In general, NH, NH_2, ND, and ND_2 deformation vibrations are very weak or not detected at all. In the spectral range of 1800–1500 cm^{-1}, identical features are observed in general for the same base moiety regardless of the substituents (ribosyl group or alkyl group) at the N(1) or N(9) position. The ribose moiety of nucleosides or -tides gives rise only to very weak Raman lines, presumably due to the absence of a double bond, while the mononucleotide bivalent phosphate ion characteristically scatters at 980 cm^{-1}. In the study of the pH effect, cation formation is always accompanied by shifts to higher frequencies of all double-bond stretching vibrations while anion formation is accompanied by a corresponding decrease in frequency. These effects can be explained only on the basis of attachment or detachment of protons (or deuterons) at ring nitrogen sites rather than at external amino groups. Attachment of protons to the ring appears to favor localization of the ring π-electrons in the various double bonds, while removal of protons appears to cause the delocalization of the π-electrons.

IR spectral studies have been made by Kyogoku et al. (*142*) on thin sections of the single crystals of 1-methylthymine, 9-methyladenine, and their 1:1 complex (A·T cocrystal) by using the polarized radiation in the spectral region of 4000–400 cm^{-1}. The absorption bands due to in-plane vibrations [such as NH stretch, CH stretch, C(2)=O stretch, C(4)=O stretch, etc., of T; NH_2 symmetric stretch of A, etc.)] and out-of-plane vibrations (CH bend, NH bend, N—CH_3 rock, C—CH_3 rock, etc., of T; NH bend, CH bend, NH_2 wag, etc., of A) can be distinguished in the spectra of the sections parallel or perpendicular to the molecular plane. Directions of some transitions have also been determined. Assignments of NH and NH_2 deformation were made on the observed N-deuteration effects.

E. Ultraviolet Absorption Spectroscopy, Fluorescence and Phosphorescence

1. Ultraviolet Absorption Spectra

Ultraviolet absorption is perhaps the most widely used of all physicochemical techniques in the study of nucleic acid. Although the collection of data is relatively easy (and there is a large volume of data), the interpretation of the data is not necessarily simple. Nevertheless, when this technique is properly used, the results can provide very useful information about structural assignments, tautomerism, ionization constants, site of protonation, electronic properties, association properties, etc.

Voet, Gratzer, Cox, and Doty (*143*) have reported the UV absorption spectra in the region of 300 to 185 $m\mu$ of the pentoses, bases, nucleosides, nucleotides, and nucleic acids in aqueous solutions at various pH's. The absorption maxima and positions of the shoulders of the natural bases and

TABLE XI

The Ultraviolet Absorption Maxima of the Bases and Ribonucleosides in Neutral Aqueous Solution[a]

	First region (4.8–4.3 eV)		Second region (5.4–4.8 eV)		Third region (6.8–6.0 eV)			
Compound	$m\mu$	$\varepsilon \times 10^{-3}$	$m\mu$	$\varepsilon \times 10^{-3}$	$m\mu$	$\varepsilon \times 10^{-3}$	$m\mu$	$\varepsilon \times 10^{-3}$
Adenine	260.5	13.4			207	23.2	<185	
Adenosine	259.5	14.9			206	21.2	190	19.8
Guanine	275	8.1	246	10.7	~205	Shoulder	196	22.1
Guanosine	276	9.0	252.5	13.7	~205	Shoulder	188	26.8
Purine	263	8.0	240	Shoulder	~200	Shoulder	188	23.6
Cytosine	267	6.1	230	6.3	~209	~10	196.5	22.5
Cytidine	271	9.1	230	8.2	~212	~10	198	23.2
Uracil	259.5	8.2			202.5	9.2		
Uridine	261.1	10.1			205	9.8		
Thymine	264.5	7.9			205	9.5		
Thymidine	267	9.7			206.5	9.8		

[a] The data were those from Voet et al. (*143*). The classification is essentially that of Tinoco and Holcomb (*144*), and Pullman and Pullman (*49*).

of the ribonucleosides (the spectra of which are very similar to the deoxynucleosides and the nucleotides) from this comprehensive report are shown in Table XI essentially in accordance with the classification of Tinoco and Holcomb (*144*) and of Pullman and Pullman (*49*). We can conclude from the data in Table XI that the replacement of a proton in a base by a sugar group (formation of nucleoside) usually causes hyperchromism and also a small red shift in the absorption maxima, except for adenine.

Yamada and Fukutome (*145*) recently reported on absorption spectra of sublimed films of adenine, guanine, cytosine, thymine, and uracil that went down to 120 mμ at room temperature. There are prominent absorption peaks in the region below 180 mμ, which are comparable in magnitude to the peaks above 180 mμ, especially for adenine, cytosine, thymine, and uracil. Comparison between the peak positions of the solid samples and those obtained from neutral aqueous solutions in the region above 185 mμ indicate that the 260 mμ and the 200 mμ bands are shifted to the red in the solid samples, probably due to a larger magnitude of polarizability in the solid samples (*145*). This red shift was also found in the crystal spectra of 9-ethyladenine, 1-methylthymine, and their dimers to be described in the next paragraph. The ratio of intensity of the 200 mμ band to the 260 mμ band is very much reduced in the solid sample as compared to the aqueous sample for adenine, cytosine, and thymine. These investigators (*145*) reported that the absorption spectra of the sublimed films change with age upon standing in vacuo at room temperature, suggesting a gradual development of ordered arrangement. In the cases of adenine, guanine, and uracil films, the absorption peaks above 160 mμ tend to decrease while the transitions below 160 mμ tend to increase in this process of changing intensities.

The vapor spectra of adenine, 9-methyladenine, 9-methylhypoxanthine, uracil, and 1,3-dimethyluracil have been reported by Clark, Peschel, and Tinoco, Jr. (*146*). These spectra are, in general, very similar in shape to those seen in aqueous solution. In 9-methylhypoxanthine, however, a new band at 280 mμ is found which is absent in the aqueous solution. The main electronic band of each compound is shifted to the blue by about 10 mμ in the vapor phase compared with the spectrum of the aqueous solution. This is commonly found for π–π^* transitions and was interpreted simply as a greater London attraction of the solvent for the excited state than for the ground state of the molecule (*146*). Clark et al. claimed that no n–π^* band was found for these compounds in vapor spectra. However, the shoulder in the 280-mμ region of the 9-methylhypoxanthine vapor spectrum is blue-shifted when dissolved in aqueous solution, suggesting an $n \rightarrow \pi^*$ transition. Nevertheless, the extinction coefficient of this shoulder appears to be rather high for an $n \rightarrow \pi^*$ transition. This possibility was not discussed by Clark et al. (*146*). The extinction coefficients for adenine and 9-methyladenine indicate that the oscillator strengths of the gas spectra are much lower than those in solution. The vacuum UV spectra of pyrimidine, pyrazine, and pyridazine vapors between 1550 Å and 2000 Å have been reported and analyzed by Parkin and Innes (*147*).

For the all-important question about the direction of the transition dipole moment, the work of Stewart and Davidson (*148,149*) and Stewart and Jensen (*150*) provide the most valuable data. They have studied the polarized absorption spectra in the molecular planes for thin sections of single crystals of 1-methylthymine, 9-methyladenine, and the 1 : 1 A·T dimer complex. For 1-methylthymine, the first absorption band (λ_{max} $\approx$ 275 mμ) is polarized 19° (inclined toward the —CH_3 group) with respect to the N(1)—C(4) axis (Fig. 4), and the second UV band ($\lambda_{max} < 230$ mμ) is polarized approximately perpendicular to the first. For the 9-methyladenine, the first transition band ($\lambda_{max} \approx 275$ mμ) consists of two components: the major component is polarized at 3° [inclined toward N(7)] with respect to the short axis of C(4)—C(5) line (Fig. 4), and the minor component (λ_{max} $\approx$ 255 mμ) is approximately perpendicular to the major component (Fig. 4) along the long axis. The dipole moment of the second transition band

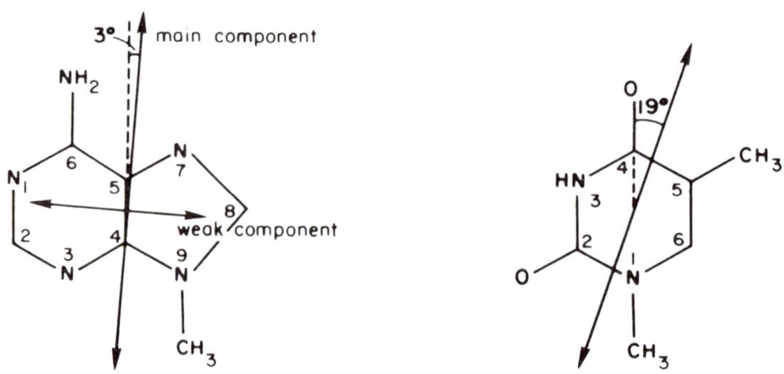

Fig. 4. Direction of the polarization of the first transition moment in 9-methyladenine and 1-methylthymine [after Stewart and Davidson (*148,149*), Stewart and Jensen (*150*)].

($\lambda_{max} < 230$ mμ) is probably also polarized along the long axis. Based on a comparative study on spectra of some mono- and poly-substituted purines, Mason earlier has suggested that the 240–280 mμ band (x band) of purine and adenine actually consists of two electronic transitions, one of which is short-axis and the other long-axis polarized (*151*). Mason, however, at that time attributed the intense absorption in adenine to a long-axis polarized band. Similarly, Clark and Tinoco (*152*) in comparing the spectra of purine and pyrimidine bases to those of acetophenone and benzimidazole, have analyzed the origin and orientation of these transition moments in relation to those of benzene. For pyrimidine, they considered that the transition at 240 mμ corresponds to $A_{1g} \rightarrow B_{2u}$ transition of benzene at 260 mμ and this transition must be polarized along the line joining the two nitrogens. Though they hesitate to assign all the absorption at the 210-mμ region to the $A_{1g} \rightarrow B_{1u}$ transition, this B_{1u} band should be polarized perpendicular to the

B_{2u} transition in accordance with the benzenoid spectral properties. They viewed that in uracil, the two keto groups are the main perturbers of the benzene system, thus the B_{2u} transition is polarized along the line joining these two keto groups, in agreement with the result of Stewart and Davidson discussed previously (*148*). In considering adenine as the $NH_2(6)$ substituted, and 4,5-imidazole attached pyrimidine, Clark and Tinoco (*152*) suggested that the B_{2u} transition is polarized along the C(4)—C(6) axis and the B_{1u} band which is assigned as a weak contributor to the adenine 260-mμ band is polarized perpendicular to the B_{2u} band. Dratz and Biltonen (*153*) have made a correlation study of the absorption band, circular dichroism, and magnetic circular dichroism of adenine and 3′-AMP. Their results clearly indicate the existence of two electronic transitions in the vicinity of 260 mμ with components perpendicular to one another. They estimated that the higher energy transition (at about 245–260 mμ) has an oscillator strength of about 20% of the known major band at 260 mμ. Callis, Rosa, and Simpson (*154*) have measured fluorescence of dilute (rigid) solutions of guanine and 9-ethylguanine in the region of 230 to 300 mμ and their results showed that the two absorption bands have approximately perpendicular transition moments. Their results on adenine were not as definitive, but they did conclude that there was a weak, oppositely polarized second transition in the main absorption band.

Hypochromism in UV absorption of nucleic acid polymers relative to the monomers is an important aspect for nucleic acid spectroscopy, since this phenomenon has been extensively used for the characterization of the conformational states of the oligomers and polymers. Stewart and Davidson (*148*) have observed that the integrated absorption coefficient for the single crystal of 1-methylthymine in the 300–260 mμ band is 25% less than that of the solution, though with a reasonably large percentage of error. The hypochromicity (comparison based on extinction coefficients at absorption maximum) attributed to the 1-methylthymine (1-methyl T) and to the 9-methyladenine (9-methyl A) in A·T dimer are 50% ($\pm$ 20%), and 75% ($\pm$ 10%), respectively. For the crystal of 9-methyl A, the hypochromicity is 20% ($\pm$ 10%). Theoretical considerations by Tinoco (*155,156*) and Rhodes (*157*) led to the conclusion that hypochromism (comparison based on the oscillator strengths of a given transition band) can be attributed to dipole–dipole interaction between transition moments in neighboring oscillators. These interactions involve the transition moment for the lowest energy π–π^* transition with those for the transitions of higher energy. Large hypochromic effects are predicted from this model when the bases are arranged parallel and above one another in a vertical stacking mode separated by a short distance as in helical DNA (*157a*) or in certain crystals. In this model, the intensity lost by the longest wavelength band is acquired by transitions further in the ultraviolet. Spectra of single-stranded stacked polynucleotides and dinucleotides which cannot form hydrogen bonds between bases qualitatively support this concept (*158,158a*, and other papers on dinucleoside monophosphates). It should be noted that both the 260-mμ band and the 190–210 band of nucleic acid are hypochromic to about the same extent with respect to those of the monomers (*143*); therefore, the predicted hyperchromic effect on transitions at higher energy (*157a*) cannot yet be confirmed or examined because of the inaccessibility of regions lower than 185 mμ in aqueous solution. Stewart and Davidson (*148*) have employed this dipole–dipole perturbation theory to calculate the hypochromicity of the crystal spectra. Their calculation for 1-methyl T in both single crystal and complex crystal is in reasonable agreement with the experimental observation, while their calculation for 9-methyl A is in very poor agreement with the observed results. On the other hand, Bolton and Weiss (*159*) and Nesbet (*160*) have attributed hypochromism to a local field effect of the induced electric dipole moments in the oscillators neighboring on the chromophore. The hypochromicity due to this resonance effect was predicted to be observable for coplanar, hydrogen-bonded base pairs even in the absence of vertical stacking. As is described in much greater detail in Section III, B, N(1) (pyrimidine) and N(9) (purine) substituted bases formed coplanar, hydrogen-bonded complementary base pairs (A to T and G to C) and self-associated base pairs in organic solvents such as CCl_4, $CHCl_3$, DMSO, etc. Thomas and Kyogoku (*161*) as well as Gratzer and McClare (*162*) simultaneously and independently reported the change of absorption in the first π–π^* transition (mostly hypochromic) due to formation of base pairs in chloroform. Thomas and Kyogoku observed the following band intensity change upon base pair formation: -10% for U at 268 mμ upon A·U formation; -15% and -5% for G at 261 mμ upon G·G formation and G·C formation respectively; -12% and $+10\%$ (hyperchromic) for C at 280 mμ upon G·C and I·C formation respectively; -10% for I at 252 mμ upon I·C formation. In the case of the A·BrU base-pair, since the absorption of 5-bromouracil is better separated from the absorption of A which shows no change in absorption (0% hypochromicity), a 20–25% hypochromicity is observed for the BrU band at 290 mμ. The results of Gratzer and McClare (*162*) qualitatively support the observation of Thomas and Kyogoku (*161*). They reported that a change of about 20% decrease in oscillator strength at the main band centered around 265 mμ upon A·U base pair formation. These observations do support the notion that hypochromicity can be brought about by the formation of a coplanar, hydrogen-bonded base pair, at least in certain solvent ($CHCl_3$) environments. The interpretation of these two papers is well supported by IR and NMR studies on these systems (see Section III, B).

This brings us to the question of solvent perturbation on the spectra of the bases. Gratzer and McClare (*162*) have studied the effect of addition of hydrogen-bonding solvents (acetic acid and ethanol) on the absorption of U and A dissolved in isooctane. They reported that the U spectrum is not perturbed by these hydrogen-bonding agents, but the A spectrum is strongly affected. The absorption of A is increased (a few per cent) and red-shifted (about 5 mμ) by the addition of ethanol and acetic acid. These changes are substantially complete at a perturber concentration of about 3% and therefore are not due to a bulk solvent effect. Based on these observations, Gratzer and McClare concluded that the hypochromicity (and hypochromism) of the 260 mμ band observed upon the formation of base-pairs is unlikely to be "... merely a hydrogen-bonding phenomenon, resulting from increased interaction of the π orbitals with non-bonding orbitals in the hydrogen-bonding substituent groups." Charney and Gellert (*163*) have found that the first transition absorption of 9-pyranosyl adenine in acetonitrile and in methylcyclohexane is hypochromic (about 10–15%) relative to the absorption of the compound in water solution, accompanied by a 2–3 mμ blue-shift in nonpolar solvents. For 5-bromouridine, little change in peak intensity and only a very small blue-shift was observed in comparing the spectrum in water to that in acetonitrile. The main form of the spectra of these two compounds, however, is the same in water and in acetonitrile. The spectra of inosine and of cytidine in acetonitrile and in water, on the other hand, are quite different, especially that of the inosine. Comparison of the inosine spectrum in acetonitrile with that in water indicates a large decrease in intensity at the 250 mμ region with a splitting into double peaks at 250 and 243 mμ and the appearance of a shoulder band at 275 mμ. In comparing the spectrum of cytidine in acetonitrile to that in water, there is a red shift of 5–10 mμ, and a large decrease in intensity at the first transition (270–280 mμ) and a small decrease at the second transition (230–240 mμ). Charney and Gellert (*163*) concluded that these changes are due to the effect of the solvent medium only and are not due to aggregation, ionization, or to change of tautomeric forms. Recent studies from our laboratory (*164*) on the spectra of inosine and 1-methylinosine in water, methanol, isopropanol, acetonitrile, and DMSO support this conclusion of Charney and Gellert that this is a solvent effect. The probable explanation for the appearance of the shoulder is a bathochromic shift of the n–π^* transition, previously buried under the π–π^* absorption region in water, when the nucleosides are placed in solvents of less hydrogen-bonding capacity. This proposition is supported by the observation that the spectrum of 9-methylhypoxanthine has a prominent shoulder in the vapor phase also (*146*). In the vapor phase, the λ_{max} of the spectrum of 9-methylhypoxanthine exhibits a 10-mμ hypsochromic shift as well, which may be related to the appearance of a 245-mμ peak in the spectra obtained in organic solvents. This shift has been discussed in the beginning of the section. Drobnik and Augenstein (*165*) have studied the effect of dielectric constant and hydrogen-bonding potential of the solvents on the absorption (and emission) spectrum of purine. They concluded that hydrogen bonding of the solvent to the N(9)H and other aza nitrogen atoms have a large effect on the absorption (and emission) spectra of purine. For 2-aminopurine and deoxyguanosine (*165*), they found that the spectra in H_2O, dioxane, and isopropanol are similar, but for 2,6-bis (diethylamino)-purine, large differences occurred in spectra when changing from organic solvents to water were observed. For the discussion on the effect of substituents, the original paper is recommended. The effect of changing from polar to nonpolar solvents in causing a bathochromic shift of the n–π^* transition is well known for heterocyclic compounds (for summary, see *166,166a*).

Quantum mechanical calculations have been made by Pullman and her associates (*167*; for summary see *49*) based on a self-consistent field approach with or without configurational mixing. The parameters shown in the results of their calculations are the transition energies, intensities, and direction of polarization. They stated, "In general, the discrepancy between the theoretical and the experimental energies of the electronic transitions, in particular as concerns the first and the second transitions, does not exceed 0.2 eV in the present calculations, in any of the compounds studied."

Ultraviolet absorption spectroscopy has been very useful for organic chemists or physicoorganic chemists in determining the site of protonation, the site of substitution in a structure proof, ionization constants, the tautomeric form, etc. For instance, it was one of the strong, early evidences to support the notion of protonation at the ring-nitrogen instead of exocyclic amino group in cytosine and adenine derivatives. Spectrophotometric comparison of the protonated base with the cationic forms of suitably methylated derivatives usually shows a similarity between a

$$\overset{|}{\underset{\oplus}{=\!\mathrm{N}}}\!-\!\mathrm{H} \qquad \text{and a} \qquad \overset{|}{\underset{\oplus}{=\!\mathrm{N}}}\!-\!\mathrm{CH_3}$$

This is because the methyl group is optically transparent in the ultraviolet and normally causes only a slight shift in λ_{max} when substituted for a hydrogen. Protonation of an exocyclic amino group will destroy the conjugation between the unpaired electron of amino-N and the π-system of the ring, resulting in a dramatic change of the spectrum as shown in the case of aniline versus aniline cation. Since there is usually a change of spectrum due to ionization, this technique has been very widely used for determination of pK_a values for the bases and nucleosides. For determination of substitution sites and tautomeric forms, a process essentially of comparative examination of spectra of structurally known compounds and of the unknown compound is required. More elaborate description on these subjects can be found in the following reviews (*168–170*) and in Section I of the Recommended General References. The comprehensive review of Lawley (*170*) is especially recommended.

2. Fluorescence and Phosphorescence

Fluorescence and phosphorescence spectra provide valuable information on the excited states of the molecules. This information can be used for identification and analytical purposes, for understanding of the mechanism of photochemical reactions, and for studying intra- and intermolecular interaction such as in the conformational analysis of polymers.

For purposes of identification and biochemical analysis to be performed in aqueous solution at room temperature, Udenfriend and Zaltzman (*171*) have reported the fluorescence of a variety of bases, nucleosides, and nucleotides. A summary of their results is presented in Table XII for reference. As shown in Table XII, guanine derivatives fluoresce much more strongly than do the other bases. This fact was utilized for a fluorometric assay of guanine in nucleic acid hydrolysates (*171*). Because of the sensitivity of the fluorescence and phosphorescence techniques, caution should be taken to ascertain that the observed spectra are not influenced by contaminants. Besides the usual care in obtaining pure compounds or excluding interfering impurities, three criteria for spectral purity should be applied whenever possible: (1) the emission spectrum should be closely related to the absorption spectrum of the compound in question; (2) the emission spectral changes should be correlated to the change of the physicochemical properties of the compound, such as ionization; (3) the characteristic exponential decay of the emission should be explicable on the basis of a single time constant.

In studying the physical chemistry of the excited states, attention is focused on the excited singlet state and triplet state of the transitions from the lowest-lying energy level. The excited singlet state is populated from the ground singlet state by direct absorption since this transition is allowed; however, the singlet-to-triplet absorption is spin-forbidden. Therefore, the excited triplet state is populated by a radiationless transition (intersystem-crossing transition) from the excited singlet state. The intersystem-crossing transition is also spin-forbidden but can take place more commonly; since the energy difference is small, there is good overlap between vibrational levels of these two excited states, and there is possible mixing with higher-lying electronic states, etc. For instance, an intersystem-crossing transition can take place more readily from a π–π^* singlet to a π–π^* triplet if there is an n–π^* state of intermediate energy lying below the lowest π–π^* state. The excited singlet state can return to the ground state through a nonradiative

TABLE XII

Fluorescence of Bases, Nucleosides, and Nucleotides in Aqueous Solutions at Room Temperature[a]

Compound[b]	Relative intensities[c]					Spectral properties at the condition for maximal emission			
	pH 13	pH 11	pH 7	pH 1	5 N H_2SO_4	Condition	Excitation max., mμ	Fluorescence max., mμ	Quantum yield %
Adenine	0	0	Negl.[d]	57	7	pH 1	265[e]	380	0.20
Adenosine	0	0	0	21	58	5 N H_2SO_4	272	390	—
Adenylic acid	0	0	0	21	51	5 N H_2SO_4	272	390	—
6-Methylaminopurine	13	0	0	Negl.	Negl.	—	—	—	—
Guanine	150	680	10	200	50	pH 1	275[e]	360	—
						pH 11	275[e]	350	3.0
Guanosine	6	16	0	200	140	pH 1	285	390	—
Guanylic acid	5	7	0	165	125	pH 1	285	390	—
1-Methylguanine	—	Negl.	0	150	—	pH 1	290	370	—
Methylaminoguanine	600	1900	20	280	0	pH 11	280[e]	355	—
Dimethylaminoguanine	2500	7100	20	740	0	pH 11	290[e]	368	—
Purine	1050	940	40	Negl.	—	pH 13	285	370	—
2,6-Diaminopurine	43	390	1400	210	230	pH 7	300	360	—
Thymine	35	35	—	0	0	pH 11	290[c]	380	0.15

[a] Data are from Udenfriend and Zaltzman (*171*).

[b] The following compounds were found to have no fluorescence: xanthine, hypoxanthine, uric acid, thymidylic acid, cytidylic acid, and other pyrimidine compounds (except free thymine) (Ref. *171*). Fluorescence of the nucleoside triphosphates are similar to those of the nucleoside monophosphates.

[c] pH 13, 0.1 N NaOH; pH 11, 1 N NH_4OH; pH 7.0, 0.1 M PO_4 buffer; pH 1, 0.1 N H_2SO_4.

[d] Negligible reading.

[e] Corrected values.

decay or a radiative decay. The radiative path is by an emission, fluorescence, with the lifetime about 10^{-9} sec. The excited triplet state can also return to the singlet ground state via a nonradiative or a radiative decay. The radiative decay path, phosphorescence, is forbidden and has large energy differences to contend with. Therefore, the lifetime of phosphorescence is long (in seconds) and can take place only at low temperature, where the nonradiative decay is retarded.

A very recommendable review recently was written by Eisinger and Shulman (*172*) about the physical chemistry of the excited states of the monomers and nucleic acid, especially from the viewpoint of the productive group from Bell Laboratories. The physicochemical properties of the excited singlet and triplet states of the monomers in glasses observed at liquid-nitrogen temperature (77°K) are summarized from the literature (*171–177*) and presented in Tables XIII and XIV. The pertinent parameters are the energy level, the quantum yields, and the radiative lifetime. Certain parameters of the excited triplet state, such as the quantum yield of the intersystem crossing, are from the electron spin resonance (ESR) measurement which provides the information about the equilibrium number of molecules in the excited triplet state at a known rate of absorption of quanta. The quantum yield of the intersystem crossing will be the same as the quantum yield of phosphorescence only when the nonradiative decay path is negligible. It can be seen from these tables (XIII and XIV) that there is reasonable agreement between various laboratories on the energy levels (location of the emission maxima) and the radiative lifetimes of these compounds. The agreement on the quantum yield values, however, is not as good, even from different publications from the same laboratory (*172,173*). This difference may be expected since the experiments are done in low-temperature glass and the amount of light emerging from the sample depends on the cracking in the glass. Best results, therefore, come from the "small sample" technique described by J. Eisinger in 1969 (*Photochem. Photobiol.* **9,** 247). The most complete results from nucleotides appear in Guéron, Eisinger, and Shulman [*J. Chem. Phys.* **47,** 4077 (1967)]. An elaborate study on the effect of temperature and viscosity on the fluorescence polarization and fluorescence quantum yields of guanine and guanosine has been made by Børresen (*178*). At 150°K, the highest yield he obtained for guanine and guanosine was 0.5 and 0.8, respectively. He has also discussed the question of the accuracy of quantum yield (*178*). Eisinger and Shulman concluded from their results on the nucleotides that the quantum yields for the nonradiative singlet decay ($1-\phi_F-\phi_{ISC}$; see Table XIII) are large, larger than those often found in aromatic compounds. Also the probability of the internal conversion for the triplets (radiationless decay) is appreciable even at 80°K, as indicated by the values of $(\phi_{ISC}-\phi_P)/\phi_{ISC}$(0.35–0.8). This is perhaps the reason for the relatively low

TABLE XIII

Properties of the Excited Singlet States (lowest-lying) of Bases, Nucleosides, and Nucleotides in Glasses at Liquid-Nitrogen Temperature (about 80°K)

Compound	Medium, glycol–H_2O	Energy (10^3 cm^{-1})	ϕF, fluorescence quantum yield	ϕISC, inter-system[a]-crossing quantum yield	1-ϕF^b – ϕISC	τR, radiative[c] lifetime (10^{-9} sec)	Ref.
Adenine	pH 1	34.4, 33.0, 32.1, 30.8	0.03	—	—	—	*173*
	pH 7	35.2, 34.0, 32.6, 31.5	0.06				
	pH 11	31.0	0.08				
Adenosine	pH 1	31.1	0.001				*173*
	pH 7	31.7	0.003				
5'-AMP	pH 7	35.2	0.01	0.02	0.97	3	*172*
5'-AMP	pH 7	35.0, 33.6, 33.2, 31.1	0.004				*173*
Guanine	pH 1	31.7	0.96				*173*
	pH 7	31.3	0.06				
	pH 11	31.1	0.20				
Guanosine	pH 1	31.1	0.16				*173*
	pH 7	30.8	0.02				
	pH 11	31.4	0.09				
5'-GMP	pH 7	34.0	0.13	0.15	0.72	12	*172*
5'-GMP	pH 1	31.0	0.32				*173*
	pH 7	30.3	0.03				
	pH 11	31.5	0.08				
CMP	pH 7	33.7	0.05		0.95	5.5	*172*
TMP	pH 7	34.1	0.16		0.84	4.5	*172*
	pH 12	34.35	0.24	0.15	0.61	5.5	*172*
UMP	pH 7	34.9	~0.01		0.99	4.5	*172*
	pH 12	35.0				4.5	*172*
Cytosine	pH 7	32.1	0.06				*173*
	pH 11	29.4, 27.3	0.10				
Thymine	pH 7	31.7					*173*
	pH 11	31.3					
Uracil	pH 7	31.7	0.0008				*173*
	pH 11	28.6	0.0013				

[a] Measured by electron spin resonance.
[b] The yield of internal conversion from the first excited singlet state.
[c] Calculated from the oscillator strength from the Einstein equation.

ϕ_F and ϕ_P. Eisinger and Shulman stated that the lowest-lying singlet and triplet states are π–π^* states based on the following arguments: (1) the absorption and fluorescence spectra of CMP and TMP are mirror images on a wave-number scale. Since the strong absorptions must be π–π^* transitions and the emission appears to come from the same state, one may conclude that the lowest excited state is a π–π^* state. The polarized fluorescence study on guanine and adenine by Callis et al. (*154*), which is mentioned in Section II, E, 1, has suggested the existence of two electronic transitions with approximately perpendicular transition moments and has established the π–π^* nature of the singlet state emission. (For other arguments on the nature of the first transition in A and G, see Section E, 1.) In his polarization fluorescence study, Børresen concluded that there is a third electronic transition within the envelope of the two major π–π^* transitions in the range between 35,000 and 42,000 cm^{-1}, and the nature of this transition is uncertain. (2) The data (Table XIV) indicate that the triplet lifetimes are of the order of one second or so for the nucleotides. From this it may be concluded that the lowest-lying triplet states of nucleotides are all π–π^* levels, since n–π^* levels would have much shorter decay times. This conclusion is supported by the values of D and E, the dipolar interaction constants for the paired electrons in the triplet states of the nucleotides, measured by the ESR technique (*179*).

Børresen (*180*) reported that the fluorescence intensity of purine, which is low as a neutral molecule, increases markedly on either acid or alkaline titration. The mono-cation of purine emits at 400 mμ with quantum efficiency of 0.008 and the anion emits at 370 mμ with the quantum efficiency of 0.045. The phosphorescence and polarized phosphorescence excitation spectra have been studied for purine, purine anion, adenine, and its derivatives by Cohen and Goodman (*175*). Their results indicate that, in all cases, the triplet–singlet transition moment is perpendicular to the molecular plane and the main phosphorescence is from a π–π^* triplet state as indicated by the long lifetimes ($\sim$ 1 sec) and the large energy difference between the absorption and phosphorescence maxima (greater than 10^4 cm^{-1}). The polarization ratio of purine at 310 mμ indicates that this transition is directed out-of-plane, consistent with the n–π^* transition. Hence, they concluded that the lowest-lying transition for purine is the n–π^* transition while the lowest-lying transition for purine anion and adenine derivatives is π–π^*.

Drobnik and Augenstein (*165*) have reported a comprehensive study on the solvent effect on the emission (as well as absorption) spectra of purine. They also concluded that the transition of the phosphorescence of purine is π–π^* in nature; they stated that the shape and the vibrational structure of purine phosphorescence is in best agreement with that of quinoline.

Børresen (*180*) has presented a general theory that the presence of a lowest excited state of n–π^* type would preclude fluorescence and possibly permit

TABLE XIV

Properties of the Excited Triplet States (lowest-lying) of Bases, Nucleosides, and Nucleotides in Glasses at Liquid-Nitrogen Temperature (about 80°K)

Compound	Medium	Energy (10^3 cm^{-1})	ϕP, phosphorescence quantum yield	τ (sec), observed lifetime	ϕISC-ϕP[a] / ϕISC	Ref.
Adenine	Ether–isopentane–ethanol	24.6		~1		*175*
Adenine	Glycol–H_2O pH 1	26.7, 25.1, 23.9	0.03	3.35		
	Glycol–H_2O pH 7	27.4, 26.0, 24.8, 23.5	0.035	2.2[b]		*173*
	Glycol–H_2O pH 11	26.2, 24.8, 23.5	0.095	2.8		
Adenosine	Glycol–H_2O pH 7	26.3, 24.9, 23.6, 22.3	0.008	2.5[b]		*173*
	Water[c]	23.0		2.3		*177*
Purine	Ether–isopentane–ethanol	25.0		1		*175*
Purine anion	Isopentane	22.7		1		*175*
5′-AMP	Glycol–H_2O pH 1	23.8	0.0008			
	Glycol–H_2O pH 7	26.2, 24.8, 23.4, 21.8	0.004	2.45[b]		*173*
	Glycol–H_2O pH 7	26.7	0.015	2.4[b]	0.35	*172*
Guanine	Glycol–H_2O pH 1	26.0, 24.7, 23.5	0.04	0.42 (40%) 1.6 (60%)		
	Glycol–H_2O pH 7	27.8, 26.3, 24.4, 24.0	0.07	1.42		*173*
	Glycol–H_2O pH 11	24.4	0.06	1.35		
Guanosine	Glycol–H_2O pH 1	23.0	0.002	0.77 (63%) 0.17 (37%)		
	Glycol–H_2O pH 7	26.3, 24.8, 23.7	0.03	1.3		*173*
	Glycol–H_2O pH 11	23.9	0.04			
	H_2O[c]	23.2		1.2		*177*
5′-GMP	Glycol–H_2O pH 1	23.0	0.002	0.64 (60%) 0.17 (40%)		*173*
	Glycol–H_2O pH 7	26.3, 24.7 23.7	0.04	1.25		
	Glycol–H_2O pH 11	23.9	0.02	1.25		
	Glycol–H_2O pH 7	27.2	0.07	1.3	0.5	*172*

TABLE XIV—continued

deoxy-GMP	Glycol–H_2O	25.0, 23.8		1.2		*176*
Uracil	DMF–ethanol–ether	23.5		<0.1		*174*
Cytosine	Ethanol	23.6		0.35		*174*
5-Hydroxy-methyl cytosine	Ethanol	24.0		0.5		*174*
Cytidine	H_2O[c]	23.0		0.6		*177*
Uridine	1 *N* ammonium[c] hydroxide	24.1		0.5		*177*
Thymidine	1 *N* ammonium[c] hydroxide	22.5		0.4		*177*
CMP	Glycol–H_2O pH 7	27.9	0.01		0.34	*172*
TMP	Glycol–H_2O pH 7	26.3			0.33	*172*
TMP	Glycol–H_2O pH 12	27.0	0.03	0.8	0.5	*172*

[a] The fraction of triplets formed which decay without emitting.
[b] In agreement with the measurements by Bersohn and Isenberg (*176*).
[c] The solute molecules tend to form crystallites in frozen water solution.

phosphorescence. Lawley (*170*) in his review has examined this proposal carefully in light of the solvent effects (*165*) and the substitution effects. In respect to purine and its substituents, he summarized that there is a low-lying n–π^* transition as well as the existence of π–π^* triplet states. Substitution of an amino group at the C(2) position [2-aminopurine (*165*)] presumably increases the energy level of the n–π^* transition relatively and causes the π–π^* state to be the lowest singlet; fluorescence rather than phosphorescence then predominates in the emission spectrum of the 2-aminopurine. Amino substituent and especially the hydroxy substituent at C(6) partially reverses the effect of the 2-amino substituent; though the lowest-lying singlet remains π–π^* in nature, the intersystem crossing from the excited singlet to the excited π–π^* triplet is enhanced (*170*).

Bell Laboratories (*172,181*) have utilized the triplet energy transfer from sensitizers, such as acetone or acetophenone, to enhance the formation of the excited triplet state of thymidylic acid. Apparently the intersystem-crossing rate is low compared to the singlet-state decay rate in the isolated TMP; therefore the phosphorescence of TMP was not observed in the absence of the sensitizer. This is an important point in the study of the photochemistry of DNA (*181*).

F. Optical Rotatory Dispersion and Circular Dichroism

Although optical rotation at a given wavelength (mostly sodium D line) has been used for a long time for the characterization of organic compounds, it is only recently that it has been possible to measure optical rotation at the absorption bands in the ultraviolet region with sufficient accuracy for the study of nucleic acids. For the nucleosides and the nucleotides, however, because of the large extinction coefficient and low optical activity, the signal-to-noise ratio is very poor (2–5) for the measurement at the absorption band. The percentage of error on the absolute measurement may be as large as 20%. Our laboratory (*182*) previously has reported the optical rotatory dispersion (ORD) of 5′-ribonucleotides, 5′-deoxynucleotides, and 2′,3′-ribonucleotides at pH 5 buffer from 320 to 589 mμ. The data were analyzed in terms of the one-term Drude equation,

$$[\alpha] = \frac{K}{\lambda^2 - \lambda_c^2}$$

where λ_c is the apparent wavelength of the optically active electronic transition, K is a constant related to the rotational strength, R, and $[\alpha]_\lambda$ is the rotation observed at the given wavelength λ. Although this equation can be shown to be applicable in describing the optical rotatory dispersion at a wavelength range distant from the absorption band of an idealized Cotton effect (*183*), the application of this equation to the experimental findings represents only a mathematical approximation and does not necessarily correspond to a real physical situation. With better instruments, Yang, Samejima, and Sarkar (*184*) have reinvestigated the ORD of both ribo- and deoxyribonucleosides and 5′-nucleotides from 600 mμ down to 200 mμ in neutral, acid, and in some cases alkaline solution. Their paper provided the most comprehensive data for the ORD of the naturally occurring monomers. Very recently, in an extensive review on ORD and circular dichroism (CD) of nucleic acids, Yang and Samejima (*185*) prepared a table summarizing the important parameters describing the Cotton effect of 16 nucleosides and nucleotides. The absorption maximum, the positions of the peak, the trough, and the crossover, as well as the amplitudes are listed. From the wavelength region above 200 mμ, the nucleosides have only one Cotton region; the ORD of the nucleosides are generally similar to those of the 5′-nucleotides within experimental error. Differences in peak positions are found for the guanosine nucleosides and nucleotides, however. For A, C, and G, the crossover position is within 2–4 mμ of the absorption maxima; the crossover positions of T and U, on the other hand, are located at a wavelength about 10 mμ longer than those of absorption maxima. As shown in Figs. 5(a) and 5(b), the two branches in the ORD curve on the red or blue side of the crossover point are not mirror-images of one another, indicating that most likely more than one optically active, electronic transition is involved in this apparently single Cotton effect. However, correlation of the optical activity to the electronic transition is a rather difficult task if the ORD data are used for this purpose; CD data are more easily correlated to the absorption spectra, a subject to be discussed in a later

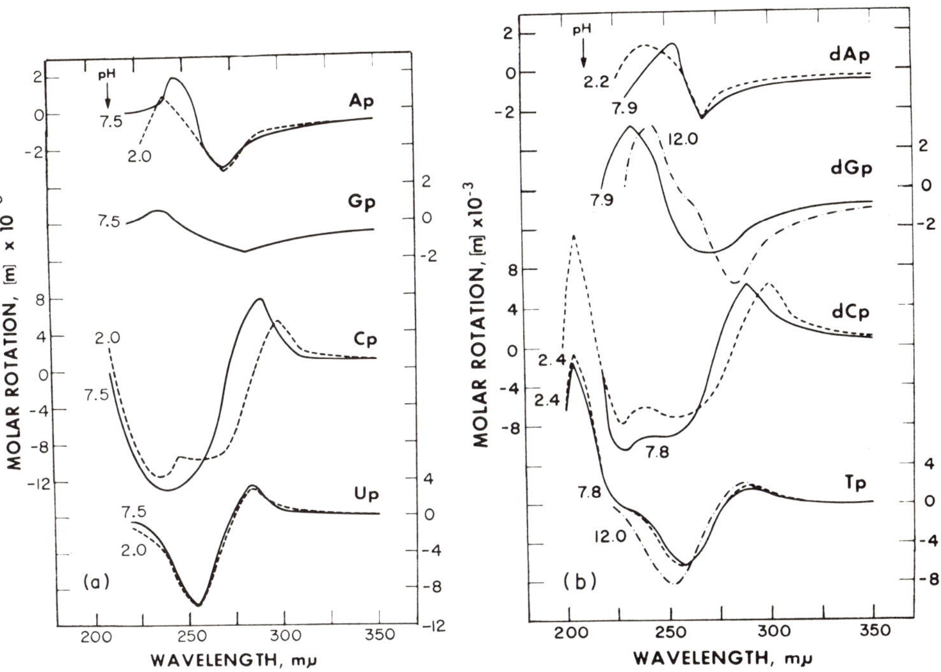

Fig. 5. (a) Ultraviolet rotatory dispersion of 5′-ribonucleotides (from Ref. *184*). (b) Ultraviolet rotatory dispersion of 5′-deoxyribonucleotides (from Ref. *184*).

paragraph. The amplitude between the peak and the trough is largest for cytidine ($[m] = +20.4 \times 10^3$), next for uridine ($+13.8 \times 10^3$), then thymidine ($+6.50 \times 10^3$), and least for adenosine and guanosine (about -5×10^3). There are small ORD differences for the deoxyribose compounds versus the ribose compounds, although no systematic correlation can be made. Our previous ORD data (*182*) on ribose and deoxyribose yielded λ_c values of 136 and 159 mμ and K values of -6.3×10^{-6} and -15.3×10^{-6}, respectively, based on the one-term Drude equation. For naturally occurring mononucleosides and nucleotides (in β form and with D-pentose), the purine compounds such as A and G have a negative Cotton effect (minimum on the long wavelength side), and the pyrimidine compounds such as T, U, and C have a positive Cotton effect (maximum on the long wavelength side).

The optical activities of the monomers were investigated in the past mainly for three purposes. The first purpose is to serve as background information for the ORD-CD stuides of dinucleotides and polynucleotides. For this purpose, Fasman et al. (*186*) have reported the ORD of CMP; Lamborg et al. (*187*), the ORD of (2′,3′)-AMP and dUMP; van Holde et al. (*188*), the CD of AMP; and Brahms et al. (*189*) the CD of various nucleotide mixtures. The second purpose is for structural study, especially for the assignment of α and β anomers. The third purpose is for the conformational analysis especially *syn* and *anti* conformations with respect to the rotation of the sugar–base torsion angle, ϕ_{cn}, about the glycosyl bond. Finally, an important, though preliminary study has appeared concerning the origin of the optical activity and correlating CD data with absorption bands (*190*). It should be mentioned that in order to determine whether a nucleoside is an L or D compound, ORD–CD data (*191,192*) provides the most direct and relevant information to this problem, a problem not readily solvable by any other spectroscopic techniques such as PMR, IR, UV absorption, etc.

The first attempt to use optical rotation data for the assignment of α and β anomers was the proposal of Hudson's isorotation rule based upon $(\alpha)_D$ values (*193*). This rule predicts that the α-D configuration of a pair of anomeric glycosides is more dextrorotatory than the corresponding β-configuration. When the α and β pyrimidine deoxynucleoside pairs became synthetically available, Fox et al. (*194,195*) and Lemieux and Hoffer (*196*) found that this rule is not applicable. In a series of short communications (*197–200*) and a full paper (*201*), Ulbricht, Emerson, and co-workers have reported an extensive survey on the ORD properties of a large number of nucleosides, especially pyrimidine nucleosides. They concluded that the purine α-nucleosides give positive Cotton effects and the β-anomers give negative Cotton effects (the same as the Hudson's isorotation rule), whereas the pyrimidine α-nucleosides give negative Cotton effects and the β-anomers give positive Cotton effects (the opposite of Hudson's isorotation rule). The question about the cyclonucleosides is discussed in the following paragraph. Fric et al. (*202*) have measured the ORD of furanosyl and pyranosyl thymine and uracil nucleosides of arabinose, lyxose, ribose, deoxyribose, and glucose. Their results supported the conclusion of Emerson et al. (*201*) on the pyrimidine compounds. Recently, Nishimura et al. (*192*) reported ORD data for about 40 anomeric nucleosides and their 5′-nucleotides, such as D- and L-ribofuranose, D- and L-arabinofuranose, D- and L-lyxofuranose and L-xylofuranose. α-D-Adenosine (or α-D-AMP) was shown to exhibit positive Cotton effects, while β-D-adenosine (or β-D-AMP) had negative Cotton effects. They also showed that β-D- and α-L-pentofuranosyl pyrimidines give single positive Cotton effects, whereas α-D- and β-L-derivatives give negative Cotton effects (the D- and L-derivatives gave exactly opposite ORD curves as expected). These observations, therefore, also support the conclusion of Ulbricht et al. stated above. All these groups agree that the configuration at the anomeric carbon determines the sign of the Cotton effect of these nucleosides. They agree also that the *cis–trans* hydroxyl substitution at C(2′) profoundly affects the amplitude of the Cotton effect. Compounds with a *cis*-hydroxyl group at the C(2′) with respect to the base such as α-D-ribosyl, α-L-ribosyl, β-D-arabinosyl, β-D-lyxosyl, and α-L-xylosyl compounds, exhibit much larger Cotton effects than compounds with a *trans*-hydroxyl group at the C(2′), such as β-D-ribose, β-L-ribosyl, α-D-arabinosyl, and α-D-lyxosyl compounds. Configurations at C(3′) and C(4′) have much less influence on the ORD data. In the review of Yang and Samejima (*185*), a list of ORD data from 13 pairs of anomeric nucleosides and nucleotides was summarized from the literature (*192,201*).

Since the bases themselves have no rotation and the sugar moiety has no absorption band located above the vacuum ultraviolet region (from 180 mμ up), the optical activity of the nucleosides observed at the absorption bands of the bases from 190 to 280 mμ must be due to an interaction between the transition dipole moment of the base and the asymmetric environment of the pentose. This interaction, therefore, should be dependent upon the geometric relationship between the base plane and the pentose moiety. As is described in minute detail in Section II, G, this geometric relationship is defined by the rotation of the glycosyl bond between the C(1′) of the pentose and the N(9) (purine) or N(1) (pyrimidine) of the base. Since this sugar–base torsion angle, ϕ_{cn}, will be defined rigorously in Section G, it is described only very briefly here. (This angle, ϕ_{cn}, was also mentioned above in Section II, A.) Owing to steric hindrance, the rotation of the glycosyl bond is restricted mainly into two ranges of the ϕ_{cn}, $-30°(\pm 45°)$ of the *anti* conformation and $+150° \pm 45°$ of the *syn* conformation. An elaborate discussion of the ϕ_{cn} angle and of these two *anti* and *syn* conformations as studied by X-ray diffraction on the crystals is presented in Section II, G. It will suffice to state that with an *anti* conformation, the H(8) proton of a purine nucleoside is in juxtaposition to the C(5′) and the H(2) protons or N(3) is the farthest distance away from the C(5′); similarly, the H(6) of a pyrimidine nucleoside is in juxtaposition to the C(5′), and the O(2) is the farthest distance away from the C(5′). With the *syn* conformation, the reverse is true: the H(2) or N(3) of the purine nucleoside, or the O(2) of the pyrimidine nucleoside is now in juxtaposition to the C(5′) of the pentose. However, before our exploration into the possible relationship between the ORD–CD data and the nucleoside conformation (in terms of the ϕ_{cn} angle), we have to understand more about the electronic transitions responsible for the optical activities.

Miles, Robins, and Eyring (*190*) have reported an important study in attempting to correlate the ORD–CD data (especially the CD data) with the absorption bands. They have labeled the absorption bands in accordance with the classification system of Clark and Tinoco (*152*). As described above in Section II, E, Clark and Tinoco have correlated the absorption bands of the bases with those of benzene. The first π–π^* transitions (220–280 mμ region) are the B_{2u} band ($A_{1g} \rightarrow B_{2u}$) and the B_{1u} band ($A_{1g} \rightarrow B_{1u}$); these two bands are polarized perpendicular to each other in the plane of the base. (The directions of these transition moments are described in Section II, E.) At the lower wavelength region (180–220 mμ), there are the two bands of E_{1u} derived from the doubly degenerate $A_{1g} \rightarrow E_{1u}$ benzene transition. In addition to these $\pi \rightarrow \pi^*$ transitions, there are the $n \rightarrow \pi^*$ transitions from the nonbonding electron pairs of N and O atoms. The $n \rightarrow \pi^*$ transitions are polarized perpendicular to the base plane. Miles et al. (*190*) attempted to correlate the CD bands with the absorption bands and to ascertain the electronic transitions responsible for both optical activity and extinction. In order to separate the $n \rightarrow \pi^*$ from the $\pi \rightarrow \pi^*$ transition, they have adopted the following premises. They noted that the nonbinding electrons on the keto oxygens contribute very little to the absorption spectra above 220 mμ [based on the study of Stewart and Davidson (*148,149*)], and the protonated N of the pyridine type is not capable of $n \rightarrow \pi^*$ excitation in the region above 220 mμ. The $n \rightarrow \pi^*$ transitions contributing to both optical activity and extinction at the spectral region above 220 mμ come from the pyridine-ring N atom which is often the site of protonation as described in Sections A, C, D, and E. Therefore, the effect of protonation on CD and UV spectra will be much larger on the bands originating from the $n \rightarrow \pi^*$ transition of the pyridine-ring N than on bands of $\pi \rightarrow \pi^*$ origin.

The assignment of the uridine and thymidine CD spectra [Fig. 6(a)] is straightforward (*190*). There is no effect of pH from pH 1 to pH 7, and, due to the absence of pyridine-ring N, there is no $n \rightarrow \pi^*$ transition at wavelengths above 220 mμ. The first CD band (positive, 265–275 mμ) is assigned to B_{2u}, the second band (negative, 240–250 mμ) to B_{1u},† the third (negative, ~ 210 mμ), and fourth (positive, 190 mμ) bands are assigned to E_{1u}, which have rotational strengths of opposite sign and nearly equal magnitude as predicted by Caldwell and Eyring (*203*). The situation for cytidine is more complex. The effect of protonation is a red-shift of the first CD and first UV absorption band from 270 to 280 mμ, a diminution of the absorption band at 240 mμ; the CD and UV bands at 215 mμ are not much affected. Miles et al. proposed that the large positive band at about 270 mμ at neutral pH comes from both B_{2u} and B_{1u} transitions in the region of 240–280 mμ

† There is a labeling error in Fig. 1 of the paper from Miles et al. (*190*) in identifying the B_{1u} band at 240 mμ as B_{2u} again for uridine.

that is shifted to 250–290 mμ upon protonation. The negative CD peak at 215 mμ is from the first E_{1u} band. The 235-mμ maximum in absorption is assigned to the $n \rightarrow \pi^*$ transition which is shifted to the far ultraviolet region upon protonation. The conclusion from the emission spectral study on CMP and TMP (*172*) is described in Section II, E, that the lowest-lying singlet and triplet states are the $\pi \rightarrow \pi^*$ transition, is in accord with these assignment of CD spectra.

For CD study on the naturally occurring purine nucleosides, Miles et al. reported that the signal-to-noise ratio is so low (about 1:1) that the quantitative measurement was not yet possible. However, the signal-to-noise ratio exhibited by the purine cyclonucleosides is 5 : 1 or better, so these compounds were studied instead. It should be noted in the beginning that all of the purine cyclonucleosides studied are linked between C(5′) of the ribose and N(3) of the purine; therefore, all of the cyclonucleosides have a fixed, typical *syn* conformation. (All of them also have the isopropylidene substitution on the 2′,3′ OH-groups, which is of no importance to the optical properties.) Again, the assignment of the CD bands and the absorption bands for 2′:3′-O-isopropylidene-xanthosine 3:5′ cyclonucleoside is straightforward, since there is no pyridine-ring N, and thus no effect of low pH [Fig. 6(b)]. The intense CD peak (positive) and absorption peak at 276 mμ are probably from the B_{2u} band; the CD peak (negative) and absorption peak at 237 mμ are likely from the B_{1u} band. The CD curve near 200 mμ indicated two close peaks of opposite sign and nearly equal magnitude, and these two peaks were assigned to the E_{1u} benzene transition with an absorption maximum at 197 mμ. The spectral properties of 2′:3′-O-isopropylidene-inosine-3:5′ cyclonucleoside *p*-tolylsulfonate are very interesting. This compound is very similar to the xanthosine cyclonucleoside except for the absence of a 2-keto group and the location of a positive charge, probably at N(3) of the six-membered ring. As shown in Fig. 6(c), the first CD peak is a negative peak at 270 mμ, followed closely by a positive CD peak at 257 mμ, which all were assigned to be of B_{2u} and B_{1u} origin, respectively. The third negative CD peak at 220 mμ consists of both an n–π^* transition and a band of E_{1u} parentage. Comparison of the structure of the xanthosine and inosine cyclonucleosides and the CD spectra of these two compounds reveals that the removal of the 2-keto group, and the presence of the positive charge at N(3) reverses the signs of the first three Cotton effects even though these two molecules have identical *syn* conformations. Another similar and equally interesting case is the CD and absorption study of the 2′:3′-O-isopropylidene-3:5′-guanosine cyclonucleoside at pH 1 and pH 7. It is rather certain that at pH 1, protonation at N(7) occurs for this compound. Miles et al. described that "At pH 1, the CD curve shows a small positive Cotton effect at 263 mμ and negative Cotton effects at 242 and 210 mμ. These Cotton

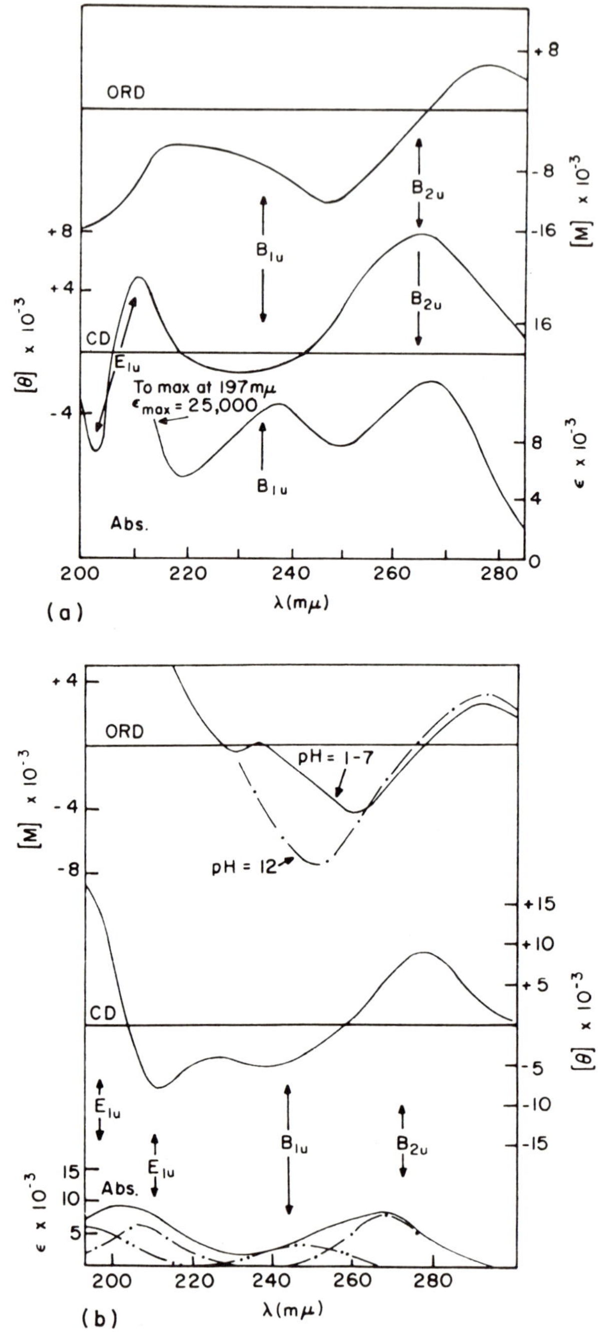
ORD
CD
Abs.
B_{1u}
B_{2u}
E_{1u}
To max at 197 mμ
ϵ_{max} = 25,000
[θ] x 10^{-3}
[M] x 10^{-3}
ε x 10^{-3}
λ(mμ)
(a)
pH = 1-7
pH = 12
(b)

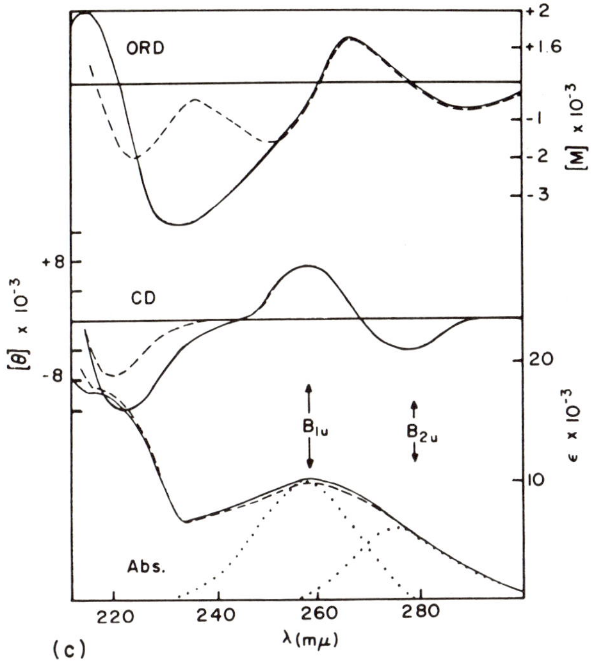

Fig. 6. Correlation of ORD, CD, and absorption spectra (after Miles, Robins, and Eyring, Ref. *190*). (a) Thymidine. (b) 2′:3′-O-isopropylidene-xanthosine-3:5′ cyclonucleoside. (c) 2′:3′-O-isopropylidene-inosine-3:5′ cyclonucleoside *p*-tolylsulfonate.

effects can quite reasonably be correlated with B_{2u}, B_{1u}, and E_{1u} transitions of benzene. This assignment requires that the broad 250 mμ absorption band at pH 1 contains both the B_{2u} and B_{1u} transitions. The very intense band at 200 mμ is likely the result of superposition of the two E_{1u} absorption bands. At pH 7, the CD curve is negative throughout the 240–270 mμ region with a maximum at 248 mμ and a prominent shoulder at 258 mμ. The change in going from pH 1 to pH 7 in this region can be attributed to a 6-mμ red shift of the return of an $n \rightarrow \pi^*$ transition to this region upon deprotonation of N(7). The position and rotational strength of the $n \rightarrow \pi^*$ transition is such as to completely hide the presence of the B_{2u} Cotton effect."† Miles et al. also have studied the spectral properties of 2′:3′-O-isopropylidene-3:5′-adenosine cyclonucleoside *p*-tolylsulfonate. At both pH 1 and pH 7, the first broad CD peak is negative (250 mμ). Hampton and Nichol (*204*) have also measured the ORD of these same compounds, 2′:3′-O-isopropylidene-3:5′-cycloadenosine, 2′:3′-O-isopropylidene 3:5′-cycloguanosine, and 3:5′-cyclo-

† Apparently, there is a mislabeling in Fig. 7 of the paper from Miles et al. (*190*) The CD spectrum labeled as pH = 7 actually is at pH 1 and the spectral curve labeled as pH = 1 actually is at pH 7.

xanthosine at several pH levels. Their spectral data are in agreement with those of Miles et al. (*190*). The work of Miles et al. (*190*) has established experimentally two very important points in regard to the optical properties of the nucleosides. First, the seemingly single Cotton effect in ORD observed for both purine and pyrimidine nucleosides at 220 to 300 mμ actually is a composite of several CD bands which can be of both π–π^* and n–π^* origin. Though the B_{2u} transition generally appears to dominate the optical properties at long wavelengths as the lowest-lying transition, there is an example (cycloguanosine) in which the rotational strength of n–π^* transition (negative) may overshadow the rotational strength of π–π^* transition (positive). Second, even with a fixed nucleoside conformation with respect to the sugar–base torsion angle, as in the case of cyclonucleosides, the sign of the Cotton effect can be reversed by substitution of the base (hypoxanthine vs. xanthine) and by protonation (guanosine vs. protonated guanosine). These examples clearly illustrate an important point (a point which is self-evident but is easily neglected because of lack of experimental data) that the Cotton effect is ultimately determined by the interaction of the transition moments of the base with the asymmetric centers of the pentose. The directions of the transition moments are not only governed by the geometry of the base in the nucleoside but are also influenced by substitution, protonation, and other chemical events.

The above discussion provides the necessary background to examine the proposal first made by Ulbricht and co-workers (the main papers are Refs. *200* and *201*). The proposal basically consists of two parts: one postulate and one "rule." They postulated "that the variation in both sign and magnitude of the Cotton effect is due to the changing conformation of the planar pyrimidine [later they extended this to purine (*200*)] ring with reference to the sugar ring, i.e., the position of the chromophore relative to the asymmetric centers" (*201*). They argued that in order to have an observable optical activity, the geometry of the chromophore must be more or less fixed with respect to an asymmetric environment; thus the rotation of the sugar–base torsion angle of the nucleosides must be restricted. In order to confirm this notion, they proposed that cyclonucleosides that have a fixed sugar–base geometry may provide valuable information. After examining an extended series of substituted uracil, thymine, and cytosine nucleosides, they suggested "that the sign of the Cotton effect in uracil, thymine, and cytosine pentofuranosides can be predicted on the basis of the following rule. The sign of the Cotton effect will be positive if: (1) the nucleoside has a preferred conformation due to restricted rotation about the glycosidic bond, and (2) a line from the C(4)=O [or C(4)—NH_2] group passing through the C(2)=O group passes from above to below the plane of the furanose ring, provided that the chromophore is not twisted to such an extent that the line passes through

C(5′) ['above' is defined as the same side of the furanose ring as C(5′)]." In formulating this "rule" originally, they did consider the direction of the B_{2u} transition dipole moment in uridine or thymidine proposed by Clark and Tinoco (*152*) and determined by Stewart and Davidson (*148,149*) (a subject discussed at some length in Section II, E). However, the direction of the transition moment of other pyrimidine nucleoside derivatives were hardly mentioned later on presumably due to lack of experimental data. The emphasis in the application of this "rule" has been put entirely on the geometric relationship between the base plane and the pentose moiety. The underlying, implicit assumption is that the transition moments of B_{2u} band of all the other modified or substituted pyrimidine nucleosides, modified so as to change the geometry of the molecule, are not affected and are still closely similar to those of thymidine or uridine. This implicit assumption is certainly difficult to justify in certain cases. Besides the α and β anomeric series (the results have been discussed in above paragraphs), the N(1)-ribosyl uracil (uridine, positive Cotton effect), and the N(3)-ribosyl uracil (negative Cotton effect, λ_{max} 261 mμ), the most important series of compounds cited as evidence for this "rule" are the cyclouridines: O(2):3′-cyclouridine (large positive Cotton effect, amplitude $+266 \times 10^2$, λ_{max} 250 mμ, 233 mμ); O(2):3′-cyclouridine (small positive Cotton effect, amplitude 43×10^{-2}, λ_{max} 246 mμ); and O(2):5′-cyclothymidine (negative Cotton effect, amplitude -299×10^{-2}, λ_{max} 250 mμ). It should be noted that the ORD curves of these cyclouridines (reported in Fig. 4, Ref. *201*) are very complex and there were not sufficient data reported about their UV absorption spectra to indicate that the transition moments are similar in all these cyclonucleosides. [Since the substitution at the O(2) is the same for all these cyclonucleosides, the only difference is in various carbon atoms of the pentose; therefore, the transition moments of the uracil base should be similar. However, changes of λ_{max} values were reported and the absorption spectra of these compounds were not presented for evaluation.] There was, however, no obvious deviation from the "rule." Therefore, Ulbricht et al. concluded that the naturally occurring uridine, thymidine, and cytidine have the *anti* conformation in aqueous solution (*201*). Ulbricht et al. also extended their argument to the purine nucleosides (*200*). Besides the α- and β-anomer pairs (results mentioned above), they built their case on the ORD of two cycloadenosine nucleosides: 8:5′-cycloadenosine with *anti* conformation (large negative Cotton effect, λ_{max} 264 mμ); and 2′:3′-isopropylidene-N(3):5′-cycloadenosine iodide with *syn* conformation (small positive Cotton effect, λ_{max} 272 mμ). They felt, therefore, that the purine nucleoside which has *anti* conformation probably exhibits a negative Cotton effect. They did mention, "however, in the case of the N(3), 5′-cyclonucleoside, the possibility that a change in the chromophore is responsible for the change in sign cannot be excluded." The ORD measurement by Miles et al. (*190*) and by Hampton and Nichol (*204*) on 2′:3′-O-isopropylidene N(3):5′-cycloadenosine, on the other hand, gave a small negative Cotton effect, contrary to the observation of Emerson, Swan, and Ulbricht (*200*). The work of Klee and Mudd (*205*) supported the results of Miles et al. and Hampton and Nichol on this N(3):5′-cycloadenosine, but they also reported the difficulty in measuring this labile compound.

Ikehara et al. (*206*) have also attempted to relate the purine nucleoside conformation with ORD data based on a series of cyclonucleosides. The reported ORD curves of these compounds are very complex, indicating multiple, optically active transitions. The sign, though not the shape of the ORD curves, is influenced by the linkage atom, sulfur or oxygen, between the base and the pentose. For instance, 8:2′-anhydro-8-mercapto-9-β-D-arabinofuranosyladenine has zero rotation at its ORD peak, and the oxygen analog has about +4300 degree rotation at its ORD peak. The absorption spectrum of the mercapto-cycloadenosine is also different from that of the oxycycloadenosine. These results again illustrate that with the same conformation (both molecules are 8:2′-anhydro-derivatives), different ORD curves are obtained presumably due to the different electronic effects of the linkage atoms, S or O, on the transition moment(s) of the base. Ikehara et al. (*206*) observed that the 8:5′-anhydro-8-mercapto-2′:3′-O-isopropylideneguanosine has a *positive* Cotton effect, and this 8:5′-anhydro molecule has an *anti* conformation. Therefore, they stated: "we are inclined to support the view that the base moiety in purine nucleoside is in *syn* conformation at least in aqueous solution."

Klee and Mudd (*205*) reported on the ORD of 5′-substituted adenosines, such as 5′-methylthioadenosine, S-adenosylmethionine, S-adenosylhomocysteine, etc. They found that all these compounds exhibit a positive Cotton effect with little change of ultraviolet absorption spectra. They proposed that "if adenosine exists in a *syn* conformation, a bulky sulfur atom on the 5′-carbon would lead to steric interference and drive the molecule more into the *anti* conformation, affording a ready explanation of the change in the sign of Cotton effect. No such ready explanation is available if adenosine usually exists in the *anti* conformation." It should be noted that while this deduction is a reasonable argument, it is not a compelling one. Stereochemical analyses about the *syn* and *anti* conformation (4 and 5) do not indicate any steric hindrance from the C(5′) atom or the OH(5′) group. Substitution of the —SCH_3 group on C(5′) for the formation of 5′-methylthioadenosine can still allow the *syn* conformation to exist, especially with the S atom *cis* to the H(4′). Klee and Mudd (*205*) expressed their view as, "however, the combined weight of the evidence is, to our view, such that the question of the *syn* versus *anti* preference of β-purine ribosides in solution be left open to await the results of further experimentation."

Lin et al. (*207*) reported an effect of pH from 4.9 to 6.1 on the ORD of AMP. To explain this effect, they proposed a hydrogen bond between the 5′-phosphate hydrogen and the N(3) of adenine. This intramolecular hydrogen bonding requires that AMP has the *syn* conformation. Our laboratory (*97*), however, was unable to confirm this observation reported by Lin et al.

In summarizing the rather uncertain situation described in the above paragraphs, this reviewer is in agreement with the basic postulate of Ulbricht and co-workers that ORD–CD data contain information about the nucleoside conformation (*syn* and *anti*) and that the observed optical activity is dependent on a more or less fixed geometry of the chromophore (the base) to the asymmetric centers (the pentose) due to rotational barriers. It cannot be overemphasized that this geometric relationship is actually for the interaction of the asymmetric center with the transition moment of the chromophore and not with the ground-state molecular framework. Any modification in order to bring about a constraint on the geometry of the pentose with respect to the base may or may not alter the direction of the transition moment. If this information is not ascertained, however, then the interpretation of the experimental data may not be meaningful. In this respect, CD data are far superior to ORD data for the purpose of identification and characterization of those electronic transitions which are responsible for the optical activities observed.

G. The Rotation of the Glycosyl Linkage—the *syn* and *anti* Conformations

As described in Section II, A, the atoms in the purine and pyrimidine rings are fixed rather firmly, and the five-membered ring of the pentose residue in the nucleoside may be considered to be a rigid unit as the first approximation (*5*). Therefore, the conformational state of the nucleoside is defined principally by the rotation of these two rigid planes, i.e., the base and the pentose, relative to each other about the axis of the C(1′)—N(9) or N(1) bond. The sugar–base torsion angle, ϕ_{cn}, was first defined by Donohue and Trueblood (*208*)† as "the angle formed by the trace of the plane of the base

† Because the bases need not necessarily be planar and because C(1′) can be significantly displaced from the least-squares plane of the base (see Section II, A), Sundaralingam and Jensen (*21*) proposed that the definition of ϕ_{cn} be modified as follows: "The torsion angle ϕ_{cn} of the bond $C_{(1')}$—N is the angle formed by the projection of $C_{(1')}$—$O_{(1')}$ relative to the projection of $N_{(1)}$—$C_{(6')}$ (in pyrimidine) and $N_{(9)}$—$C_{(8)}$ (in purine) when viewed along $C_{(1')}$—N. This angle is taken as zero when $O_{(1')}$ is antiplanar to $C_{(2)}$ of the pyrimidine or $C_{(4)}$ of the purine ring, and positive angles are taken as those measured in a clockwise direction when viewing from $C_{(1')}$ to N. It is seen that ϕ_{cn} in both these definitions will be the same if the atoms comprising the bases are coplanar and $C_{(1')}$ lies in this plane." At present, however, most of the existing data on the ϕ_{ce} angle are those following the definition of Donohue and Trueblood. Therefore, the older definition is followed here.

with the projection of the $C_{1'}$—$O_{1'}$ bond of the furanose ring when viewed along the $C_{1'}$—N bond. This angle will be taken as zero when the furanose-ring oxygen atom is antiplanar to C_2 of the pyrimidine or purine ring, and positive angles will be taken as those measured in a clockwise direction when viewing from $C_{1'}$ to N."

A graphic definition of the angle is shown in Fig. 7, adopted from Haschemeyer and Rich (*4*). Donohue and Trueblood further concluded that there are two ranges of ϕ_{cn} for the nucleosides: ca. $-30°$ for the *anti* conformation and ca. $+150°$ for the *syn* conformation. The range for each conformation is about $\pm$ 45°. This torsion angle, ϕ_{cn}, is a very important parameter in building molecular models of polynucleotides (*4,6*) and of dinucleotides (*119*) from the monomers. At present, there are three experi-

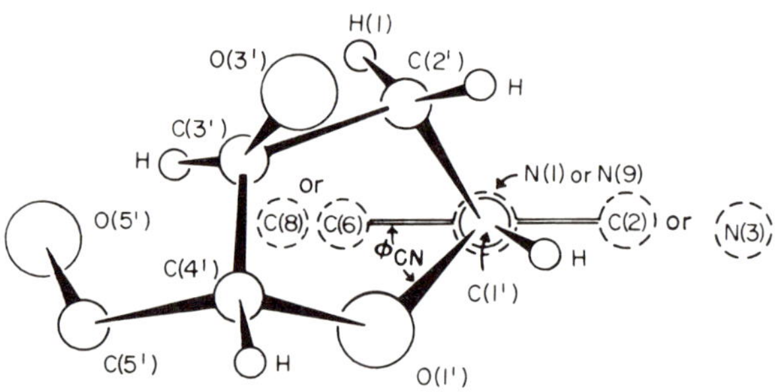

Fig. 7. Schematic illustration of the torsion angle in a pyrimidine or purine nucleoside (after Haschemeyer and Rich, Ref. *4*). The plane of the base is viewed end-on with the glycosyl bond between C(1′) and N(1) [or N(9)] of the base perpendicular to the paper. The torsion angle (ϕ_{cn}) is the dihedral angle between the plane of the base and the plane formed by the C(1′)—O(1′) bond of the furanose ring and the C(1′)—N(1′) bond. The angle is zero when O(1′) lies directly in front of C(6) [or C (8) for a purine], and positive angles are measured when C(1′)—O(1′) is rotated in a clockwise direction when viewing from C(1′) to N.

mental approaches in obtaining information about this ϕ_{cn} angle of the nucleoside (or nucleotide): X-ray diffraction studies on the crystal, ORD–CD studies, and PMR studies on the compounds in solution.

1. X-Ray Diffraction Studies

The results of X-ray studies on 16 pyrimidine nucleosides or nucleotides and on 11 purine compounds for the ϕ_{cn} torsion angle are presented in Table XV. For the pyrimidine compounds, all the angles observed are in the *anti* conformation range with the variation from $-5°$ to $-72°$; 12 of them have the angle from $-40°$ to $-60°$. For the purine compounds, the variation is much larger; six of them in a typical *anti* conformation ranging from $-3°$

TABLE XV

The Sugar–Base Torsion Angle (ϕ_{cn}) of the Nucleosides and Nucleotides Determined by X-Ray Diffraction

	Torsion angle (ϕ_{cn})	Reference
A. Pyrimidine compounds		
Thymidine	−39°	*209*
Calcium thymidylate	−48°	*11*
5′-Bromo-5′-deoxythymidine	−56°	*210*
5-Bromodeoxyuridine	−43°	*31*
5-Bromouridine	−56°	*31*
5-Bromouridine in (5-bromouridine·adenosine complex)	−20°	*32*
5-Fluorodeoxyuridine	−62°	*33*
5-Iododeoxyuridine	−65°	*211*
Uridine in (β-adenosine-2′-β-uridine-5′-phosphate)	−5°	*25*
Barium uridine 5′-phosphate	−43°	*13*
4-Thiouridine disulfide	−17° and −18°	*212*
Uridine 3′:5′-cyclic phosphate[a]	−72° and −56°	*22*
Cytidine	−18°	*213*
Cytidine 3′-phosphate[b]		
orthorhombic	−42°	*21*
monoclinic	−39°	*19*
5-Bromodeoxycytidine in (deoxyguanosine·5-bromodeoxycytidine)	−61°	*35*
B. Purine compounds		
Deoxyadenosine	−3°	*10*
Adenosine in (5-bromouridine·adenosine complex)	−10°	*32*
Adenosine 5′-phosphate	−18°	*18*
Adenosine 3′-phosphate[b]	−4°	*12*
Adenosine 3′:5′-cyclic phosphate[a]	−50° and +102°	*23*
Adenosine in (β-adenosine-2′-β-uridine-5′-phosphate)	−54°	*25*
8-Aza-9-carboadenosine (formycin)	+148°	*214*
Guanosine[c]	−120° and −45°[d]	*17*[e]
Inosine[c]	−120° and −45°[d]	*17*[e]
Inosine 5′-phosphate	−44°	*14*
Deoxyguanosine in (5-bromodeoxycytidine·deoxyguanosine complex)	+138°	*35*

[a] Two crystallographically independent molecules in an asymmetric unit cell.

[b] ϕ_{cn} angle redefined by Sundaralingam and Jensen (*21*) (See footnote, p. 113).

[c] Two crystallographically independent molecules in an asymmetric unit cell. Guanosine and inosine crystals are isomorphous.

[d] This molecule has an unusual furanose conformation (See text, Section II, A).

[e] C. E. Bugg, private communication.

two of them in the *syn* conformation; formycin is a special case discussed to $-54°$, two of them located between the *syn* and the *anti* range ($-120°$), later in this section. Davies (*6*) in his survey of 16 cases considered that the angle has some dependence on whether C(2′) or C(3′) is the out-of-plane atom (most are *endo*, see Section II, A); for C(2′) out of plane, ϕ_{cn} averages $-51°$ compared with $-16°$ for C(3′) out of plane. He concurred with the conclusion reached in Section II, A, however, that the energy difference between conformations with C(2′) or C(3′) displacement is probably small. No dependence on the ribose or the deoxyribose can be observed.

As described in Section II, A, four crystals were found to contain two crystallographically independent molecules in an asymmetric unit cell. In the case of uridine 3′:5′-cyclic phosphate (*22*), the ϕ_{cn} of the two molecules are quite similar, $-72°$ and $-56°$. In the case of guanosine and inosine (their crystal structures are isomorphous), the ϕ_{cn} of one molecule in the unit cell is $-120°$ and the ϕ_{cn} of the other molecule in the unit cell is $-45°$. In the case of adenosine 3′:5′-cyclic phosphate, it is most interesting that one molecule is in the *anti* range ($-50°$) and the other molecule is in the *syn* range ($+102°$) in the same unit cell. These data suggest that the energy difference in favor of the *anti* conformation between the *syn* and *anti* conformation is larger for the pyrimidine compounds than for the purine compounds.

As to the all-important question about the rotational barrier between the *anti* conformation and the *syn* conformation, Haschemeyer and Rich (*4*) have done an extensive analysis on the allowed conformations and the steric barriers between these allowed states. The crystallographic data from 15 compounds were analyzed on the basis of a hard-sphere model with accepted van der Waal's radii (3 levels) for H, O, N, and C atoms. The excluded volume of a pair of atoms ranged from about 2 Å for H···H to about 2.8 Å for C···N. Therefore, an interatomic distance of 1.6–2.2 Å found at a given torsion angle (ϕ_{cn}) represents an unallowed conformational state or a steric barrier at that angle. Situations for different conformations of the furanose ring, i.e., C(3′)-*endo*, C(2′)-*endo*, etc., were studied. For the pyrimidine compounds, Haschemeyer and Rich have concluded that the barriers between the *syn* and *anti* conformation are fairly great. These barriers are due to the H(6) and the O(2) atoms: on one side the H(6) runs into close contact with H(2′) (~ 1.7 Å), with C(2′) (~ 2.3 Å), and with O(1′) (~ 2.1 Å); on the other side, the O(2) of the *oxo* group runs into a close contact with H(2′) [~ 1.7 Å for C(3′)-*endo* and ~ 1.5 for C(2′)-*endo*], with H(3′) [~ 1.9 Å for C(3′)-*endo*], with C(2′) (~ 2.3 Å), and with O(1′) [~ 2.1 Å for C(3′)-*endo*]. As for the purine compounds, the steric barriers are caused by the N(3) and the H(8) atoms. For the C(3′)-*endo* conformation, the N(3) runs into H(2′) (~ 1.9 Å), into H(3′) (~ 1.6 Å), and into O(1′) (~ 2.5 Å); the

H(8) runs into H(2′) (~ 1.9 Å). A similar situation also exists for the C(3′)-*exo* conformation, except the contacts of N(3) to H(2′), and H(8) to H(2′) are even closer. As for the C(2′)-*endo* conformation, however, the steric barrier for the N(3) atom appears to be much less: N(3) to H(2′) is only 2.2 Å; N(3) to O(1′) is only 2.6 Å. The reason for not analyzing the H(8) contacts for the C(2′)-*endo* conformation were not stated (*4*). Haschemeyer and Rich concluded that the steric barriers to the interconversion between *syn* and *anti* conformation is not high for the C(2′)-*endo* purine nucleosides, but is higher for C(3′)-*endo* or C(3′)-*exo* purine nucleosides and is much higher for the pyrimidine nucleosides. The allowed conformational states defined by the ϕ_{cn} (the *syn* and the *anti* range) calculated by Haschemeyer and Rich are consistent with the experimental observation. Sasisekharan et al. have reached essentially the same conclusion in their stereochemical analysis (*5*).

As mentioned in the above paragraph and in Table XV, formycin has a *syn* conformation with the ϕ_{cn} at +148° (*214*). Formycin [7-amino-3-(β-D-ribofuranosyl) pyrazolo (4,3-*d*) pyrimidine], has a carbon-to-carbon linkage between the C(1′) of the ribose and the C(3), or C(9) in previous adenine numbering system of the base. The C-to-C linkage is expected to be about 0.07 Å longer than the C-to-N linkage in the natural nucleoside. In addition, the C(8) of the adenosine now becomes an aza nitrogen [2, or N(8) in the previous adenine numbering system] in formycin. Therefore, the steric hindrance of H(8) in adenosine (or other purine nucleosides) now vanishes since no hydrogen atom is located on the corresponding nitrogen (8, adenine numbering system) in the formycin. It appears, therefore, that the steric hindrance to the rotation of the sugar–base torsion angle (ϕ_{cc} instead of ϕ_{cn}) will be substantially reduced for formycin because of the increase in length of the C–C linkage between the base and the pentose, and because of the absence of an hydrogen atom at the former 8 position of the base-ring. Recently, Beynon and Fuller (*215*) have made stereochemical calculations based upon the molecular structures known from the X-ray diffraction study and on "Lennard-Jones 6–12 potential function" with the van der Waal's radii of 1.6, 1.5, 1.0, and 1.4 Å, respectively, for C, N, H, and O atoms. Their results indicated that the rotational barrier of the C–N glycosyl bond is much greater for the pyrimidine nucleosides than for the purine nucleosides. Two other theoretical calculations have been made by Tinoco et al. (*216*) and by Jordan and Pullman (*217*). The computed results from these two groups both agree that while the pyrimidine nucleosides clearly prefer the *anti* conformation, the purine nucleosides do not have a large energy difference (estimated to be 1 kcal/mole by Jordan and Pullman) between these two conformational states. The results of these two groups also indicate that guanosine prefers the *syn* conformation over the *anti* conformation.

2. ORD–CD Studies

ORD–CD studies on nucleosides with respect to the problem of the *syn* and *anti* conformation have been discussed extensively in Section II, F; therefore, they will not be repeated here. Generally, the results support the notion that pyrimidine nucleosides have the *anti* conformation and have a relatively large rotational barrier between these two conformations. The results in the purine nucleosides are not conclusive. Part of the reason is that the optical activity of the purine nucleosides is much less than that of the pyrimidine nucleosides. The optical activities observed presumably are due to the interaction between the transition moment(s) of the chromophore (the base) and the asymmetric centers (the pentose). The information about the direction of the transition moment(s) of the bases is very scarce. In addition, ORD data are not in a convenient form to be correlated to the absorption bands responsible for the optical activity. Therefore, the interpretation of the results usually relies on empirical correlations. When a constraint is imposed upon the nucleoside to adopt a given conformation (*syn* or *anti*) by chemical modification (such as the synthesis of cyclonucleosides) it is often uncertain whether such a modification has affected the electronic transition of the base sufficiently to render the interpretation rather tenuous.

3. PMR Studies

In Section II, C, 2, the work of Cushley, Watanabe, and Fox (*95*) on the assignment of anomeric structures for the pentofuranosyl and the pyranosyl pyrimidine nucleosides by the PMR method was described. These authors have observed the anisotropic effect of the double bond in the pyrimidine aglycon on the chemical shifts of the acetyls substituted at the C(2′) position of the sugars. This anisotropic field effect can be clearly illustrated by the comparison between the chemical shifts of the acetyl groups before and after the catalytic hydrogenation of the 5,6 double bond of the pyrimidine base. With 2′-acetylated pentofuranosyl nucleosides, hydrogenation of the double bond caused a downfield shift of the acetyl resonance when a *cis*-C(1′)–C(2′) relationship exists between the base and the acetyl group (for example, α-anomers of ribosyl and xylosyl and β-anomers of arabinosyl and lyxosyl compounds). The same hydrogenation treatment causes small highfield shifts of the acetyl resonance for nucleosides having a *trans* relationship (such as the β-anomers of ribosyl and xylosyl and the α-anomers of arabinosyl and lyxosyl compounds). In order for the anisotropic field of the 5,6 double bond of the pyrimidine base to be effective on the acetyl groups on the C(2′) (shielding for *cis* relationship), pyrimidine nucleosides have to assume an *anti* conformation at least part of the time. Without a quantitative calculation, these observations do not provide definitive information about the rotational barrier between the *syn* and the *anti* conformation of the pyrimidine nucleosides. If one accepts the conclusion from the computation based on the X-ray data that there is a large rotational barrier between the *syn* and *anti* conformation of the pyrimidine nucleosides, then pyrimidine nucleosides must assume an *anti* conformation in DMSO solution based on the deduction from the data of Cushley et al. (*95*).†

The next approach to this problem is based on the effect of the phosphate group on the chemical shifts of the base protons. Our recent paper (*97*) has

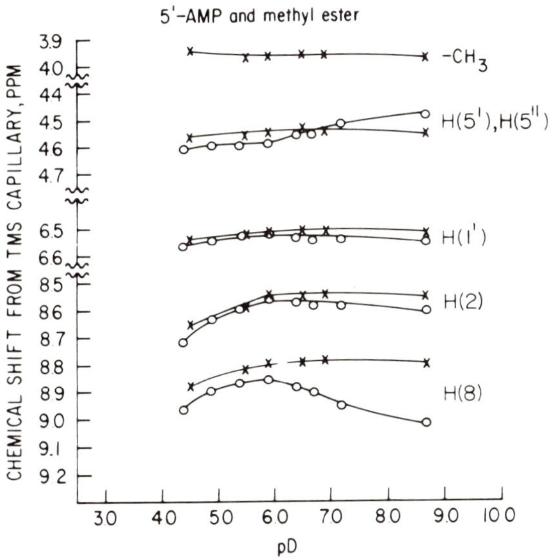

Fig. 8. Chemical shift vs. pD for protons of 5′-AMP (—O—O—, 35°, 0.08 *M*) and 5′-AMP monomethyl ester (—X—X—, 30°, 0.08 *M*) (from Ref. *97*).

described these results in great detail. For example, the effect of pD on the chemical shifts of 5′-AMP and 5′-AMP monomethyl ester at 0.09 *M* is shown in Fig. 8. In the region of pD 4.5 to 6.0, the effect of pD is on the deprotonization of the base. The loss of the proton from the adenine moiety causes an upfield shift of the base protons both by the removal of the intrinsic deshielding effect of the positive charge and by an increase in the intermolecular association of the bases. This interpretation is confirmed by the

† In Ref. *95*, Cushley et al. used the description of "5,6 double bond of the aglycon is *endo* to the five-membered sugar ring" which is equivalent to the *anti* conformation.

identical results obtained from both 5′-AMP and 5′-AMP monomethyl ester. In the region of pD 6.0 to 8.7, the secondary phosphate is being ionized and thus increases its electron density. This is illustrated by the increase in shielding of the H(5′), H(5″) protons of 5′-AMP at higher pD, while the H(5′) proton of the methyl ester is unchanged as expected. In this region, the H(8) proton of 5′-AMP is specifically and substantially deshielded (~ 0.1 ppm), while the H(8) and H(2) protons of the corresponding methyl ester are unaffected. There is a slight downfield shift (~ 0.02 ppm) of the H(2) proton of 5′-AMP. This shift is most likely due to the small effect of the ionization of the secondary phosphate on the stacking of 5′-AMP and to the fact that the H(2) proton is most sensitive to the degree of stacking as is described in Section III. Therefore, since the deshielding effect due to the

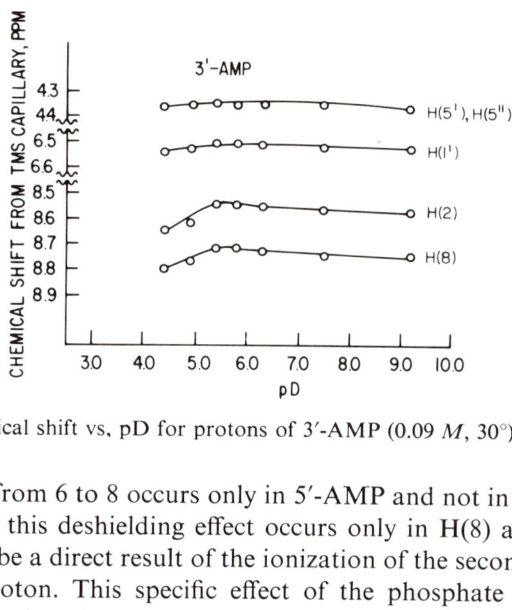

Fig. 9. Chemical shift vs. pD for protons of 3′-AMP (0.09 *M*, 30°) (from Ref. *97*).

increase of pD from 6 to 8 occurs only in 5′-AMP and not in its monomethyl ester, and since this deshielding effect occurs only in H(8) and not in H(2), this effect must be a direct result of the ionization of the secondary phosphate on the H(8) proton. This specific effect of the phosphate group on H(8) suggests strongly that 5′-AMP is in an *anti* conformation as depicted in Fig. 1. In this conformation, the H(8) proton is in juxtaposition to the 5′-phosphate group. In Fig. 9, the effect of pD on the chemical shifts of 3′-AMP is shown. Now, when the phosphate group is moved from the 5′-position to the 3′-position, the specific effect of ionization of the secondary phosphate essentially disappears. The small deshielding effect upon increase in pD from 6 to 9 observed for the H(2), H(8), and H(1′) protons (0.01–0.02 ppm) is probably due to a slight decrease in stacking because of the increase in the charge of the phosphate group as explained above for a similar effect on the H(2) proton of 5′-AMP. The shielding effect, upon the increase of pD from

4.5 to 6 due to the deprotonation of the adenine ring, is also observed here. The ionization of the secondary phosphate group in 2′-AMP has virtually no effect on the chemical shift of the H(8) protons; the ionization has a small effect on the H(2) protons, mainly due to an indirect effect on the stacking of 2′-AMP (see Section III, C, or Ref. *97*). Table XVI shows the effect of the secondary phosphate ionization (from pD 5.9, 10% dianion to pD 7.4, 90% dianion) on the chemical shift of various mononucleotides. For the purine

TABLE XVI

The Effect of the Secondary Phosphate Ionization (from pD 5.9 to pD7.4) on the Chemical Shift of Various Mononucleotides ($\Delta\delta = \delta$pD 5.9–δpD 7.4)[a]

A. Purine nucleotides		δ^aH(8)			δ^aH(2)		
	Conc., *M*	pD 5.9	pD 7.4	$\Delta\delta$	pD 5.9	pD 7.4	$\Delta\delta$
5′-AMP	(0.08)	8.853	8.950	0.097	8.567	8.588	0.021
5′-AMP	(0.25)	8.792	8.895	0.103	8.425	8.442	0.071
5′-MeAMP	(0.08)	8.795	8.795	0	8.550	8.553	0.003
3′-AMP	(0.09)	8.717	8.737	0.020	8.547	8.562	0.015
2′-AMP	(0.09)	8.733	8.780	0.047	8.508	8.608	0.100
2′-MeAMP	(0.09)	8.758	8.758	0	8.628	8.625	+0.003
5′-IMP	(0.09)	8.888	9.000	0.112	8.645	8.662	0.017
5′-GMP[b]	(0.02)	8.500	8.613	0.113			
5′-MeGMP	(0.09)	8.462	8.465	0.003			
3′-GMP[c]	(0.02)	8.393	8.423	0.030			
5′dAMP	(0.08)	8.788	8.883	0.095	8.533	8.562	0.029
Tubercidin 5′-phosphate	(0.1)	8.000	8.008	0.008	8.510	8.513	0.003

B. Pyrimidine nucleotides		δ^aH(6)			δ^aH(5)		
	Conc., *M*	pD 5.9	pD 7.4	$\Delta\delta$	pD 5.9	pD 7.4	$\Delta\delta$
5′-CMP	(0.02)[d]	8.424	8.504	0.080	6.588[e]	6.600	0.012
5′-MeCMP	(0.09)	8.383	8.383	0		*f*	
3′-CMP	(0.03)	8.303	8.287	+0.016		*g*	
2′-CMP	(0.03)	8.241	8.233	+0.008		*g*	
5′-UMP[h]	(0.02)	8.403	8.533	0.130	6.370	6.398	0.028
5′-MeUMP	(0.1)	8.352	8.357	0.005	6.372	6.378	0.006
5′-dUMP	(0.09)	8.443	8.527	0.084	6.400	6.430	0.030
5-TMP	(0.095)	8.190	8.240	0.050	2.361[i]	2.371[i]	0.010
5,6 dihydro-5′-UMP	(0.1)	4.097	4.115	0.018	3.220	3.235	0.015
5′-dCMP	(0.09)	8.411	8.472	0.061	6.551	6.554	0.003

TABLE XVI—continued

C. Nucleosides		δ^aH(6)			δ^aH(5)		
	Conc., *M*	pD 5.9	pD 7.4	$\Delta\delta$	pD 5.9	pD 7.4	$\Delta\delta$
Uridine	(0.1)	8.370[j]	8.368[k]	+0.002	6.398[j]	6.393[k]	+0.005
			H(8)			H(2)	
Inosine	(0.05 *M*)	8.762	8.755[m]	+0.007	8.653[l]	8.642[m]	+0.011

[a] Measured from TMS capillary; $\Delta\delta$ values negative unless indicated.
[b] In mixture with 0.02 *M* 3′-AMP.
[c] In mixture with 0.02 *M* 5′-AMP.
[d] In mixture with 0.02 *M* 2′ (3′)-CMP. H(5) not discernible.
[e] In 0.09 *M* concentration measured with a 60-Mc instrument. The absolute δ value is not as accurate as that quoted in Table XVII, but $\Delta\delta$ is accurate.
[f] H(5) was not unambiguously determined.
[g] Difficult to measure shifts due to broad resonances.
[h] In mixture with 0.02 *M* 3′ (2′)-UMP.
[i] Methyl protons.
[j] 0.2 *M* NaH_2PO_4.
[k] 0.2 *M* Na_2HPO_4.
[l] Na-acetate buffer.
[m] Adjusted by NaOD checked before and after run.

nucleotides, the change of pD from 5.9 to 7.4 has virtually no effect on the H(2) protons save that for 2′-AMP, as has been discussed above. As a control, this change of pD has no effect on the H(2) or H(8) protons of all the monomethyl esters of 5′-AMP, 2′-AMP, and 5′-GMP nor on the nucleoside, inosine. Little or no effect on H(2) can also be observed for the dianion formation of the 3′-phosphate of AMP. The ionization of the secondary phosphate has a large effect (0.1 ppm) on the H(8) protons of 5′-AMP, 5′-dAMP, 5′-IMP, and 5′-GMP. This ionization has no effect on the H(8) or H(7) protons of tubercidin 5′-phosphate which is discussed later in relation to the mechanism.

As for the pyrimidine nucleotides, this change of the pD from 5.9 to 7.4 has virtually no effect on the H(5) protons. As a control, this increase of pD has no effect on the H(5) or H(6) protons of the methyl esters of 5′-CMP and 5′-UMP, nor on the protons of the nucleoside, uridine, which was dissolved in 0.2 *M* phosphate buffer. This ionization of the secondary phosphate has a specific and substantial effect (~ 0.08 ppm) on the H(6) protons of 5′-CMP, 5′-UMP, 5′-dUMP, 5′-dTMP, and 5′-dCMP. The ionization has no effect on the H(6) of 3′-CMP or 2′-CMP, nor on the H(6) proton(s) of 5,6 dihydro-5′-UMP as is discussed later. It should be noted also that the chemical shifts of the base protons in pyrimidine nucleosides are not influenced by the degree of their self-association as is discussed in Section III, C. We may conclude, therefore, by analogy with the 5′-purine nucleotides, that the 5′-phosphate group of the pyrimidine nucleotides also has a direct and specific effect on the H(6) proton and not on the H(5) proton. Again, in order to produce this intramolecular deshielding, the 5′-pyrimidine phosphate must be in the *anti* conformation. In this conformation, the H(6) proton is in juxtaposition to the 5′-phosphate group.

The effect of the phosphate group on the chemical shifts of the base protons is again shown in Table XVII. In Table XVII, comparison is made between the nucleoside and the corresponding nucleotide in monoanion form (the effect of intermolecular stacking of adenine derivatives has been eliminated by extrapolation to infinite dilution). First of all, if the monoanionic phosphate group is substituted at the 3′-position, no difference in chemical shift can be found for the base protons [H(8) and H(2) of purine, H(5) and H(6) of pyrimidine] between the nucleoside and the corresponding 3′-nucleotide as exemplified by adenosine versus 3′-AMP and cytidine versus 3′-CMP. If the phosphate group is substituted at the 5′-position, the H(8) protons of the purine nucleotides, such as 5′-AMP, 5′-dAMP, 5′-GMP, and 5′-IMP, are deshielded specifically by about 0.12 ppm, while the H(2) protons are unaffected. In an analogous manner and with a similar magnitude, the H(6) protons of the pyrimidine nucleotides such as 5′-CMP, 5′-UMP, and 5′-TMP are deshielded to a much greater extent than the H(5) protons. The substitution of the phosphate group at 2′-position as in the case of 2′-AMP, has caused the shielding of H(8) and especially the H(2) protons (0.05 to 0.08 ppm). This effect of the 2′-phosphoryl group on the base protons is most likely due to an indirect effect of the removal of the OH(2′) group which can form a hydrogen bond with N(3) of the base as discussed in Section H, which follows. Therefore, we may conclude from the data presented in Table XVII that the 5′-phosphate group does exert a specific effect on the H(8) of the purine nucleotides and the H(6) of the pyrimidine nucleotides.

The mechanism of the deshielding effect of 5′-phosphate on the H(8) proton of purine nucleotides or on the H(6) proton of the pyrimidine nucleotides deserves further inquiry. One possible mechanism is the paramagnetic effect of the anisotropy of the phosphoryl group which may be spatially oriented in a proper manner toward the sensitive protons [H(8) or H(6)] in order to produce the deshielding effect. It should be noted, however, that the phosphoryl group presents a complex case with regard to its magnetic anisotropy, because not only does the P=O bond resonate between the

TABLE XVII

Influence of the Phosphate Group as Shown by the Comparison of the Chemical Shifts of the Nucleosides and Nucleotides (pD 5.9). Chemical Shifts Measured from TMS Capillary

		H(8)	H(2)		H(6)	H(5)
5′-AMP[a]		8.957	8.705			
Adenosine		8.820	8.730			
	Δ	0.137	−0.025			
Adenosine 5′-methyl hydrogen phosphate[a]		8.883	8.717			
Adenosine[a]		8.820	8.730			
	Δ	0.063	−0.013			
5′-dAMP[a]		8.861	8.640			
2′-deoxyadenosine[a]		8.756	8.686			
	Δ	0.105	−0.046			
3′-AMP[a]		8.805	8.697			
Adenosine[a]		8.820	8.730			
	Δ	−0.015	−0.033			
2′-AMP[a]		8.773	8.653			
Adenosine[a]		8.820	8.730			
	Δ	−0.047	−0.077			
5′-GMP[b]		8.544				
1-methylguanosine[b]		8.418				
	Δ	0.126				
5′-IMP[c]		8.888	8.662			
Inosine[c]		8.762	8.653			
	Δ	0.126	0.009			
5′-CMP					8.424	6.546
Cytidine[c]					8.292	6.513
				Δ	0.132	0.033
Cytidine 5′-methyl hydrogen phosphate[c]					8.383	*d*
Cytidine					8.292	6.513
				Δ	0.091	

TABLE XVII—continued

5′-UMP[b]		8.403	6.370
Uridine[c]		8.275	6.327
	Δ	0.128	0.043
Uridine 5′-methyl hydrogen phosphate		8.352	6.372
Uridine[c]		8.275	6.327
	Δ	0.077	0.045
3′-CMP[c]		8.303	6.533
Cytidine[c]		8.292	6.513
	Δ	0.011	0.023
5′-TMP[c]		8.190	2.350[e]
Thymidine[c]		8.087	2.318[e]
	Δ	0.103	0.032

[a] Extrapolated infinite dilution shifts.
[b] 0.02 *M*, in mixture with 0.02 *M* 3′ (2′)-UMP. 1-methylguanosine was used because of solubility problem of guanosine. Since the chemical shift of the H(8) proton is the same for inosine and 1-methylinosine, this substitution was considered justifiable (see Ref. *64*).
[c] 0.09 *M*.
[d] These resonances not discernible.
[e] Methyl protons.

three oxygen atoms, but the P—O ester bond can also rotate freely. This consideration leads us to expect that the anisotropic effect of the phosphoryl group on the H(8) or H(6) proton may be minimized on a time average basis. Another possible mechanism is the direct electrostatic field effect of the negatively charged phosphate. This mechanism hypothesizes that the electrostatic field of the negative charge tends to attract the positive proton and thus polarize the C—H bond to a greater extent, resulting in lower electron density at the proton, and hence a deshielding effect. In other words, the former mechanism emphasizes the influence of the *magnetic* field effect of the phosphate while the latter emphasizes the *electrostatic* field effect of the phosphate on the proton. One way to distinguish these two proposals is to compare the deshielding effect of the phosphate group on the sensitive protons that have different degrees of "acidity." It is also important for the success of this experiment that the spatial relationship of the phosphate group to the sensitive protons of these various analogs remain essentially constant.

Following this line of reasoning, one can examine the data on 5′-dTMP and 5′-dUMP in Table XVI. As expected, 5′-dTMP, bearing an electron-releasing methyl group at the C(5) position, gives a chemical shift of the H(6) proton about 0.25 ppm upfield in comparison with the H(6) of 5′-dUMP. The less "acidic" H(6) of 5′-dTMP is shifted only 0.05 ppm downfield over the pD range 5.9 to 7.4, compared with 0.084 ppm for the more "acidic" H(6) of 5′-dUMP. If the 5,6 double bond of 5′-UMP is saturated as in 5,6 dihydro-uridine-5′-phosphate, the effect of pD change is practically eliminated (0.018 ppm, Table XVI). One cannot be sure, however, in the case of 5,6-dihydro-UMP that the spatial relationship of the 5′-phosphate group to the H(6) is the same as the other unsaturated nucleotides.

As for the 5′-purine nucleotides, the H(8) proton is well known for being "acidic," judging from its rate of exchange with deuterium (*63,79*). With regard to the change of the "acidity" of the H(8) proton, we have studied the pD shift of the tubercidin-5′-phosphate [4-amino-7-β-D-ribofuranosyl-5′-phosphate-7H-pyrrolo-(2,3-d) pyrimidine], which is a 7-deazaadenosine-5′-phosphate (Table XVI). The "acidity" of the H(8) (purine numbering system) of this molecule is decreased as compared with the "acidity" of the H(8) of adenine due to the replacement of the electronegative N(7) with C—H in the 5-membered ring of adenine. As clearly shown in Table XVI, neither the pyrole ring protons H(8) and H(7) nor H(2) in the pyrimidine ring are deshielded upon dianion formation of the phosphate. We expect that the general conformation of tubercidin-5′-phosphate is similar to that of 5′-AMP since the 5-membered rings of the base remain aromatic (therefore, planar) in both molecules.

The foregoing experiments on analogs of 5′-pyrimidine and 5′-purine nucleotides do provide some reliable evidences that this deshielding effect of the phosphate is reduced when the "acidity" of the sensitive protons is decreased. Unfortunately, no experimental data are yet available to prove or to disprove the reverse situation. Nevertheless, we feel that these data favor the "electrostatic mechanism" over the "paramagnetic mechanism," since the latter mechanism should not be concerned with the "acidity" of the sensitive protons.

There have been other observations on the deshielding effect of a phosphoryl group on the proton chemical shifts through space. In a recent paper, Lira and Huffman (*218*) have reported a deshielding influence of a phosphonate group upon H(8) in N(9)-bis (β-chloroethyl)-β-adenine-9-ylethyl phosphonate. These authors suggested that the deshielding arose either as a result of the P=O anisotropy or the hydrogen bonding of H(8) to the phosphoryl oxygen. More direct evidence for the phosphate deshielding effect is forthcoming from the data on arabinosyl cytosine-2′:5′-cyclic phosphate in which the phosphate anion is held in a position very close to H(3′) (*219*). It is found that H(3′) of this cyclic phosphate is about 1 ppm to lower field than in arabinosyl cytosine.

Danyluk and Hruska (*220*) have studied the pD dependence of the chemical shifts of the base protons and the H(1′) proton of the nucleotides of adenosine, guanosine, xanthosine, and thymidine over the pD range of 1–8.2. They reached a conclusion similar to that of our work, i.e., the 5′-nucleotides are in an *anti* conformation.

It should be strongly emphasized here that although we favor the proposal of the electrostatic field effect of the 5′-phosphate on the sensitive protons in

TABLE XVIII

Difference in Chemical Shift ($\Delta\delta$ 5′–3′, ppm) between 5′-Nucleotides and 3′-Nucleotides at 30° and 60°—Indication of "*anti*" Conformation $\Delta\delta$ 5′–3′ = (δ 5′-nucleotide) – (δ 3′-nucleotide), ppm[a]

Nucleotide	Conc., molal	Temp.	H(8)	H(2)	H(6)	H(5)	H(1′)
AMP	0[b]	30°	0.148	0	—	—	0.011
	0	60°	0.141	0.01	—	—	0.063
GMP	0[b]	30°	0.116	—	—	—	0.038
	0	60°	0.094	—	—	—	0.018
CMP	0.045	30°	—	—	0.143	0.047	0.044
	0.045	60°	—	—	0.134	0.040	0.055
UMP	0.045	30°	—	—	0.130	0.060	0.047
	0.045	60°	—	—	0.133	0.049	0.044
TMP[c]	0.430	30°	—	—	0.133	0.03 ($-CH_3$)	0.070
	0.430	60°	—	—	0.116	0	0.052

[a] Values of the δ5′-nucleotide and δ3′-nucleotides are from Ref. *119*.
[b] Extrapolation to infinite dilution.
[c] δ value of the thymidine is used as substitute for the δ value of 3′-TMP.

order to cause the deshielding phenomenon, we do not mean that the phosphate group is as strongly influenced by these protons as by formation of a hydrogen bond to the phosphoryl oxygen atom. We feel that our data and our proposal are in essential agreement with the conclusion of Phillips et al. (*221*). They have studied the pK_a of the secondary phosphate group of various 5′-nucleoside mono-, di-, and triphosphates by potentiometric titration. They found that the pK_a of the secondary phosphate is not very sensitive to the nature of the bases attached to the nucleotides and concluded that there is no "ring-chain interaction" in nonmetal complexed mononucleotides. As a corollary, we do not consider that this phosphate–proton interaction provides much of the stabilization energy for the *anti* conformation.

While the PMR studies described above provide a clear indication that these 5′-nucleotides must be in an *anti* conformation, these studies do not give information about the rotational barrier and the energy level between these two conformations. These data do not exclude the possibility that for part of the time the 5′-nucleotides are in *syn* conformation unless there is a large rotational barrier. The observation that the phosphate in 2′-AMP does not deshield the H(8) proton tends to minimize the possibility for the *anti* conformation from the PMR point of view. Recently, we (*36*) have studied the effect of temperature on the nucleotide conformation. As described in the above paragraphs, the differential in chemical shift between the H(8) proton of the 5′-nucleotides and that of the 3′-nucleotides in case of a purine compound, or the differential between the chemical shift of the H(6) proton of the 5′-nucleotide to that of the 3′-nucleotide in case of a pyrimidine compound is an index of the sugar–base torsion angle, ϕ_{cn}, which defines the *anti* or *syn* conformation. As shown in Table XVIII (from Ref. *36*), the same differentials in the values for the H(8) or H(6) protons between the 5′-nucleotides and the 3′-nucleotides are obtained at both 30° and 60°. We conclude from these data, therefore, the sugar–base torsion angle, ϕ_{cn}, of these 5′-nucleotides must be very much the same at both 30° and 60°; thus, the 5′-nucleotides still assume the same *anti* conformation (at least to the same extent) at 60° as at 30°.

H. Intramolecular Hydrogen Bonding between the 2′-Hydroxyl Group of the Ribose and the Base in Nucleosides and Nucleotides

The discussion in Section II, G indicates that the nucleosides and nucleotides have an *anti* conformation in aqueous solutions. In this *anti* conformation (but not in the *syn* conformation), the OH(2′) group of the pentose in the ribosyl nucleosides and nucleotides appears to be capable of forming a hydrogen bond with the N(3) of purines or the O(2) (keto) of pyrimidines. The stereochemical considerations of this intramolecular hydrogen bonding for all four ribosyl nucleosides have been recently calculated by Beynon and Fuller (*215*), based on a computer program described in Section II, G. With the ribose in a C(2′) *endo* conformation and with the most favorable torsion angle (ϕ_{cn}) in the *anti* range (−50° to −70°), the hydrogen-bond length [O(2′) to N(3) in purine, or O(2′) to O(2) in pyrimidine nucleoside] can be as close as 2.8 Å with the hydrogen-bond angle deviating only about 10° from linearity. As suggested previously from our PMR study on 3′-deoxyadenosine (*64*), this intramolecular hydrogen-bond formation is more favorable when the C(2′) is in the *endo* conformation and less favorable when the C(3′) is in the *endo* conformation. The computation (*215*) showed that though the hydrogen-bond length is about the same for both C(2′) *endo* and C(3′) *endo* conformations, the deviation from linearity of the hydrogen-bond angle is much greater for the C(3′) *endo* conformation (about 25°–30°). Discussion of X-ray studies in Section II, A indicates that the furanose conformation is subjected to the influence of the environment and that there is no large energy barrier between these conformational states. Therefore, we may conclude that this intramolecular hydrogen bonding between the OH(2′) group of the ribose and the N(3) of purine or O(2) of pyrimidine nucleosides can exist as far as the stereochemical requirement is concerned. This intramolecular hydrogen bonding, as far as this reviewer knows, has not yet been observed in crystal structures studied by X-ray diffraction. The OH(2′) group has been found to form a hydrogen bond with N(3) *inter*molecularly between two neighboring molecules in the 3′-AMP crystal structure (*12*).

To study this intramolecular hydrogen bonding in solution is not an easy experimental task. In organic solvents such as chloroform, the OH frequency can be directly examined by the IR technique. By this method, Pitha et al. (*133*) concluded that they detected the existence of an intramolecular bonding of the OH(5′) group of 2′,3′ di-substituted adenosine and pyrimidine nucleoside derivatives with the base moiety. In a review, Michelson (*222*) mentioned unpublished data concerning the infrared spectra of 2′,3′-di-O-acetyladenosine and 3′,5′-di-O-acetyladenosine. These spectra indicated hydrogen bonding in the latter compound; presumably the OH(2′) group is hydrogen bonded to the adenine in the organic solvent.

The much more important issue is whether such an intramolecular hydrogen bond can exist in an aqueous environment, Currently, there are only three indirect experimental approaches to this problem. The first two approaches are concerned with the effect of the OH(2′) group on the pK_a of the base moiety in the nucleosides and on the UV absorption spectra. Fox and Shugar (*223*) first noticed these two effects on the pyrimidine nucleosides in 1952 and subsequently Fox and his co-workers (*224,225*) have utilized these effects for purposes of structural assignments. For instance, Fox et al. (*223,224*) had noted the pK_a values for the following cytosine compounds: 4.11 for cytidine, 4.25 for deoxycytidine, 4.16 for cytidine PO_4(3′), and 4.30 for cytidine PO_4(2′). These data are consistent with the notion that the OH(2′) group in cytidine and cytidine PO_4(3′) forms a hydrogen bond with the cytosine and thereby lowers the pK_a of the base for protonation. Preliminary potentiometric titration experiments done in our laboratory (0.05 *M* nucleoside, 0.15 *N* NaCl, 25°, Ref. *226*) indicate that 2′-deoxyadenosine has a pK_a 0.1 pH unit higher than that of adenosine. This conclusion is supported by a PMR-titration experiment. We have found that

H(2) and H(8) of 2′-deoxyadenosine are shifted 10 cps downfield going from a neutral solution to pH 3.6, whereas H(2) and H(8) of adenosine are shifted only 5 cps to lower field over the same pH range. Thus the base of 2′-deoxyadenosine is more readily protonated, indicating a higher pK_a. Grossman et al. (*227*) reported the pK_a value of 2.15 and 2.80, respectively, for guanosine and deoxyguanosine as determined by spectrophotometric titration. Spectrophotometric titration experiments done in our laboratory (*228*) indicated pK_a values of 2.2 and 2.4 for the guanosine and deoxyguanosine, respectively. The pK_a of uridine (9.25), however, is quite close in comparison to that of deoxyuridine (9.30), both determined spectrophotometrically (*223*). It is not immediately apparent why the difference is smaller than that found for the cytosine derivatives, even though this pK_a value (near 9) is closer to the pK_a of the OH(2′) group (about 12). The pK_a value of uridine determined titrimetrically was reported to be 9.17 by Levene and Bass (*229*). We may safely conclude that, at least for adenine, guanine, and cytosine nucleosides, the ribosyl compound has a pK_a value about 0.1 to 0.2 lower than that of the corresponding deoxyribosyl compound. This observation supports the notion of intramolecular hydrogen bonding between the OH(2′) group and the base which will decrease the pK_a through the readily delocalized electronic system.

The second effect which Fox et al. (*223–225*) had observed is the difference in spectra between the ribosyl and the deoxyribosyl nucleosides at alkaline pH due to the ionization of the OH(2′) group. The real mechanism of this effect is not very clear, though it is related to ionization of the pentose; the observed difference is dependent on the type of pentosyl substituents such as ribose, deoxyribose, arabinose, lyxo, etc., and not observed with an alkyl substituent. Though it is clearly indicative of an interaction between the base and the pentose, the real nature of this interaction has yet to be elucidated. On the basis of this data and the results on the influence of the 3′-nucleotide bases upon the hydrolysis rate of dinucleoside phosphates at all pH ranges, Witzel (*231*) proposed that there is an intramolecular hydrogen bonding between the OH(2′) group and the bases [N(3) of purine and 2-keto of pyrimidine] of the 3′-nucleotidyl unit in the dinucleoside phosphate.

The third type of experimental data available presently is the difference in the PMR properties of the ribosyl monomer versus those of deoxyribosyl monomer. In Table XIX (*97*) the chemical shifts of the base protons of adenine nucleosides and nucleotides are shown. The chemical shifts of H(2) and the H(8) of the ribosyl compounds are located 0.04–0.06 ppm downfield from those of the corresponding deoxyl compounds. Similarly, the base protons of 3′-AMP are located at a lower field than those of 2′-AMP which does not have a OH(2′) group. On the other hand, the chemical shifts of the base proton of O(2′) methyl adenosine are identical to those of 2′-deoxyadenosine (*64*). All these results are compatible with the explanation that an intramolecular hydrogen bond exists between the OH(2′) group and the N(3) of adenine, and this hydrogen bond brings about a low-field shift of the base protons. Interestingly, there is an apparent exception to this reasoning: The chemical shifts of the base protons of 3′-deoxyadenosine (Cordycepin) are the same as those of 2′-deoxyadenosine (*64*). However, it was shown that the value of $J_{H(1')-H(2')}$ of 3′-deoxyadenosine (2.1–2.3 cps) is much smaller than that of adenosine (6.1–6.4 cps) (*64*), indicating a difference in the pentose conformation between these two nucleosides. As discussed in our paper (*64*), it appears that the OH(2′) group of 3′-deoxyadenosine is not in a favorable conformation to form a hydrogen bond with the N(3) of the adenine (*64*).

TABLE XIX

Chemical Shifts of the Adenine Nucleosides and Nucleotides at Infinite Dilution

	H(8)	H(2)
Adenosine	8.820	8.730
2′-deoxyadenosine	8.756	8.686
5′-AMP·Na	8.957	8.705
5′-dAMP·Na	8.861	8.640
5′-MeAMP·Na	8.883	8.717
5′-AMP·Na_2	9.055	8.717
5′-dAMP·Na_2	8.960	8.672
3′-AMP·Na	8.805	8.697
2′-AMP·Na	8.773	8.653

As for the pyrimidine nucleotides, the H(6) and H(5) of 5′-CMP appear to be slightly less shielded than those of 5′-dCMP; similarly, the H(6) and H(5) of 3′-CMP are located at a slightly lower field position than those of 2′-CMP (Table XVI). These values, however, are very small and such a difference could not be found between 5′-UMP and 5′-dUMP, nor between 3′-UMP and 2′-UMP (Table XVI and Ref. *97*). It appears that such a weak intramolecular hydrogen bonding to the *exo*cyclic oxygen is unlikely to affect the protons of the double bonds in the ring especially since the ring is not extensively conjugated. On the other hand, the deshielding influence of the 5′-phosphoryl group on the H(6) proton is dependent on the *anti* conformation which in turn is likely to be stabilized by a hydrogen bond from the OH(2′) group to the 2-keto group. Thus, it might be expected that the loss of these stabilizing forces would reduce the proportion of the molecules in a proper *anti* conformation and consequently the phosphate influence would be less.

Data in Section B of Table XVI show the deshielding effect on H(6) protons of the ionization of the secondary phosphate (from pD 5.9 to 7.4) of various pyrimidine nucleotides. The deshielding due to this secondary ionization is indeed less for the deoxynucleotides, only to the extent of about 60–70% of the corresponding ribonucleotides, such as 5′-dUMP (0.084 ppm) versus 5′-UMP (0.13 ppm) and 5′-dCMP (0.062 ppm) versus 5′-CMP (0.08 ppm). Thus, the deshielding effect of the 5′-phosphate group on the H(6) protons is dependent to a certain extent on the presence of the OH(2′) group in the ribose. However, such a dependence appears to be negligible for the adenine nucleotides. In summarizing the PMR results, there is evidence to support the existence of this intramolecular hydrogen bonding: deshielding of the base protons by the OH(2′) group and augmentation of the 5′-phosphate effect on the H(6) protons of pyrimidine nucleotides by the OH(2′) group.

In examining all the data from the studies on the monomers in solutions, including both the pK_a and the PMR studies, one feels that while the effect of the presence of the OH(2′) group in each individual case is small, all the evidence as a whole does give a clear indication of the existence of such an intramolecular hydrogen bond between the OH(2′) group and the bases in the nucleosides and 5′-nucleotides. The crucial problem now is the question of the 3′-nucleotides, which have a phosphate group adjacent to the OH(2′) group. The hydrogen of the OH(2′) group can interact with the negative oxygen of the neighboring phosphate group, or the oxygen of the OH(2′) group can interact with the positive phosphorus atom of the phosphate group, a reaction which can lead to the formation of a cyclic phosphate. All these possible interactions are of importance, since they may be closely related to the properties of RNA and the mechanism of nucleases (such as pancreatic RNAse and T_1-RNAse). In a comparative study of the ORD and absorption spectra of ribosyl polynucleotides versus deoxyribosyl polynucleotides (poly A vs. poly dA, poly C vs. poly dC, and poly U vs. poly dT), we have found a large influence of the 2′-hydroxyl group of ribose on the conformation and the helix-coil transition of these polymers (*232*). The same conclusion was reached by Chamberlin and co-workers on these and other polynucleotides (*233–235*). Similarly, differences in optical properties have been found for ribosyl dinucleoside monophosphates versus deoxyribosyl dimers (*189,236–238*). Therefore studies on the possible interaction between the OH(2′) group and the base as well as the interaction with the neighboring phosphate group in the ribosyl 3′-mononucleotides are of considerable importance to the problem of conformation of dinucleotides and nucleic acid.

III. INTERACTION OF THE MONOMERS

A. Interaction between the Base-Pairs in the Solid State Molecular Complexes in Crystals

The first successful co-crystallization of purine and pyrimidine derivatives was reported by Hoogsteen (*239,240*). He observed that 9-methyladenine and 1-methylthymine form an interlocking structure in which the two bases are hydrogen bonded in a planar complex. Since that time, this interesting phenomenon has been investigated extensively by Professor Rich and his associates (or former associates) at M.I.T. by X-ray diffraction. The research on these complexes has provided us with more elaborate information about hydrogen bonding and packing phenomenon.

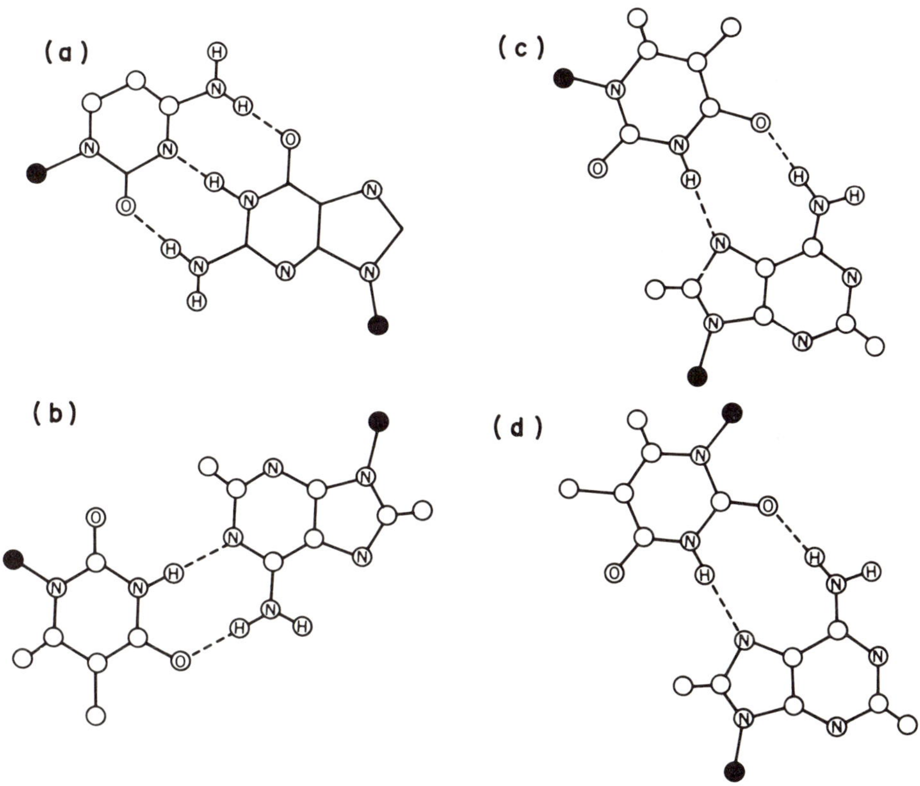

Fig. 10. Base-pairing scheme of guanine·cytosine and of adenine·uracil. (a) Guanine·cytosine in a Watson–Crick pair. (b) Adenine·uracil in a Watson–Crick pair. (c) Adenine·uracil in a Hoogsteen pair involving N(7) of adenine. (d) Adenine·uracil in a reversed Hoogsteen pair. Related to C by rotation of uracil about N(3)—C(6). The pairing involves C(2)=O of uracil instead of C(4)=O of uracil.

TABLE XX

Hydrogen Bond Lengths (Å) between the Base-Pairs in the Crystalline Complexes

A. Guanine to cytosine

Crystalline complexes	References	C(4)—NH_2 of C to C(6)═O of G	N(3) of C to N(1)—H of G	C(2)═O of C to C(2)—NH_2 of G
9-Ethylguanosine + 5-Bromo-1-methyl-cytosine	*241*	2.86	2.95	2.91
9-Ethylguanine + 1-Methylcytosine	*242*	2.943	2.924	2.806
9-Ethylguanosine + 1-Methyl-5-fluoro-cytosine	*244*	2.964	2.942	2.823
Deoxyguanosine + 5-Bromodeoxy-cytidine	*243*	2.82	2.91	2.78

B. Adenine to thymine or uracil

Crystalline complexes	References	N(3)—H of T(U) to N(7) of A	C(4)═O of T(U) to C(6)—NH_2 of A (Hoogsteen type)	C(2)═O of T(U) to C(6)—NH_2 of A (Reversed Hoogsteen type)
1-Methyl T + 9-Methyl A	*239,240*	2.924	2.846	
1-Methyl U + 9-Ethyl A	*246*	2.83	2.98	
1-Methyl-5-bromo U + 9-Methyl A	*249*	2.86	2.97	

TABLE XX—continued

1-Methyl-5-fluorouracil + 9-Ethyladenine	*250*	2.78	2.96	
5-Bromouridine + Adenosine	*251*	2.80		3.10
1-Methyl-5-bromouracil + 9-Ethyladenine	*247* *248*	2.80		3.04

These intermolecular complexes in the crystals are essentially planar, with the dihedral angle between the two bases less than 6.5° (mostly about 3°–4°). Four G·C complexes have been investigated: 9-ethylguanine + 5-bromo-methyl cytosine (*241*); 9-ethylguanine + 1-methylcytosine (*242*); deoxyguanosine + 5-bromodeoxycytidine (*243*); and 9-ethylguanine + 1-methyl-5-fluorocytosine (*244*). All these G·C complexes have the Watson–Crick type of hydrogen-bonding scheme with three pairs of bonds (Fig. 10), i.e., C(4)—NH_2 of C bonded to C(6)═O of G; N(3) of C bonded to N(1)—H of G; and C(2)═O of C bonded to C(2)—NH_2 of G. Among these three pairs of H-bonds, the distance between the C(2)═O of C to the C(2)—NH_2 of G is shorter (0.04–0.14 Å) than those of the other two H-bonds in three out of the four complexes investigated (see Table XX). The shortest distance among these G·C H-bonds in these 4 crystals is that of the C(2)═O of C to C(2)—NH_2 of G (2.78 Å) in the crystal of deoxy G + 5-bromodeoxy C; while the longest distance is that of C(4)—NH_2 of C to C(6)═O of G (2.964 Å) in the crystal of 9-ethyl G + 1-methyl-5-fluoro C (Table XX). In addition, the N(7) of G is also bonded to the C(4)—NH_2 of C in the crystals of 9-ethyl G + 1-methyl C and 9-ethyl G + 1-methyl-5-fluoro C. As mentioned in Section II, the conformation of deoxyguanosine in the molecular complex with 5-bromodeoxycytidine (*243*) has two interesting features: the rotation of ϕ_{cn} angle is 138° with the *syn* orientation, and there is an intramolecular hydrogen bond between the N(3) and the C(5′)—OH (2.82 Å) of the deoxyguanosine. The crystal structure of 9-ethyl-2-aminopurine and 1-methyl-5-fluorouracil has also been investigated (*245*). It is interesting to note that this complex has two pairs of hydrogen bonds which follow the Watson–Crick scheme for the G·C pair (Fig. 10), i.e., C(2)═O of fluoro U bonded to C(2)—NH_2 of the aminopurine, and N(3)—H of fluoro U bonded to N(1) of the aminopurine. These observations suggest that the 5-fluorouracil nucleosides are able to form base pairs with both adenine and guanine nucleosides. In a biological system, 2-aminopurine is a well-known mutagenic agent, and 5-fluorouracil promotes phenotypic reversion and inhibits DNA and RNA synthesis.

Since the original study of Hoogsteen on the crystalline complex of 9-methyladenine + 1-methylthymine (*239,240*), four other complexes of bases, and one other complex of nucleosides consisting of adenine and uracil (or thymine) derivatives have been investigated. These are: 9-ethyladenine + 1-methyluracil (*246*); 9-ethyladenine + 1-methyl-5-bromouracil (*247,248*); 9-methyladenine + 1-methyl-5-bromouracil (*249*); 9-ethyladenine + 1-methyl-5-fluorouracil (*250*); and adenosine + 5-bromouridine (*251*). The most important common feature among these six crystalline complexes is that N(7) and *not N(1)* of the adenine moiety is involved in the hydrogen bonding with the N(3)—H of the uracil or thymine moiety in the base-pairing scheme. Therefore, none of the crystalline complexes has the base-pairing scheme of the Watson–Crick type or of the DNA type, which involves the N(1) of adenine moiety instead of N(7) (Fig. 10). There are two types of base-pairing schemes for A+T (or U) involving the N(7) of adenine. The first type is the original "Hoogsteen" type (Fig. 10), in which the N(3)—H of T (or U) is bonded to N(7) of A and the C(4)═O of T (or U) is bonded to C(6)—NH_2 of A. The base-pairing scheme of 1-methyl T + 9-methyl A (*240*), 1-methyl U + 9-ethyl A (*246*), 1-methyl-5-bromo U + 9-methyl A (*244*), and of 1-methyl-5-fluoro U + 9-ethyl A (*250*) belongs to the first type (Table XX). The second type is the "reversed Hoogsteen" type (Fig. 10), in which the N(3)—H of T (or U) is still bonded to N(7) of A, but now C(2)═O of T (or U) is bonded to C(6)—NH_2 of A. The relationship between the original "Hoogsteen" type and the "reversed Hoogsteen" type is a rotation of the pyrimidine of 180° about the line joining N(3) and C(6), thereby substituting the C(4)═O in the original "Hoogsteen" type with the C(2)═O in the "reversed Hoogsteen" type. The base-pairing scheme of 5-bromouridine + adenosine, and of 1-methyl-5-bromo U + 9-ethyl A, belongs to the second type (Table XX). The distance for the hydrogen bond is usually shorter for the bond between N(3)—H of U or T to the N(7) of A, in five out of the six complexes investigated, and is usually longer for the bond between C(2 or 4)═O of U and T to the C(6)—NH_2 of A (Table XX). The shortest distance among these A·T(U) H-bonds in these 6 crystals is that of the N(3)—H of U to N(7) of A (2.78 Å) in the crystal of 1-methyl-5-fluoro U + 9-ethyl A, while the longest distance is that of C(2)═O of U to C(6)—NH_2 of A (3.10 Å) in the crystal of 5-bromouridine + adenosine (Table XX).

It is especially interesting to note that the complex of 1-methyl-5-bromo U + 9-methyl A has the "Hoogsteen" type of base-pairing scheme, while the complex of 1-methyl-5-bromo U + 9-ethyl A has the "reversed Hoogsteen" type of scheme. Tomita, Katz, and Rich (*250*) suggested that " . . . the crystal form with the greatest stability is one which involves not only favorable hydrogen bonding but also a stabler type of vertical packing, especially involving a significant overlap of the planar adenine molecules." Bugg, Thewalt, and Marsh (*17*) have also emphasized the preponderant importance of the stacking of the bases in determining the structure of the crystals, since they found that the guanosine crystal is isomorphous to the inosine crystal, even though guanosine has one more hydrogen-bonding site (the 2-amino group) than inosine.

Thin sections of the 1-methylthymine + 9-methyladenine crystal have been studied by IR spectroscopy with polarized radiation (*142*). A marked difference was observed between the spectra of the crystalline complex and an equimolar mechanical mixture of 1-methylthymine and 9-methyladenine. The out-of-plane vibrations of NH and NH_2 are sensitive to the type of hydrogen bonding engaged in.

Donohue (*252*) and Donohue and Trueblood (*253*) have shown that 29 base pairs linked with two or three hydrogen bonds could be formed between two of the four nucleosides present in the nucleic acids, if the only restriction for the base-pair formation is the geometrical requirement for the formation of single hydrogen bonds between nitrogen and oxygen atoms, i.e., approximate colinearity of the three atoms A—H· · ·B involved and interatomic distance between A and B ranging from about 2.80 to 3.00 Å. It was reasonable to believe, therefore, in the past, that the specificity of the base pairing in DNA resides mainly on the stereochemical constraints imposed on the purines and the pyrimines by the polynucleotide backbone in the double helix. As stated in a recent, comprehensive review on these crystalline complexes by Hoogsteen (*254*) "Numerous attempts have been made to cocrystallize purine and pyrimidine derivatives in all possible combinations. The failure of these experiments so far to produce in polar solvents coplanar complexes in combinations other than the Watson–Crick complementary pair is indicative for the specificity of their interactions in these solvents, but does not constitute proof in itself." It is difficult to evaluate quantitatively the effect of the packing forces in determination of the specificity of the formation of these crystalline, coplanar complexes. It is logical, therefore, to examine the formation of these complexes in solution.

B. Interaction of the Monomers in Organic Solvents

In studying the interaction of solute molecules through hydrogen bonding, a liquid without (or with the least) hydrogen-bonding capacity naturally should be selected as solvent in order to minimize the competitive interaction between the solute and the solvent. Usually, the main obstruction to such an experiment is the problem of solubility.

Independently and within the span of one year, five communications appeared reporting the interaction of bases through hydrogen bonding in organic solvents. Hamlin, Lord, and Rich at M.I.T. (*255*) and Kuchler and Derkosch (*256*) at Vienna reported the interaction of adenine and uracil derivatives, and Pitha, Jones, and Pithova (*257*) at Ottawa, reported the interaction of guanine and cytosine derivatives. All these three observations were based on IR spectroscopic studies on substituted bases dissolved in $CHCl_3$ and CCl_4. Shoup, Miles, and Becker (*258*) at N.I.H. reported the interaction of guanosine, cytosine, adenine, thymine, and 2-amino adenine derivatives dissolved in DMSO and DMF as studied by PMR; similarly, Katz and Penman (*259*) situated at Albert Einstein at the time also reported PMR studies on the interaction of derivatives of the four bases in DMSO. Subsequently, this problem was investigated extensively with the IR technique by Kyogoku, Lord, and Rich (*260–262*), as well as by Miller and Sobell (*263*). Katz has continued the study with the PMR technique on the adenine–uracil interaction (private communication) and recently, Newmark and Cantor (*264*) have studied the guanosine–cytosine interaction by PMR.

Thomas and Kyogoku (*161*) as well as Gratzer and McClare (*162*) have reported on the hypochromic effect on UV absorption due to base pairing.

In order to overcome the solubility problem of the bases or nucleosides in organic solvents, lipophylic substituents were added to the bases or the nucleosides. For the bases, the N(9) (purine) or N(1) (pyrimidine) was substituted by alkyl, cyclohexyl groups, etc.; for the nucleosides, the OH groups of the pentose were substituted by acetyl, isopropylidenyl, trityl, benzylidinyl groups, etc.

The IR measurements were usually taken at the region of 3500–2800 cm^{-1}, the region of NH and NH_2 symmetrical and antisymmetrical stretching vibrations. The frequency of these bonds is dependent on the state of hydrogen bonding of these groups. For instance, the 3395 cm^{-1} band of 1-cyclohexyluracil is assigned to be the free NH stretching, and upon hydrogen bonding, this bond is shifted to 3184 cm^{-1} (*261*). For 9-ethyladenine, the antisymmetric and symmetric stretching vibrations of the nonbonded NH_2 group are located at 3527 and 3416 cm^{-1}, respectively; these vibrations of the H-bonded NH_2 group are located at 3485 and 3312 cm^{-1} in concentrated solution (*261*). Deuteration experiments are very helpful for the identification of these bonds. Some studies on the carbonyl stretching frequency have also been made (*256,261*). In dilute solution, the stretching bonds of the C(2) carbonyl group and the C(4) carbonyl group of cyclohexyluracil in chloroform solution are assigned to the 1710 cm^{-1} and the 1687 cm^{-1} bands, respectively. In concentrated solution, these bands are shifted to about 1700 and 1670 cm^{-1} (*261*). Therefore, formation of hydrogen bonds, which generally lower the frequency of these groups can then be used for thermodynamic measurements. Kyogoku et al. (*261*) have given a complete assignment of the bands for the 9-ethyl A, 1-cyclohexyl U, 1 + 1 mixture of A and U, in hydrogen-form, in deutero-form, in solution, and in solid. They have also shown that the thermodynamic quantities calculated from two different bonds are in essential agreement with each other, For the IR studies on the interaction of guanosine and cytosine derivatives, the spectral data in the 3500–3200 cm^{-1} region of the NH stretching vibration were used again (*257,260*).

The PMR studies on the association of the bases and nucleosides in organic solvents were dependent on the downfield shift of the —NH proton of G and U and the —NH_2 proton of A and C upon hydrogen bonding (*258,259,264*). As may be expected, the observed effect on —NH is usually larger than that for the —NH_2 in the association of the base pair (A + T, for instance). The concentration employed in the PMR studies was usually 10^2-fold higher than those for the IR studies. DMSO (dimethylsulfoxide) is the solvent most commonly used, even though mixtures of DMSO–DMF (dimethylformamide) (*258*) and $CDCl_3$ (with a small amount of DMSO) (*259*) have been employed. Since DMSO can serve as a powerful hydrogen-bond acceptor, the interaction of the bases in this solvent is much less than that observed in chloroform. In fact, the interaction of adenine and uracil derivatives is hardly detectable in DMSO but can be well-demonstrated in $CDCl_3$ (with a small amount of DMSO) by the PMR technique (*258,259*). Only one line was observed for both free and bonded protons of the —NH or —NH_2 group with a narrow linewidth, indicating there is sufficiently rapid exchange between the free form and the bonded form.

Interactions of many compounds under a variety of experimental conditions, such as total concentrations, ratios of interactants, temperature, etc., have been investigated. The most important conclusion obtained is that for the interaction of two compounds, only the interaction of the Watson–Crick complementary pairs [A·U(T) and G·C] can be observed. Self-association of these compounds can be observed but usually with the association constants 10–50 times less than those of the complementary pairs (see Table XXI). A careful study was carried out to investigate the possible interaction of 5-bromo U with G and 2,6-diaminopurine with cytosine (*262*), the compounds with the best chance to interact in violation of the Watson–Crick complementary scheme because of their strong tendency to associate. Interaction of these compounds was not found.

Based on a variety of arguments including some pertinent spectroscopic data, all the investigators concluded that these compounds associate to form cyclic dimers of 1 : 1 stoichiometry, at least in dilute solution. In the

TABLE XXI

Formation of Hydrogen-Bonded Molecular Complexes by the Bases and Nucleosides in Organic Solvents[a]

Compounds		Solvent	Association constants (liter/mole) Pair association	Self association	Temperature	Methods of investigation	Ref.
Guanosine	Cytosine, 5 aza-C 6 aza-C	$CHCl_3$	10^4–10^5[b]	—	25°	IR	*257, 260*
Adenosine	Uridine	CCl_4	600	A(80) U(115)	20°	IR	*256*
Adenine	3-Methyluracil	$CDCl_3$	1	A(3.1) 3-MU(0)	25°	IR	*261,262*
	5,6-Dihydrouracil	$CDCl_3$	30	DHU(2.9)	25°	IR	*262*
	4-Thiouracil	$CDCl_3$	90	4TU(2.7)	25°	IR	*262*
	Uracil	$CDCl_3$	100[c]	U(6.1)	25°	IR	*261,262*
	Thymine	$CDCl_3$	130	T(3.2)	25°	IR	*262*
	5-Bromouracil	$CDCl_3$	240[d]	5BrU(4.1)	25°	IR	*262*
	5-Iodouracil	$CDCl_3$	220	5IU(5.7)	25°	IR	*262*
Uracil	6-Dimethylaminopurine	$CDCl_3$	1.5	U(6.1) DMP(0)	25°	IR	*261,262*
	6-Methylaminopurine	$CDCl_3$	50	MAP(1.5)	25°	IR	*262*
	2-Aminopurine	$CDCl_3$	45	2AP(2.0)	25°	IR	*262*
	2,6-Diamino purine	$CDCl_3$	170[e]	DAP(2.0)	25°	IR	*262*
	8-Bromoadenine	$CDCl_3$	140	8BrA(120)	25°	IR	*262*
2-Aminopurine	5,6-Dihydrouracil	$CDCl_3$	10		25°	IR	*262*
	Thymine	$CDCl_3$	55		25°	IR	*262*
	5-Bromouracil	$CDCl_3$	75		25°	IR	*262*
2,6-Diaminopurine	5,6-Dihydrouracil	$CDCl_3$	100		25°	IR	*262*
	Thymine	$CDCl_3$	210		25°	IR	*262*
	5-Bromouracil	$CDCl_3$	550[f]		25°	IR	*262*

TABLE XXI—continued

6-Methylamino purine	5,6-Dihydro-uracil	$CDCl_3$	15		25°	IR	*262*
	Thymine	$CDCl_3$	70		25°	IR	*262*
	5-Bromo-uracil	$CDCl_3$	100		25°	IR	*262*
Guanosine	Cytidine	DMSO	3.6–4.0	G = ?[g] C = 0.1	32°	PMR	*264*

[a] Calculations of all the association constants are based on the formation of 1:1 stoichiometry of cyclic dimer (see text).

[b] The value of 1×10^5 was estimated by Pitha et al. (*257*) and the value of 1×10^4 was estimated by Kyogoku et al. (*260*).

[c] Estimated by Miller and Sobell (*263*) to be 150 ± 50.

[d] Estimated by Miller and Sobell (*263*) to be 350 ± 75.

[e] Estimated by Miller and Sobell (*263*) to be 300 ± 100.

[f] Estimated by Miller and Sobell (*263*) to be 600 ± 150.

[g] This association constant was not evaluated because of the uncertainty about the stoichiometry (*264*).

concentrated solution of A and U derivatives, Katz has good evidence for the formation of a complex with 1A : 2U stoichiometry (*259*, private communication). The association constants calculated on the basis of 1 : 1 cyclic dimer formation are shown in Table XXI. It is evident that blocking of the —NH or —NH_2 groups by methyl substitution results in a large reduction of association constants. Bromo-substitution of uracil [C(5)] and adenine [C(8)] increases the association constants, presumably by increasing the acidity of the —NH and the —NH_2 protons due to the electron-attracting power of the bromine atom (*262*). It is interesting to note that the association constant of 2,6-diaminopurine with uracil is only 1.7 times that of adenine (6-aminopurine). There is no definitive information about which nitrogen atom, N(7) or N(1), of adenine participates in the cyclic dimer formation. Kyogoku et al. suggested that these two nitrogen atoms can be equally involved (*262*). Similarly, no detailed information is known about the extent of participation of the C(4)═O group versus the C(2)═O group of uracil in dimer formation.

Thermodynamic quantities for self- and cross-association of 9-ethyladenine and 1-cyclohexyluracil in chloroform (*261*) and for the association of guanosine and cytidine in DMSO (*264*) have been obtained by measurement of association constants at varying temperatures. All log K vs. 1/T plots are reasonably linear. The $-\Delta H^\circ$ (kcal/mole) values for the formation of U·U dimer, A·A dimer, and A·U dimer are, respectively, 4.3 ± 0.4, 4.0 ± 0.8, and 6.2 ± 0.6, averaging about 2 kcal per hydrogen bond in the self-association and 3 kcal per hydrogen bond in the base pairing (or cross-association). The

$-\Delta S$ (e.u.) values for the formation of all these three types of dimers (U·U, A·A, and A·U) are essentially the same (respectively, 11.0 ± 1, 11.4 ± 2, and 11.8 ± 1.2). For the G·C system in DMSO, ΔH was found to be -5.8 kcal/mole, averaging about 2 kcal per hydrogen bond, and ΔS was found to be -16 e.u. (*264*). Thermodynamic quantities of these two different base pairs measured by two different spectroscopic techniques in two different solvent systems are in accord with each other. As mentioned in Section II, changes of UV absorption spectra (mostly hypochromic) due to base-pair formation in chloroform have been reported simultaneously and independently by Thomas and Kyogoku (*161*) as well as by Gratzer and McClare (*162*). These observations are important, though their significance is more pertinent to the problem of hypochromism in UV absorption spectroscopy than to the phenomenon of base pairing. Nevertheless, the interpretations of these UV absorption data in many respects rely upon IR and PMR data, and conclusions based on these data.

Kinetic studies of hydrogen bonding between 1-cyclohexyluracil and 9-ethyladenine in chloroform have been made by Hammes and Park by ultrasonic attenuation measurements (*265*). They found that the association rate constants of U·U formation ($1.5 \times 10^9\ M^{-1}\ \text{sec}^{-1}$ at 25°), of A·A formation ($\geqslant 2 \times 10^9\ M^{-1}\ \text{sec}^{-1}$ at 25°) and of A·U formation ($4.0 \times 10^9\ M^{-1}\ \text{sec}^{-1}$ at 20°) are similar even though their equilibrium constants for formation are quite different, They interpreted the magnitudes of these rate constants to indicate that in all three cases the association is essentially diffusion controlled. The dissociation rate constants were reported to be $25 \times 10^7\ \text{sec}^{-1}$, $\geqslant 60 \times 10^7\ \text{sec}^{-1}$, and $3.2 \times 10^7\ \text{sec}^{-1}$, respectively, for U·U pair, A·A pair, and A·U pair. These authors suggested that these dissociation rate constants are directly related to the strength of the hydrogen bonds of these pairs.

The discourse in this section clearly indicates that the specificity in the Watson–Crick base-pairing scheme for DNA does not rely on the constraint of the polynucleotide backbone in the double helix nor on the packing force in a crystalline solid. The specificity in base pairing resides in the bases themselves as part of their intrinsic properties which can be displayed in their association in solution. Theoretical calculations have been made by Bradley and Nash (*51,52*, for a summary see Ref. *266*) and by Pullman's group (*267,268*, for a summary see Ref. *49*) in attempts to explain these experimental observations. These efforts have been rather successful to a certain extent. Pullman and Pullman (*49*) have suggested that the origin of the specificity of the base pairing resides in the relatively high stability of the self-association process especially those of C·C and of G·G. They proposed that the G·C and A·T pairs can be formed only because these are the only two pairs which have higher interaction energy to overcome the stability of G·G and C·C pairs in solution, or the A·A and T·T pairs in solution, respectively. For instance, in their calculations, the A·C pair (−7.8 kcal/mole) and the A·G pair (−7.5 kcal/mole) have interaction energies higher than the A·T pair (−7.0 kcal/mole). However, they wished to point out that only the A·T pair has higher energies than the A·A pair (−5.8 kcal/mole) and the T·T pair (−5.2 kcal/mole), while the A·C pair has lower energy than the C·C pair (−13 kcal/mole), and the A·G pair has lower energy than the G·G pair (−14.5 kcal/mole). Therefore, only A·T pairs can be formed in the mixture of A and T, while no A·C pair or A·G pair can be formed in the mixture of A and C or mixture of A and G, respectively. Similarly, the energy of the G·C pair (−19.2 kcal/mole) is sufficiently large to overcome that of the C·C pair and G·G pair. This explanation appears to be applicable for the monomer–monomer interaction; however, it appears not to be adequate for polymer–oligomer interaction or polymer–polymer interaction. It is possible that stereochemical contributions to the specificity of interaction should be considered at the polymer level, although they can be temporarily ignored at the monomer level.

C. Interaction of the Monomers in Water

Studies of the interaction of bases, nucleosides, and nucleotides in aqueous solutions provide some most important information about the properties of nucleic acid. Generally, the research on this problem can be separated into thermodynamic studies and spectroscopic studies.

1. Thermodynamic Studies

a. *Osmotic Coefficients, Activity Coefficients, and Free Energy of Association*

Through measurements of vapor pressure osmometry (VPO), the osmotic coefficients, ϕ, of 16 bases and nucleosides in aqueous solutions at 25° and at varying concentrations have been determined in our laboratory (*64,86, 269,270*), as shown in Table XXII. Very recently, Solie and Schellman (*271*) have published osmotic coefficient measurements of deoxyadenosine, inosine, ribosylpurine, deoxythymidine, deoxycytidine, and uridine, as well as ϕ values of equimolal mixtures of deoxy (C+T, A+T, A+C, G+C) and of deoxyadenosine + uridine, Their ϕ values of the pyrimidine compounds such as thymidine, deoxycytidine, and uridine are in satisfactory agreement with those of thymidine, cytidine, and uridine listed in Table XXII. Small differences were observed between the ϕ values of ribosylpurine reported from these two laboratories. However, substantial deviations were found in the ϕ values of deoxyadenosine and inosine obtained by Solie and Schellman as compared to those reported from our laboratory. For deoxyadenosine, at

TABLE XXII

Molal Osmotic Coefficients (ϕ) at 25°C

	Molal Concentration															
	0.025	0.05	0.10	0.15	0.20	0.25	0.30	0.35	0.40	0.45	0.50	0.60	0.70	0.80	0.90	1.00
Uridine[a]	—	0.969	0.943	0.931	0.901	0.883	0.866	0.849	0.833	0.817	0.801	0.773	0.775	—	—	—
Cytidine[a]	—	0.967	0.935	0.905	0.876	0.850	0.826	0.804	0.785	0.768	0.752	0.724	0.695	—	—	—
Thymidine[b]	—	—	0.905	—	0.865	—	—	—	—	—	—	—	—	—	—	—
5-Bromouridine[c]	—	0.948	0.894	0.844	0.801	0.766	0.738	0.715	0.693	—	—	—	—	—	—	—
Inosine[b]	0.994	0.957	0.888	0.830	—	—	—	—	—	—	—	—	—	—	—	—
1-Methylinosine[b]	0.962	0.926	0.860	0.800	0.750	—	—	—	—	—	—	—	—	—	—	—
7-Methylinosine[e]	—	0.975	0.875	0.850	—	—	—	—	—	—	—	—	—	—	—	—
Ribosylpurine[b]	0.965	0.930	0.860	0.810	0.770	—	0.710	—	—	—	—	—	—	—	—	—
Purine[a]	—	0.917	0.849	0.794	0.749	0.714	0.685	0.662	0.643	0.627	0.614	0.590	0.567	0.544	0.522	0.505
Adenosine[b]	0.915	0.836	0.740[d]	—	—	—	—	—	—	—	—	—	—	—	—	—
2′-O-Methyl-adenosine[b]	0.908	0.828	0.723	0.658	0.611	—	0.550	—	—	—	—	—	—	—	—	—
2′-Deoxy-adenosine[b]	0.900	0.800	0.668	0.598	—	—	—	—	—	—	—	—	—	—	—	—
6-Methylpurine[c]	—	0.786	0.682	0.624	0.582	0.544	0.510	0.484	0.469	0.461	0.456	0.427	0.410	—	—	—
N-6-Methyl-adenosine[b]	0.805	0.685	0.558	0.480	—	—	—	—	—	—	—	—	—	—	—	—
N-6-Methyl-2′-deoxyadenosine[b]	0.790	0.680	0.540	0.468	—	—	—	—	—	—	—	—	—	—	—	—
N-6-Dimethethyl-adenosine[b]	0.712	0.608	0.470	0.408	0.378	—	—	—	—	—	—	—	—	—	—	—

[a] From Ref. *269*.
[b] From Ref. *64*.
[c] From Ref. *270*.
[d] At solubility limit of 0.085 *M*.
[e] From Ref. *86*.

0.05 M, $\phi = 0.72 \pm 0.03$ was reported by Solie and Schellman (*271*) and $\phi = 0.80 \pm 0.005$ listed in Table XXII (see the references cited in the table for the deviation of the measurement). For inosine at 0.05 M, $\phi = 0.873 \pm 0.03$ was reported by Solie and Schellman and $\phi = 0.957 \pm 0.02$ is listed in Table XXII. Unfortunately no elemental analysis was reported in the paper of Solie and Schellman (*271*) and their calculations were apparently based on the assumption that the compounds were anhydrous. If the commercial deoxyadenosine used by them is indeed in monohydrate form as in our case, then their value of ϕ recalculated on the basis of 2′-deoxyadenosine monohydrate is in accord with ours, i.e., 0.05 M, $\phi = 0.77 \pm 0.03$. It should be noted that our data on 2′-deoxyadenosine is very close to that for adenosine and 2′-O-methyladenosine, as may be expected. In the case of inosine, our compound is anhydrous. If the inosine used by Soiie and Schellman was the monohydrate (this form is also available commercially), then the recalculated value of Solie and Schellman (0.936 ± 0.036) is again in agreement with ours. The real reason for the discrepancies is not certain at present.

Compounds such as purine, 6-methyl purine, uridine, 5-bromouridine, and cytidine are more soluble than others and therefore they can be studied over a larger concentration range. Activity coefficients for these more soluble compounds at 25° were calculated from the osmotic coefficients by the Gibbs–Duhem relationship using a computer which performed a numerical integration on the fitted polynomials and related molal concentration to ϕ (Table XXIII). The data in Table XXIII clearly indicate that the properties of the bases and nucleosides in solution are far from ideal. Values of both osmotic coefficients and activity coefficients are much below unity. These results establish the concept that purine and pyrimidine nucleosides do interact extensively in aqueous solution.

After further analysis for their congruence to different models for multiple equilibria, the thermodynamic data were found to be incompatible with the model which assumes that only dimers are formed (Ts'o et al., *269*). Thus the degree of association of these compounds may go beyond the dimer stage to a higher degree of polymerization. Most of the results are consistent with the model which assumes that the association process continues through many successive steps (at least more than five steps) with the same equilibrium constant (*269,270*). The apparent equilibrium constant, K, for the association at various steps can be obtained by the following equation:

$$K = \frac{1-\phi}{m\phi^2} \qquad (7)$$

where m is the concentration in molality. For more complicated situations, a knowledge of activity coefficients is also required for the mathematical analysis of the multiple equilibria (*270*). Comparison of the equilibrium

TABLE XXIII

Molal Activity Coefficients[a] at 25° Computed from the Fitted Osmotic Coefficients[b]

Molal conc.	Purine	6-Methyl-purine	Uridine	5-Bromo-uridine	Cytidine
0.05	0.844	0.626	0.939	0.902	0.936
0.10	0.728	0.469	0.888	0.811	0.878
0.15	0.641	0.385	0.845	0.732	0.824
0.20	0.575	0.329	0.808	0.666	0.776
0.25	0.522	0.287	0.775	0.613	0.733
0.30	0.480	0.255	0.744	0.569	0.695
0.35	0.446	0.230	0.716	0.533	0.661
0.40	0.418	0.211	0.690	0.502	0.631
0.45	0.394	0.196	0.665		0.604
0.50	0.374	0.185	0.641		0.580
0.55	0.355	0.173	0.620		0.558
0.60	0.339	0.162	0.600		0.537
0.65	0.324	0.152	0.582		0.518
0.70	0.311	0.146	0.568		0.499
0.75	0.297				
0.80	0.286				
0.85	0.275				
0.90	0.264				
0.95	0.255				
1.00	0.247				
1.05	0.240				
1.10	0.235				

[a] Data from Refs. *269* and *270*.
[b] See Table XXII.

constants and thus the standard free energy changes of these bases, nucleosides, and urea are given in Table XXIV. One immediate general conclusion is that the tendency of purine to associate is much greater than that of pyrimidine nucleosides, which in turn is greater than that of urea. Solie and Schellman (*271*) have also applied Eq. (7) (this equation was first derived by Schellman, Ref. *272*) for the analysis of their data. The results are similar to those in Table XXIV, except that their K value for deoxyadenosine is higher for above discussed reasons concerning the osmotic coefficient measurements.

Helmkamp and Kondo (*273*) have reported the osmotic coefficients of alkyl-substituted purines. They have also applied Eq. (7) to analyze their VPO data and obtained K values (molal^{-1}) for the following compounds:

TABLE XXIV

Equilibrium Constants and the Standard Free Energy Change of the Association for the Following Compounds at 25°C

	K (molal^{-1})	ΔF° ($RT \ln k$, cal)
Urea[a]	0.041	+1190
Uridine[b]	0.61	+290
Cytidine[b]	0.86	+80
Thymidine[c]	0.91	+60
5-Bromouridine[d, e]	$K_1 = 1.0$	0
	$K = 2.9$	−630
Ribosylpurine[f]	1.91	−380
1-Methylinosine[f]	1.8–2.0	−360 to −410
Purine[f]	2.1	−440
Adenosine[f]	4.5	−900
2′-O-Methyladenosine[f]	5.1	−970
6-Methylpurine[f]	6.7	−1120
2′-Deoxyadenosine[f]	7.5–4.7	−(1195 to 920)
N-6-Methyladenosine[f]	14.9–11.8	−(1600 to 1460)
N-6-Methyl-2′-deoxyadenosine[f]	15.9	−1640
N-6-Dimethyladenosine[f]	22.2	−1840

[a] Cited from Ref. *272*.
[b] Cited from Ref. *269*.
[c] Cited from Ref. *271*.
[d] Cited from Ref. *270*.
[e] The treatment of multiple equilibria of this compound requires two equilibrium constants. K_1 is for the first step and $\bar{K}$ is for the successive steps. See original paper.
[f] Cited from Ref. *64*.

9-methylpurine (1.9); 9-ethylpurine (2.1); 9-isopropylpurine (2.3); 9-*t*-butylpurine (2.5); 6-*t*-butylpurine (8.7); 2-*t*-butylpurine (9.2); and 8-isopropylpurine (4.4). These K values are to be compared with the K value (2.1) of the unsubstituted purine (Table XXIV). The significance of this comparison will be discussed in later paragraphs of this section.

What is the mode of association of these molecules in aqueous solution? Do they associate with each other vertically through hydrophobic and stacking interactions, or do they associate horizontally through hydrogen bonding? The thermodynamic data of these compounds conclusively do not support the hypothesis of horizontal association through hydrogen bonding because of the following reasons:

(1) All these bases and nucleosides associate much more extensively than urea which is already known to be one of the best hydrogen-bonding agents in water.

(2) Methylation enhances association. From the values of K of association (Table XXIV) and ϕ (Table XXII), the order of association tendencies with respect to the degree of methylation can be listed as follows: 2′-deoxy-adenosine < N-6-methyl-2′-deoxyadenosine; adenosine < N-6-methyl-adenosine < N-6-dimethyladenosine; and inosine < 1-methylinosine. In every case examined, substitution of a hydrogen of the bases by a methyl group removes a hydrogen-bond donor and the association tendency is enhanced significantly. 1-Methylinosine and N-6-dimethyladenosine, in spite of the fact that the hydrogen-bond donor sites of those two compounds have been completely removed by methylation, do associate substantially more than inosine or adenosine, respectively. It is interesting to note that in order to obtain the promoting effect of methylation, the methylation has to take place at the base and not at the pentose. Thus, the association tendency of 2′-O-methyladenosine is about the same as that of adenosine and may be slightly less than that of 2′-deoxyadenosine. The 9-alkylpurines do not have a hydrogen available as hydrogen-bond donor, and yet these compounds all have association constants, K, substantially higher than that of urea.

The self-association of purine in aqueous solution has been studied by van Holde and Rossetti (*274*) using the sedimentation equilibrium technique. Their results are in agreement with the VPO data in that a simple reversible association is involved, with a constant free energy increment for the addition of each successive purine molecule. The value of the association constant that they reported (2.8 M^{-1}) is slightly higher than that (2.1 M^{-1}) obtained from the VPO data (Table XXIV). Solie and Schellman (*272*) have also studied the self-association of deoxyadenosine, ribosylpurine, inosine, and thymidine by sedimentation equilibrium. The K values for the self-association obtained from their analysis of the data are in reasonable agreement with those from their VPO data.

The investigation detailed above concerns solutions containing a single kind of solute, a two-component system of self-association. What are the interactions between different compounds, especially between purine and pyrimidine? To obtain precise information from a three-component system is more difficult. The increase in solubility of the sparingly soluble adenine and thymine caused by the presence of the highly soluble purine or nucleoside was adopted in our laboratory as the method for investigation of interactions of this type (*269*). It was found that the solubility of adenine is much enhanced by the presence of purine. The enhancement is moderate in the presence of cytidine, uridine, and pyrimidine, and is practically nil in the presence of cyclohexanol, adonitol, and urea. Similarly, the solubility of thymine is enhanced by purine and to a lesser extent by uridine and cytidine. These data were also analyzed by multiple equilibria treatment. The assumption in the treatment is that the base interacts to the same extent with the free and the associated forms of the interactants. Equilibrium constants for such interactions between different compounds have been calculated and the general conclusions can be summarized as follows: Interactions between purines (i.e., 6-methylpurine with 6-methylpurine, purine with purine, purine with

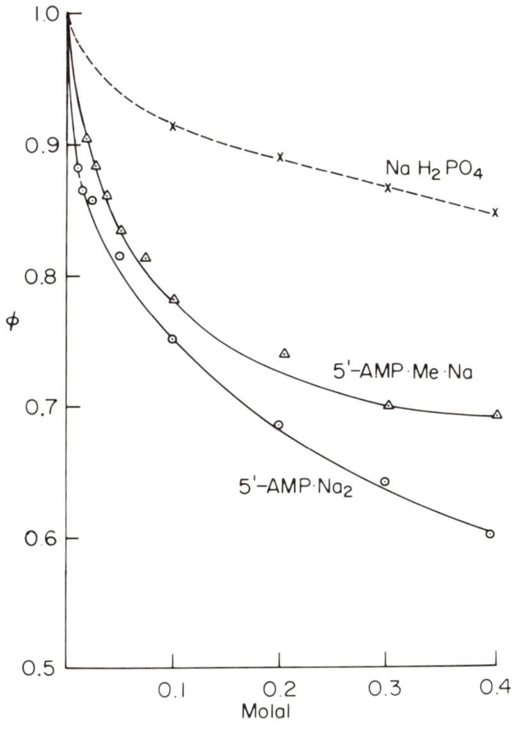

Fig. 11. The osmotic coefficient, ϕ, vs. molal concentration (m) for NaH_2PO_4 (···×···×···), 5′-AMP monomethyl ester sodium salt (—△—△—), and 5′-AMP disodium salt (—⊙—⊙—) (from Ref. *97*).

adenine, etc.) are stronger than interactions of purine with pyrimidine (purine with thymine; uridine or cytidine with adenine, etc.) which are in turn stronger than the interactions of pyrimidine with pyrimidine (cytidine with cytidine, uridine with uridine, thymine with uridine, etc.). These compounds all interact much more extensively, however, than does urea.

As mentioned above, Solie and Schellman (*272*) have studied the solution properties of equimixtures of purine nucleoside and pyrimidine nucleoside such as dA + T. Their qualitative conclusion on these mixtures is the same as cited above, i.e., interaction of purine–purine is greater than the interaction of purine–pyrimidine, which, in turn, is greater than the interaction of pyrimidine–pyrimidine.

The studies described so far, concern only the interaction of the uncharged molecules, such as the neutral bases and nucleosides in aqueous solutions. How about the ionized monomers such as the nucleotides? In Fig. 11, the osmotic coefficients (ϕ) of adenosine 5′-phosphate monomethylester, 5′-disodium AMP, and monosodium dihydrogen phosphate in various molal concentrations are presented (*97*). The values for the nucleotides were obtained with the assumption that these compounds are fully ionized. Clearly, the ϕ value of the 5′-AMP methyl ester is considerably lower than that of NaH_2PO_4. This is strong evidence of the association of the 5′-AMP methyl ester. Unfortunately, these data cannot be treated quantitatively without making an assumption about the ionization of the sodium phosphate part of 5′-AMP with respect to the change of concentration. Recently, by the method of sedimentation equilibrium, Rossetti and van Holde (*275*) have found an increase of weight-average "apparent molecular weight" of 5′-AMP (monoanion) for concentration of 0.05 *M* (40% increase) and higher (up to about 110% increase). It is therefore virtually certain that AMP does stack in aqueous solution as does adenosine. The effect of phosphate substitution in the association has been studied by the PMR method and discussed in later paragraphs.

The effects of a large variety of salts on the solubility of bases and nucleosides have been studied by Robinson and Grant (*276*). They have analyzed the data by the use of the Setschenow equation:

$$\log \frac{S_i^\circ}{S_i} = \log \frac{f_i}{f_i^\circ} = K_s C_s \ldots \tag{8}$$

where S_i° and S_i are the solubilities of the nonelectrolyte, *i*, in water or in salt solution, respectively; f_i° and f_i are the molar activity coefficients in water and in salt solutions, respectively; K_s is the salting-out constant, and C_s is the salt concentration in moles per liter. The effect on the solubilities observed with different salts is mainly due to the anions, and the order of effectiveness of different salts on all of the bases is nearly the same. Anions such as tosylate, trichloroacetate, thiocyanate, perchlorate, etc., increase the solubility of the bases and nucleosides (K_s values ranging from -0.12 to -0.6) and anions such as sulfate and phosphate, etc., decrease the solubility of the bases and nucleosides (K_s values ranging from $+0.04$ to $+0.45$). There is a good correlation between the effect of concentrated salt solutions on the activity coefficients (or solubilities) of the bases and their effect on the denaturation of DNA. Sodium salts of anions such as Cl_3COO^-, SCN^-, ClO_4^-, etc., are potent denaturants of DNA. These results indicate that the denaturation of DNA by these concentrated salt solutions can be accounted for by their effects on the activity coefficients of the bases and is attributable to a greater degree of exposure of the bases to solvent in denatured DNA than in the native form.

b. *Enthalpy and Entropy of Association*

Enthalpy and entropy of the self-association of seven bases and nucleosides have been determined by Gill and co-workers (*277–279*) by microcalorimetry in conjunction with the osmotic coefficients obtained from the VPO data. Van Holde and Rossetti (*274*) have also measured the enthalpy and entropy of self-association of purine by the use of the van't Hoff equation which

TABLE XXV

Thermodynamic Parameters of Self-Association of Bases and Nucleosides in Water

Compound	$\Delta H°$, kcal	$\Delta F°$, kcal	$\Delta S°$, eu	$K°$, m^{-1}
Purine[a, b]	-4.2 ± 0.2	-0.44	-13	2.1
6-Methylpurine[a]	-6.0 ± 0.4	-1.12	-16	6.7
Ribosylpurine[c]	-2.5 ± 0.1	-0.38	-7	1.9
Deoxyadenosine[d]	-6.5 ± 1.0	-1.00	-18	$4.7-7.5$
Caffeine[a]	-3.4 ± 0.2	-1.5	-6	13 ± 1
Cytidine[e]	-2.8 ± 0.1	0.08	-10	0.87
Uridine[e]	-2.7 ± 0.1	0.29	-10	0.61
Thymidine[f]	-2.4 ± 0.3	0.06	-9	0.91

[a] Data from Refs. *277* and *269*.

[b] The values reported by van Holde and Rossetti (*274*) based on van't Hoff relationship are the following: ΔH, -4.9 kcal/mole; $\Delta F = -0.61$ kcal/mole; and $\Delta S = -14$ cal/degree/mole.

[c] Data from Refs. *278* and *64*.

[d] The values reported by Gill et al. originally were based on the osmotic coefficient data of Solie and Schellman (*271*). The values reported here are obtained by a recalculation of the calorimetry data of Gill et al. based on the VPO data reported in Tables XXII and XXIV. The discrepancy between the VPO data of Solie and Schellman and those from our laboratory have been discussed elsewhere.

[e] Data from Refs. *278* and *269*.

[f] Calculated from data in Ref. *278* and in Table XXIV.

relates the equilibrium constant to $1/T$. The value of $\Delta H^\circ \approx -4$ kcal/mole for the association of purine was obtained by both procedures, in satisfactory agreement with each other. A summary of the data from the work of Gill and co-workers and from the VPO studies is presented in Table XXV. The significance of these results are discussed later in this section, after the presentation of the spectroscopic data. Heats of vaporization of several bases have been determined by Clark et al. (*146*) as described in Section II, E. These values are in the range of 20–30 kcal/mole. The heats of solution for 9-methyladenine and 1-methylthymine measured by Gill et al. are 6.9 and 5.3 kcal/mole, respectively. Clark et al. (*146*) have calculated the heat of solution of adenine to be 8 kcal/mole based on our solubility data at two temperature levels (*269*). Therefore, Clark et al. pointed out that the process of "purine or pyrimidine in gas phase → purine or pyrimidine in aqueous solution" has a large negative ΔH of about 20 kcal/mole.

2. Spectroscopic Studies and other Measurements

The extensive interaction of GMP in gel formation was first investigated by Gellert, Lipsett, and Davies (*280*). They showed that the optical density and the optical rotation of 3′-GMP and 5′-GMP, but not 2′-GMP, are very sensitive to temperature in the range of 0° to 40°. The decrease in negative rotation and the increase in absorbancy in response to the elevation of temperature from 0° upward indicates the transition of states from an ordered structure to a disordered one. X-Ray diffraction patterns from fibers drawn from the gels of 5′-GMP and of 3′-GMP confirmed the above conclusion and provided more elaborate information about the helical structure of the GMP gel which involves both hydrogen bond pairing and stacking of bases. Recently, a refinement of the originally proposed structure was reported by Davies (*6*). 3′-GMP forms square coplanar stacks involving four molecules in its helical complex with a fourfold rotation axis parallel to the helical axis, while 5′-GMP forms a structure with a continuous hydrogen-bonded helix with 15 bases in four turns of the helix. In both cases, each GMP has four pairs of hydrogen bonds with its two nearest planar (or nearly so) neighbors. Each molecule serves both as donor and acceptor of the following two pairs of hydrogen bonds: NH_2(2) to N(7); and N(1)—H to 6-keto. Sarkar and Yang (*281*) reported the ORD pattern and the UV absorption spectrum of the 5′-GMP gel near 0° and at room temperature. The transition of ordered structure to disordered state can be clearly demonstrated. Most interestingly, the ORD curve of 5′-GMP gel is similar to that of poly I in showing a negative Cotton effect while all other nucleic acids and polynucleotides show a positive Cotton effect. The explanation for this difference is not yet certain though it may be related to the fact that the GMP helix has a left-handed helical turn.

Helical formation of isoguanosine (2-keto-6-amino-9-β-D-ribofuranosylpurine) has been reported by Ravindranathan and Miles (*282*) as indicated by viscosity measurements, UV and IR spectra, as well as by optical rotation studies. The ordered structure of this nucleoside exhibits a positive Cotton effect.

Solie and Schellman (*271*) cited in their paper, the concentration dependency of the optical rotatory properties of deoxyadenosine. However, no optical activity was detected even for a 1 *M* purine solution at 5° (unpublished results of Ts'o and Sander). Lord and Thomas (*283*) have studied the Raman spectra of aqueous solutions of complementary base pairs at

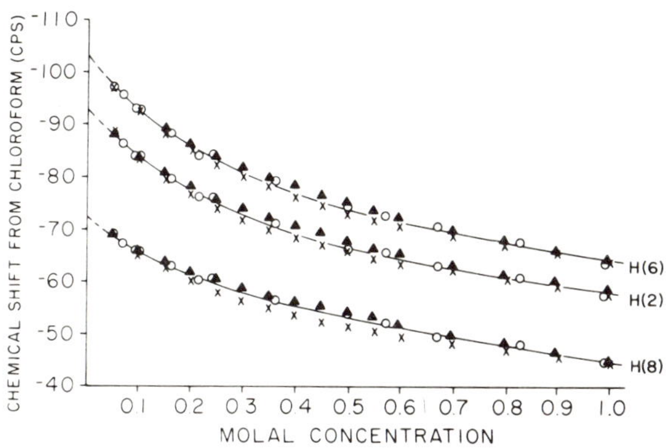

Fig. 12. Concentration dependence of the proton chemical shifts for purine in aqueous solution at 25° (corrected for bulk susceptibility); shifts measured from external chloroform reference: —○—, experimental values; —×—, calculated values from overall average model; —▲—, calculated values from statistical partial-overlapping model (from Ref. *84*.)

neutral pH or pD up to 1 *M* concentration. No evidence for specific base-pairing interactions have been found in these aqueous solutions, even though, as described in Section III, B, specific base-pairing takes place readily in organic solvents.

By far, the most informative results came from PMR studies which provided elaborate information about the extent and the mode of association of purine nucleosides. As discussed in Section II, C, nuclear magnetic shielding is a very sensitive probe of inter- and intramolecular interactions. In this case, vertical stacking interactions are easily distinguished from hydrogen bonding interactions since these interactions manifest themselves differently in the concentration dependence of the PMR spectra. The first extensive study was the PMR spectra of purine over the concentration range of 0.05 to 1 *M* (*84*). Chemical shifts of the three protons in purine versus the concentration are shown in Fig. 12. A pronounced concentration effect has been observed. Proton resonances in purine are all shifted to higher fields as the solute concentration is increased. Shifts to high fields with concentration are well known for aromatic systems and are generally attributed to the magnetic anisotropy associated with the ring currents in neighboring molecules. Because of the mobile electrons, a large diamagnetic current is induced in the plane of the ring by an external magnetic field when the field is perpendicular to the plane of the molecule. This ring current gives rise to a small secondary

TABLE XXVI

Concentration Dependence of Chemical Shifts for 11 Purine Nucleosides [(0.0–0.2 M) in D_2O]

Compound	Temp. (C°)	$\Delta\delta$, cps[e]				
		H(2)	H(8)	H(6)	H(1′)	—CH_3
Inosine[b]	32	6.4	5.3		7.1	
1-Methylinosine[a]	33	8.9	6.4		6.8	5.3
Ribosylpurine[b]	30	10.7	6.4	13.1	8.8	
Purine[c]	25–27	12.6	9.6	14.2		
2′-O-Methyladenosine[b]	31	13.7	7.5		8.8	
6-Methylpurine[c]	25–27	19.4	13.3			17.0
2′-Deoxyadenosine[b]	30	19.8	13.0		13.6	
N-6-Methyl-2′-deoxyadenosine[b]	32	26.0	15.8		14.0	15.2
N-6-Dimethyladenosine[b]	28	27.2	14.5		14.4	25.5
N-6-Methyladenosine[b]	26	32.6	17.5		12.6	18.1
2′-Deoxyadenosine[b, d]	30	14.8	10.0		9.8	
Adenosine[b, d]	32	14.8	8.3		6.9	
3′-Deoxyadenosine[b, d]	25	15.8	9.0		9.6	

[a] Peak positions of H(8) and H(2) are reversed with respect to the other 6-substituted nucleosides studied.

[b] Cited from Ref. *64*.

[c] PMR data of the two compounds are from Ref. *84*.

[d] Differences measured over the concentration range 0–0.10 M because of solubility limitation.

[e] From a 60-Mc instrument.

magnetic field which reinforces the primary field at the peripheral protons in the plane of the ring. In the region directly above and below the molecular plane, however, the two fields are opposed. As the concentration of a solution of aromatic molecules is increased, the average distance between molecules decreases and the protons of a given molecule will feel the secondary magnetic fields produced by the ring current of neighboring molecules. Since it is much more probable to find the molecules somewhere above or below the molecular plane of another aromatic molecule due to the dish-shaped nature of the aromatic molecules, this magnetic anisotropy of the ring-current effect will lead to a high-field shift with concentration or to a low-field shift upon dilution. At higher temperatures, or when the purine is dissolved in organic solvents such as dimethylformamide, such concentration-dependent chemical shifts for the purine protons are greatly reduced (*84*). Furthermore, when the purines are protonated by hydrochloride so that they cannot associate because they carry a positive charge, such concentration-dependent chemical shifts are again practically eliminated (*84*). These data clearly suggest that the mode of association of purine is by the vertical stacking of rings in a partially overlapping fashion. As described above, the osmotic coefficients and activity coefficients of purine have been interpreted in terms of multiple equilibria and on this basis, populations of various associated species at varying concentrations were computed. Based on these population distributions of the associated species, we can calculate the concentration dependence of the chemical shifts, which is also given in Fig. 12 (*84*). It can be seen that the calculated value and the experimental values are in satisfactory agreement. Therefore, a numerical correlation between the PMR data and the osmotic data has been successful in the sense that they reinforce and support the interpretations of the other.

Similar results have been obtained from the purine nucleosides (*64*), especially the adenine nucleoside series (Table XXVI). The close correlation between the magnitudes of $\Delta\delta$, the concentration dependence of the chemical shifts and the values of ϕ and K (Tables XXII and XXIV) for this series of purine nucleosides in solution not only establishes that stacking is the mode of association of these solutes but also verifies the usefulness of PMR as a tool for studies of association. The differential in the magnitude of $\Delta\delta$ for various protons of a given nucleoside may provide additional information about the average geometry of the stacks in solutions. For instance, in the case of 2′-deoxyadenosine (Fig. 13), the concentration dependence of the chemical shift is even larger than that of purine (Fig. 13). In all these cases, the H(2) proton of the six-membered ring of the adenine is shifted to the higher field than the H(8) proton in the five-membered ring. This indicates that the six-membered ring of the adenine does participate to a greater extent in the stacks than the five-membered ring of the adenine nucleosides. The H(1′) pentose protons are also considerably shifted to higher fields when the concentration is increased while the H(5′) pentose protons are hardly affected. As one proceeds around the pentose ring from C(1′) to C(5′), there is a progressive drop or decrease in the magnitude of these concentration-dependent chemical shifts. This indicates that adenine nucleoside interaction is preferentially localized at the purine base of the nucleoside so that the

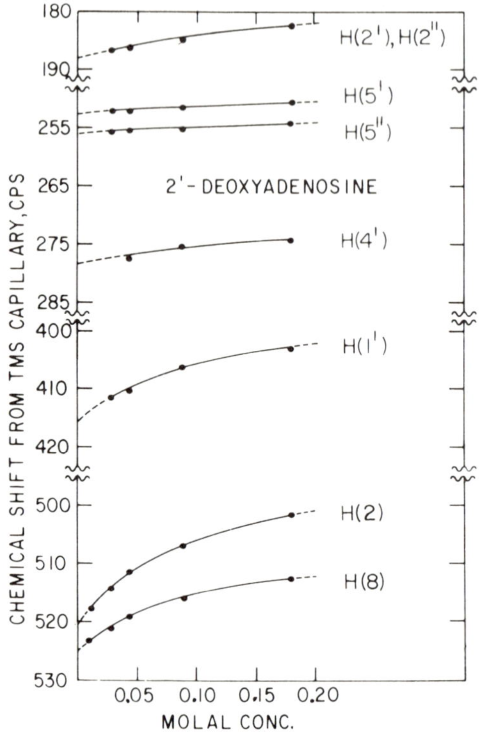

Fig. 13. Concentration dependence of the proton chemical shifts for 2′-deoxyadenosine (from Ref. *64*).

ring-current magnetic anisotropy is principally felt by the base protons. From this type of study, therefore, not only can we obtain a general picture about the mode of association, but we can even get down to the detailed molecular structure of the stacks. Two models [Figs. 14(a) and 14(b)] for the preferred average orientation of the nucleoside bases in the stacks have been proposed by our laboratory to account for the observed results (*64*). In these dimer models, one can see that the H(2) proton will be shielded strongly by the six-membered ring of the neighboring bases most of the time. The H(8) proton and the H(1′) proton will be shielded mainly by the five-membered ring. In addition, according to these models, the H(8) and H(1′) protons will spend less time on the average in the proximity of the five-membered ring than the H(2) will in the proximity of the six-membered ring. These two models are related to each other by symmetry considerations. The base-stacking arrangement for model Fig. 14(a) is face-to-back and that for Fig. 14(b) is face-to-face (or back-to-back), one of the nucleosides being rotated by 180° along an axis in the plane bisecting the C(4)—C(5) bond.

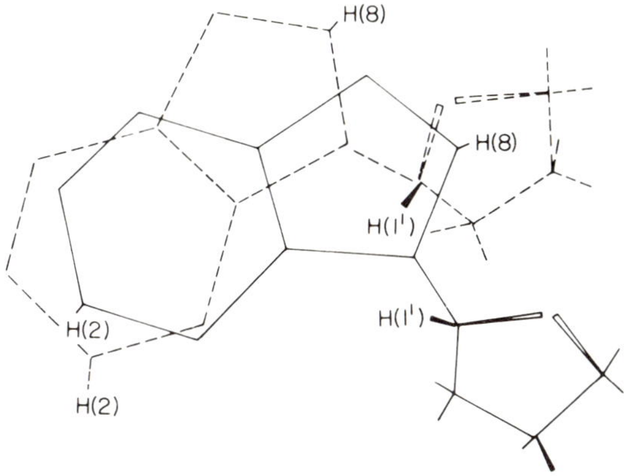

(a)

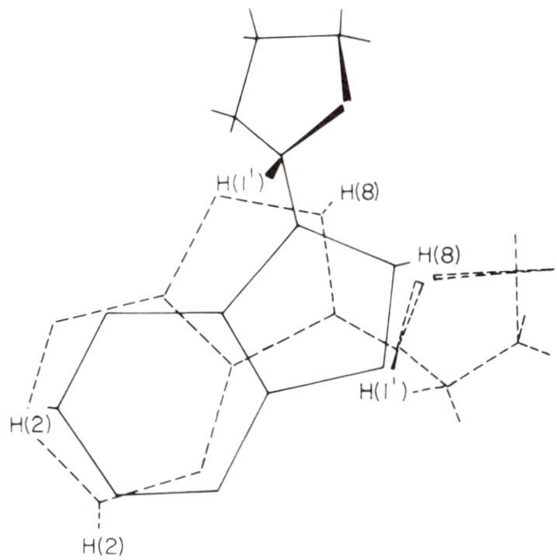

(b)

Fig. 14. Models illustrating the proposed arrangements for two nucleosides in stacks. (a) Straight stack (face-to-back); the ribosyl substituents at N(9) are on the same side of the stack. (b) Alternate stack (face-to-face or back-to-back): one of the nucleosides is rotated by 180° along an axis in the plane bisecting the C(4)—C(5) bond. The ribosyl substituents are opposite to each other in the stack (from Ref. *64*).

Thus, in model Fig. 14(a) the two ribosyl substituents at N(9) will be on the *same side* of the dimer (straight stack). On the other hand, in model Fig. 14(b), the two ribosyl moieties at N(9) will be *opposite* to each other in the dimer resulting in an alternating arrangement (alternate stack). The alternate

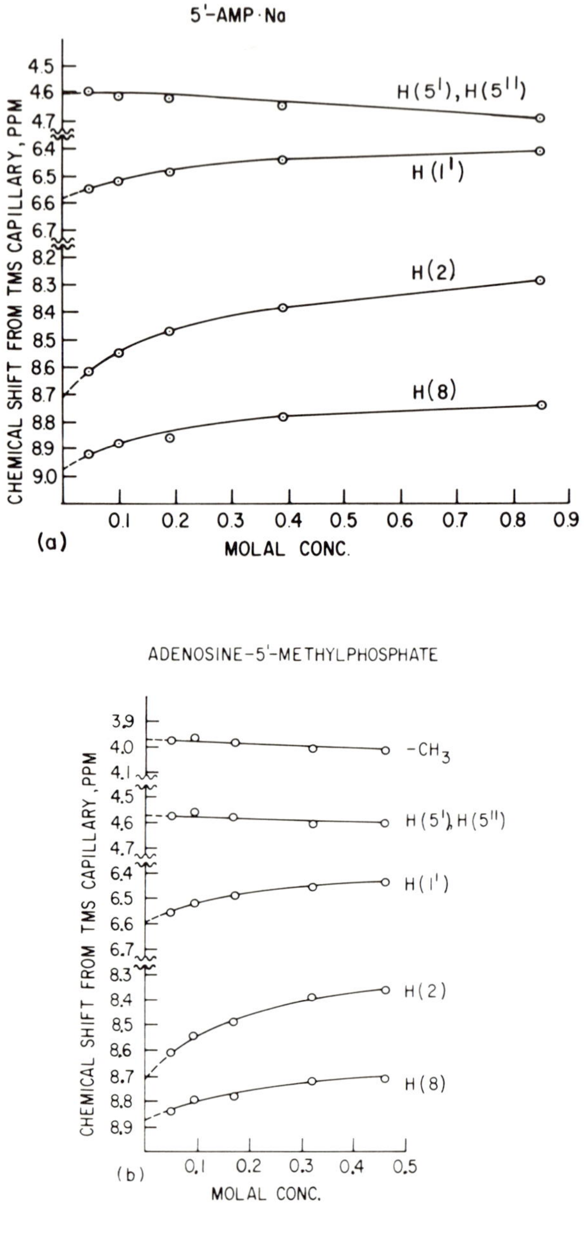

Fig. 15. (a) Chemical shift vs. molal concentration for 5′-AMP(Na) protons; 30°, D_2O, pD 5.9. (b) Chemical shift vs. molal concentration for protons of adenosine 5′-methyl phosphate [5′-MeAMP(Na)]; 31°, D_2O, pD 5.9 (from Ref. *97*).

stack [Fig. 14(b)] may be favored based on three considerations: (i) the steric hindrance of the ribosyl group is reduced; (ii) this arrangement reduces the repulsion between the dipole moments of the adjacent bases expected in the straight stack; (iii) finally, the concentration dependencies, $\Delta\delta$, of the chemical shifts of H(2), H(8), and H(1′) of monosodium (pD 5.9) 5′-AMP [Fig. 15(a)], of monomethyl-5′-AMP [Fig. 15(b)], and of disodium 5′-AMP (pD 7.4) were found to be about the same as described in later paragraphs, indicating the negligible effect of electrostatic repulsion in the stack. This observation suggested that the phosphate groups of the 5′-nucleotides are as far apart as they would be in the case of the model of Fig. 14(b). In addition, the chemical shifts of the methyl protons of adenosine 5′-methyl phosphate [Fig. 15(b)] are not sensitive to concentration. This is another indication that the phosphate groups of the nucleotide stacks are far away from any diamagnetic moieties such as the purine rings and the hydroxyl groups. These two observations, therefore, strongly favor the "face-to-face" model of Fig. 14(b), at least for the stacks of 5′-AMP.

Two qualifying statements should now be mentioned in the discussion of these models. (a) The standard free energy change for the association is of the order of the thermal energy kT, and no significant line broadening or separate shifts for stacked and free species were observed in the PMR studies; therefore, these stacks must break and reform rapidly. What we observe and describe is an overall average phenomenon. (b) The preferred orientation of the bases in the stack must vary to a certain extent depending upon the nature of the bases in the nucleosides. Substitution of an oxo group, amino group, methyl or dimethylamino group, etc., should exert an influence on the geometric arrangement of the stacks. Nevertheless, these stacks will have a partially overlapped orientation so that the H(8) and the H(1′) can both be shielded to a similar extent.

The chemical shifts of the base protons [H(5),H(6) or CH_3(5)] of the pyrimidine nucleosides are not concentration dependent (Fig. 16) even though these compounds have all been shown to associate in aqueous solutions by a lowering of osmotic coefficients (Table XXII). Therefore, we should interpret this result on the basis that these pyrimidine nucleosides do not support ring currents as indicated in Section II, C, so that their interactions cannot be monitored by proton magnetic resonance via the effect of the ring-current magnetic anisotropy (*284*). Interaction of pyrimidine nucleosides with purine or purine nucleosides, however, can be monitored by PMR as shown in Fig. 16 for the interaction of thymidine and purine (*284*). Marked upfield shifts are noticed, particularly for the thymidine base protons [CH_3(5) and H(6)] and the anomeric proton H(1′). As one proceeds around the deoxyribose ring from C(1′) to C(5′), however, there is a progressive drop off or decrease in the magnitude of these upfield shifts. This

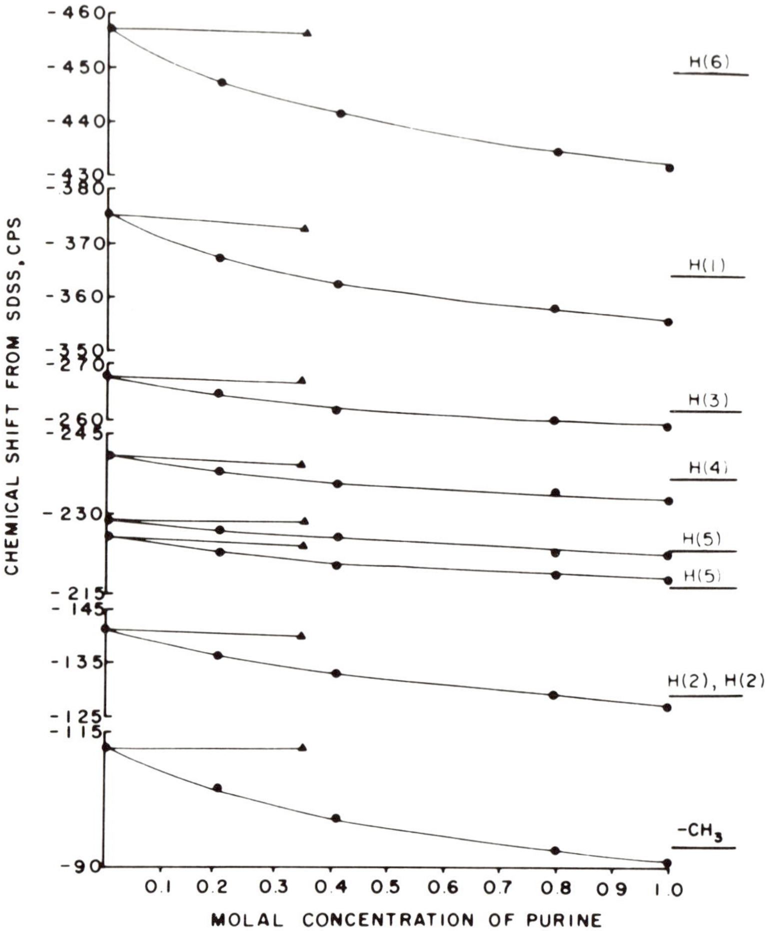

Fig. 16. Chemical shift dependence of thymidine protons upon thymidine (—▲—▲—) and purine concentration (—●—●—) at 35° in D_2O. Only two thymidine concentrations were measured, 0.1 *M* (placed at origin of the abscissa) and 0.35 *M* (placed at 0.35 *M* scale of the abscissa). Shifts measured from external SDSS. Magnetic field increases from top to bottom along the ordinate. Spectra obtained at 60 Mc (from Ref. *284*).

phenomenon, which is similar to that observed for the self-interaction of purine nucleosides, again indicates that the purine–pyrimidine nucleoside interaction is preferentially localized at the pyrimidine base of the nucleoside and the mode of interaction is vertical ring stacking of the bases (*284*). The same effect can be observed when a purine nucleoside is used in place of purine. For example, a mixture of 0.2 *m* 2′-O-methyl adenosine and 0.1 *m* thymidine results in the following upfield shifts (in cps) of the thymidine protons: 8.2 (—CH_3), 8.9 [H(6)], 7.1 [H(1′)], 4.3 [H(2′), H(2″)] (*284*). Similar effects were noted for N-6-methyladenosine and 2′-deoxyadenosine in thymidine solutions, whereas ribosylpurine is only 70% as effective as the adenosine derivatives.

The above studies describe mainly the stacking of the uncharged bases and nucleosides. It is of importance to investigate the association of nucleotides such as AMP in aqueous solution in order to evaluate the charge effect on stacking. We have carefully studied the concentration dependence of the PMR spectra of various nucleotides (*97*). As shown in Table XXVII and Figs. 15(a) and 15(b) (*97*), the concentration dependence of various protons of these 5′-adenine nucleotides is very similar to that reported for the adenine nucleosides. The δ value of the sugar protons located far away from the adenine base, such as the H(5′), H(5″), and the 5′-methyl protons are not sensitive to concentration changes. For the base protons, the magnitude of the high-field shift ($\Delta\delta$) in response to the increase of concentration is largest for H(2). The $\Delta\delta$ values for H(8) are about the same as those for the anomeric proton, H(1′). This comparison is dependent on individual nucleotides and is further discussed below. These results not only suggest the association of 5′-AMP in water by vertical stacking but also suggest that the geometry for the mode of association of 5′-AMP is probably very similar to that proposed for the association of adenine nucleosides [Fig. 14(b)] which is discussed above. The data in Figs. 15(a) and 15(b) and Table XXVII provided the experimental evidence to indicate that diionization and methylation of 5′-phosphate have very little effect on the $\Delta\delta$ of these protons and that the 5′-methyl protons are not concentration sensitive. These observations support the "face-to-face" model of Fig. 14(b) as described previously.

Comparison of the $\Delta\delta$ values of 5′-dAMP (*97*) at a range of 0.0–0.2 *M* with that of 2′-deoxyadenosine (*64*) does indicate a reduction in the extent of interaction, by about 40% at H(8) and about 30% at H(2) and H(1′). Comparison of the $\Delta\delta$ of the protons of 2′-AMP and 3′-AMP with the $\Delta\delta$ of the protons of 5′-AMP reveals some very interesting effects of the position of the charged phosphate group. When the charged phosphate group is moved from the 5′-position to the 2′-position, which is much closer to the five-membered ring and H(1′), the $\Delta\delta$ values of H(8) and H(1′) were reduced by about 50% (Table XXVII). The $\Delta\delta$ value for the more distant H(2) proton located at the six-membered ring remains unaffected by this 2′-phosphate group until the phosphate is doubly ionized. Then there is a reduction of $\Delta\delta$ of about 30% for the H(2), 50% for the H(8), and 70% for the H(1′) as compared to the $\Delta\delta$ value of 5′-AMP (Table XXVII). Thus, placement of the phosphate near the site of the interaction results in a different stacking orientation for 2′-AMP·Na, chiefly involving overlap of the six-membered

TABLE XXVII

Effect of Concentration upon the Chemical Shifts of Various Protons of the Mononucleotides[a, b]

Mononucleotides	Conc., m		δ, ppm from TMS capillary H(8)	H(2)	H(1′)	H(5′), H(5″)	H(2′), H(2″), CH_3	
5′-AMP·Na	0.0		8.957	8.705	6.578	4.602		
	0.3		8.805	8.412	6.463	4.658		
		Δδ	0.153	0.293	0.115	−0.056		
5′-AMP·Na_2	0.0		9.055	8.717	6.587	4.483		
	0.3		8.897	8.392	6.472	4.580		
		Δδ	0.158	0.325	0.115	−0.097		
5′-MeAMP·Na	0.0		8.883	8.717	6.605	4.575	3.990	(CH_3)
	0.3		8.730	8.408	6.447	4.585	4.020	
		Δδ	0.153	0.309	0.158	−0.010	−0.030	
5′-dAMP·Na	0.0		8.861	8.640	6.922			
	0.3		8.720	8.375	6.730			
		Δδ	0.141	0.265	0.192			
5′-dAMP·Na_2	0.0		8.960	8.672	6.945			
	0.3		8.803	8.395	6.753			
		Δδ	0.157	0.277	0.192			
3′-AMP·Na	0.0		8.805	8.697	6.530			
	0.3		8.613	8.347	6.407			
		Δδ	0.192	0.350	0.123			
2′-AMP·Na	0.0		8.773	8.653	6.610			
	0.3		8.697	8.328	6.555			
		Δδ	0.076	0.325	0.055			
2′-AMP·Na_2	0.0		8.810	8.700	6.597	4.323		
	0.3		8.730	8.462	6.562	4.365		
		Δδ	0.080	0.238	0.035	−0.042		
5′-dGMP·Na_2	0.049		8.568		6.712	4.373	3.133	
	0.90		8.520		6.598	4.480	3.073	
		Δδ	0.048		0.104	−0.110	0.060	
			H(6)	CH_3	H(1′)	H(5), H(5″)	H(2′), H(2″)	
5′-dTMP·Na_2	0.052		8.272	2.392	6.812	4.518	2.838	
	0.98		8.280	2.438	6.830	4.580	2.912	
		Δδ	−0.008	−0.046	−0.018	−0.062	−0.074	

[a] Temperature 28–31°

[b] Δδ values are all positive unless otherwise noted, indicating upfield shifts with increase in concentration. Values of chemical shifts at 0.0 m were obtained from extrapolations. Data from Ref. *97*.

rings. When the phosphate is moved from the 5′-position to the 3′-position, the per cent increase of $\Delta\delta$ values appear to be higher especially for the H(8) proton. Examination of the molecular model indicates that now the 3′-phosphate group is distant from the base rings regardless of the rotation around the glycosyl bond [N(9)—C(1′)]. Thus, the specific interaction of the 5′-phosphate group with the H(8) proton, which occurs when the nucleotide is in an "anti" conformation, is abolished. This is undoubtedly the reason for the enhancement of the $\Delta\delta$ value of the H(8) proton of 3′-AMP as compared to the H(8) proton of the 5′-AMP. The H(8) proton of 3′-AMP can now be overlapped closer by the neighboring molecules in the stack.

Deoxyribose 5′-AMP also stacks to a similar extent as ribose 5′-AMP. The $\Delta\delta$ for the H(1′) of deoxyribose AMP, however, is even larger by about 0.06 ppm (Table XXVII) than the $\Delta\delta$ for the H(1′) of the ribose AMP. Evidently, the replacement of the OH(2′) group by the hydrogen atom allows the neighboring base rings to approach the H(1′) region more readily. Similar behavior was noted for deoxyribose versus ribose adenine nucleosides (Table XXVI).

The $\Delta\delta$ value of 5′-TMP·Na_2 is very small even over a large concentration range (Table XXVII). This is due to the absence of ring-current magnetic anisotropy of this nonaromatic pyrimidine as discussed above with the $\Delta\delta$ values of the pyrimidine nucleosides. Similarly, the $\Delta\delta$ value for 5′-dGMP·Na_2 is very small even at a large concentration range (0.05–0.90 *m*, Table XXVII). At high concentration, the nucleotide solution is viscous indicating the formation of helical structure discussed above (*6,280,281*). The six-membered ring of guanine has very little ring current and is "non-aromatic," but the five-membered ring of guanine (the imidazole) has approximately the same calculated ring current as that of adenine (*72*). Furthermore, association of GMP in aqueous solution is caused not only by the stacking of the bases but also by hydrogen bonding of the bases [including N(7)] to yield multiple strands (*6,280*). It is possible that the noticeable $\Delta\delta$ value for the H(1′) (0.10 ppm) of 5′-dGMP·Na_2 at this large concentration range (Table XXVII) originates from the ring-current effect of the five-membered ring in the stacks. This $\Delta\delta$ value for the H(1′) is substantially larger than the $\Delta\delta$ value for the H(8) proton (0.05 ppm). The only explanation we can propose at the present is the possible occurrence of intermolecular hydrogen bonding at N(7) which will deshield the C(8) protons and offset the high-field shift caused by stacking of the five-membered ring.

Jardetzky (*90*) has also studied the concentration dependence of purine, 5′-AMP, and some other nucleosides and nucleotides. The results are similar to those described above.

3. Nature of the Stacking Association

From the above thermodynamic and spectroscopic studies on the extensive association of bases and nucleosides in water, three main concepts have been developed in describing the nature of this interaction. First, the mode of association is by vertical stacking of the bases and not by horizontal hydrogen bonding [Fig. 14(b)]. The degree of association certainly proceeds much beyond the dimer stage (Table XXIV). Second, the association is exothermic (Table XXV). The enthalpy for the association is about -2 to -6 kcal

TABLE XXVIII

Correlation between the Osmotic Coefficients (ϕ)[a] of the Nucleosides, and the Polarizabilities (α)[b] and Dipole Moments (μ)[c] of the Bases

	ϕ (0.1 *m*, 25°C)	α, Å^3	μ, D
Uridine	0.943	10.2	(3.9)[d]
Cytidine	0.935	11.0	7.2
Thymidine	0.905	12.0	3.6
Inosine	0.888	13.0	5.2
Ribosylpurine	0.860	12.5	4.3 (4.2)[d]
2′-O-Methyladenosine[e]	0.723	13.9	3.0 (3.2)[d]
2′-Deoxyadenosine[e]	0.688		

[a] From Table XXII.
[b] From Pullman (*48,285*).
[c] From Pullman (*48,285*). See Section II, B for discussion.
[d] Experimental value from Ref. *47*.
[e] Adenosine is not sufficiently soluble to give a 0.1 *M* solution.

and the overall entropic contribution is always negative, ranging from -5 to -18 e.u. Third, the extent of association of the purine derivatives is greater than that of the pyrimidine derivatives. Methylation at the base moiety usually enhances the association while methylation at the pentose moiety has little effect. The range of the free energy for the association at 25° is from $+300$ cal (uridine) to -1.9 kcal [N-6-dimethyladenosine] (Table XXIV). Comparison of the tendencies of association of several nucleosides (as indicated by their osmotic coefficients) with certain parameters of the molecular properties of their bases provides additional understanding about this problem. In Table XXVIII, the correlation among the osmotic coefficients (ϕ) of the nucleosides, and the polarizabilities (α), and the dipole moments (μ) of the bases is listed. Comparison of the order of ϕ and the order of μ among the four purine nucleosides clearly indicates that these two quantities relate to each other in a reverse manner. Comparison of the order of ϕ and μ among the three pyrimidine nucleosides again shows no correlation between these two quantities. Thymine has the smallest dipole moment, but thymidine has the greatest tendency to associate, while cytosine has the greatest dipole moment among the six compounds in Table XXVIII, but cytidine has the next to the lowest tendency to associate. Therefore, it can be concluded with a considerable degree of certainty that as far as these nucleosides are concerned, permanent dipole moment attraction is not the most important driving force for their self-association.

This conclusion receives decisive support from our recent comparative studies on the solution properties of inosine, 1-methylinosine, and 7-methylinosine (*164*). 7-Methylinosine, which has a betaine structure, is highly polar because of the deficiency of electrons in the five-membered ring and the excess of electrons in the six-membered ring. VPO data (Table XXII) and PMR results (*164*) both indicate that this polar 7-methylinosine does not associate in water more than inosine or 1-methylinosine. This comparative study demonstrates that the electrostatic interaction of monopoles and dipoles of the bases do not contribute significantly to the free energy of stacking interaction of nucleosides in aqueous solution. Though the contribution of electrostatic interaction derived from theoretical considerations is very substantial in the vacuum state ($F_{\rho\rho}$, $F_{\rho\mu}$, $F_{\mu\mu}$) (*47*), the present experimental conclusion is perhaps not too surprising. If electrostatic interaction of fixed charges is of primary importance, and if the solute–solute interaction is not dominantly influenced by the strong solvent–solvent interaction, then the self-association of the bases in organic solvent of low dielectric constant should be much greater than that in water, which it is not. In fact, bases do not self-associate in organic solvent by stacking but form hydrogen bonds with the specific, complementary bases (Table XXI). Also, positive and negative ions such as Na^+ and Cl^- do not associate to a great extent in water in the range of 0.01–0.1 *m* concentration. It should be noted, that it is entirely possible that the charge properties of 7-methylinosine have made a contribution to the negative enthalpy for the association. If so, then this contribution is counterbalanced by the negative entropy with no resultant increase in the negative free energy for the association. This important possibility hopefully will soon be investigated. On the other hand, in the correlation between the order of ϕ and the order of α, the polarizability is of interest (Table XXVIII). The only problem is that although hypoxanthine was calculated to have a larger polarizability than purine, the association tendency of inosine appears to be slightly less than that of ribosylpurine. This anomaly may be due to the negative influence of the large permanent dipole moment of hypoxanthine which may hinder the stacking either directly or indirectly by more extensive hydration. Other than this exception, the rather good correlation here is in accord with the recent conclusion, based upon a correlation of polarizabilities and the effectiveness of a variety of denaturing agents, that London dispersion forces are responsible for the stability of the DNA helix (*286*).

Though the nature of the molecular force(s) providing the negative enthalpy for the association is not well understood at present, these forces undoubtedly are the cause for the mode of association stacking of the planar bases. Some indications do exist that there is an entropic contribution to the association though the overall entropic value is negative, opposing the association. For instance, the free energy of association for caffeine (−1.5 kcal) is definitely larger than that of purine (−0.44 kcal) or 6-methylpurine (−1.12 kcal), yet the enthalpy for purine (−4.2 kcal) or 6-methylpurine (−6.0 kcal) is substantially larger than that of caffeine (−3.4 kcal) (Table XXV). The reason for a larger negative free energy for the self-association of caffeine comes from a smaller negative entropy (−6 e.u.) versus that of purine (−13 e.u.) or that of 6-methylpurine (−16 e.u.). Similarly, it appears that the larger negative free energy for the self-association of ribosylpurine (−0.4 kcal) over that for the pyrimidine nucleosides (−0.08 to +0.3 kcal) comes from the smaller negative entropy (−7 e.u.) for the purine nucleosides versus that of the pyrimidine nucleosides (−10 e.u.), since the negative enthalpy for ribosyl purine association (−2.5 kcal) is actually smaller than that of the pyrimidine nucleosides (about −2.75 kcal) (Table XXV). These two comparisons strongly suggest the existence of a positive entropic contribution to the association, hidden in the overall negative entropic value. Such a positive entropic contribution most likely is derived from the competition between solute–solvent (water) interaction and solvent–solvent (water) interaction. Very recently, Crothers and Ratner (*287*) have described a thermodynamic study on the association system of actinomycin with deoxyguanosine which may be relevant for the discussion here. The association process of actinomycin with deoxyguanosine has a negative enthalpy and a negative entropy, which is similar to the situation for nucleoside association. A hidden, *positive* entropic contribution in the aqueous system can be demonstrated, since the entropy value becomes more negative upon increase of the methanol content of the system. They concluded that this is a demonstration of the "hydrophobic bonding" through the ordering water molecules around the dissolved particles (*287*). They further viewed that "the results favor the standard view of hydrophobic effects as arising from an ordering of solvent molecules around the solute over the description in terms of varying solvent surface tension [Sinanoglu, O., and Abdulnur, S. (1965), *Federation Proc.* **24**, 5]." This notion of the hidden positive entropy contribution in the aqueous system and the fact that the self-association of bases does not take place in organic solvents but only in water, indicates that the early description of this interaction as "hydrophobic-stacking interaction" (*269*) may be rather appropriate. As a working hypothesis, we view that the bases are clustered in aqueous solution because of the strong self-interaction of water molecules. With the bases in a close proximity to each other, short-ranged, attractive forces (forces proportional to $1/r^6$ or even to $1/r^3$) begin to become dominant, which most likely involve π-electron polarizabilities and energies through electrostatic interaction.

The recent paper by Helmkamp and Kondo (*273*) is of significance in providing further understanding about the nature of the molecular forces participating in the association. They reported that the apparent free energy for self-association (ΔF, cal/mole) of 9-methylpurine in aqueous solution at 25° is -380, while the ΔF for purine and 6-methylpurine are -440 and -1100, respectively (Table XXV). The concentration dependence of the chemical shifts ($\Delta\delta$) of the protons of 6-methylpurine (*84*) correspondingly is much higher than those of 9-methylpurine and purine (*84,273*). This difference between C(6) methylation versus N-methylation is unlikely due to the loss of contribution of hydrogen bonding to the association by the removal of the hydrogen in N(9) methylation. The reasons are that these molecules associate mainly through stacking, as shown by the upfield shift observed, and that they do not associate in organic solvents. At present, there is also no obvious reason why this difference would be caused by a steric effect. The best explanation we can propose is that the N(9) alkylation has blocked the tautomerization of the hydrogen between N(9) and N(7) at the five-membered ring of the purine derivatives. The blockage of this tautomerization causes a reduction in the allowable patterns of electron distribution in the system and should, therefore, decrease the polarizability of the molecule. This decrease may lead to a lowering of the tendency of association in stacking. Comparison of the thermodynamic qualities of the association with the polarizability of these compounds will be of significant value. Helmkamp and Kondo (*273*) have also reported that the substitution of isopropyl or *t*-butyl at the N(9) position will have a much smaller effect on the extent of association of the alkylated compound than when substitution by these groups is at C(8), C(2), or C(6) positions. However, the concentration dependence of the chemical shifts ($\Delta\delta$) observed for these N(9) alkylated compounds is slightly higher than for those substituted at various carbon atoms. These data imply the possibility that these N(9) substituted compounds may have a different mode of association than that of the carbon substituted compounds, besides having comparatively lower association constants.

For the solvent effects on molecular associations, Sinanoglu and his coworkers (*288–290*) have developed an interesting theory. The more quantitative treatment developed recently (*288*) indicates mainly two usually opposing solvent effects which may be classified as solvation (or inverse volume forces) and surface forces. He considered that " 'Solvophobic forces' with 'hydrophobic' as an especially strong special case arise from the surface forces and act on not only nonpolar, hydrocarbon-like groups, but also on polar groups (like purine, pyrimidine bases). It is just that if the groups are polar, they may act in a way to counteract some of the surface effect." This theory offers a clear conceptual explanation for various solvent effects and can be tested experimentally [see above discussion of the paper by Crothers and Ratner (*287*)]. The difference in association tendencies of the isomeric pairs, such as the 6-methylpurine versus 9-methylpurine just discussed above in paper by Helmkamp and Kondo (*273*), and such as 2-O-methyladenosine versus N-6-methyladenosine (Table XXIV), is unlikely to be explained only on the basis of the surface tension effect, since these molecules in the isomeric pair presumably have very similar volume properties.

D. Interaction of Monomers with Polymers in Aqueous Solution

1. Interaction of Bases and Nucleosides with Nucleic Acids

a. *Unspecific and Noncooperative Interaction*

The unspecific and noncooperative interaction of bases and nucleosides were studied in the early sixties by our laboratory by two approaches (*291*, *292*). The first approach is to study the effect of the monomers on the conformation of nucleic acids by means of optical rotation. Optical rotation has been shown to be valuable because it possesses the following advantages: One, optical rotation measured at wavelengths above 350 mμ is not obstructed by the highly UV absorbing monomers which have little or no optical activity. Second, the change of optical rotation caused by a transition in nucleic acid conformation is very large while the optical rotation of the nucleoside is not sensitive to temperature. The second approach is the direct measurement of the binding of the monomer to the nucleic acid polymer, usually by the method of equilibrium dialysis. The effects of purine, purine nucleosides, pyrimidine, pyrimidine nucleosides, and analogs and derivatives of these compounds on the properties of nucleic acids in helical form have been studied (*291*). The presence of these compounds causes no change of optical rotation or viscosity of helical nucleic acid, such as native DNA or helical poly A in acidic solution. Nevertheless, the helix-coil transition temperature, T_m, of both native DNA and of helical poly A is significantly reduced in the presence of these compounds. For instance, in the presence of 0.3 M purine, the T_m of thymus DNA and the T_m of helical poly A is reduced as much as 9° and 14°, respectively. A series of compounds has been tested for their effectiveness in lowering the melting temperature of poly A and of thymus DNA. The order of effectiveness was found to be: adonital, methyl–riboside (both negligible) < cyclohexanol < phenol, pyrimidine, uridine, < cytidine, thymidine < purine, adenosine, inosine, deoxyguanosine < caffeine, coumarin, 2,6-dichloro-7-methylpurine. Urea was found to be completely ineffective with poly A and only slightly effective with DNA. The relationship between the structures of these molecules and their effectiveness in reducing T_m of native DNA or helical poly A indicates that their denaturing effects do not appear to depend upon hydrogen bonding but rather upon hydrophobic stacking interaction. For instance, coumarin, 2,6-dichloro-7-methylpurine, and caffeine have much less hydrogen-bonding abilities than purine, yet these compounds are very effective denaturants. These results are understandable in light of the above discussion in Section III, C. The effects of purine and related compounds on the physical properties of helical nucleic acids are not noticeable in the temperature range below T_m. In contrast, the effect of purine on the physical properties of denatured DNA and on the coil form of poly A is quite striking. In the presence of 0.3 *M* purine, the specific rotation (at 20°C and at 436 mμ) of denatured DNA drops from 240° to 170°, and in the presence of 0.6 *M* purine, to 50°. Concurrent with the decrease of specific rotation, there is a significant increase of 30 to 60% in reduced viscosity. Similar effects of purine have also been observed for the coil form of poly A. From these facts, a coherent picture is emerging. Compounds such as purine and nucleosides do not appear to interact noticeably with the helical forms of nucleic acids. They do, however, interact significantly with the coil form. These interactions bring about further uncoiling and extension of the polymer with concomitant reduction of the base–base interaction of the nucleic acid itself. It is this preferential interaction of compounds such as purine with the coil rather than with the helical form of nucleic acid, that causes the helix-coil transition of nucleic acid to be shifted to a lower temperature by their presence.

This picture is further confirmed by direct measurements of the affinities of purine and pyrimidine bases and related compounds to nucleic acids by equilibrium dialysis (*292*). The affinity constants (*nk*), where *n* is the number of the binding sites and *k* is the average association constant of each site, indicated that the order of values of (*nk*) is thymine < adenine, caffeine < naphthalene < steroids < phenanthrene. In addition, the (*nk*) affinity constants of the nucleic acids in coil form for these small compounds were found to be much higher than those of nucleic acids in helical form. The (*nk*) values of adenine and caffeine (about 7–9 M^{-1}) with denatured DNA is about 10 times higher than those with native DNA. This fact clearly indicates that the affinity of a substrate for nucleic acid depends greatly on the secondary structure of the nucleic acid. This conclusion substantiates the picture proposed above; namely, that the mechanism by which the T_m

of nucleic acids is reduced is based on the preferential binding of the substrate to the coil form of the nucleic acid.

More recently, this problem was reinvestigated by Chan and co-workers by the PMR technique (*293,294,295*). First, they studied the PMR spectra of purine [H(2), H(6), and H(8)] and of the TpT, TpdU, and dUpT (T= thymine, dU′=2′-deoxyuridine) in the D_2O solution influenced by each other (*293*). Upon addition of purine, the resonance of the base protons and H(1′) protons of the dimers are shifted upfield significantly. In addition, the base protons of TpT are split into two sets of resonances of equal intensity in the presence of purine, indicating that the two sets of thymine base protons of TpT have become much more magnetically nonequivalent. Binding of purine to the dimers also induces broadening of the linewidth of the purine molecule, especially the H(6) protons. The influence of the three dimers on the linewidth of the purine protons is the largest for TpdU, less for TpT, and least for dUpT. Based on these observations, Chan et al. proposed that the purine is inserted between the stack of the dimers in a sandwich-like fashion, in addition to the possible external ring stacking. The broadening of the purine spectral lines is attributed to the nuclear spin relaxation in the dynamic binding process induced by the fluctuations of the dipolar field of the 2′-, 3′-, and 5′-ribose protons in the "bend" of the dimer cage. Later on, Bangerter and Chan extended their investigation to the binding to poly U (*294*). The poly U proton resonances were shifted to higher fields by the added purine, indicating that purine binds to the uracil bases of the polymer by base stacking. Severe broadening of the purine proton resonance was also observed, and this observation is interpreted as strong evidence for the intercalation of purine between adjacent uracil bases of poly U. The linewidths of the poly U proton resonances were not noticeably broadened in the presence of purine, thus, the binding of purine to poly U does not result in a more rigid or ordered structure for the polymer.

b. *Specific and Cooperative Interaction of Nucleosides with the Complementary Polynucleotides*

This type of interaction was investigated independently by an NIH group (*296,297*) and by our laboratory (*298,299*) at the same time from different approaches. The results from our laboratory will be described first.

The specific and cooperative interaction of poly U and adenosine was investigated (*298*). The binding of adenosine to poly U was first studied by equilibrium dialysis at 5°C in 0.4 *M* NaCl. When the fraction of the occupied poly U binding sites is plotted as a function of free adenosines (Fig. 17), the resulting adsorption isotherm shows a very steep transition. No binding was detectable until a critical threshold concentration of adenosine was reached.

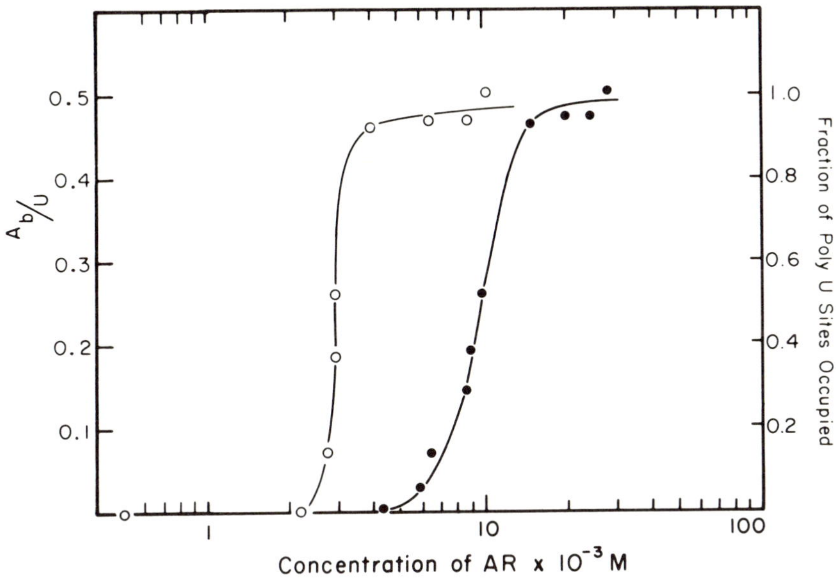

Fig. 17. Adenosine bound per UMP of the poly U (1.5×10^{-2} M) vs. adenosine input concentration at 5°C, 0.4 M NaCl, 0.01 M phosphate (HMP)(—●—). The fraction of poly U sites occupied versus free adenosine concentration is also shown (—○—) (from Ref. *298*).

This steep curve of adsorption isotherm is analyzed by the following equation derived from lattice statistics based on the nearest-neighbor interaction (*300*):

$$\left(\frac{\partial \theta}{\partial \ln Y}\right)_{\theta=\frac{1}{2}} = \frac{\exp(-W/2kT)}{4}$$

where θ is the fraction of sites occupied, W is the interaction energy of the nearest neighbor, and Y is the function of absolute activity of the adsorbate ($\lambda = e^{u/hT}$) and the partition function for a molecule of bound adsorbate, q; λ_q at dilute solution is equivalent to $K_0 M$, where K_0 is the intrinsic association constant for one molecule of adsorbate with a single site, and M is the molar concentration of free adsorbate (*301*). Therefore:

$$\left(\frac{\partial \theta}{\partial \ln M}\right)_{\theta=\frac{1}{2}} = \frac{\exp(-W/2kT)}{4} \tag{8}$$

Estimation of the slope of the curve (Fig. 17, open circle) yields a value of 30–60. In Eq. (8), W is calculated to be -5 to -6 kcal/mole, which is the stacking energy of adenosine upon pairing with 2U residues of poly U. This is comparable to the values of -4.8 kcal/mole and -7.5 kcal/mole calculated for the stacking energy of poly d(A–T) and poly dT·poly dBrC, respectively, by Crothers and Zimm (*302*).

Similar experiments were also performed using cytidine or inosine as the dialyzable components. No detectable binding was found even at an input nucleoside concentration as high as 2×10^{-2} *M*. Therefore, this interaction has the same specificity as the system of long-chain polymers, i.e., the base-pairing scheme of Watson–Crick. The stoichiometry of this binding reaction was studied by using solubility measurements, and it was found that at low temperature that the stoichiometry is 2U to 1A, while at 20° the stoichiometry becomes 1A to 1U. The physical properties of this poly U·adenosine complex were further analyzed by sedimentation, viscosity, and optical rotation measurements.

The formation of poly U·adenosine (AR) complex can be demonstrated by analytical ultracentrifugation. Sedimentation coefficients (S) of poly U in the absence (control) and presence of nucleosides are given in Table XXIX. When N-6-methyladenosine, cytidine, or inosine was mixed with

TABLE XXIX

Sedimentation of Poly U in Nucleoside Solutions[a]

Nucleosides	Buffer	Temp., °C	S_{20} control	S_{20} complex	% increase
Adenosine	0.4 *M* NaCl	5	4.68	6.21	33
Adenosine	0.4 *M* NaCl	10	4.03	5.78	43
Adenosine	0.4 *M* NaCl	19	4.00	6.11	53
Adenosine	0.02 *M* $MgCl_2$	5	4.70	6.45	37
L-Adenosine	0.4 *M* NaCl	5	4.77	6.69	40
N-6-methyladenosine	0.4 *M* NaCl	5	4.77	4.77	0
Cytidine	0.4 *M* NaCl	5	4.06	4.06	0
Inosine	0.4 *M* NaCl	5	4.68	4.68	0

[a] Poly U conc. = 1.5×10^{-2} *M*; Nucleoside conc. = 1.5×10^{-2} *M*, from Ref. *218*.

poly U in equal amounts (1.5×10^{-2} *M* each) at 5°C and 0.4 *M* NaCl, no change in either the pattern or the S value was found. When adenosine was mixed with poly U under identical conditions, a 33% increase in S value and a sharpening of the boundary was observed as compared with the control. Similar results were obtained in 0.02 *M* $MgCl_2$ with the same mixture. In 0.4 *M* NaCl, as the temperature was raised, the percentage change in S value also increased to 43% at 10°C and 53% at 19°C, but it was accompanied by a decrease in sharpness of the boundary. The specific viscosities of poly U (1.5×10^{-2} *M*) and poly U·AR complex (1.5×10^{-2} *M* of each) in 0.4 *M* NaCl at 5° were, respectively, 0.602 and 1.05. As previously stated a parallel increase in S value (33%) has also been observed. The concurrent increase in both specific viscosity and the sedimentation coefficient of the poly U·AR complex as compared with those of poly U, unambiguously showed that there is a molecular weight increase in the polymer resulting from the complex formation.

Optical rotation measurement at 350 mμ was used to determine the conformation and stability of the poly U·AR complex. Poly U in 0.4 *M* NaCl gave a small positive rotation at low temperature (at 1.5×10^{-2} *M*, the observed rotation was about 0.2 deg at 5°C). The rotation decreased with increasing temperature, finally becoming temperature insensitive beyond 12°C as shown in the control curve in Fig. 18. On the other hand, 1.5×10^{-2} *M* adenosine alone gave an observed rotation of $-0.09°$ calculated from the

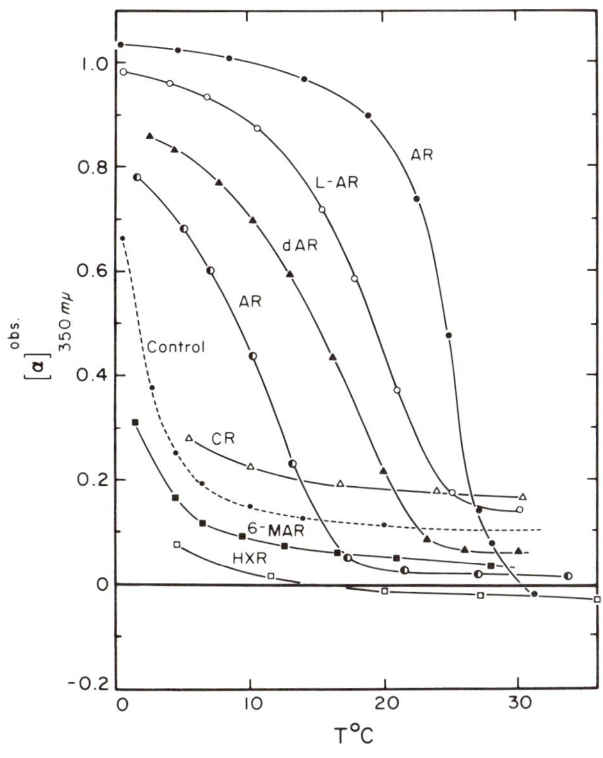

Fig. 18. Observed rotation of poly U (1.5×10^{-2} *M*) nucleoside mixtures vs. temperature in 0.4 *M* NaCl, HMP. Concentrations of the nucleosides are: adenosine (AR), 1.5×10^{-2} *M*, [— ● — and 9.3×10^{-3} *M* (— — ○ — —)]; L-adenosine (L-AR), 9.3×10^{-3} *M*; deoxyadenosine (dAR), 7.8×10^{-3} *M*; cytidine (CR), 1.1×10^{-2} *M*; N(6)-methyladenosine (6=MAR), 7×10^{-3} *M*; inosine (H × R), 1.5×10^{-2} *M* (from Ref. *298*).

rotation at 1.2×10^{-2} *M* which was temperature independent. Nevertheless, when the two were mixed, a large increase in positive rotation was observed, $+1.03°$ at 5°C. In the temperature-insensitive region beyond 30°C, the rotation of the mixture was the algebraic sum of its constituents. In 0.4 *M* salt, the optical rotation measurements remained essentially invariant in a temperature range from 0.5° to 20°C and then decreased abruptly from 20° to 30°C. We took these results to mean that the poly U·AR complex formed an ordered structure in 0.4 *M* NaCl and that its stability was reflected by its melting behavior in response to the temperature variation. When poly U was mixed with cytidine, inosine, or methylated adenosines, no complex formation was observed (Fig. 18). Formation of poly U·AR complex and its thermostability were highly dependent on adenosine concentration. Various analogs of adenosine were also tested for their binding capacity to poly U by the optical rotation and sedimentation methods with the expectation of obtaining information about an involvement of binding sites and the role of the sugar moiety. The following compounds were tested: deoxyadenosine, L-adenosine (the pentose was L-ribose instead of D-ribose), 9-γ-hydroxypropyladenine and 9-hydroxypentyladenine (long-chain alcohols replacing the sugar moiety). Complexing with poly U was found for all these four compounds. When the point of attachment of the purine ring was changed from the 9-position to the 3-position as in the case of 3-isoadenosine, complex formation could still take place. All these observations indicate that the sugar moiety of the adenosine does not play an important role in the binding. Optical rotation studies of mixtures of poly U with N(6)-methyladenosine, with 1-methyladenosine and tubercidin (7-deazaadenosine) revealed that no interaction took place. Therefore, the N(6)-amino group and N(1) position of the adenine appear to be definitely involved in binding with poly U. The other possible binding site is the N(7) position of the adenine.

The two important aspects of the participation of adenosine in the interaction are its concentration dependence and its specificity. The complex formation is undetectable in low nucleoside concentration. After a threshold concentration of adenosine is reached, the binding increased rapidly in a cooperative manner until saturation. The key to the understanding of the interaction resides in the properties of nucleosides in solutions of moderate concentration as detailed in the section on monomer–monomer interactions. From these studies, we know that the stacking of adenosine occurs when the concentration increases. These stacks behave as the oligonucleotides do, and therefore have much greater affinity to poly U than does free adenosine. At moderate concentrations, these associated stacks may serve as initiators for the subsequent binding of the adenosine molecule to poly U by a cooperative mechanism. In fact, the stability of the completely interacting complexes measured in our experiments is comparable to that obtainable for the poly

U-trimer or tetramer (oligonucleotides) interaction. The forces responsible for stacking energy are short ranged. Calculation based on consideration of the nearest-neighbor interaction gave an estimation of approximately 5 or 6 kcal/mole as the free energy of stacking for this poly U·AR system. The results clearly indicated that hydrogen bonding cannot be the sole force responsible for the binding, since in dilute solution no binding is detected, even though hydrogen-bonding capacity is still present. On the other hand, hydrophobic stacking forces alone do not allow the interaction to occur.

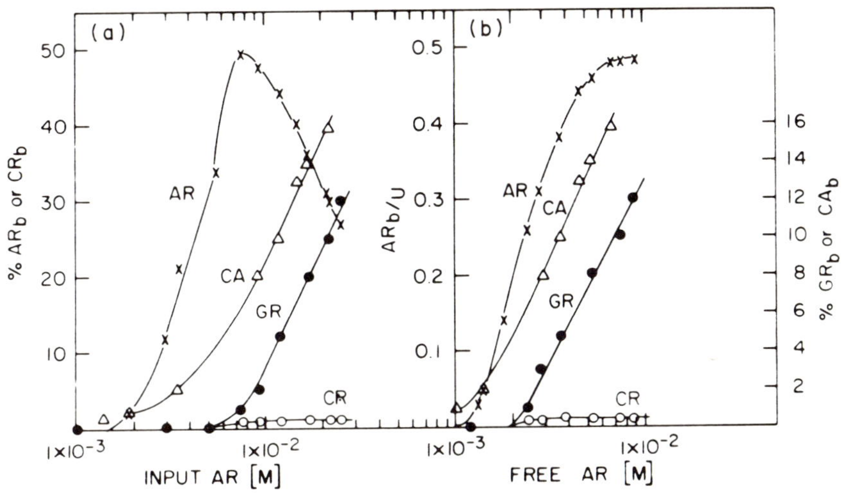

Fig. 19. Binding of nucleosides and caffeine to $r(U)_n$ in equilibrium dialysis in 0.01 *M* tris, 0.4 *M* NaCl, pH 7.0 at 5°. (a) Per cent of bound adenosine, AR (×—×); bound caffeine, CA (△—△); bound guanosine, GR (●—●); or bound cytidine, CR (○—○) versus input concentration of adenosine. (b) Amount of adenosine bound per UMP of $r(U)_n$; per cent bound caffeine; per cent bound guanosine; or per cent bound cytidine versus concentration of free adenosine. Symbols are the same as in (a) (from Ref. *299*).

Inosine, methylated adenosines, and other adenine analogs (Table XXVI) all form stacks, yet they fail to bind to poly U. It appears, therefore, the hydrogen-bonding and hydrophobic stacking forces are both essential, with the former related to specificity and the latter related to stability. The combined action of the hydrophobic stacking forces and the hydrogen bonding in the polymer–monomer system can be evaluated separately, and the results further indicate the importance of the stacking forces in this specific interaction. This was achieved in our laboratory by studying the binding of guanosine, caffeine, and cytidine to poly U as shown in Fig. 19 (*299*). No binding of guanosine, caffeine, and cytidine to poly U was detected by equilibrium dialysis and this conclusion was reinforced by solubility and optical-rotation measurements. This is to be expected from the Watson–Crick base-pairing scheme. Upon the formation of adenosine · poly U complex in the presence of adenosine, however, binding of guanosine and caffeine starts to take place, whereas the binding of cytidine does not. Caffeine has no hydrogen-bond donor site, but it has a large tendency for stacking (*269*). The binding of guanosine and caffeine, but not cytidine, to the adenosine · poly U complex indicates that the mechanism of this unusual binding process is the co-stacking of the guanosine or caffeine with the adenosine which in turn is hydrogen bonded to poly U through the base-pairing scheme. These results show that mispairing and intercalation in nucleic acid interaction are caused by co-stacking of these compounds with the bases in nucleic acid.

The association of adenosine and poly U has been studied by PMR methods by our group (*303*) and by Bangerter and Chan (*295*). Extensive broadening of the linewidths of the adenosine and poly U protons, especially of the base protons, was found when conditions are favorable for the formation of the soluble complexes. The formation of the complex is governed by the temperature (*295*), ionic strength, divalent ions (*303*), and monomer concentration.

The major source of the line broadening must come from the anisotropic nuclear dipole–dipole interaction of the protons with their environment due to the loss of their freedom of motion when the binding occurs. This conclusion is supported by two lines of evidence (*303*). (1) There exists a gradient of line broadening among various protons of the adenosine starting from the base to the 5′ end of the ribose; as compared with free adenosine, the base protons of adenosine in the interaction system have the highest degree of broadening, next the H(1′) proton, and least the H(5′, 5″) protons. This relaxation gradient suggests that the base protons of adenosine lose more freedom of motion in the binding process than do the H(1′) protons, and in turn the H(1′) protons lose more motional freedom than do the H(5′), H(5″) protons. This is the predicted result if the binding of adenosine to the $(U)_n$ takes place at the bases. (2) Within the range of $\pm$ 0.03 ppm, the chemical shifts of the adenosine protons are the same in the mixture as in its own individual solution. This observation is not unexpected in light of our previous work on the extensive association of adenosine in solution (*64*). Thus adenosine already partially self-aggregates into stacks, before binding. The chemical shifts of the H(2) and H(8) protons are already 0.1 ppm upfield from the values at infinite dilution. Furthermore, the hydrogen bonding which takes place in the complex formation should cause a downfield shift of the base protons, and therefore may compensate for the additional shielding due to further stacking in the complex. A similar situation was also found for the $(U)_n$. Since the chemical shifts of the bound form and the free form appear to be very close, the broadening observed cannot be due to an exchange of two forms having very different chemical shifts. Therefore, we may conclude that the main cause of the linewidth broadening comes from the dipolar interaction of the protons with their environment as a result of loss of motional freedom upon binding. The line shapes of the resonances further suggest that there is a rapid exchange between the bound form and the free form. Finally, the proton magnetic resonance results on the 0.029 M inosine, 0.027 M adenosine, and 0.03 M $(U)_n$ mixture in 0.04 M $MgSO_4$ provided the most direct demonstration of base-pairing specificity. In this mixture, the spectra indicated that extensive interaction between adenosine and $(U)_n$ occurs, while the inosine is left untouched in solution.

Howard et al. from NIH (*296*) have studied extensively the interaction of poly U, poly C, poly I, and poly A with a variety of the bases, nucleosides, and nucleotides, mainly by the IR technique but by solubility, optical rotation, and enzymic phosphorolysis as well. Their results demonstrated the formation of purine monomer complex with the complementary pyrimidine polynucleotides, but no pyrimidine monomer–purine polynucleotide complex could be found. The melting curves of their complexes constructed from the IR spectra (usually in the 1600–1650 cm^{-1} region) show a steep transition indicating the cooperative nature of the binding. Stoichiometry of the interaction, concentration dependence, and ionic strength effects on T_m have also been investigated. In general, the results of this work (*296*) are similar to those reported from our laboratory (*298*). Maxwell et al. (*297*) have indicated that the formation of this adenosine or 2-aminoadenosine complex with poly U and poly (U, C) inhibits these pyrimidine polynucleotides from serving as messenger RNA for polyphenylalanine synthesis in the ribosome system of *E. coli* in rat liver.

2. Interaction of Nucleotides with Nucleic Acids

Interaction of nucleoside monophosphates with polynucleotides was first reported by Howard et al.(*304*)on the system of guanosine-5′-mononucleotides with poly C. By the IR technique, the melting curve (in the 1600–1700 cm^{-1} region) of the complex was constructed which shows a sharp transition, indicating a cooperative process of melting. They reported (*304*) the formation of two-stranded and three-stranded 5′-dGMP·poly C complex.† Subsequently, Miles et al. (*305*) reported in a short communication on the interaction of 5′-GTP, 5′-ATP, and 5′-AMP with poly C and poly U, respectively, as studied by the IR technique. While the T_m of these complexes were reported,

† Most of the experimental results reported are for 5′-dGMP; they stated that "The same reactions are observed for both 5′-GMP and 5′-dGMP. The advantage of the latter is that it has a much weaker self-interaction."

no information, however, on the physical states of the complexes were described (*304,305*).

Independently, our laboratory has investigated the interaction of nucleotides (ribosyl and deoxyribosyl monophosphates and triphosphates) with polynucleotides (*306*). Insoluble complexes of adenine nucleotides (AMP, dAMP, ATP, and dATP) and poly U as well as guanine nucleotides (GMP, dGMP, and GTP) and poly C are formed reversibly when the concentration of the polynucleotides and especially of the nucleotides is sufficiently high and the temperature is sufficiently low. At moderate concentration of the interactants (0.01–0.02 *M*), Mg^{2+} ion is usually required for the interaction. These insoluble complexes have a definite stoichiometry, generally 2U to 1A or 1C to 1G and, in one case, 2C to 1G was found. The base-pairing specificity is the same as that of the polynucleotide interaction.

TABLE XXX

Specificity of the Interaction of Poly U (0.01 *M*) and Nucleotides at 0°C, pH 5.7–6.5

			Precipitation	
	(*M*)	$MgCl_2$, *M*	Occurrence	A/U
3′-AMP	0.02	0.03	−	—
	0.04	0.03	+	0.53
2′-AMP	0.02	0.03	−	—
	0.02	0.05	−	—
	0.02	0.07	+	0.56
5′-AMP	0.02	0.02	+	0.50
5′-TuMP	0.02	0.03	−	—
		0.07	−	—
5′-CMP	0.04	0.05	−	—
5′-IMP	0.04	0.05	−	—
5′-GMP	0.04	0.05	+	no poly U
5′-ATP[a]	0.02	0.04	+	0.52
5′-CTP	0.02	0.05	−	—
5′-ITP	0.02	0.05	−	—

[a] pH 8.5.

The specificity and the stoichiometry of the A · poly U interaction is shown in Table XXX. These stoichiometric, insoluble complexes form and redissolve reversibly in response to temperature changes as illustrated in Table

XXXI. On the other hand, a soluble complex between the mononucleotides and their complementary polynucleotides cannot be detected by optical rotatory dispersion, by sedimentation, or by viscosity, and PMR techniques which were employed successfully before to characterize the adenosine·$(U)_n$ system (*303,306*). The observations from the PMR investigation are particularly convincing: When the spectral positions and the linewidths of the base protons and H(1′) protons of the mononucleotides and the polynucleotides are the same, despite their being dissolved together in a mixture or individually in separate solutions, it may be safely concluded that no appreciable interaction exists among these compounds in the mixture under these conditions. Soluble complexes *cannot* be found in the mixtures of mononucleotides and

TABLE XXXI

Formation of Insoluble Complex of Adenosine Mono- and Triphosphates with Poly U (0.01 *M*)

Nucleotide, 0.02 *M*	$MgCl_2$, *M*	pH	T_p, Dissoc. °C	T_p, Assoc. °C	A/U, (precipitate)
5′-AMP	0.03	7.0	17	12.5	0.46
5′-dAMP	0.03	7.0	24.5	20.5	0.53
	0.03	8.4	—	—	—
5′-ATP	0.04	8.4	7	2	—
	0.05	8.4	10	5	0.52
5′-dATP	0.04	8.4	15	10	—
	0.05	8.4	18	13.5	0.57

the polynucleotides, though *insoluble* complexes will form in these same mixtures at temperatures 10–20° lower than those for the spectral measurement. These data indicate that interaction in these mixtures takes place only when accompanied by a phase transition, such as precipitation or gelation. The following mechanism is proposed. When the conditions are sufficiently favorable, the monomer–polymer interaction occurs through hydrogen bonding between the base pairs and through the cooperative stacking of the mononucleotides alongside the complementary polynucleotides. In the complex, mononucleotides become "polymer-like." Thus, a phase transition occurs, since the polymer–polymer complexes are insoluble under these conditions. This transition of physical state provides the additional driving force needed for the interaction to proceed by removing the complex into another phase of the system.

IV. CONCLUSION

We may conclude, after reviewing the rather voluminous amount of information presented above, that there is a great deal of interesting and yet rather complex physicochemistry about these relatively simple monomeric units of nucleic acids. A large variety of techniques has been applied and many fundamental principles are involved in the interpretation of these data.

It is self-evident that the physicochemical studies on nucleic acids and the interpretation of the data so collected must rely on the fundamental knowledge about the monomeric units, i.e., the bases, nucleosides, and nucleotides. The more precise and the more thorough our understanding about the physicochemical data of the monomers, the better we know how to investigate the polymer.

Another conclusion is that many important properties of nucleic acids are the intrinsic or inherent properties of the monomers. These monomers form hydrogen-bonded base pairs in crystalline solids and in organic solvents with almost identical specificity and bonding scheme as the polymer. These monomers form stacks in aqueous solutions in a fashion very similar to the stacked conformation of the polymers. When interacting with the complementary polymer, the monomers (at least the purine monomers) form both hydrogen-bonded base pairs and costack along the complementary strand, showing "polymer-like" properties. These observations have provided a sound basis for the comprehension of the physicochemical properties of nucleic acids.

This review has demonstrated clearly that the research on nucleic acid done through a thorough study on its monomers is a rewarding approach.†

Acknowledgment. This author wishes to express his gratitude for the support given to his laboratory by AEC, NSF, and NIH for the research in the area of the nucleic acid monomers as well as in the preparation of this manuscript.

He also wishes to thank the following scientists for the communication of their unpublished data which were very helpful in writing this manuscript: Dr. E. D. Bergmann, Dr. C. E. Bugg, Dr. W. Fuller, Dr. P. D. Lawley, Dr. J. H. Lister, and Dr. J. T. Yang.

He wishes to thank the following colleagues for reading the review and providing helpful suggestions: Dr. R. Biltonen, Dr. W. Fuller and Dr. W. Cohn.

Special thanks are due Dr. Lloyd Stempel, of our laboratory, who has proofread this paper in its entirety and has provided very valuable advice and corrections.

This author wishes to thank Mrs. M. Groenheim for her assistance in preparing this paper.

† The literature survey of this review ended in December, 1968.

Recommended General References

I. Synthesis, Structure Proof, Reactivities, and Other Organic and Physico-organic Properties

"Pyrimidine Nucleosides," J. J. Fox and I. Wempen, *Advan. Carbohydrate Chem.*, **14**, 283 (1959).

"Purine Nucleosides," J. A. Montgomery and H. J. Thomas, *Advan. Carbohydrate Chem.*, **17**, 301 (1962).

The Chemistry of Nucleosides and Nucleotides, A. M. Michelson, Academic Press, New York, 1963.

"Physicochemical Aspects of the Chemistry of Purines," J. H. Lister, *Advan. Heterocyclic Chem.*, **6**, 1 (1966).

"The Mononucleotides," T. Ueda and J. J. Fox, *Advan. Carbohydrate Chem.*, **22**, 307 (1967).

"Neighboring-group Participation in Sugars," L. Goodman, *Advan. Carbohydrate Chem.*, **22**, 109 (1967).

"The Purines and Related Ring Systems," R. K. Robins, in *Heterocyclic Compounds*, Vol. 8, Chap. 3 (Elderfield, ed.), Wiley, New York, 1967, p. 162.

II. Physical Properties

"Nucleic Acids," I. Tinoco and D. N. Holcomb, *Ann. Rev. Phys. Chem.*, **15**, 371 (1964).

"The Physical and Chemical Properties of Nucleic Acids," G. Felsenfeld and H. T. Miles, *Ann. Rev. Biochem.*, **36** (Part II), 407 (1967).

"Some Recent Developments in the Quantum-Mechanical Studies on the Electronic Structure of the Nucleic Acids," B. Pullman, *J. Chem. Phys.*, **43**, S233 (1965).

"Aspects of the Electronic Structure of the Purine and Pyrimidine Bases of the Nucleic Acids and of Their Interactions." A. Pullman and B. Pullman, *Advan. Quantum Chem.*, **4**, 267 (1968).

"Polarographic Behavior of Nucleosides and Nucleotides of Purines, Pyrimidines, Pyridines, and Flavins," B. Janik and P. J. Elving, *Chem. Rev.*, **68**, 295 (1968).

"X-ray Diffraction Studies of Macromolecules," D. R. Davies, *Ann. Rev. Biochem.*, **36** (Part I), 321 (1967).

"Nucleoside Conformations: An Analysis of Steric Barriers to Rotation about the Glycosidic Bond," A. E. V. Haschemeyer and A. Rich, *J. Mol. Biol.*, **27**, 369 (1967).

"Stereochemistry of Nucleic Acids and Polynucleotides. I. Theoretical Determination of the Allowed Conformations of the Monomer Unit," V. Sasisekharan, A. V. Lakshminarayanan, and G. N. Ramachandran, in

Conformation of Biopolymers, Vol. 2 (G. N. Ramachandran, ed.), Academic Press, New York, 1967.

"The Optical Rotation and Circular Dichroism of Nucleic Acids," J. T. Yang and T. Samegima, *Progr. Nucleic Acid Res. Mol. Biol.*, **9** (1969), in press.

"Infrared Spectra of Nucleic Acids and Related Compounds," T. Shimanouchi, M. Tsuboi, and Y. Kyogoku, *Advan. Chem. Phys.*, **7**, 435 (1964).

"Raman Spectral Studies of Nucleic Acids and Related Molecules. I. Ribonucleic Acid Derivatives," R. C. Lord and G. J. Thomas, Jr., *Spectrochim. Acta*, **23A**, 2551 (1967).

"Mass Spectra of Nucleic Acid Derivatives-Pyrimidine," J. M. Rice, G. O. Dudek, and M. Barber, *J. Am. Chem. Soc.*, **87**, 4569 (1965).

"Mass Spectra of Nucleic Acid Derivatives. II. Guanine, Adenine, and Related Compounds," J. M. Rice and G. O. Dudek, *J. Am. Chem. Soc.*, **89**, 2719 (1967).

"Mass Spectrometry of Nucleic Acid Components. Trimethysilyl Derivatives of Nucleotides, Nucleosides, and Bases," J. A. McCloskey, A. M. Lawson, K. Tsuboyana, P. M. Kueger, and R. N. Stillwell, *J. Am. Chem. Soc.*, **90**, 4182 (1968).

"Excited Electronic States of DNA," J. Eisinger and R. G. Shulman, *Science*, **161**, 1311 (1968).

III. Special Topics

"Nucleic Acids—Selected Topics Related to their Enzymology and Chemistry," C. A. Dekker, *Ann. Rev. Biochem.*, **29**, 453 (1960).

"Mechanism of Action and Application of Azapyrimidines," J. Skoda, *Progr. Nucleic Acid Res. Mol. Biol.*, **2**, 197 (1963).

"5-Hydroxymethyl Pyrimidines and their Derivatives," T. L. V. Ulbricht, *Progr. Nucleic Acid Res. Mol. Biol.*, **4**, 189 (1965).

"Fluorinated Pyrimidines," C. Heidelberger, *Progr. Nucleic Acid Res. Mol. Biol.*, **4**, 2 (1965).

"Amino Acid Esters of RNA, Nucleosides, and Related Compounds," H. G. Zachau and H. Feldmann, *Progr. Nucleic Acid Res. Mol. Biol.*, **4**, 217 (1965).

"The Chemistry of Pseudouridine," R. W. Chambers, *Progr. Nucleic Acid Res. Mol. Biol.*, **5**, 350 (1966).

"Introduction to the Biochemistry of D-arabinosyl Nucleosides," S. S. Cohen, *Progr. Nucleic Acid Res. Mol. Biol.*, **5**, 2 (1966).

"Chemistry of Guanine and Its Biologically Significant Derivatives," R. Shapiro, *Progr. Nucleic Acid Res. Mol. Biol.*, **8**, 73 (1968).

REFERENCES

1. B. F. C. Clark, B. P. Doctor, K. C. Holmes, A. Klug, K. A. Marcker, S. J. Morris, and H. H. Paradies, *Nature*, **219**, 1222 (1968).
2. S. H. Kim and A. Rich, *Science*, **162**, 1381 (1968).
3. A. Hampel, M. Labanauskas, P. G. Connors, L. Kirkegard, U. T. Rajbhandary, P. B. Sigler, and R. M. Bock, *Science*, **162**, 1384 (1968).
4. A. E. V. Haschemeyer and A. Rich, *J. Mol. Biol.*, **27**, 369 (1967).
5. V. Sasisekharan, A. V. Lakshminarayanan, and G. N. Ramachandran, in *Conformation of Biopolymers*, Vol. 2 (G. N. Ramachandran, ed.), Academic, New York, 1967, p. 641.
6. R. A. Davies, *Ann. Rev. Biochem.*, **36**, 321 (1967).
7. J. M. Broomhead, *Acta Cryst.*, **4**, 92 (1951).
8. G. A. Jeffrey and Y. Kinoshita, *Acta Cryst.*, **16**, 20 (1963).
9. R. Gerdil, *Acta Cryst.*, **14**, 333 (1961).
10. D. G. Watson, D. J. Sutor, and P. Tollin, *Acta Cryst.*, **19**, 111 (1965).
11. K. N. Trueblood, P. Horn, and V. Luzzati, *Acta Cryst.*, **14**, 965 (1961).
12. M. Sundaralingam, *Acta Cryst.*, **21**, 495 (1966).
13. E. Shefter and K. N. Trueblood, *Acta Cryst.*, **18**, 1067 (1965).
14. S. T. Rao and M. Sundaralingam, *Chem. Commun.*, **16**, 955 (1968).
15. K. Hoogsteen, *Acta Cryst.*, **16**, 28 (1963).
16. D. G. Watson, R. M. Sweet, and R. E. Marsh, *Acta Cryst.*, **19**, 573 (1965).
17. C. E. Bugg, U. T. Thewalt, and R. E. Marsh, *Biochem. Biophys. Res.* Commun., **33**, 436 (1968).
18. J. Kraut and L. H. Jensen, *Acta Cryst.*, **16**, 79 (1963).
19. C. E. Bugg and R. E. Marsh, *J. Mol. Biol.*, **25**, 67 (1967).
20. D. L. Barker and R. E. Marsh, *Acta Cryst.*, **17**, 1581 (1964).
21. M. Sundaralingam and L. H. Jensen, *J. Mol. Biol.*, **13**, 914 (1965).
22. C. L. Coulter, *Science*, **159**, 888 (1968).
23. K. Watenpaugh, J. Dow, L. H. Jensen, and S. Furberg, *Science*, **159**, 206 (1968).
24. E. Shefter and H. G. Mautner, *J. Am. Chem. Soc.*, **89**, 1249 (1967).
25. E. Shefter, M. Barlow, R. Sparks, and K. N. Trueblood, *J. Am. Chem. Soc.*, **86**, 1872 (1964).
26. G. N. Reeke, Jr., and R. E. Marsh, *Acta Cryst.*, **20**, 703 (1966).
27. D. Rohrer and M. Sundaralingam, *Chem. Commun.*, 746 (1968).
28. D. G. Watson, D. J. Sutor, and P. Tollin, *Acta Cryst.*, **18**, 313 (1965).
29. L. Katz, K. Tomita, and A. Rich, *J. Mol. Biol.*, **13**, 340 (1965).
30. S. Furberg, C. S. Petersen, and C. Rømming, *Acta Cryst.*, **18**, 313 (1965).
31. J. Iball, C. H. Morgan, and H. R. Wilson, *Nature*, **209**, 1230 (1966).
32. A. E. V. Haschemeyer and H. M. Sobell, *Acta Cryst.*, **18**, 525 (1965).
33. D. R. Harris and W. M. Macintyre, *Biophys. J.*, **4**, 203 (1964).
34. A. E. V. Haschemeyer and H. M. Sobell, *Nature*, **202**, 969 (1964).
35. A. E. V. Haschemeyer and H. M. Sobell, *Acta Cryst.*, **19**, 125 (1965).
36. P. O. P. Ts'o, N. S. Kondo, M. P. Schweizer, and D. P. Hollis, *Biochemistry*, **8**, 997 (1969).
37. D. L. Barker and R. E. Marsh, *Acta Cryst.*, **17**, 1581 (1964).
38. D. G. Watson, R. M. Sweet, and R. E. Marsh, *Acta Cryst.*, **19**, 573 (1965).
39. B. Pullman, H. Berthod, and J. Caillet, *Theoret. Chim. Acta* (Berlin), **10**, 43 (1968).
40. W. Cochran, *Acta Cryst.*, **4**, 81 (1951).
41. R. F. Bryan and K. I. Tomita, *Acta Cryst.*, **15**, 1174 (1962).

42. M. Sundaralingam and L. H. Jensen, *J. Mol. Biol.*, **13,** 930 (1965).
43. R. E. Marsh, R. Bierstedt, and E. L. Eichhorn, *Acta Cryst.*, **15,** 310 (1962).
44. E. O. Akinrimisi, C. Sander, and P. O. P. Ts'o, *Biochemistry*, **2,** 340 (1963).
45. R. Langridge and A. Rich, *Nature*, **198,** 725 (1963).
46. K. A. Hartman, Jr., and A. Rich, *J. Am. Chem. Soc.*, **87,** 2033 (1965).
47. H. DeVoe and I. Tinoco, Jr., *J. Mol. Biol.*, **4,** 500 (1962).
48. B. Pullman, *J. Chem. Phys.*, **43,** No. 10, S233 (1965).
49. A. Pullman and B. Pullman, *Advan. Quantum Chem.*, **4,** 267 (1968).
50. P. Claverie, in *Molecular Associations in Biology* (B. Pullman, ed.), Academic, New York (1968), p. 115.
51. A. Nash and D. F. Bradley, *Biopolymers*, **3,** 261 (1965).
52. A. Nash and D. F. Bradley, *J. Chem. Phys.*. **45,** 1360 (1966).
53. C. Lifschitz, E. D. Bergmann, and B. Pullman, *Tetrahedron Letters*, **46,** 4583 (1967).
54. I. Omura, H. Baba, and K. Higasi, *Bull. Chem. Soc. Japan*, **30,** 633 (1957).
55. E. D. Bergmann and C. Lifschitz, in *Physicochemical Mechanism of Carcinogenesis*, (E. D. Bergmann and B. Pullman, eds.). The Israel Academy of Sciences and Humanities, Jerusalem (1969), p. 72.
56. H. Weiler-Feilchenfeld and E. D. Bergmann, *Israel J. Chem.*, **6,** 823 (1969).
57. C. P. Smyth, *Dielectric Behavior and Structure*, McGraw-Hill, New York (1955).
58. H. Lumbroso, C. Pigenet, R. Nasielski-Hinskens, and R. Promel, *Bull. Soc. Chim. France*, **1967,** 1833.
59. B. Pullman, E. D. Bergmann, H. Weiler-Feilchenfeld, and Z. Neiman, *Int. J. Quant. Chem.*, **35,** 103 (1969).
60. C. D. Jardetzky and O. Jardetzky, *J. Am. Chem. Soc.*, **82,** 222 (1960).
61. C. D. Jardetzky, *J. Am. Chem. Soc.*, **83,** 2919 (1960).
62. C. D. Jardetzky, *J. Am. Chem. Soc.*, **82,** 229 (1960).
63. M. P. Schweizer, S. I. Chan, G. K. Helmkamp, and P. O. P. Ts'o, *J. Am. Chem. Soc.*, **86,** 696 (1964).
64. A. D. Broom, M. P. Schweizer, and P. O. P. Ts'o, *J. Am. Chem. Soc.*, **89,** 3612 (1967)
65. P. O. P. Ts'o, M. P. Schweizer, and D. P. Hollis, *Ann. N. Y. Acad. Sci.*, **158,** 256 (1969).
66. M. P. Schweizer, Ph.D. thesis, The Johns Hopkins University (1968).
67. T. Schaefer and W. G. Schneider, *Can. J. Chem.*, **41**, 966 (1963).
68. A. Veillard, *J. Chim. Phys.*, **59,** 1056 (1962).
69. B. M. Lynch and H. J. Dou, *Tetrahedron Letters*, **1966**, 2627.
70. A. D. Buckingham, *Can. J. Chem.*, **38,** 300 (1960).
71. J. I. Musher, *J. Chem. Phys.*, **37,** 34 (1962).
72. C. Giessner-Prettre and B. Pullman, *Compt. Rend.*, **261,** 2521 (1965).
73. S. Gronowitz and R. A. Hoffman, *Arkiv. Kemi*, **16,** 459 (1960).
74. R. L. Miller, P. G. Lykos, and H. N. Schmeising, *J. Am. Chem. Soc.*, **84,** 4623 (1962).
75. A. Veillard and B. Pullman, *Compt. Rend.*, **253,** 2277 (1961).
76. J. A. Pople, W. G. Schneider, and H. J. Bernstein, *High Resolution Nuclear Magnetic Resonance*, McGraw-Hill, New York, 1959, p. 178.
77. V. M. S. Gil and J. N. Murrell, *Trans. Faraday Soc.*, **60,** 248 (1964).
78. S. Matsuura and T. Goto, *Tetrahedron Letters*, **1963**, No. 22, 1499.
79. F. J. Bullock and O. Jardetzky, *J. Org. Chem.*, **29,** 1988 (1964).
80. G. S. Reddy, L. Mandell, and J. H. Goldstein, *J. Chem. Soc.*, **1963,** 1414.
81. A. Veillard and B. Pullman, *J. Theoret. Biol.*, **4,** 37 (1963).
82. B. Pullman and A. Pullman, *Quantum Biochemistry*, Wiley (Interscience), New York, 1963, p. 764.
83. J. I. Fernandez-Alonso, *Advan. Chem. Phys.*, **7,** 1 (1964).

84. S. I. Chan, M. P. Schweizer, P. O. P. Ts'o, and G. K. Helmkamp, **86,** 4182 (1964).
85. F. E. Hruska, C. L. Bell, T. A. Victor, and S. S. Danyluk, *Biochemistry*, **7,** 3721 (1968).
86. P. O. P. Ts'o, N. S. Kondo, R. K. Robins, and A. D. Broom, *J. Am. Chem. Soc.*, **91,** 5625 (1969).
87. R. J. Pugmire, D. M. Grant, R. K. Robins, and G. W. Rhodes, *J. Am. Chem. Soc.*, **87,** 2225 (1965).
88. J. R. Fox, Ph.D. thesis, Univ. of Illinois, 1965.
89. W. C. Coburn, Jr., M. C. Thorpe, J. A. Montgomery, and K. Hewson, *J. Org. Chem.*, **30,** 1114 (1965).
90. O. Jardetzky, *Biopolymer Symp.*, **1,** 501 (1965).
91. J. P. Kokko, L. Mandel, and J. H. Goldstein, *J. Am. Chem. Soc.*, **84,** 1042 (1962).
92. E. D. Becker, H. T. Miles, and R. B. Bradley, *J. Am. Chem. Soc.*, **87,** 5575 (1965).
93. J. Ladik and K. Appel, *Theoret. Chim. Acta*, **4,** 132 (1966).
94. M. J. Robins and R. K. Robins, *J. Am. Chem. Soc.*, **87,** 4934 (1965).
95. R. J. Cushley, K. A. Watanabe, and J. J. Fox, *J. Am. Chem. Soc.*, **89,** 394 (1967).
96. H. P. M. Fromageot, B. E. Griffin, C. B. Reese, J. E. Sulston, and D. R. Trentham, *Tetrahedron*, **22,** 705 (1966).
97. M. P. Schweizer, A. D. Broom, P. O. P. Ts'o, and D. P. Hollis, *J. Am. Chem. Soc.*, **90,** 1042 (1968).
98. H. T. Miles, F. B. Howard, and J. Frazier, *Science*, **142,** 1458 (1963).
99. J. M. Read and J. H. Goldstein, *J. Am. Chem. Soc.*, **87,** 3440 (1965).
100. H. T. Miles, R. B. Bradley, and E. D. Becker, *Science*, **142,** 1569 (1963).
101. O. Jardetzky, P. Pappas, and N. G. Wade, *J. Am. Chem. Soc.*, **85,** 1657 (1963).
102. J. C. N. Ma and E. W. Warnhoff, *Can. J. Chem.*, **43,** 1849 (1965).
103. C. D. Jardetzky, *J. Am. Chem. Soc.*, **84,** 62 (1962).
104. M. Karplus, *J. Chem. Phys.*, **30,** 11 (1959).
105. R. V. Lemieux, J. D. Stevens, and R. R. Fraser, *Can. J. Chem.*, **40,** 1955 (1962).
106. R. V. Lemieux and J. D. Stevens, *Can. J. Chem.*, **44,** 249 (1965).
107. M. Tsuboi, F. Kuriyagawa, K. Matsuo, and Y. Kyogoku, *Bull. Chem. Soc. Japan*, **40,** 1813 (1967).
108. M. Tsuboi, M. Kainosho, and A. Nakamura, *Recent Developments of Magnetic Resonance in Molecular Biology* (S. Fujiwara and L. H. Piette, eds.), Hirokawa, Tokyo, 1968.
109. R. R. Shoup, H. T. Miles, and E. D. Becker, *J. Am. Chem. Soc.*, **89,** 6200 (1967).
110. G. L. Eichhorn, P. Clark, and E. D. Becker, *Biochemistry*, **5,** 245 (1966).
111. S. M. Wang and N. C. Li, *J. Am. Chem. Soc.*, **88,** 4592 (1966).
112. S. M. Wang and N. C. Li, *J. Am. Chem. Soc.*, **90,** 5069 (1968).
113. M. Cohn and T. R. Hughes, Jr., *J. Biol. Chem.*, **237,** 176 (1962).
114. R. L. Ward and J. A. Happe, *Biochem. Biophys. Res. Commun.*, **28,** 785 (1967).
115. R. G. Shulman, H. Sternlicht, and B. J. Wyluda, *J. Chem. Phys.*, **43,** 3116 (1965).
116. H. Sternlicht, R. G. Shulman, and E. W. Anderson, *J. Chem. Phys.*, **43,** 3123 (1965).
117. H. Sternlicht, R. G. Shulman, and E. W. Anderson, *J. Chem. Phys.*, **43,** 3133 (1965).
118. H. Sternlicht, D. E. Jones, and K. Kustin, *J. Am. Chem. Soc.*, **90,** 7110 (1968).
119. P. O. P. Ts'o, N. S. Kondo, M. P. Schweizer, and D. P. Hollis, *Biochemistry*, **8,** 997 (1969).
120. F. E. Hruska and S. S. Danyluk, *J. Am. Chem. Soc.*, **90,** 3266 (1968).
121. R. L. Sinsheimer, R. L. Nutter, and G. R. Hopkins, *Biochim. Biophys. Acta*, **18,** 13 (1955).
122. H. T. Miles, *Biochim. Biophys. Acta*, **30,** 328 (1958).

123. C. L. Angell and R. L. Werner, *J. Chem. Soc.*, **1952,** 2911.
124. C. L. Angell, *J. Chem. Soc.*, **1961,** 504.
125. M. Tsuboi, *J. Am. Chem. Soc.*, **79,** 1351 (1957).
126. E. R. Blout and H. Lenormant, *Biochim. Biophys. Acta*, **17,** 325 (1955).
127. G. B. B. M. Sutherland and M. Tsuboi, *Proc. Roy. Soc.* (London), **A239,** 446 (1957).
128. G. R. Wilkinson, W. C. Price, and E. M. Bradbury, *Spectrochim. Acta*, **14,** 284 (1959).
129. R. S. Morgan and E. R. Blout, *J. Am. Chem. Soc.*, **81,** 4625 (1959).
130. Y. Kyogoku, M. Tsuboi, T. Shimanouchi, and I. Watanabe, *Nature*, **189,** 120 (1961).
131. E. M. Bradbury, W. C. Price, and G. R. Wilkinson, *J. Mol. Biol.*, **3,** 301 (1961).
132. J. Pitha and J. Beranek, *Coll. Czech. Chem. Commun.*, **28,** 1507 (1963).
133. J. Pitha, S. Chládek, and J. Smrt, *Coll. Czech. Chem. Commun.*, **28,** 1622 (1963).
134. J. Pitha and J. Zemlicka, *Coll. Czech. Chem. Commun.*, **29,** 410 (1964).
135. J. Pitha and S. Vasickova, *Coll. Czech. Chem. Commun.*, **30,** 1792 (1965).
136. T. Shimanouchi, M. Tsuboi, and Y. Kyogoku, *Advan. Chem. Phys.*, **7,** 435 (1964).
137. M. Tsuboi, Y. Kyogoku, and T. Shimanouchi, *Biochim. Biophys. Acta*, **55,** 1 (1962).
138. T. Sato, Y. Kyogoku, S. Higuchi, Y. Mitsui, Y. Iitaka, and M. Tsuboi, *J. Mol. Biol.*, **16,** 180 (1966).
139. H. T. Miles, F. B. Howard, and J. Frazier, *Science*, **142,** 1458 (1963).
140. F. B. Howard and H. T. Miles, *J. Biol. Chem.*, **240,** 801 (1965).
141. R. C. Lord and G. J. Thomas, Jr., *Spectrochim. Acta*, **23A,** 2551 (1967).
142. Y. Kyogoku, S. Higuchi, and M. Tsuboi, *Spectrochim. Acta*, **23A,** 969 (1967).
143. D. Voet, W. B. Gratzer, R. A. Cox, and P. Doty, *Biopolymers*, **1,** 193 (1963).
144. I. Tinoco, Jr., and D. N. Holcomb, *Ann. Rev. Phys. Chem*,, **15,** 371 (1964).
145. T. Yamada and H. Fukutome, *Biopolymers*, **6,** 43 (1968).
146. L. B. Clark, G. G. Peschel, and I. Tinoco, Jr., *J. Phys. Chem.*, **69,** 3615 (1965).
147. J. E. Parkin and K. K. Innes, *J. Mol. Spectry.*, **15,** 407 (1965).
148. R. F. Stewart and N. Davidson, *J. Chem. Phys.*, **39,** 255 (1963).
149. R. F. Stewart and N. Davidson, *Biopolymers Symp.*, **1,** 465 (1964).
150. R. F. Stewart and L. H. Jensen, *J. Chem. Phys.*, **40,** 2071 (1964).
151. S. F. Mason, *J. Chem. Soc.*, **1965,** 2071.
152. L. B. Clark and I. Tinoco, Jr., *J. Am. Chem. Soc.*, **87,** 11 (1965).
153. E. A. Dratz and R. L. Biltonen, data to be published, 1969.
154. P. R. Callis, E. J. Rosa, and W. T. Simpson, *J. Am. Chem. Soc.*, **86,** 2292 (1964).
155. I. Tinoco, Jr., *J. Chem. Phys.*, **33,** 1332 (1960); **34,** 1087 (1961).
156. I. Tinoco, Jr., *J. Am. Chem. Soc.*, **82,** 4785 (1961).
157. W. Rhodes, *J. Am. Chem. Soc.*, **83,** 3609 (1961).
157a. H. DeVoe and I. Tinoco, Jr., *J. Mol. Biol.*, **4,** 518 (1962).
158. J. Brahms, K. E. van Holde, and A. M. Michelson, *J. Mol. Biol.*, **15,** 467 (1966).
158a. A. M. Michelson, *J. Chem. Soc.*, **1959,** 1371.
159. H. C. Bolton and J. J. Weiss, *Nature*, **195,** 666 (1962).
160. R. K. Nesbet, *Biopolymers Symp.*, **1,** 129 (1964); *Mol. Phys.*, **7,** 211 (1964).
161. G. J. Thomas, Jr., and Y. Kyogoku, *J. Am. Chem. Soc.*, **89,** 4170 (1967).
162. W. B. Gratzer and C. W. F. McClare, *J. Am. Chem. Soc.*, **89,** 4224 (1967).
163. E. Charney and M. Gellert, *Biopolymers Symp.*, **1,** 169 (1964).
164. P. O. P. Ts'o, N. S. Kondo, R. K. Robins, and A. D. Broom, *J. Am. Chem. Soc.*, **91,** 5625 (1969).
165. J. Drobnik and L. Augenstein, *Photochem. Photobiol.*, **5,** 13, 83 (1966).
166. M. Kasha, *Light and Life* (W. B. McElroy and B. Glass, eds.), The Johns Hopkins Press (1961), p. 31.
166a. S. F. Mason, *Quart. Rev.*, **3,** 287 (1961).

167. H. Berthod, C. Giessner-Prettre, and A. Pullman, *Theoret. Chim. Acta*, **5,** 53 (1966); *Compt. Rend. Acad. Sci.*, **262,** 2657 (1966).
168. C. A. Dekker, *Ann. Rev. Biochem.*, **29,** 453 (1960).
169. R. Shapiro, *Progr. Nucleic Acid Res. Molecular Biol.*, **8,** 73 (1968).
170. P. Lawley, "Ionization and Ultraviolet Spectra of Purines," in *Purine* (J. H. Lister, ed.). Interscience Publishers, New York and London, 1971.
171. S. Udenfriend and P. Zaltzman, *Anal. Biochem.*, **3,** 49 (1962).
172. J. Eisinger and R. G. Shulman, *Science*, **161,** 1311 (1968).
173. J. W. Longworth, R. O. Rahn, and R. G. Shulman, *J. Chem. Phys.*, **45,** 2930 (1966).
174. A. Haug and P. Douzou, *Z. Naturforsch.*, **206,** 509 (1965).
175. B. J. Cohen and L. Goodman, *J. Am. Chem. Soc.*, **87,** 5487 (1965).
176. R. Bersohn and I. Isenberg, *J. Chem. Phys.*, **40,** 3175 (1964).
177. C. Helene, *Biochem. Biophys. Res. Commun.*, **22,** 237 (1966).
178. H. C. Børresen, *Acta Chem. Scand.*, **21,** 920 (1967).
179. R. O. Rahn, J. W. Longworth, J. Eisinger, and R. G. Shulman, *Proc. Natl. Acad. Sci. U.S.*, **51,** 1299 (1964).
180. H. C. Børresen, *Acta Chem. Scand.*, **17,** 921 (1963).
181. A. A. Lamola, M. Guéron, T. Yamane, J. Eisinger, and R. G. Shulman, *J. Chem. Phys.*, **47,** 2210 (1967).
182. P. O. P. Ts'o, G. K. Helmkamp, and C. Sander, *Biophys. Biochim. Acta*, **55,** 584 (1962).
183. A. Moscowitz, *Optical Rotatory Dispersion* (C. Djerassi, ed.), McGraw-Hill, New York, 1960, p. 150.
184. J. T. Yang, T. Samejima, and P. K. Sarkar, *Biopolymers*, **4,** 623 (1966).
185. J. T. Yang and T. Samejima, *Progr. Nucleic Acid Res. Mol. Biol.*, **9** (1969), in press.
186. G. D. Fasman, C. Lindblow, and L. Grossman, *Biochemistry*, **3,** 1015 (1964).
187. M. R. Lamborg, P. C. Zamecnik, T. K. Li, J. Kagi, and B. L. Vallee, *Biochemistry*, **4,** 63 (1965).
188. K. E. van Holde, J. Brahms, and A. M. Michelson, *J. Mol. Biol.*, **12,** 726 (1965).
189. J. Brahms, J. C. Maurizot, and A. M. Michelson, *J. Mol. Biol.*, **25,** 481 (1967).
190. D. M. Miles, R. K. Robins, and H. Eyring, *Proc. Natl. Acad. Sci. U.S.*, **57,** 1138 (1967).
191. E. M. Acton, K. J. Ryan, and L. Goodman, *J. Am. Chem. Soc.*, **86,** 5352 (1964).
192. T. Nischimura, B. Shimizu, and I. Iwai, *Biochim. Biophys. Acta*, **157,** 221 (1968).
193. C. S. Hudson, *J. Am. Chem. Soc.*, **31,** 66 (1909).
194. J. J. Fox, N. C. Yung, I. Wempen, and M. Hoffer, *J. Am. Chem. Soc.*, **83,** 4066 (1961).
195. J. J. Fox and I. Wempen, *Advan. Carbohydrate Chem.*, **14,** 283 (1959).
196. R. V. Lemieux and M. Hoffer, *Can. J. Chem.*, **39,** 110 (1961).
197. T. L. V. Ulbricht, J. P. Jenning, P. M. Scopes, and W. Klyne, *Tetrahedron Letters*, **1964,** 695.
198. T. L. V. Ulbricht, T. R. Emerson, and R. J. Swan, *Biochem. Biophys. Res. Commun.*, **19,** 643 (1965).
199. T. L. V. Ulbricht, T. R. Emerson, and R. J. Swan, *Tetrahedron Letters*, **1966,** 1561.
200. T. R. Emerson, R. J. Swan, and T. L. V. Ulbricht, *Biochem. Biophys. Res. Commun.*, **22,** 505 (1966).
201. T. R. Emerson, R. J. Swan, and T. L. V. Ulbricht, *Biochemistry*, **6,** 843 (1967).
202. I. Fric, J. Smejkal, and J. Farkas, *Tetrahedron Letters*, **1966,** 75.
203. D. J. Caldwell and H. Eyring, *Ann. Rev. Phys. Chem.*, **15,** 281 (1964).
204. A. Hampton and A. W. Nichol, *J. Org. Chem.*, **32,** 1688 (1967).
205. W. A. Klee and S. H. Mudd, *Biochemistry*, **6,** 988 (1967).

206. M. Ikehara, M. Kaneko, K. Muneyama, and H. Tanaka, *Tetrahedron Letters*, **1967,** 3977.
207. C. Y. Lin, D. W. Urry, and H. Eyring, *Biochem. Biophys. Res. Commun.*, **17,** 642 (1964).
208. J. Donohue and K. N. Trueblood, *J. Mol. Biol.*, **2,** 363 (1960).
209. P. Tollin, H. R. Wilson, and D. W. Young, *Nature*, **217,** 1148 (1968).
210. P. M. Huber, *Acta Cryst.*, **10,** 129 (1957).
211. N. Camerman and J. Trotter, *Science*, **144,** 1348 (1964).
212. E. Shefter and T. I. Kalman, *Biochem. Biophys. Res. Commun.*, **32,** 878 (1968).
213. S. Furberg, C. S. Petersen, and C. H. R. Rømming, *Acta Cryst.*, **18,** 313 (1965).
214. G. Koyama, K. Maeda, H. Umezawa, and Y. Iitaka, *Tetrahedron Letters*, **1966,** 597.
215. M. E. Beynon and W. Fuller, personal communication.
216. I. Tinoco, Jr., R. C. Davis, and S. R. Jaskunas, *Molecular Associations in Biology* (B. Pullman, ed.), Academic, New York, 1968.
217. F. Jordan and B. Pullman, *Theoret. Chim. Acta* (Berlin), **9,** 242 (1968).
218. P. Lira and C. W. Huffman, *J. Org. Chem.*, **31,** 2188 (1966).
219. W. J. Wechter, *J. Org. Chem.*, **34,** 244 (1969).
220. S. S. Danyluk and F. E. Hruska, *Biochemistry*, **7,** 1038 (1968).
221. R. Phillips, S. J. P. Eisenberg, P. George, and R. J. Rutman, *J. Biol. Chem.*, **240,** 4393 (1965).
222. A. M. Michelson, *Ann. Rev. Biochem.*, **30,** 133 (1961).
223. J. J. Fox and D. Shugar, *Biochem. Biophys. Acta*, **9,** 369 (1952).
224. J. J. Fox, L. F. Cavalieri, and N. Chang, *J. Am. Chem. Soc.*, **75,** 4315 (1953).
225. J. J. Fox, J. F. Codington, N. C. Yang, L. Kaplan, and J. O. Lampen, *J. Am. Chem. Soc.*, **80,** 5155 (1958).
226. J. Pitha and P. O. P. Ts'o, unpublished data. 1967.
227. L. Grossman, S. S. Levine, and W. S. Allison, *J. Mol. Biol.*, **3,** 47 (1961).
228. L. M. Stempel and P. O. P. Ts'o, unpublished data, 1968.
229. P. A. Levene and L. W. Bass, *Nucleic Acid*, Chemical Catalog Co., New York (1931).
230. H. Witzel, *Ann. Chem.*, **635,** 182 (1960).
231. H. Witzel, *Progr. Nucleic Acid Res.*, **2,** 221 (1963).
232. P. O. P. Ts'o, S. A. Rapaport, and F. J. Bollum, *Biochemistry*, **5,** 4153 (1966).
233. M. J. Chamberlin, *Federation Proc.*, **24,** 1446 (1965).
234. M. J. Chamberlin and D. L. Patterson, *J. Mol. Biol.*, **12,** 410 (1965).
235. M. Riley, B. Maling, and M. J. Chamberlin, *J. Mol. Biol.*, **20,** 359 (1966).
236. A. Adler, L. Grossman, and G. D. Fasman, *Proc. Natl. Acad. Soc. U.S.*, **57,** 423 (1967).
237. A. J. Adler, L. Grossman, and G. D. Fasman, *Biochemistry*, **7,** 3836 (1968).
238. J. N. Vournakis, D. Poland, and H. A. Scheraga, *Biopolymers*, **5,** 403 (1967).
239. K. Hoogsteen, *Acta Cryst.*, **12,** 822 (1959).
240. K. Hoogsteen, *Acta Cryst.*, **16,** 907 (1963).
241. H. M. Sobell, K. Tomita, and A. Rich, *Proc. Natl. Acad. Sci. U.S.*, **49,** 885 (1963).
242. E. J. O'Brien, *J. Mol. Biol.*, **7,** 107 (1963).
243. A. E. V. Haschemeyer and H. M. Sobell, *Acta Cryst.*, **19,** 125 (1965).
244. E. J. O'Brien, *J. Mol. Biol.*, **22,** 377 (1966).
245. H. M. Sobell, *J. Mol. Biol.*, **18,** 1 (1966).
246. F. S. Mathews and A. Rich, *J. Mol. Biol.*, **8,** 89 (1964).
247. L. Katz, K. Tomita, and A. Rich, *J. Mol. Biol.*, **13,** 340 (1965).
248. L. Katz, K. Tomita, and A. Rich, *Acta Cryst.*, **21,** 754 (1966).
249. Yu. G. Baklagine, M. V. Volkenshtein, and Yu. D. Kondrasher, *Zh. Struk. Khim.*, **7,** 399 (1966).

250. T. Tomita, L. Katz, and A. Rich, *J. Mol. Biol.*, **30,** 545 (1967).
251. A. E. V. Haschemeyer and H. M. Sobell, *Acta Cryst.*, **18,** 525 (1965).
252. J. Donohue, *Proc. Natl. Acad. Sci. U.S.*, **42,** 60 (1956).
253. J. Donohue and K. Trueblood, *J. Mol. Biol.*, **2,** 362 (1960).
254. K. Hoogsteen, *Molecular Associations in Biology* (B. Pullman, ed.), Academic, New York, 1968, p. 21.
255. R. M. Hamlin, Jr., R. C. Lord, and A. Rich, *Science*, **148,** 1734 (1965).
256. E. Kuchler and J. Derkosch, *Z. Naturforsch.*, **21b,** 209 (1966).
257. J. Pitha, N. Jones, and P. Pithova, *Can. J. Chem.*, **44,** 1045 (1966).
258. R. R. Shoup, H. T. Miles, and E. D. Becker, *Biochem. Biophys. Res. Commun.*, **23,** 194 (1966).
259. L. Katz and S. Penman, *J. Mol. Biol.*, **15,** 220 (1966).
260. Y. Kyogoku, R. C. Lord, and A. Rich, *Science*, **154,** 518 (1966).
261. Y. Kyogoku, R. C. Lord, and A. Rich, *J. Am. Chem. Soc.*, **89,** 496 (1967).
262. Y. Kyogoku, R. C. Lord, and A. Rich, *Proc. Natl. Acad. Sci. U.S.*, **57,** 250 (1967).
263. J. H. Miller and H. M. Sobell, *J. Mol. Biol.*, **24,** 345 (1967).
264. R. A. Newmark and C. R. Cantor, *J. Am. Chem. Soc.*, **90,** 5010 (1968).
265. G. G. Hammes and A. C. Park, *J. Am. Chem. Soc.*, **90,** 4151 (1968).
266. D. F. Bradley and H. A. Nash, *Molecular Associations in Biology* (B. Pullman, ed.), Academic, New York, 1968.
267. B. Pullman, P. Claverie, and J. Caillet, *Proc. Natl. Acad. Sci. U.S.*, **55,** 904 (1966).
268. B. Pullman, P. Claverie, and J. Caillet, *J. Mol. Biol.*, **22,** 373 (1967).
269. P. O. P. Ts'o, I. S. Melvin, and A. C. Olson, *J. Am. Chem. Soc.*, **85,** 1289 (1963).
270. P. O. P. Ts'o and S. I. Chan, *J. Am. Chem. Soc.*, **86,** 4176 (1964).
271. T. N. Solie and J. A. Schellman, *J. Mol. Biol.*, **33,** 61 (1968).
272. J. A. Schellman, *Compt. Rend. Trav. Lab. Carlsberg*, **29,** 223 (1956).
273. G. K. Helmkamp and N. S. Kondo, *Biochim. Biophys. Acta*, **157,** 242 (1968).
274. K. E. van Holde and G. P. Rossetti, *Biochemistry*, **6,** 2189 (1967).
275. G. P. Rossetti and K. E. van Holde, *Biochem. Biophys. Res. Commun.*, **26,** 717 (1967).
276. D. P. Robinson and M. E. Grant, *J. Biol. Chem.*, **241,** 4030 (1966).
277. S. J. Gill, M. Downing, and G. F. Sheats, *Biochemistry*, **6,** 272 (1967).
278. E. L. Farquhar, M. Downing, and S. J. Gill, *Biochemistry*, **7,** 1224 (1968).
279. S. J. Gill, D. B. Martin, and M. Downing, *J. Am. Chem. Soc.*, **85,** 706 (1963).
280. M. Gellert, M. N. Lipsett, and D. R. Davies, *Proc. Natl. Acad. Sci. U.S.*, **48,** 2013 (1962).
281. P. K. Sarkar and J. T. Yang, *Biochem. Biophys. Res. Commun.*, **20,** 346 (1965).
282. R. V. Ravindranathan and H. T. Miles, *Biochim. Biophys. Acta*, **94,** 603 (1965).
283. R. C. Lord and G. J. Thomas, Jr., *Biochem. Biophys. Acta*, **142,** 1 (1967).
284. M. P. Schweizer, S. I. Chan, and P. O. P. Ts'o, *J. Am. Chem. Soc.*, **87,** 5241 (1965).
285. B. Pullman, in *Molecular Biophysics* (B. Pullman and M. Weissbluth, ed.), Academic, New York, 1965, pp. 154–157.
286. S. Hanlon, *Biochem. Biophys. Res. Commun.*, **23,** 861 (1966).
287. D. M. Crothers and D. I. Ratner, *Biochemistry*, **7,** 1823 (1968).
288. O. Sinanoglu, in *Molecular Associations in Biology* (B. Pullman, ed.), Academic, New York, 1968, p. 427.
289. O. Sinanoglu and S. Abdulnur, *Federation Proc.*, **24,** No. 2, Part III, S-12 (1965).
290. O. Sinanoglu, S. Abdulnur, and N. R. Krestner, in *Electronic Aspects of Biochemistry* (B. Pullman, ed.), Academic, New York, 1964, p. 301.
291. P. O. P. Ts'o, G. K. Helmkamp, and C. Sander, *Proc. Natl. Acad. Sci. U.S.*, **48,** 686 (1962).

292. P. O. P. Ts'o and P. Lu, *Proc. Natl. Acad. Sci. U.S.*, **51,** 17 (1964).

293. S. I. Chan, B. W. Bangerter, and H. H. Peter, *Proc. Natl. Acad. Sci. U.S.*, **55,** 720 (1966).

294. B. W. Bangerter and S. I. Chan, *Biopolymers*, **6,** 983 (1968).

295. B. W. Bangerter and S. I. Chan, *Proc. Natl. Acad. Sci. U.S.*, **60,** 1144 (1968).

296. F. B. Howard, J. Frazier, M. F. Singer, and H. T. Miles, *J. Mol. Biol.*, **16,** 415 (1966).

297. E. S. Maxwell, L. M. Barnett, F. B. Howard, and H. T. Miles, *J. Mol. Biol.*, **16,** 440 (1966).

298. W. H. Huang and P. O. P. Ts'o, *J. Mol. Biol.*, **16,** 523 (1966).

299. P. M. Pitha, W. M. Huang, and P. O. P. Ts'o, *Proc. Natl. Acad. Sci. U.S.*, **61,** 332 (1968).

300. T. L. Hill, *Introduction to Statistical Thermodynamics*, Chap. 14, Addison-Wesley, Reading, Mass., 1960.

301. R. F. Steiner and R. F. Beers, Jr., *Polynucleotides*, Chap. 10, Elsevier, Amsterdam, 1960.

302. D. M. Crothers and B. H. Zimm, *J. Mol. Biol.*, **9,** 1 (1964).

303. P. O. P. Ts'o and M. P. Schweizer, *Biochemistry*, **7,** 2963 (1968).

304. F. B. Howard, J. Frazier, M. N. Lipsett, and H. T. Miles, *Biochem. Biophys. Res. Commun.*, **17,** 93 (1964).

305. H. T. Miles, F. B. Howard, and J. Frazier, *Federation Proc.*, **25,** 519 (1966).

306. P. O. P. Ts'o and W. M. Huang, *Biochemistry*, **7,** 2954 (1968).

CHAPTER 3

CIRCULAR DICHROISM OF NUCLEIC ACIDS

J. Brahms and S. Brahms**

CENTRE OF MOLECULAR BIOPHYSICS (C.N.R.S.)
ORLEANS, FRANCE

I. INTRODUCTION

The high sensitivity of optical rotation to the three-dimensional structure of molecules makes it particularly suitable to the investigation of macromolecular conformation. The progress in instrumentation and in theoretical

* Present address: Institut de Biologie Moléculaire, Faculté des Sciences de Paris, Paris V^{e}, France.

and experimental development allows one to anticipate a more extensive application in the area of physical biochemistry.

The purpose of this review is a presentation of both theoretical and experimental results which allow the interpretation of the spectra and facilitate the correlation with macromolecular conformation. Thus, it is not a general review with compilation of results and references.

We prefer to consider circular dichroism rather than optical rotatory dispersion. This method has the advantage of high resolving power and simplifies the analysis to the restricted region of an absorption band. We cannot consider here the description of basic parameters of circular dichroism and of instrumentation since it can be found in several fundamental books and reviews, particularly in Volume 1 of this series in the review by Beychok (1967) on "Circular Dichroism of Poly-α-Amino Acids and Proteins" and in the book by Veluz, Legrand, and Grosjean (1965). We consider the experimental results on polynucleotides since the circular dichroism of polypeptides was reviewed in Volume 1. Furthermore, in contrast to polypeptide and protein investigations, the studies of optical activity of polynucleotides are rather restricted. Only recently have theoretical advances and investigations in the region of polynucleotide absorption allowed the correlation of spectra with structure.

Basic Expressions

The intensity of the absorption of monochromatic light by a molecule is expressed by the dipole strength D_{0a} which is a scalar product of electric dipole moment $\boldsymbol{\mu}$ for the transition $0-a$ induced by the light wave.

$$D_{0a} = (0|\boldsymbol{\mu}|a)\cdot(a|\boldsymbol{\mu}|0) \tag{a}$$

The magnitude of the dipole strength can be obtained experimentally from the integrated area of an absorption band according to the expression

$$D_{0a} \simeq 3hc\,(2.303)/8\pi^3 N \int (\varepsilon/v)\,dv \tag{b}$$

in which h is the Planck constant, c the velocity of light, N the Avogadro number, and ε the molecular extinction coefficient in liters/mole cm. Thus, the magnitude of an induced electric dipole moment of a transition may be determined from the measurements of the absorption spectra.

The corresponding quantity in circular dichroism is the rotational strength, R_{0a}. The rotational strength is determined from the area of the circular dichroic band

$$R_{0a} \simeq 3hc\,(2.303)/32\pi^3 N \int [(\varepsilon_L-\varepsilon_R)/v]\,dv \tag{c}$$

where $(\varepsilon_L-\varepsilon_R)$ is the measured circular dichroism and expresses the difference of the molecular extinction coefficients for left and right circularly polarized light. Alternatively, this may be expressed in terms of molecular ellipticity $[\theta]$, which is given as

$$[\theta] = 2.303\,(4500/\pi)(\varepsilon_L - \varepsilon_R). \tag{d}$$

The circular dichroism band can be approximated by a Lorentzian (Condon, 1937) or a Gaussian (Moscowitz, 1960). If an isolated band is considered Gaussian, then the expression for $(\varepsilon_L - \varepsilon_R)$ is

$$(\varepsilon_L - \varepsilon_R) = (\varepsilon_L - \varepsilon_R)_{\max}\, e^{-(\nu - \nu_{\max})^2/\Gamma^2} \tag{e}$$

where Γ is the band width at half-maximum and $(\varepsilon_L - \varepsilon_R)_{\max}$ is the maximum value of $(\varepsilon_L - \varepsilon_R)$ occurring at frequency $\nu_{\max}$.

The theoretical expression defines the rotational strength R_{0a} associated with the transition from the ground state 0 to the excited state a as the scalar product of the electric transition dipole moment $(0|\boldsymbol{\mu}|a)$ and of the magnetic dipole transition moment $(a|\mathbf{m}|0)$.

$$R_{0a} = \operatorname{Im}\,(0|\boldsymbol{\mu}|a)\cdot(a|\mathbf{m}|0) \tag{f}$$

This definition leads to two observations: (1) R_{0a} will express the handedness of the dissymmetrical path of a molecule or of a group in which the perturbing electromagnetic wave induces electric and magnetic dipole transition moments. The magnitude and the positive or negative sign of R_{0a} will depend on the magnitude of the electric and magnetic transition moments and the cosine of the angle between them; (2) the conditions for optical activity are that the induced electric and magnetic transition moments must be different from zero and cannot be oriented at right angles to each other. The optimal stereochemical requirement for optical activity is fulfilled when the electric and magnetic moments are oriented parallel to each other.

An important rule was derived by Kuhn (1929) classically and by Condon (1937) on the basis of quantum mechanics that shows that the sum of the rotational strengths of all electronic transitions in a molecule vanishes.

$$\sum_a R_{0a} = 0 \tag{g}$$

The very general reciprocal relations between dispersive and absorptive phenomena are the Kronig–Kramers equations. They can be applied in order to transform ORD spectra into circular dichroism, demonstrating the intimate connection between the two phenomena. Moscowitz (1960) proposed the following relation:

$$\theta(\nu) = \left(\frac{-2}{\pi}\right)\int_0^{\infty} \frac{[m](\nu)\nu'}{(\nu'^2 - \nu^2)}\, d\nu \tag{h}$$

which allows one to calculate θ as a function of ν knowing the molecular rotation $[m]$. However, one has to realize that the relation implies a complete knowledge of ORD spectra since the integral should be taken for ν from 0 to ∞.

More recently, numerical procedures were developed by Emeis, Oosterhoff, and de Vries (1967) and by Thiéry (1968) which show the possibility of computation of circular dichroism from the measured ORD spectra, limited to a certain spectral region.

II. THEORIES OF OPTICAL ROTATORY POWER

A. Kuhn–Kirkwood Dipole-Coupling Mechanism

This theory is based on the coupled oscillator model developed by Born (1933) and Kuhn (1929) within the framework of the classical theory of optical rotation. The essential feature of Kuhn's simple model is that two oscillators located at different positions in the molecule can produce optical rotation. The necessary conditions based on asymmetric requirements for this mechanism are:

(1) The two oscillators must not be coplanar nor colinear; that is, the direction of the vibration of two oscillators must form an angle different from 180° or 0°;

(2) The two oscillators are coupled and, as a result, symmetric or antisymmetric oscillations will occur (symmetric when both electrons are directed in a positive direction along the x,y coordinates and antisymmetric when one of the directions is positive and the other is negative).

Kuhn was able to show that this model can predict the sign of optical rotation, which depends on phase conditions; that is, whether the vibration of one of the oscillators is in phase with the other or whether it is 180° out of phase. According to this model, the coupling will produce a splitting of the absorption band, which Kuhn and Rotmetsch (1944) demonstrated experimentally for substituted diphenyls.

According to Kuhn, an inherently symmetrical chromophore will be optically active by the virtue of the "vicinal effect," that is, by the presence of nearby groups.

The quantum mechanical theory of optical rotation was developed by Rosenfeld (1928). The general expression for the rotational strength associated with an electronic transition from ground state 0 to excited state a of a molecule is given by the imaginary part of the scalar product of electric and magnetic transition moments; that is

$$R_{0a} = -\operatorname{Im}(\boldsymbol{\mu}_{0a} \cdot \mathbf{m}_{a0}) \tag{1}$$†

† The expression (1) for rotational strength R_{0a} is identical to that given by equation (f). The brackets in (f) indicate integration over configuration space.

where

$$\boldsymbol{\mu}_{0a} = \int \psi_0^* \boldsymbol{\mu} \psi_a \, d\tau$$
$$\mathbf{m}_{a0} = \int \psi_a^* \mathbf{m} \psi_0 \, d\tau \tag{2}$$

ψ_0 and ψ_a are the electronic wave functions of the ground and excited states, respectively. $\boldsymbol{\mu}$ and $\mathbf{m}$ are electric and magnetic dipole operators.

Kirkwood's (1937) polarizability theory is the quantum mechanical formulation of the coupled oscillator model. Kirkwood calculated the molecular rotatory parameter g which is related to the rotational strength as

$$g = \frac{c}{3\pi h} \sum_a \frac{R_{0a}}{\nu_{a0}^2 - \nu^2} \tag{3}$$

The basic hypothesis for calculation of rotatory power is that the molecule can be divided into N atomic groups, all optically inactive. It is assumed that there is no electron exchange between these groups and that there is no resonance phenomena. For a molecule with N groups, the zero-order wave functions are products of the group functions. The ground state is expressed by

$$\psi_0^0 = \varphi_{01}\varphi_{02}\cdots\varphi_{0N}. \tag{4}$$

The excited state wave function, in which a single state in group j is excited, is defined by

$$\psi_{ja}^0 = \frac{\varphi_{ja}}{\varphi_{j0}} \varphi_{01}\cdots\varphi_{0N} \tag{5}$$

and which gives the sum over all transitions in the molecule. The electric and magnetic transition moments can be expressed as the sum of the contributions of transition moments of various groups from the ground state 0 to the excited state a

$$\boldsymbol{\mu}_{0a} = \sum_i^N \boldsymbol{\mu}_{i0a}$$
$$\mathbf{m}_{a0} = \sum_i^N \left(\mathbf{m}_{ia0} + \frac{e}{2mc} \mathbf{R}_i \times \mathbf{p}_{ia0} \right) \tag{6}$$

where

$$\boldsymbol{\mu}_{i0a} = e \int \psi_{i0}^* \left(\sum_j \mathbf{r}_{ij}\right) \psi_{ia} \, d\tau_i$$
$$\mathbf{m}_{ia0} = \frac{e}{2mc} \int \psi_{ia}^* \sum_j (\mathbf{r}_{ij} \times \mathbf{p}_{ij}) \psi_{i0} \, d\tau_i \tag{7}$$
$$\mathbf{p}_{ia0} = \int \psi_{ia}^* \left(\sum_j \mathbf{p}_{ij}\right) \psi_{i0} \, d\tau_i$$

$\mathbf{p}_{ij}$ and $\mathbf{r}_{ij}$ are the momentum and the position of electron j in a group i. The second term in the expression for the magnetic moment is due to the angular momentum. $\mathbf{R}_i$ is the position of group i in an arbitrary coordinate system.

The substitution of Eq. (6) into (2) and then (1) into (3) gives the expression for the molecular rotatory parameter. From an expression with many terms, Kirkwood had retained only one term, neglecting the others. He assumed the coupling of electric transition moments within neighboring groups to be the principal mechanism contributing to the optical activity; that is

$$g = \frac{1}{3h}\sum_a \sum_i \sum_{j>1} \frac{\nu_{a0}(\mathbf{R}_i - \mathbf{R}_j)\cdot \boldsymbol{\mu}_{i0a} \times \boldsymbol{\mu}_{j0a}}{\nu_{a0}^2 - \nu^2} \tag{8}$$

As it follows from Eq. (7), a knowledge of ground and excited wave functions is necessary for the calculation of the parameter g. Kirkwood has applied first-order perturbation theory in order to obtain ψ. The potential energy part of the Hamiltonian expressing the perturbation is assumed to have the form

$$\sum_i \sum_{j>1} V_{ij}$$

where V_{ij} is the Coulombic energy of interaction between the groups i and j, and the sum is taken over each pair of groups. The first-order wave functions for the ground and the excited states are, respectively,

$$\begin{aligned} \psi_0 &= \psi_0^0 - \sum_a \sum_{b\neq a} \sum_i \sum_{j\neq i} \frac{V_{i0a;j0b}\,\psi_{ia,jb}^0}{h(\nu_{ia0}+\nu_{jb0})} \\ \psi_{ia} &= \psi_{ai}^0 - \sum_{j\neq i} \sum_{b\neq a} \frac{V_{i0a;j0b}\,\psi_{jb}^0}{h(\nu_{jb0}-\nu_{ia0})} \end{aligned} \tag{9}$$

where

$$V_{i0a;j0b} = \int \psi_{i0}^{0*}\psi_{j0}^{0*} V_{ij} \psi_{ia}^0 \psi_{jb}^0 \, d\tau$$

and $\psi_{ia,jb}^0$ is a wave-function corresponding to a doubly-excited state; that is, the group i is excited to state a and the group j to state b. These potential matrix elements express the interaction of the electric dipole moment of the 0–a transition in group i with that of 0–b transition in group j. Then the resulting equation for the molecular rotatory parameter is

$$g = \frac{2}{3h^2}\sum_i \sum_{j\neq i} \sum_a \sum_b \frac{\nu_{ia0}\nu_{jb0} V_{i0a;j0b} \mathbf{R}_{ij}\cdot \boldsymbol{\mu}_{i0a} \times \boldsymbol{\mu}_{j0b}}{(\nu_{ia0}^2 - \nu^2)(\nu_{jb0}^2 - \nu^2)} \tag{10}$$

The potential energy terms are expanded in a multipole expansion in R_{ij}^{-1}, and only the dipole–dipole term is retained.

$$V_{i0a;j0b} = \boldsymbol{\mu}_{i0a}\cdot \tilde{T}_{ij}\cdot \boldsymbol{\mu}_{j0b} \tag{11}$$

where $\tilde{T}_{ij}$ is the dipole–dipole interaction tensor given by

$$\tilde{T}_{ij} = \frac{1}{R_{ij}^3}\left[1 - \frac{3\mathbf{R}_{ij}\mathbf{R}_{ij}}{R_{ij}^2}\right] \tag{12}$$

and $\mathbf{R}_{ij} = \mathbf{R}_i - \mathbf{R}_j$ is the distance between the groups i and j. Previously de Malleman (1924) applied the polarizability approximation for the calculation of optical rotation. The polarizability tensor $\tilde{\alpha}$ for a group i is related to the electronic transition in this group by

$$\tilde{\alpha}_i = \frac{2}{3h}\sum_a \frac{\mu_{ia0}\cdot\mu_{ia0}\nu_{ia0}}{\nu_{ia0}^2 - \nu^2} \tag{13}$$

This equation has allowed Kirkwood to avoid the calculation of wave functions by expressing the transition moments in terms of polarizabilities. Polarizability values are available from experimental data, e.g., from the measurements of refractivity of atoms which constitute the group.

For cylindrically symmetrical groups, the elements of the polarizability tensor in the direction of principal axes will be $(\alpha_{11})_i = (\alpha_{22})_i \neq (\alpha_{33})_i$. Then the polarizability tensor will be expressed in terms of the anisotropy ratio β_i and the average polarizability α_i

$$\beta_i = \frac{(\alpha_{33})_i - (\alpha_{11})_i}{\alpha_i} \tag{14}$$

where α_{33} denotes the polarizability along the axis of cylindrical symmetry and α_{11} in the direction normal to it.

$$\alpha_i = \tfrac{1}{3}[(\alpha_{11})_i + (\alpha_{22})_i + (\alpha_{33})_i] \tag{15}$$

Using these equations in (10), one obtains the polarizability expression for the rotational strength of a system of coupled oscillators. The final expression obtained by Kirkwood has the form

$$g = \tfrac{1}{6}\sum_i \sum_{j<i} \alpha_i\alpha_j\beta_i\beta_j G_{ij}\mathbf{R}_{ij}\,\mathbf{e}_i\times\mathbf{e}_j \tag{16}$$

$$G_{ij} = \frac{1}{R_{ij}^3}\left[\mathbf{e}_i\mathbf{e}_j - 3\,\frac{(\mathbf{e}_i\mathbf{R}_{ij})(\mathbf{e}_j\mathbf{R}_{ij})}{R_{ij}^2}\right] \tag{17}$$

$\mathbf{e}_i$ and $\mathbf{e}_j$ are unit vectors in the direction of the symmetry axis of groups i and j. For electric dipole allowed transitions with strong oscillator strength ($\varepsilon > 1000$), dipole coupling gives a dominant contribution to the optical activity of a molecule. However, the theory is not able to deal with weakly electrically allowed transitions, particularly with dipole magnetic allowed transitions.

B. The One-Electron Theory

The one-electron theory takes into consideration each single chromophore individually. The rotational strengths are due to single absorption bands in a molecule and therefore are related to individual electronic transitions. It was shown by Condon, Altar, and Eyring (1937) that optical activity will be generated by a single electron if a perturbation due to the asymmetric force field of surrounding atoms will act on the electron.

The essential feature of the theory is the static nature of the perturbing field. This may be due to the permanent or induced dipole moments in nearby groups that act on the electron or to the static field of ions in the surrounding groups.

Moffitt and Moscowitz (1956) have extended the "one-electron" theory, and Moscowitz (1962) has used Condon, Altar, and Eyring concepts for the calculation of the rotational strength of $n \rightarrow \pi^*$ transitions in ketones and steroids. Some aspects of the "one-electron" theory is discussed in relation to Schellman's contribution. A particular chromophore i in a molecule or system containing $j = 1, \cdots, N$ residues is considered. The chromophore is characterized by two transitions $0 \rightarrow a$ and $0 \rightarrow b$. Their unperturbed wave functions may be approximated by hydrogen-like wave functions. Then it is assumed that the $0 \rightarrow a$ electrically forbidden transition of energy $h\nu_a$ is a $2p \rightarrow 2p$ transition, e.g., in a C=O amide group; that is, the nonbonding electron on the oxygen of the carbonyl chromophore is excited to an anti-bonding π^* state. This transition is magnetically allowed and its magnetic transition moment is $\mathbf{m}_{ia0}$. On the other hand, the $0 \rightarrow b$ transition is a $2p \rightarrow 3d$ transition of energy $h\nu_b$, which is electrically allowed, and its transition moment is $\boldsymbol{\mu}_{i0b}$. The isolated chromophoric group is supposed to have a plane or a center of symmetry, and it is not optically active. However, the static field of the rest of the molecule produced by all N residues will behave as a perturbation altering the symmetry of the group and gives rise to an induced optical activity.

Applying first-order perturbation theory, the wave functions for the ground state and the excited states are, respectively,

$$\psi_0 = \psi_0^0 - \sum_{i \neq j} \frac{V_{i0b;\,j00}}{h\nu_b} \psi_{ib}^0 \tag{18}$$

$$\psi_a = \psi_a^0 - \sum_{i \neq j} \frac{V_{iab;\,j00}}{h(\nu_b - \nu_a)} \psi_{ib}^0 \tag{19}$$

Similar equations hold for the $0 \to b$ transition. By putting the perturbed wave functions in expression (1), one obtains the rotational strengths of the $0 \to a$ and $0 \to b$ transitions.

$$R_{ia} = -\text{Im} \sum_{j \neq i} \frac{V_{iab;\,j00}}{h(\nu_b - \nu_a)} \boldsymbol{\mu}_{i0b} \cdot \mathbf{m}_{ia0} \tag{20}$$

$$R_{ib} = -\text{Im}\ \boldsymbol{\mu}_{i0b} \cdot \sum_{j \neq i} \frac{V_{iab;\,j00}}{h(\nu_a - \nu_b)} \mathbf{m}_{ia0} \tag{21}$$

R_{ia} and R_{ib}, occurring at energies $h\nu_a$ and $h\nu_b$, respectively, have equal magnitude and opposite sign.

The potential terms $\Sigma_{j \neq i} V_{iab;\,j00}$ are the electrostatic interaction of a charge density associated with transition a–b between the two excited states, with the asymmetric permanent field of all residues j. They are defined by

$$V_{iab;\,j00} = \int \psi_{ia}^{*0} \psi_{j0}^{*0} V_{ij} \psi_{ib}^{0} \psi_{j0}^{0}\, d\tau_i\, d\tau_j \tag{22}$$

The perturbation causes a mixing of the two excited states. The $2p$ excited state will acquire some $3d$ character and makes the transition $0 \to a$ weakly electrically allowed. Similarly in the case of the $o \to b$ transition a magnetic moment is acquired due to the external static field.

Schellman and Oriel (1962) have applied first-order perturbation theory to the calculation of the rotational strength of the magnetically allowed $n \to \pi^*$ transition in helical polypeptides. These authors demonstrated that the main contribution to the optical activity of $n \to \pi^*$ transitions is due to the "one-electron" terms. The justification of this procedure, which does not take into account intergroup coupling, is the smallness of the Kirkwood term for an $n \to \pi^*$ transition, which Schellman and Oriel (1962) calculated explicitly. The $n \to \pi^*$ transition was previously ignored in theoretical considerations of polypeptide optical activity. The experimental measurements of the polypeptide Cotton effect occurring at 222 mμ (Simmons and Blout, 1960; Simmons et al., 1961) showed its significance for conformation studies of polypeptides and proteins. Schellman and Oriel assigned this optically active band as due to the $n \to \pi^*$ transition and have shown the origin of this Cotton effect. The $n \to \pi^*$ transition of the planar amide group will give rise to optical activity when it is surrounded by the asymmetric field such as the electrostatic field of the helical polymer. The mechanism of this induced optical activity involves mixing, as a result of which the electrically forbidden $n \to \pi^*$ transition acquires a small electric component from the electrically allowed $\pi \to \pi^*$ transition moments and from the "charge transfer" associated with the $n \to \pi^*$ transition. Schellman and Oriel considered two of the electrically allowed transitions of the amide group namely NV_1 and NV_2 transitions whose energies are $h\nu_{NV_1}$ and $h\nu_{NV_2}$ and their electric transition moments are μ_{iNV_1} and μ_{iNV_2}. The rotational strength of the $n \rightarrow \pi^*$ transition of an amide group i in a helical polypeptide is

$$R_{in\pi^*} = -\operatorname{Im}\left[\sum_{i\neq j}\frac{V_{in\pi^*;\,j00}\,(\boldsymbol{\mu}_{in\pi^*-n\pi^*}-\boldsymbol{\mu}_{i00})}{h\nu_{n\pi^*}}\right.$$
$$-\sum_{i\neq j}\frac{V_{in\pi^*-NV_1;\,j00}\,\boldsymbol{\mu}_{iNV_1}}{h(\nu_{NV_1}-\nu_{n\pi^*})} \tag{23}$$
$$\left.-\sum_{i\neq j}\frac{V_{in\pi^*-NV_2;\,j00}\,\boldsymbol{\mu}_{iNV_2}}{h(\nu_{NV_2}-\nu_{n\pi^*})}\right]\cdot\mathbf{m}_{in\pi^*}$$

Since the $n\pi^*$ transition is electrically forbidden, the terms in brackets can be defined as a first-order expression for the electric dipole transition moment. $\mathbf{m}_{in\pi^*}$ is the zero-order magnetic transition moment. In the first term, $(\boldsymbol{\mu}_{in\pi^*-n\pi^*}-\boldsymbol{\mu}_{i00})$ is the difference in the permanent electric dipole moment between the excited and the ground state. Schellman has assimilated this moment into an intragroup charge-transfer vector associated with the $n\pi^*$ transition, which will interact with $\mathbf{m}_{in\pi^*}$ in a way similar to a transition electric moment in the static field of the polymer. The potential factors $\Sigma_{i\neq j}\,V_{in\pi^*;\,j00}$ arise from the interaction of the $n\pi^*$ transition charge density of the amide group i with the permanent field of all other groups in their ground state. The next two terms involve the mixing of NV_1 and NV_2 with $n\pi^*$ states, due to the permanent field. In their calculation, Schellman and Oriel have shown that if the helical perturbing potential is expanded in a multipole expression, the $n\pi^*$ states will mix with other states only by virtue of the quadrupole term. This result follows from symmetry considerations. Therefore, only a potential of restricted symmetry can induce a Cotton effect, by mixing the $n\pi^*$ states belonging to a determined irreducible representation with states NV_1 and NV_2 of a different irreducible representation.

C. Symmetry Considerations for Optical Activity Arising from Static Perturbations

Schellman (1966) has applied the method of group theory to the matrix elements of the one-electron equations. The electronic states are classified according to the symmetry of the unperturbed chromophoric group. The symmetry properties of the perturbation are considered, and it is shown that in order to produce optical activity they must depend on the symmetry of the chromophore itself. The general form and the simplest example of perturbing potentials for each symmetry group are given.

The zero-order wave functions belong to the representations of the symmetry group of the isolated chromophore. Those wave functions whose representations transform like the components of a polar vector correspond to excited states of electric dipole allowed transitions. The symbol β denotes these electronic states. Those excited states to which magnetic dipole transitions are allowed are symbolized by α, and their representations have different transformation properties, i.e., of an axial vector. Applying first-order perturbation theory the one-electron expressions for rotational strengths of $0 \rightarrow \alpha$ and $0 \rightarrow \beta$ transitions will have the following forms in bracket notation:

$$R_{0\alpha} = \text{Im}\left\{\left[\sum_{\beta} \frac{(0|\boldsymbol{\mu}|\beta)(\beta|V|\alpha)}{h(\nu_\beta - \nu_\alpha)} + \sum_{\rho} \frac{(\rho|\boldsymbol{\mu}|\alpha)(\rho|V|0)}{h(\nu_\rho - \nu_0)}\right]\cdot(\alpha|\mathbf{m}|0)\right\} \tag{20a}$$

$$R_{0\beta} = \text{Im}\,(0|\boldsymbol{\mu}|\beta)\cdot\left[\sum_{\alpha} \frac{(\alpha|\mathbf{m}|0)(\beta|V|\alpha)}{h(\nu_\alpha - \nu_\beta)} + \sum_{\rho} \frac{(\beta|\mathbf{m}|\rho)(\rho|V|0)}{h(\nu_\rho - \nu_0)}\right] \tag{21a}$$

where $|0)$ is the ground state and $|\rho)$ stands for higher excited states. The higher energy states are omitted in Eqs. (20) and (21).

The external perturbation potential is usually due to polar groups of the rest of the molecule or to an ion, and this perturbing group is situated outside the chromophore. It will not normally belong to any of the representations of the symmetry group of the chromophore. However, the potential can be expressed as a sum of functions, each of which are representations of the symmetry of the chromophore. It is shown that only the pseudoscalar functions will have the proper symmetry requirements to give rise to optical activity.

One can suppose that α and β have one-dimensional representations. It follows from Eqs. (20a) and (21a) that the potential matrix elements of the first summations will differ from zero only if V belongs to the same representation as the product $\alpha_x\beta_x$ and therefore the same representation as $m_x\mu_x$. The scalar product of an axial and a polar vector is a pseudoscalar whose transformation properties are those of a scalar except that it changes sign under reflection or inversion. There is always a representation for pseudoscalar functions in any point group other than the pure rotational groups. Thus the perturbation V should belong to the pseudoscalar representation of the chromophore. Schellman constructed a table containing the potential pseudoscalar functions for common point-symmetry groups.

In the case of an electrostatic potential the charge distribution of the perturbant can be approximated by point charges placed at the position of nuclei. The Coulombic potential can be approximated in terms of a multipole expansion.

$$V(\mathbf{r}) = \sum \frac{q_i}{R_i} + \left[\left(\frac{\Sigma q_i X_i}{R_i^3}\right)x + \left(\frac{\Sigma q_i Y_i}{R_i^3}\right)y + \left(\frac{\Sigma q_i Z_i}{R_i^3}\right)z\right]$$

$$+ \left[3\left(\frac{\Sigma q_i X_i Y_i}{R_i^5}\right)xy + 3\left(\frac{\Sigma q_i Y_i Z_i}{R_i^5}\right)yz + 3\left(\frac{\Sigma q_i Z_i X_i}{R_i^5}\right)zx\right.$$

$$+ \frac{3}{4}\left(\frac{\Sigma q_i (X_i^2 - Y_i^2)}{R_i^5}\right)(x^2 - y^2) + \frac{1}{4}\left(\frac{\Sigma q_i (2Z_i^2 - X_i^2 - Y_i^2)}{R_i^5}\right)$$

$$\left.\cdot(2z^2 - x^2 - y^2)\right] + \text{octopole terms} + \text{higher terms}$$

where R_i is the distance from the center of the chromophore to the external perturbing source.

The electronic states α and β of a chromophore of C_s symmetry belong to A″ and A′ representations, respectively, while the pseudoscalar representation is A″ and the simplest form of the pseudoscalar potential is z. The XY symmetry plane of the chromophore divides the space surrounding the chromophore into two regions. When the perturbing group moves from one side of the plane to the other, the perturbation-generated optical activity will change sign. These considerations indicate a planar rule for C_s symmetry. A quadrant rule holds for the $n\pi^*$ transition of the peptide group since the local symmetry on the oxygen atom is C_{2v}. The n orbitals have a nodal plane, XZ, perpendicular to the XY nodal plane of the π system. The space is divided into four regions of alternating positive and negative signs for the Cotton effect. If the perturbation is approximated by an expansion, the dominant term is an xy quadrupole term. Similarly, the regional rule for ketones is the well-known octant rule. The ketone group itself has a C_{2v} symmetry. However, if the electronic orbitals involved in the $n\pi^*$ transition are considered, there will be three planes of symmetry instead of two, raising the symmetry to D_{2h}. The pseudoscalar term in the expansion is an octopole term xyz.

In general, the higher the symmetry of the chromophoric group, the greater is the number of nodal planes, and the smaller is the effectiveness of a Coulombic perturbing field. This explains the decrease in magnitude of the Cotton effect when a proton is removed from a COOH group, raising its symmetry from C_s to C_{2v}. For a chromophore of a known symmetry, if the sign of Cotton effect is defined from experiment for one region, then the regional symmetry rules allow one to determine whether there will be a positive or a negative contribution to rotational strength, depending upon which spatial region the perturbing group lies in.

Recently Schellman and co-workers have developed another approach to the calculation of rotational properties of helical polypeptides by investigation of simple model compounds such as a pair of peptide units (Bayley, Nielson, and Schellman, 1969; Schellman and Nielson, 1967; Litman and Schellman, 1965). The theoretical procedure differs from the general treatment of Tinoco (1962) in that a matrix method is preferred rather than the first-order perturbation theory. The polymeric system is divided into N groups (monomers). The separate groups are optically inactive and there is no electron exchange between groups. However, additional assumptions restrict the theory to systems for which the following two conditions are fulfilled. First, the number of considered transitions is limited to those observable in the experiments. In particular, high-energy transitions in the far UV are neglected. Obviously this assumption is valid only if the sum of rotational strengths of observable transitions is approximately zero. The second condition is that only singly excited states are considered. Interactions involving the ground state and simultaneously excited states are omitted. This will cause only small errors if the levels are degenerate or if the considered transitions are situated in a narrow energy range. The theory has been applied to the calculation of optical rotation properties of diamides. The system is assumed to be dominated by the low-energy transitions; that is, two magnetically allowed $n\pi^*$ and two electrically allowed $\pi\pi^*$ transitions, and both are degenerate or nearly so.

Each amide in the dimer system has two excited states α_1 and β_1, α_2 and β_2, where 1 and 2 symbolize group 1 and group 2, while α and β designate $n\pi^*$ and $\pi\pi^*$ transitions, respectively. The transition energies from ground state 0 to these excited states are E_{α_1}, E_{α_2}, E_{β_1}, E_{β_2}. The Hamiltonian for this four-state system is expressed in the form of a secular determinant containing four submatrices, each of which involves the specific group–group interactions.

$$\begin{pmatrix} (1\text{—}1) & (1\text{—}2) \\ (2\text{—}1) & (2\text{—}2) \end{pmatrix} \tag{24}$$

The submatrices can be divided into two types. The first type (1—1) and (2—2) will characterize the interactions which lead to a mixing of states within the same group. The second type (1—2) and (2—1) represents the coupling of states in different groups. A more detailed examination of the different submatrices shows that despite the simplification of the problem all mechanisms producing optical rotation are involved. The (1—1) and (2—2) submatrices are defined as follows:

$$\begin{aligned} (1\text{—}1) &= \begin{pmatrix} E_{\alpha_1} & \langle \alpha_1 | V | \beta_1 \rangle \\ \langle \beta_1 | V | \alpha_1 \rangle & E_{\beta_1} \end{pmatrix} \\ (2\text{—}2) &= \begin{pmatrix} E_{\alpha_2} & \langle \alpha_2 | V | \beta_2 \rangle \\ \langle \beta_2 | V | \alpha_2 \rangle & E_{\beta_2} \end{pmatrix} \end{aligned} \tag{25}$$

The off-diagonal elements involve the mixing of $n\pi^*$ and $\pi\pi^*$ states within one group due to the static field developed by the other group. The intergroup submatrices (1—2) and (2—1) are defined by

$$\begin{pmatrix} \langle\alpha_1|V|\alpha_2\rangle & \langle\alpha_1|V|\beta_2\rangle \\ \langle\beta_1|V|\alpha_2\rangle & \langle\beta_1|V|\beta_2\rangle \end{pmatrix}$$

and

$$\begin{pmatrix} \langle\alpha_2|V|\alpha_1\rangle & \langle\alpha_2|V|\beta_1\rangle \\ \langle\beta_2|V|\alpha_1\rangle & \langle\beta_2|V|\beta_1\rangle \end{pmatrix} \tag{26}$$

The two elements of the kind $\langle\beta_i|V|\beta_j\rangle$, where $i = 1,2; j = 1,2; i \neq j$ represent the dipole–dipole coupling mechanism (Kuhn–Kirkwood coupled oscillator theory). The four elements of the kind $\langle\alpha_i|V|\beta_j\rangle$ involve the interaction of quadrupolar and dipolar charge distributions. The other terms can be neglected since they correspond to quadrupole–quadrupole interaction of $n\pi^*$ transitions. The electric and magnetic transition moments are given by the equations

$$\boldsymbol{\mu} = C\boldsymbol{\mu}^0, \quad \mathbf{m} = C^*\mathbf{m}^0 \tag{27}$$

$\boldsymbol{\mu}^0$ and $\mathbf{m}^0$ are row vectors and contain only transitions from the ground state. C is the transformation matrix. The matrix coefficients are obtained from the solution of the secular equation. The theory has the advantage of dealing with very well-known states observable from spectroscopy, and offers the possibility of being confronted with experimental data.

D. Tinoco Contribution

Tinoco's theoretical treatment of optical rotation (1962, 1963) is based on Kirkwood's polarizability theory as extended by Moffitt (1956). The effects of static field as well as interaction among identical groups are considered. As in the Kirkwood theory, the system is divided into groups, and electrons are assigned to these particular groups. Both electron delocalization and electron exchange interaction are neglected. The zero-order wave functions for the system are products of group functions [Eq. (3)]. The system can be a molecule or a polymer. When a polymer is considered, identical groups are present. A polymer consisting of N identical units has a unique ground state and N-fold degenerate excited states in zero order. The zero-order wave function of a polymer excited state A is expressed by a linear combination of wave functions of the system in which the excitation is located on a single unit a.

$$\psi_{AK}^0 = \sum_{j=1}^{N} C_{jak}\psi_{ja}^0 \qquad k = 1,2\cdots N \tag{28}$$

The values of coefficients C_{jak} depend on the geometry of the system. They all have the following properties:

$$\sum_{i=1}^{N} C_{iak}C_{iak}^{*} = 1$$
$$\sum_{k=1}^{N} C_{iak}C_{jak}^{*} = 0 \qquad \text{where} \quad i \neq j \tag{29}$$

The perturbed first-order approximated wave functions for the ground state and the excited state of the polymer are, respectively,

$$\psi_0^1 = \psi_0^0 - \sum_{i}\sum_{i\neq j}\sum_{a} \frac{V_{i0a;\,j00}\psi_{ia}^0}{h\nu_a} - \sum_{i}\sum_{j>i}\sum_{a}\sum_{b} \frac{V_{i0a;\,j0b}\psi_{ia,\,jb}^0}{h(\nu_a-\nu_b)} \tag{30}$$

$$\psi_{AK}^1 = \sum_{i=1}^{N} C_{iak}\left[\psi_{ia}^0 + \sum_{j\neq 1} \frac{V_{i0a;\,j00}\psi_0^0}{h\nu_a} - \sum_{j\neq i}\sum_{a\neq b} \frac{V_{i0a;\,j0b}\psi_{jb}^0}{h(\nu_b-\nu_a)} - \sum_{j\neq i}\sum_{b\neq a} \frac{V_{iab;\,j00}\psi_{ib}^0}{h(\nu_b-\nu_a)}\right] \tag{31}$$

These wave functions and the properties of C_{iak} coefficients were used by Tinoco to calculate the electric and magnetic transition moments and to obtain the rotational strength for a transition from ground state 0 to the excited state A.

$$R_{AK} = \text{Im}\ \boldsymbol{\mu}_{0AK}\cdot \mathbf{m}_{A0K} \tag{32}$$

If the polymer excited state is N-fold degenerate then the total rotational strength will be

$$R_A = \sum_{i=1}^{N} R_{AK} \tag{33}$$

or

$$R_A = \sum_{i=1}^{N}\Bigg[\text{Im}\ \boldsymbol{\mu}_{i0a}\cdot \mathbf{m}_{ia0} \tag{34a}$$

$$-2\sum_{j\neq i}\sum_{b\neq a} \frac{\text{Im}\ V_{i0a;\,j0b}(\boldsymbol{\mu}_{i0a}\cdot \mathbf{m}_{jb0}\nu_a + \boldsymbol{\mu}_{j0b}\cdot \mathbf{m}_{ia0}\nu_b)}{h(\nu_b^2-\nu_a^2)} \tag{34b}$$

$$-\sum_{j\neq i}\sum_{b\neq a} \frac{\text{Im}\ V_{iab;\,j00}(\boldsymbol{\mu}_{i0a}\cdot \mathbf{m}_{ib0} + \boldsymbol{\mu}_{i0b}\cdot \mathbf{m}_{ia0})}{h(\nu_b-\nu_a)} \tag{34c}$$

$$-\sum_{j\neq i}\sum_{b\neq a}\frac{\text{Im } V_{i0b;j00}(\boldsymbol{\mu}_{i0a}\cdot\mathbf{m}_{iab}+\boldsymbol{\mu}_{iab}\cdot\mathbf{m}_{ia0})}{h\nu_b} \tag{34d}$$

$$-\sum_{j\neq i}\frac{\text{Im } V_{i0a;j00}(\boldsymbol{\mu}_{iaa}-\boldsymbol{\mu}_{i00})\cdot\mathbf{m}_{ia0}}{h\nu_a} \tag{34e}$$

$$\left.-(2\pi/c)\sum_{j\neq i}\sum_{b\neq a}\frac{V_{i0a;j0b}\nu_a\nu_b(\mathbf{R}_j-\mathbf{R}_i)\cdot(\boldsymbol{\mu}_{j0b}\times\boldsymbol{\mu}_{i0a})}{h(\nu_b^2-\nu_a^2)}\right] \tag{34f}$$

The first term is the contribution of an isolated group which is, for a monomer, very often small or zero if it possesses a center or a point of symmetry, unless the monomer itself is dissymmetric.

The second term arises from the interaction of the transition electric dipole moment on one group with the transition magnetic dipole moment in another group. The perturbation potential due to this interaction produces an electromagnetic coupling. The electronic motion on one group is coupled to the electron motion associated with the magnetic moment of the other group. This term was very often neglected in the calculation of optical rotatory power, but recently it was taken into consideration by Woody and Tinoco (1967), and Schellman (1968). Its contribution is not negligible in the case of a large magnetic transition moment.

The next three terms show the effect of the static field of the polymer. The third and fourth terms are due to the interaction of electric and magnetic transition moments within the same group and involve transitions between excited states. They are the "one-electron" terms in which the potential factors come from the interaction of transition charge densities on one group with the permanent charge distribution of all other groups in the polymer. The third term may have an appreciable value if its denominator is small. The contribution of the fourth term will be somewhat smaller owing to the large value of $h\nu_b$. The fifth term comes from an electric moment arising from the difference between the permanent dipole moment of the excited state $\boldsymbol{\mu}_{i0a}$ and the permanent dipole moment of the ground state $\boldsymbol{\mu}_{i00}$, which interacts with the magnetic transition moment within the same group.

The last term is the Kuhn–Kirkwood coupled oscillators term, which is expected to be large for electric dipole allowed transitions. This term was extensively discussed in section IIA.

Tinoco's theory of optical rotation has the advantage of presenting in one general unified equation all previously considered theories and includes all possible mechanisms for the generation of optical rotation. Moreover, it shows the additive character of different mechanisms, which is of fundamental importance for the theory of optical rotation. Equations (34) give the contribution of a single transition in a particular group to the total rotation of the polymer. The calculation is quite complicated but the predominance of some terms over others will lead to a simplification of the problem. In the treatment of an actual molecule or polymer, it is generally assumed that the terms of importance are the second, third, and sixth. One of them or all three terms may dominate the Cotton effect of a specific transition. The two other terms (the fourth and the fifth), although less significant because of their large denominators, merit an evaluation especially in the case of a weak Cotton effect.

In order to illustrate the use of the general equation, we briefly outline the Woody and Tinoco (1967) calculation of the rotational strengths associated with $n\pi^*$ and NV_1 transitions of poly-L-alanine in an α-helical conformation. The $n\pi^*$ rotational strength of a polypeptide was calculated previously by Schellman and Oriel (1962) [Eq. (23)]; however, the Woody and Tinoco treatment (1967) is more rigorous since it accounts for all possible sources of optical activity, and it considers the contribution of all transitions including those in the far ultraviolet region. A detailed evaluation of each term is reported.

1. Rotational Strength of an $n\pi^*$ Transition

In an isolated amide group, the $n\pi^*$ transition is electrically forbidden; thus by setting $\boldsymbol{\mu}_{i0a} = 0$ the general equation will have the following form:

$$\begin{aligned} R_{n\pi^*} = \sum_{i=1}^{N} \Bigg[& -2 \sum_{j \neq i} \sum_{b \neq a} \frac{\operatorname{Im} V_{i0a;\, j0b}(\boldsymbol{\mu}_{j0b} \cdot \mathbf{m}_{ia0} \nu_b)}{h(\nu_b^2 - \nu_a^2)} \\ & - \sum_{j \neq i} \sum_{b \neq a} \frac{\operatorname{Im} V_{iab;\, j00}(\boldsymbol{\mu}_{i0b} \cdot \mathbf{m}_{ia0})}{h(\nu_b - \nu_a)} \\ & - \sum_{j \neq i} \sum_{b \neq a} \frac{\operatorname{Im} V_{i0b;\, j00}(\boldsymbol{\mu}_{iab} \cdot \mathbf{m}_{ia0})}{h\nu_b} \\ & - \sum_{j \neq i} \frac{\operatorname{Im} V_{i0a;\, j00}(\boldsymbol{\mu}_{iaa} - \boldsymbol{\mu}_{i00}) \cdot \mathbf{m}_{ia0}}{h\nu_a} \Bigg] \end{aligned} \tag{35}$$

$R_{n\pi^*}$ is the contribution of the $n\pi^*$ transition of the amide group to the rotational strength of the polymer which contains $i - N$ residues. For the determination of potential factors a monopole expansion is used. Thus the matrix elements $V_{ia0;\, j0b}$ in the first term arising from the interaction of $n\pi^*$ transition monopoles in groups i with those in groups j are expressed by Tinoco (1962):

$$V_{ia0;\, j0b} = \sum_s \sum_t \frac{\rho_s^{ia0} \rho_t^{j0b}}{r_{st}} \tag{36}$$

Transitions $0 \rightarrow b$ refer to electrically allowed transitions of the amide group available from spectral measurements, that is NV_1, $n'\pi^*$, $n\sigma^*$, and NV_2.

The contribution of interaction of $n\pi^*$ transitions with those in the far ultraviolet is calculated by using the polarizability approximation. Similarly, the other potential factors involve the interaction of transition monopoles in group i with permanent monopoles of the rest of the polymer.

An account of the geometrical features of transition moments will show that the $n'\pi^*$ and $n\sigma^*$ electric transition moments are perpendicular to $\mathbf{m}_{ia0}$. Therefore the second term will include only interactions of $n\pi^*$ with NV_1 and NV_2 transition moments and the sum $\Sigma_{i \neq j} V_{iab;j00}$ which accounts for the electrostatic interaction of charge densities associated with $n\pi^* \to NV_1$ and $n\pi^* \to NV_2$ transitions between excited states with the permanent charge distribution of all other groups in the polymer. This contribution is expressed by the last two terms of Eq. (23) (Schellman and Oriel, 1962).

The third term will vanish because $\boldsymbol{\mu}_{iab} = 0$ for $n\pi^* \to NV_1$ and $n\pi^* \to NV_2$ transitions which are electrically forbidden, while $(\boldsymbol{\mu}_{iab} \cdot \mathbf{m}_{ia0}) = 0$ for $n\pi^* \to n'\pi^*$ and $n\pi^* \to n\sigma^*$ since their transition electric moments are perpendicular to $n\pi^*$ magnetic moments. Finally, the fourth term corresponds to the first term of Eq. (23). Woody and Tinoco confirmed the correctness of the Schellman and Oriel results that showed that the major contribution to $R_{n\pi^*}$ comes from mixing of NV_1 and $n\pi^*$ transitions owing to the static field produced by the rest of the polymer.

2. Rotational Strength for the NV_1 Transitions

In the case of electrically allowed transitions, by setting $\mathbf{m}_{ia0} = 0$ one obtains the vanishing of the fifth term in Eq. (34). Furthermore one assumes that static field effects are negligible if compared to the large contribution due to the coupled oscillator terms. Thus Eq. (34) will be reduced to the second and last terms. We first consider the rotational strengths produced by interactions of NV_1 transitions with $n\pi^*$, $n'\pi^*$, $n\sigma^*$ and of far UV transitions. The contribution of $n\pi^*$ transitions to R_{NV_1} will have this simple form if static fields are neglected.

$$(R_{NV_1})_{n\pi^*} = -\sum_{i=1}^{N} 2 \sum_{j=1} \frac{\nu_{NV_1} \operatorname{Im} V_{iNV_1;jn\pi^*} \boldsymbol{\mu}_{iNV_1} \mathbf{m}_{jn\pi^*}}{h(\nu_{n\pi^*}^2 - \nu_{NV_1}^2)} \tag{37}$$

The coupled oscillator contribution due to $n'\pi^*$ and $n\sigma^*$ is obtained from equations

$$(R_{NV_1})_{n'\pi^*} = -\sum_{i=1}^{N} \left[\frac{2\pi}{c} \sum_{j \neq i} \frac{V_{iNV_1;jn'\pi^*} \nu_{NV_1} \nu_{n'\pi^*} (\mathbf{R}_j - \mathbf{R}_i) \cdot \boldsymbol{\mu}_{jn'\pi^*} \boldsymbol{\mu}_{iNV_1}}{h(\nu_{n'\pi^*}^2 - \nu_{NV_1}^2)} \right]$$

$$(R_{NV_1})_{n\sigma^*} = -\sum_{i=1}^{N} \left[\frac{2\pi}{c} \sum_{j \neq i} \frac{V_{iNV_1;jn\sigma^*} \nu_{NV_1} \nu_{n\sigma^*} (\mathbf{R}_j - \mathbf{R}_i) \cdot \boldsymbol{\mu}_{jn\sigma^*} \boldsymbol{\mu}_{iNV_1}}{h(\nu_{n\sigma^*}^2 - \nu_{NV_1}^2)} \right] \tag{38}$$

For interaction with transitions occurring in the far UV, the polarization equation is used.

$$R_A = \sum_{i=1}^{N} \frac{\pi}{c} \left[\frac{\nu_a \nu_0^2 (\mu_{iNV_1})^2 (\alpha_{33} - \alpha_{11}) G_{ij} (\mathbf{e}_{3i} \times \mathbf{e}_{3j}) \cdot R_{ij}}{\nu_0^2 - \nu_{NV_1}^2} \right] \tag{39}$$

ν_0 is an average frequency for the higher-energy far UV transitions. It is assumed that the ellipsoid polarizability has cylindrical symmetry about direction 3 perpendicular to the amide plane. Denoting by α_{33} the polarizability along this axis, α_{11} is the polarizability perpendicular to it. G_{ij} is the interaction factor from Eq. (17).

The presence in the polymer of identical amide groups leads to consideration of exciton contribution. A detailed discussion of exciton effects is given in the next section. For the interaction among NV_1 transitions, the contribution to the rotational strength of each of the $k = 1, 2 \cdots N$ excited levels is:

$$(R_{NV_1})_k = \frac{\pi \nu_0}{c} \sum_i \sum_{j \neq i} C_{ik} C_{jk}^* (\mathbf{R}_j - \mathbf{R}_i) \boldsymbol{\mu}_{jNV_1} \cdot \boldsymbol{\mu}_{iNV_1} \tag{40}$$

where C_{ik} are the mixing coefficients [Eq. (29)] and $\mathbf{R}_i$ is a vector from an arbitrary origin to residue *i*. The Woody and Tinoco calculation has shown that *N* rotational strengths tend to an aggregation around three distinct groups of wavelengths.

(1) A negative rotational strength centered at the lower energy of the monomer absorption ν_0 polarized parallel to the helix axis.

(2) A group of positive rotational strengths of large amplitude centered in the neighborhood of ν_0 with perpendicular polarization.

(3) A second group situated on the high-energy side of the monomer frequency with negative rotational strengths and polarized perpendicularly to the helix axis.

Despite discrepancies in the magnitude of the Cotton effect, the Woody and Tinoco calculations provide a basis for the understanding of the sources of optical rotatory power of helical polypeptides in the UV region. It should be underlined that this theory gives the right signs and the extrema location of the Cotton effect. These results allowed the following band assignments of observed CD spectra; the band at 225 mμ is due to $n\pi^*$ transitions while the 205 and 109 mμ CD bands are associated with transitions of a split NV_1 excited state giving a parallel polarization for the low-energy band and a perpendicular polarization relative to the helix axis for that of higher energy. This is in quite good agreement with experimental results on polarized absorption spectra of polypeptides of Gratzer, Holzwarth, and Doty (1961), Jaffé, Jaffé, and Rosenheck (1967), and Brahms et al. (1968).

III. CORRELATION OF THE SPECTRA WITH STRUCTURE

A. Nearest Neighbor (Exciton) Interaction Theory

The exciton theory seems to be particularly simple and suitable for the understanding of optical properties of a polymer or of an aggregate composed of N identical residues with strong electrically allowed transitions. This theory is based on the exciton idea of Davydov (1948), applied to polymers by Moffitt (1956). In an oligomer or polymer, a light quantum is not only absorbed by a residue but migrates and leads to the excitation of any of N residues. The transfer of excitation energy occurs by resonance from residue i to the residue j and is characterized by the term V_{ij}, which is very similar to the dipole–dipole interaction energy. This resonance interaction will give rise to the splitting of N degenerate, originally singly excited states into $1, 2, 3 \cdots N$ nondegenerate discrete states. Therefore, the excited state of a polymeric array can be considered as an exciton band of many closely spaced energy levels. A transition of energy $E_K = h\nu_K$ to each of these exciton levels will give rise to a rotational strength R_K

$$R_K = \frac{\pi \nu_0}{c} \sum_{i=1}^{N} \sum_{j>i} C_{ik} C_{jk}^* (\mathbf{R}_j - \mathbf{R}_i) \cdot \boldsymbol{\mu}_j \times \boldsymbol{\mu}_i \tag{41}$$

$$h\nu_K = h\nu_0 + \sum_{i=1}^{N} \sum_{j>i} C_{ik} C_{jk}^* V_{ij} \tag{42}$$

$\boldsymbol{\mu}_j$ is the transition electric dipole moment of residue j and $\mathbf{R}_j$ is a vector from an arbitrary origin to residue j. The potential term V_{ij} is the interaction energy of residues i and j, while $h\nu_0$ is the transition energy of the monomer. These equations have been derived by Tinoco, Woody, and Bradley (1963). The zero-order wave function of a polymer excited state [Eq. (28)] was used in their calculation of R_K for each transition in the exciton band. Applying first-order perturbation theory to these wave functions leads to the following secular equation:

$$\sum_{i=1}^{N} (C_{ik} V_{ij} - E_K \delta_{ij}) = 0 \tag{43}$$

The first-order transition energies E_K are obtained from the solution of the determinant

$$|V_{ij} - E_K \delta_{ij}| = 0 \tag{44}$$

Substitution for E_K values in Eq. (43) gives the values of coefficients C_{ik}. Consequently in a helix containing N residues, the shape of the circular dichroism curve arises from many rotational strengths. If the shape of a CD curve is assumed to be Gaussian, the contribution to CD of a single transition ν_K is related to R_K,

$$[\theta_K] = 0.811 \times 10^{42} \left(\frac{\nu_K}{\Gamma_K}\right) R_K \exp\left[-\frac{(\nu_K - \nu)^2}{\Gamma_K^2}\right] \tag{45}$$

where Γ_K is the width of the band at one-half of its intensity.

The absorption can be either perpendicular or parallel to the helix axis. In the case of exciton bands polarized perpendicular to the helix axis, the dichroism band arises from the overlapping of many transitions near ν_0, the frequency of the monomer absorption maximum. The sign of the CD band is determined by

$$\sum_{j>i}^{N} (j-i) V_{ij} \sin \frac{2\pi(j-i)}{P} \tag{46}$$

where P is the number of groups per turn. According to Tinoco (1964), the expression for the CD spectrum polarized perpendicularly to the helix axis for a single-strand helix is

$$[\theta] = A \left\{\left[2\frac{(\nu - \nu_0)\nu_0}{\Gamma^2} + 1\right] \exp\left[-\frac{(\nu_0 - \nu)^2}{\Gamma^2}\right]\right\} \tag{47}$$

where

$$A = \frac{48\pi^{5/2} N_0 z \mu_\perp^2 \nu_0}{h^2 c^2 \Gamma} \sum_{j>i}^{N} (j-i) V_{ij} \sin 2\pi \frac{(j-i)}{P} \tag{48}$$

where N_0 is the Avogadro's number, z is the distance between residues measured along the helix axis, $\mu_\perp$ is the component of the transition moment perpendicular to the helix axis, and $[\theta]$ is the molar ellipticity.

Since the same phenomenon of resonance interaction will occur in an oligomer, one can consider first the simplest case of a dimer. The consideration of a dimer, e.g., diadenylic acid ApA, presents the advantage of a particular simplicity in calculation and will allow correlation with experimental results.

The two excited states of a dimer composed of two identical units are expressed by eq. 28 and are

$$\psi_\pm = \frac{1}{\sqrt{2}} (\psi_{1a}\psi_{20} \pm \psi_{10}\psi_{2a}) \tag{49}$$

where ψ_1 and ψ_2 are the wave functions of the monomers (C_{jk} are equal to $\pm 1/\sqrt{2}$). The energies of the split states in the dimer composed of two identical units are

$$E_\pm = E_0 \pm V_{12} \tag{50}$$

where E_0 is the energy level of the monomer excited state and V_{12} is the potential energy of interaction between groups 1 and 2. One can expect that this will be reflected in the spectrum by the splitting of the monomer band into two bands with frequencies

$$\nu_{\pm} = \nu_0 \pm V_{12}/h \tag{51}$$

(where ν_0 is the frequency of the monomer band). Thus, the spectrum of the dimer will consist of two bands in which transition energies $h\nu_+$ and $h\nu_-$ are separated by the interaction energy V_{12}. The splitting energy V_{12} can be evaluated by theoretical computation of transition monopoles placed at the carbon and nitrogen nuclei in each purine (Bush, 1965; Warshaw, Bush, and Tinoco, 1965). The interaction energies are

$$V_{1a;2a} = \langle \psi_{1a}\psi_{20} | V_{12} | \psi_{10}\psi_{2a} \rangle \tag{52}$$

Bush expresses ψ_{10} and ψ_{2a}, etc., as molecular orbitals that are sums of atomic orbitals centered on each atom of each purine and/or pyrimidine ring. The Coulomb interaction, V_{12}, is

$$\sum_{i,j} e^2/r_{ij} \tag{53}$$

where i sums over all the atoms on monomer 1, j sums over the atoms of monomer 2, and r_{ij} is the distance between the electrons.

Another way of calculating V_{12} is to apply the dipole–dipole approximation, where the dipoles involved are simply the dipole transition moments according to the expression

$$V_{12} = \frac{1}{R^3}\left[\boldsymbol{\mu}_1 \cdot \boldsymbol{\mu}_2 - \frac{3(\mathbf{R}_{12}\cdot\boldsymbol{\mu}_1)(\mathbf{R}_{12}\cdot\boldsymbol{\mu}_2)}{R^2}\right] \tag{54}$$

R_{12} is the vector distance between the two point dipoles $\mathbf{R}_{12} = (\mathbf{R}_2 - \mathbf{R}_1)$.

A third approach for determination of the band position is semiempirical (Van Holde et al., 1965; Brahms et al., 1966). If one assumes that the dimer (ApA) represents the beginning of a single-stranded helix, the interaction energy, V_{12}, should be approximately the same in the dimer and in the polymer. Very often this shift is small, but sufficient to be experimentally determined. On the basis of the nearest neighbor theory (Bradley et al., 1963) the shift of the maximum in the polymer perpendicularly polarized band $(\nu_\perp - \nu_0)$ is related to that of the oligomer $(\nu_K - \nu_0)$ by

$$\frac{\nu_K - \nu_0}{\nu_\perp - \nu_0} = \frac{\cos[\pi K(N+1)]}{\cos\gamma}, \qquad (K = 1, 2, 3 \ldots N) \tag{54}$$

where γ is the angle of rotation of successive residues (bases) about the helix axis. One can thus calculate the absorption frequencies of any oligomer, ν_K, from the observed shift in the polymer spectrum.

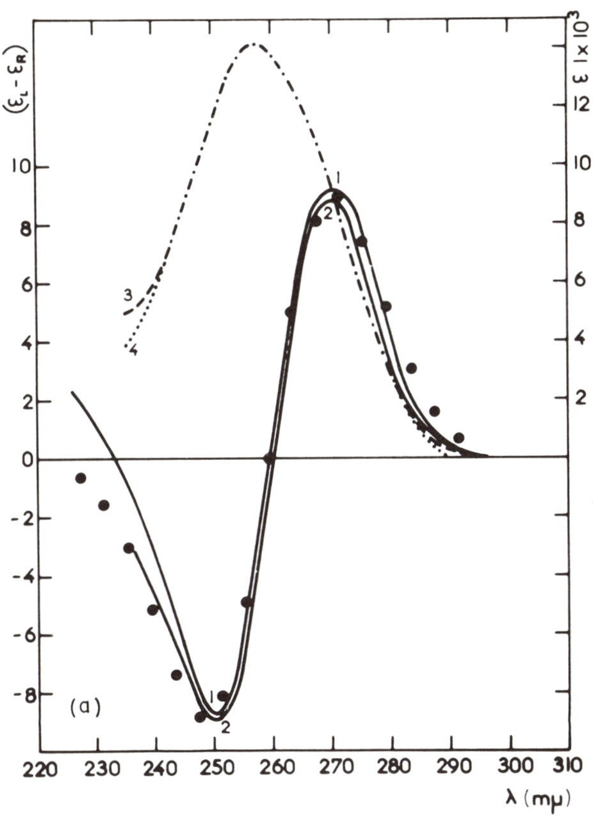

Fig. 1*a*. The circular dichroism and absorption spectra of diadenylic acid. Curves 1 and 2 show the CD at pH 7.4 and 4.9, respectively, at 5.5°C. The absorption spectra at these two pH are shown as curves 3 and 4. The points (● ●) correspond to the theoretical curve (see text). (After Van Holde, Brahms, and Michelson, 1965.) $(\varepsilon_L - \varepsilon_R)$ is calculated per nucleotide residue.

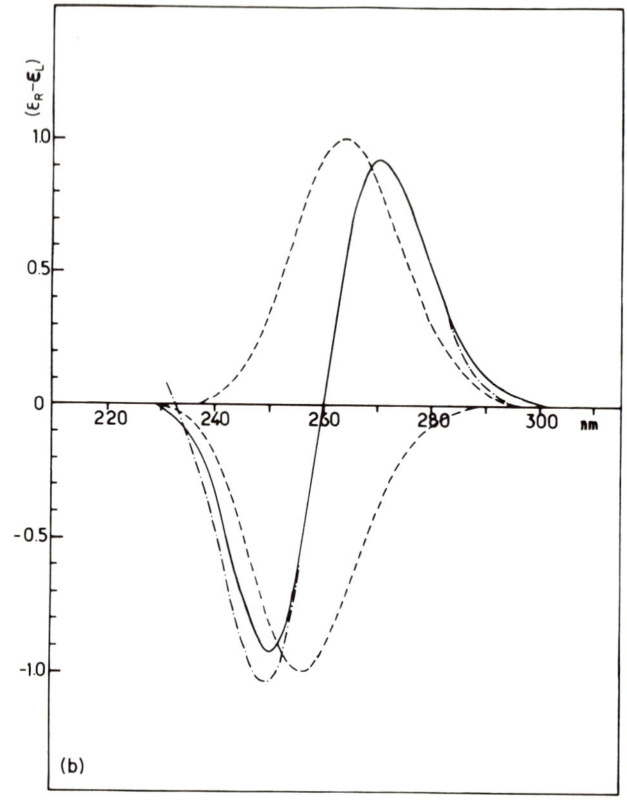

Fig. 1*b*. Resolution of a measured CD spectrum of stacked ApA (–·–·–) in arbitrary units, showing the contribution of two bands of equal rotational strengths (– – –) and opposite sign, situated at different wavelengths. The obtained separation of these two bands $\Delta\nu \sim 1200\ \text{cm}^{-1}$, is in agreement with experimental results of the derivative absorption spectrum of ApA (Brahms, Sellini, and Brahms, 1969).

Since $\Delta\nu \ll \Gamma$, one can assume that the shape of the CD curve can be approximated by the derivative of a Gaussian (——); then the value of Γ, the half band width, can be obtained from the intersection point and one of the extrema $\nu_{\text{extr.}}$ of the experimental curve according to the expression

$$\Gamma = (\nu_{\text{extr.}} - \nu_0)\sqrt{2 \log 2}$$

It is also possible to obtain the position of split bands by the decomposition of a measured CD spectrum. As shown in Fig. 1*b*, the CD of a stacked dimer is resolved into two bands assuming their shape is Gaussian. For the dimer, the expression for the rotation strength [Eq. (41)] is reduced to

$$R_{\pm} = \mp \frac{\pi \nu_0}{2c} [\mathbf{R}_{12} \cdot \boldsymbol{\mu}_1 \times \boldsymbol{\mu}_2] \tag{55}$$

It predicts that the circular dichroic spectrum of the dimer will be composed of two bands equal in magnitude and of opposite sign. Figure 1*a* shows the circular dichroic spectrum of diadenylic acid experimentally determined and calculated on the basis of the nearest neighbor interaction theory (Van Holde, Brahms, and Michelson, 1965). The calculated circular dichroism curve was obtained assuming that both components are of Gaussian shape, similar to that of the monomer AMP absorption band. The agreement between the calculated and observed spectrum is reassuring both for the theory and for the model of a stacked base dimer, similar to the beginning of a helix. The higher oligomers are considered later.

B. Nondegenerate Case

For the dimer composed of different residues, the expression for rotational strength obtained from Eq. 34f will be (Bush and Tinoco, 1967):

$$R_{0A} = \frac{\pi}{c}\left[\frac{\nu_a \nu_b V_{1a;2b}}{h(\nu_b^2 - \nu_a^2)} (\mathbf{R}_{12} \cdot \boldsymbol{\mu}_{10a} \times \boldsymbol{\mu}_{20b}) \right. \tag{56a}$$

$$\left. + \frac{\nu_a \nu_b V_{1b;2a}}{h(\nu_b^2 - \nu_a^2)} (\mathbf{R}_{12} \cdot \boldsymbol{\mu}_{10b} \times \boldsymbol{\mu}_{20a}) \right] \tag{56b}$$

This equation allows the calculation of the CD spectrum when residue one absorbs at ν_a and residue two at ν_b (56a) while the second part (56b) gives the rotational strength for the dimer in which the residue 1 absorbs at ν_b and residue 2 at ν_a. Consequently this equation predicts that the rotational strength R_{0A} will be equal in magnitude and opposite in sign to R_{0B}. In other words, the predicted CD spectrum will be again composed of positive and negative bands. This is in good agreement with the observed CD spectra of some heterodimers. From Figures 2 and 3, it can be seen that the CD of dinucleotides such as ApU, ApC, CpA is composed of positive and negative bands in the spectral region from 320 to 220 mμ (Brahms, Maurizot, and Michelson, 1967b). Since the local sum rule is obeyed in the spectral region

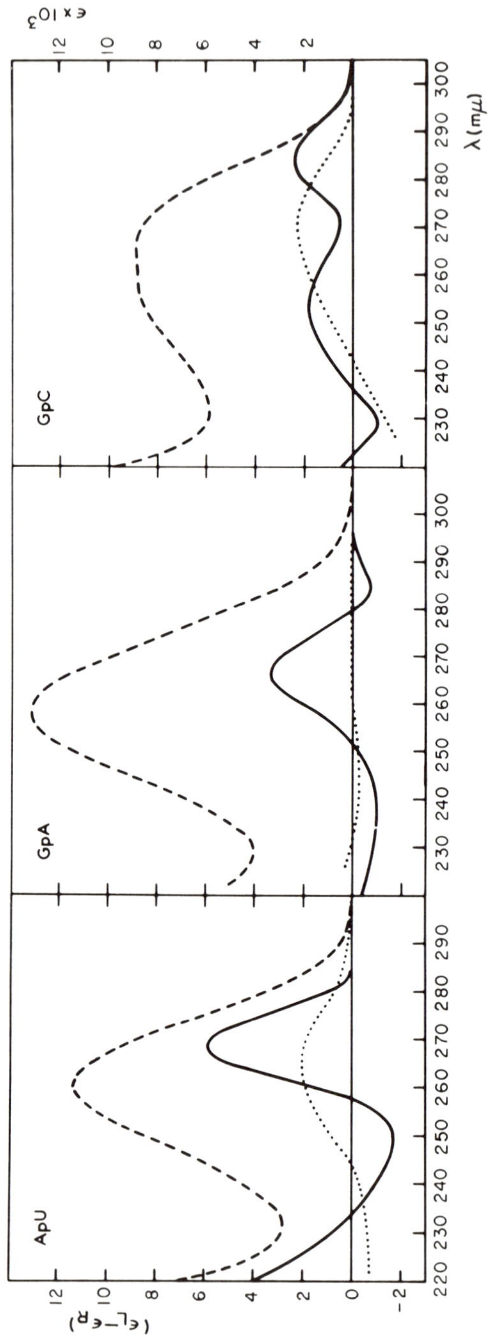

Fig. 2. Circular dichroic spectra of 3′ → 5′ dinucleotide phosphates at neutral pH (in 4.7 *M* KF, 0.01 *M* tris, and at low temperatures of −16° to −20°C) full line; (— — — —) the absorption spectra. For comparison the CD of the sum of the monomer constituents (· · · ·) is shown. (After Brahms, Maurizot, and Michelson, 1967b.)

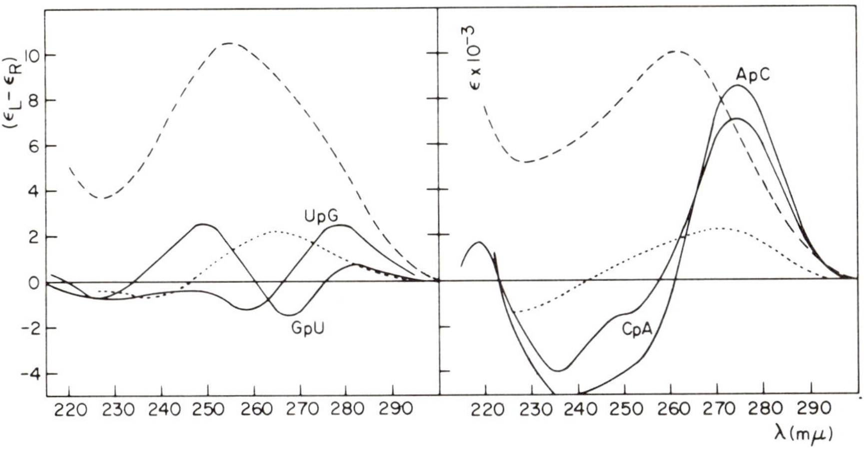

Fig. 3. Circular dichroic spectra of 3′ → 5′ dinucleoside phosphates isomers (————) at low temperature (−18° to −20°C) in 4.7 *M* KF, 0.01 *M* tris at neutral pH. The absorption spectra are left (— —) UpG, right (— — —) CpA. Dotted lines are the circular spectra of monomer constituents. (After Brahms et al., 1967b.)

under consideration, this type of spectrum is called conservative by Bush and Brahms (1967); that is

$$\sum_{K}^{N} R_K = 0$$

Figure 4 shows the corresponding ORD of dinucleotides, shown for comparison and studied by Warshaw and Tinoco (1965, 1966) and Bush and Tinoco (1967).

C. Influence of Chain Length (Homoadenylates)

The studies of circular dichroism of a series of homo-oligomers allow one to correlate the spectra with structure. This is particularly important for single-stranded synthetic polynucleotides which were believed to exist in an unknown, probably disordered randomly coiled form.

The same nearest neighbor theory can be extended to a series of longer oligomers composed of identical residues. The rotational strengths are given by Bradley, Tinoco, and Woody (1963). For an oligomer with a perpendicularly polarized band, their expression for R_K can be simply reduced to

$$R_K = \left(\frac{z\pi}{c}\right)(\nu_0 \mu_\perp^2)(S_2 S_4 - S_1 S_5) \tag{57}$$

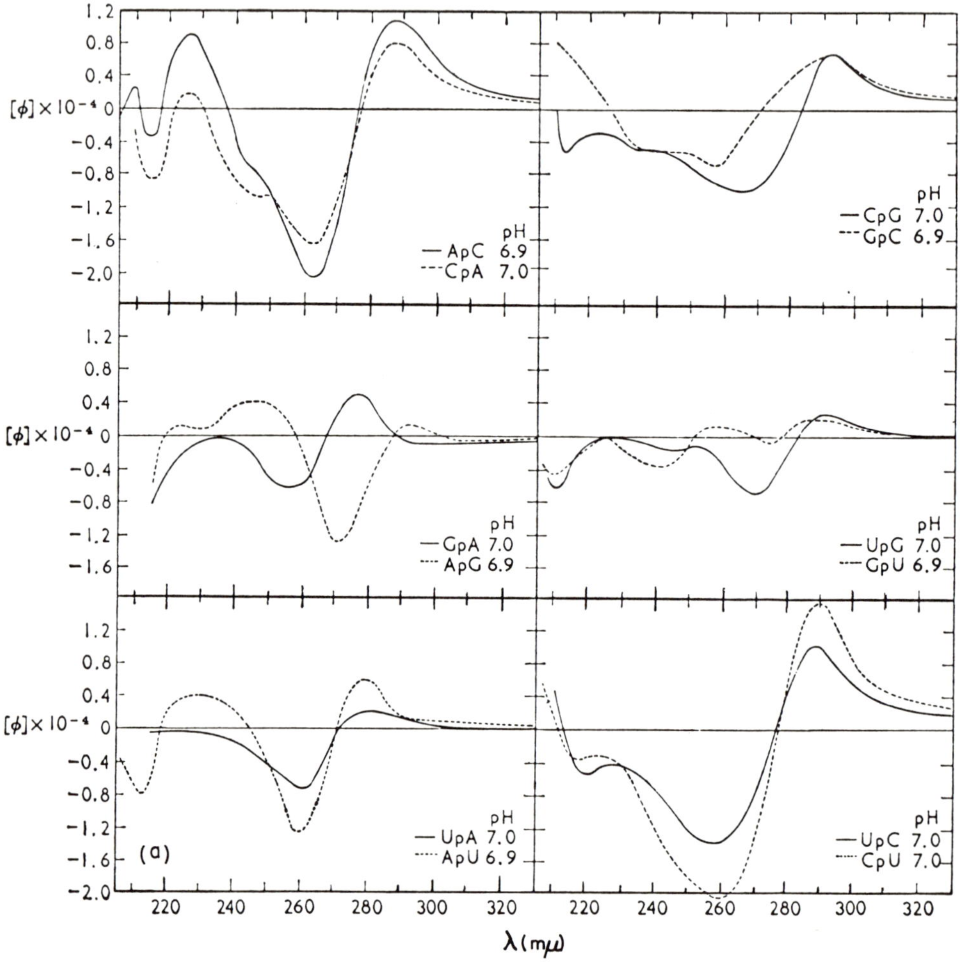

Fig. 4*a*. Optical rotatory dispersion of six dinucleoside diphosphates at pH 6.9 (in 0.08 *M* $KClO_4$ and 0.01 *M* phosphate buffer). The molar rotation [ϕ] is calculated per base. (After Warshaw and Tinoco, 1966.)

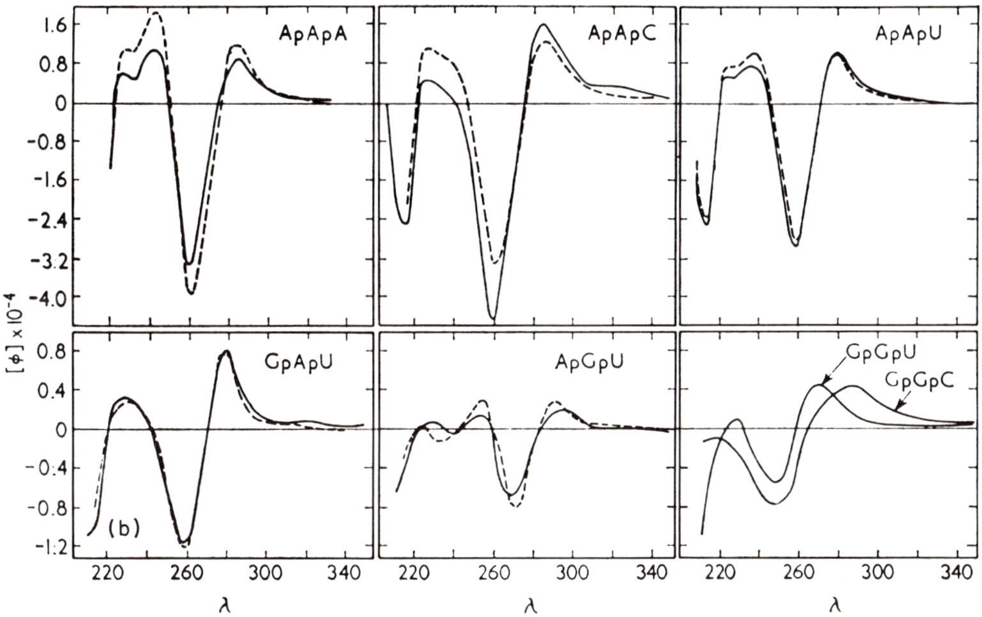

Fig. 4*b*. Molar rotation of trinucleoside diphosphates (in the same solvent). (After Cantor and Tinoco, 1965.)

where z is the displacement of successive bases in the direction of the helix axis, α is the rotation of successive bases about that axis, and $\boldsymbol{\mu}_\perp$ is the electric transition moment,

$$S_1 = \sum_{i=1}^{N} C_{iK} \cos i\alpha \qquad S_2 = \sum_{i=1}^{N} C_{iK} \sin i\alpha$$
$$S_4 = \sum_{i=1}^{N} iC_{ik} \cos i\alpha \qquad S_5 = \sum_{i=1}^{N} iC_{ik} \sin i\alpha \tag{58}$$

The mixing coefficients are given by

$$C_{ik} = \sqrt{\left(\frac{2}{N+1}\right)} \sin\left(\frac{\pi ik}{N+1}\right) \tag{59}$$

The explicit expressions for R_k for a series of small oligomers have been evaluated (see Brahms, Michelson, and Van Holde, 1966) are shown in

TABLE I

Rotational Strength for Adenylate Oligomers

K	$N = 2$	$N = 3$	$N = 4$	$N = 5$
1	$-\frac{B^{a}}{2}\sin\alpha$	$-B\left(\frac{1}{\sqrt{2}}\sin\alpha + \frac{1}{2}\sin 2\alpha\right)$	$-\frac{2B}{5}[(b^2+2ab)\sin\alpha+4ab\sin 2\alpha + 3a^2\sin 3\alpha]^{b}$	$-\frac{B}{3}\left[\frac{3\sqrt{3}}{2}\sin\alpha + \frac{7}{2}\sin 2\alpha + \frac{3\sqrt{3}}{2}\sin 3\alpha + \sin 4\alpha\right]$
2	$+\frac{B}{2}\sin\alpha$	$+B\sin 2\alpha$	$-\frac{2B}{5}[(2ab-b^2)\sin\alpha-4ab\sin 2\alpha - 3b^2\sin 3\alpha]$	$-\frac{B}{3}\left[\frac{3}{2}\sin\alpha - \frac{3}{2}\sin 2\alpha - \frac{9}{2}\sin 3\alpha - 3\sin 4\alpha\right]$
3		$-B\left(-\frac{1}{\sqrt{2}}\sin\alpha + \frac{1}{2}\sin 2\alpha\right)$	$-\frac{2B}{5}[(a^2-2ab)\sin\alpha-4ab\sin 2\alpha + 3b^2\sin 3\alpha]$	$-\frac{B}{3}[-4\sin 2\alpha+4\sin 4\alpha]$
4			$-\frac{2B}{5}[(-b^2-2ab)\sin\alpha + 4ab\sin 2\alpha+3a^2\sin 3\alpha]$	$-\frac{B}{3}\left[-\frac{3}{2}\sin\alpha - \frac{3}{2}\sin 2\alpha + \frac{9}{2}\sin 3\alpha - 3\sin 4\alpha\right]$
5				$-\frac{B}{3}\left[-\frac{3\sqrt{3}}{2}\sin\alpha+\frac{7}{2}\sin 2\alpha-\frac{3\sqrt{3}}{2}\sin 3\alpha+\sin 4\alpha\right]$

[a] $B = \frac{\pi z \mu_\perp{}^2}{1}$

[b] $a = \sin 36^\circ = 0.5878;\ b = \sin 72^\circ = 0.9511.$

Table I. The position of the maximum of the oligomer bands are related to monomer frequency ν_0 by

$$\nu_k = \nu_0 + \frac{2V}{h} \cos \frac{\pi k}{N+1}, \qquad (k = 1, 2, 3 \ldots N) \tag{60}$$

Experimental results have confirmed this theory, at least for adenylates at neutral pH. The circular dichroism curves of ApA (Fig. 1) and of a series of

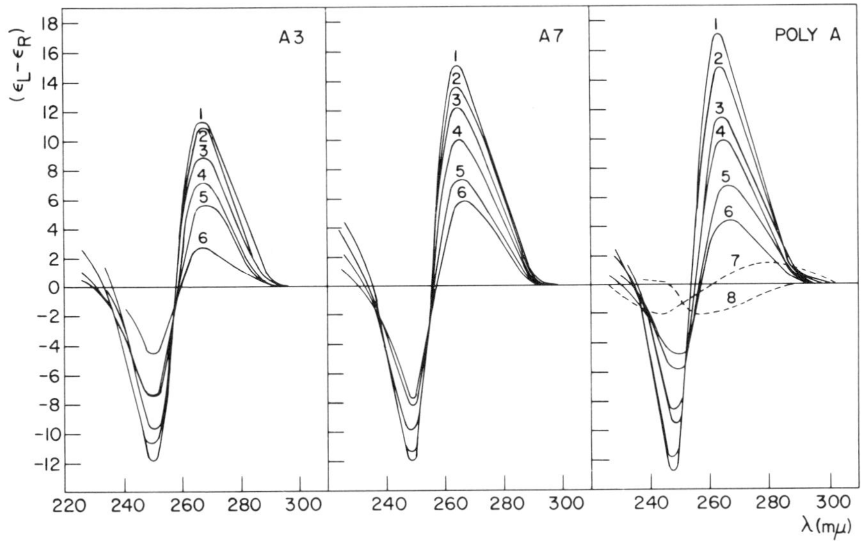

Fig. 5. Circular dichroic spectra of adenylate oligonucleotides at pH 7.4 at various temperatures. Left, the trimer in 0.1 *M* NaCl–0.01 *M* tris at: (1) −2°C; (2) +0.5°C; (3) 4.5°C; (4) 18°C; (5) 25°C; (6) 47°C. Center, the heptamer at: (1) −2°C; (2) +0.5°C; (3) 8°C; (4) 18.5°C; (5) 32°C; (6) 40°C. Right, poly A at: (1) temperature range from −2°C to +6°C; (2) +17°C; (3) 34°C; (4) 42°C; (5) 57°C; (6) 74°C; (7) poly A quaternary ammonium salt in 98% ethanol at 0°C; (8) AMP at 0°C. (After Brahms, Michelson, and Van Holde, 1966.)

adenylates (Fig. 5) at neutral pH indicate clearly that a pair of bands of opposite sign are present in the region of adenine absorption (230–300 mμ). The two CD bands of ApA are spaced above and below the position of the monomer absorption band which confirms the splitting, in agreement with the theory of the exciton coupling. Thus, while the difference in sign allows resolution in the circular dichroism spectrum, the bands cannot at present be resolved in the absorption spectrum. This CD spectrum is completely different in form from that of the mononucleotide AMP (Van Holde et al., 1965; Warshaw et al., 1965). The ApA circular dichroic and ORD curves are particularly intense at low temperatures. The circular dichroism spectra shown in Fig. 5 indicate a similarity in the general shape for all oligoadenylates from the dimer to the high polymer. This Cotton effect arises from the dissymmetrical base–base interaction which is discussed below.

The intensity of these circular dichroic bands increases as a function of degree of polymerization, i.e., with increase of the number of residues, N. The rotational strength per mole of residue increases slowly as the chain

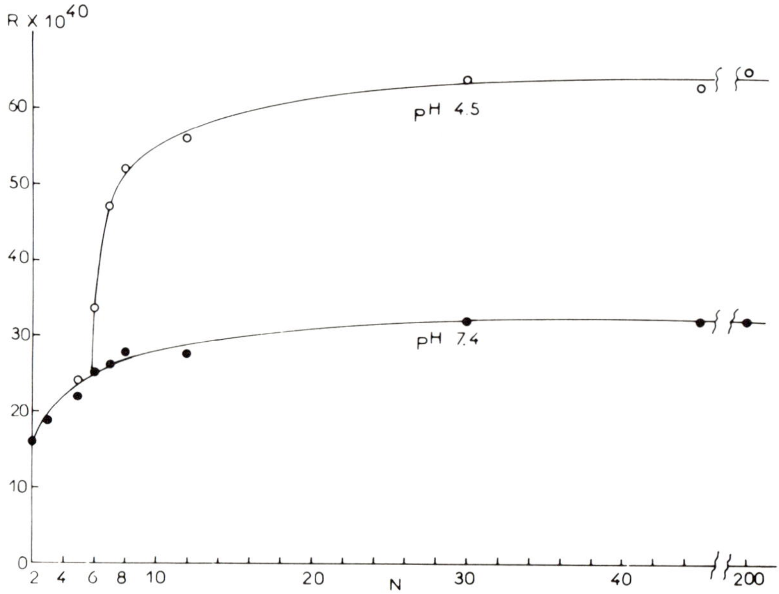

Fig. 6. Rotational strength (R) as a function of the chain length N of adenylate oligomers (positive CD main band) at 0°C. Open circles at pH 4.5; filled circles at pH 7.4. (After Brahms, Michelson, and Van Holde, 1966.)

becomes longer and reaches an asymptotic maximum for the high polymer, poly A (see Fig. 6). Brahms et al. (1966) also observed the broadening of the positive band as a function of the number of residues in the oligomeric chain. There is a general shift toward lower wavelengths with increasing N, that is the position of the maximum, minimum, and crossing point are blue shifted. This is accompanied by a corresponding decrease in absorbance hypochromicity (Brahms et al., 1966; Leng and Felsenfeld, 1966).

The quantitative calculation of these circular dichroism properties was made using the outlined exciton theory and an appropriate model of dissymmetric ApA and adenylate structure. The geometry assumed for ApA and adenylates is that of parallel stacked purines, rotated by an angle of about 30° to 45°. This can represent the start of a single-strand helix quite similar in dimension to half of a double-strand DNA helix, i.e., a single strand of DNA. Another model of a coupled oscillator in which the bases are not coplanar but perpendicular (see substituted diphenyl model, Kuhn and

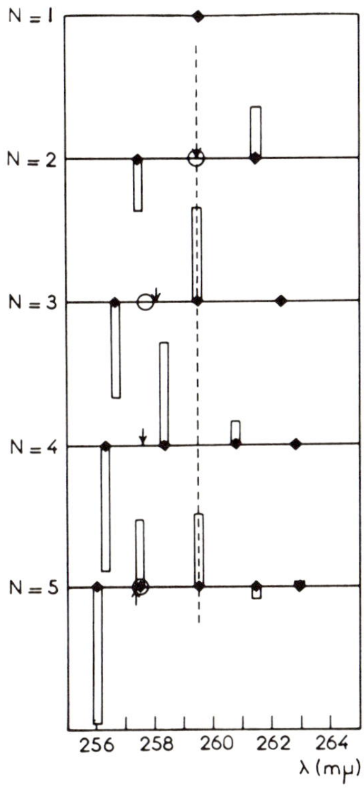

Fig. 7. Schematic distributions of rotational strength predicted for small adenylates oligomers for chain length $N = 1$ to 5. Predicted crossing points are shown by arrows observed crossing points by circles. (After Brahms, Michelson, and Van Holde, 1966.)

Rometch, 1944) will not give the splitting. Van Holde et al. (1965) calculated R using the stacked base model assuming an angle between the bases of 30° to 45°, a vertical distance of 3.4 Å, and the electric transition moment $\mu_{\perp}$ of this $\pi \rightarrow \pi^*$ band perpendicular to the helix axis of 4.1 D. The obtained value of 15×10^{-40} or 20×10^{-40} for an angle of 30° or 45°, respectively, is in good agreement with the experimentally obtained value of about 21×10^{-40}. Bush (1965) and Bush and Tinoco (1967) also obtained excellent accord between calculated and experimental CD and ORD curves, using a similar model.

Calculations for higher oligomers are more difficult because of a large number of bands to be taken into account with resulting overlapping and cancellation. Brahms, Michelson, and Van Holde (1966) calculated the distribution of rotational strength for a series of adenylates for N from 1 to 5, which are shown in Fig. 7. The predicted form of the circular dichroic spectra will have a longer wavelength component of positive rotational strength and a second, shorter wavelength component of negative rotational strength. With increasing N, the predicted crossing point will shift to the blue (Fig. 7). Figure 7 shows that the agreement between predicted and observed CD parameters (see Fig. 5) is satisfactory. This calculation was made by applying the "nearest neighbor" exciton theory outlined before. The agreement supports the model, similar to one strand of Watson–Crick DNA, where oligoadenylates and poly A are single-chain stacked-base helices (Langridge et al., 1960).

Qualitatively similar results were obtained for oligocytidylates at neutral pH by Brahms, Maurizot, and Michelson (1967a). With increasing degree of polymerization, there is a progressive increase of the rotational strength; however, this investigation revealed the contribution of other effects reflected in the nonconservative spectrum.

D. Other Dinucleotides: Nonconservative Spectra

All previously examined oligoadenylate circular dichroic spectra were in good agreement with the exciton theory (Tinoco, 1964; Tinoco, Woody, and Bradley, 1963). Warshaw and Tinoco (1965, 1966) examined ORD spectra of almost all 16 dinucleoside phosphates which were theoretically calculated by Bush and Tinoco (1967) using the expression for R outlined previously for degenerate and nondegenerate cases. In general, the stacked base structure gives rise to the exciton type ORD spectra. (One may guess that the only exception is the ORD curve of GpA.) Therefore, one would expect that in general the circular dichroic spectra will be composed of a pair of adjacent positive and negative bands with an intersection point situated at a wavelength very near that of the absorption maximum.

Brahms, Maurizot, and Michelson (1967b) have found two types of circular dichroic spectra for a variety of $3' \rightarrow 5'$ dinucleotides at low temperatures.

The first type, described previously, is called the conservative type, since the sum rule of R is conserved. In this category are the CD spectra of ApU, CpA, CpU, and the typical ApA and adenylates spectra (Figs. 1, 2 and 3). However, a second type was found for CD spectra that essentially consist of one or two positive bands shown in Fig. 2 for GpC and GpA. One can clearly see that the sum of rotational strengths is not equal to zero in the spectral region under consideration. This type of spectra is not predicted by the exciton theory of rotatory power.

Very interesting features of these nonconservative spectra were revealed by the analysis of CpC, cytidylate oligomers, and poly C circular dichroism and absorption spectra at neutral pH (Fig. 8). Some of the conclusions

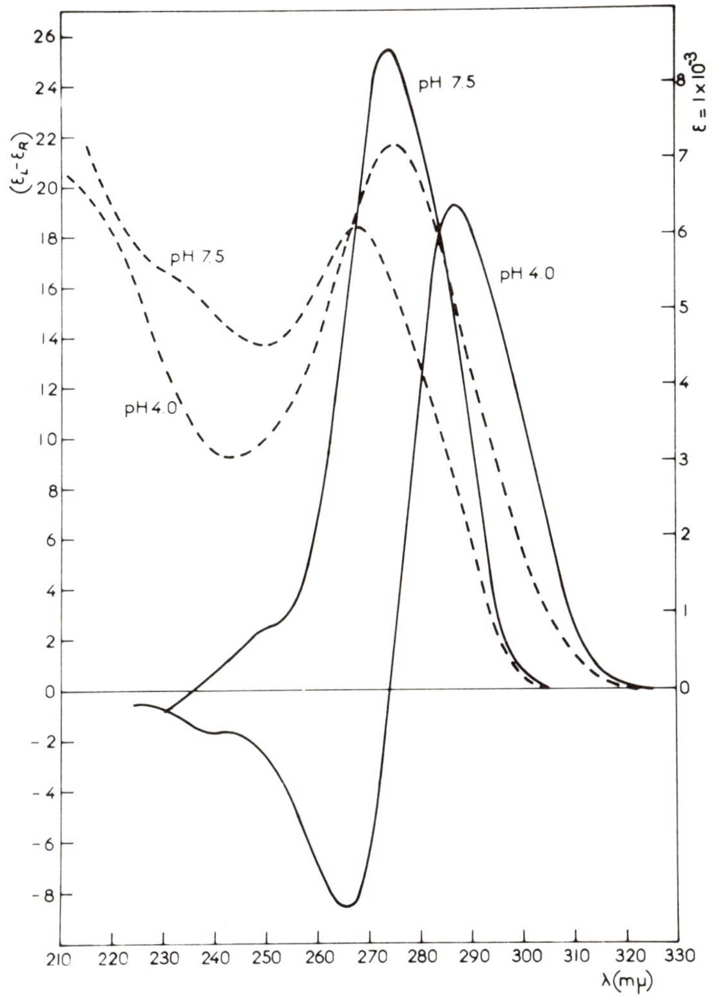

Fig. 8. Circular dichroic and absorption spectra of poly C at pH 4.0 (in 0.1 *M* NaCl, 0.05 *M* acetate) and at pH 7.5 (in 0.1 *M* KF, 0.01 *M* tris) both circular dichroic spectra (———) at about 0°C and absorption spectra at 25°C (— — —). (After Brahms, Maurizot, and Michelson, 1967a.)

TABLE II

Circular Dichroism and Absorption Spectral Parameters of 3′ → 5′ and 2′ → 5′ Cytidylics[a,b]

	Absorption		Circular dichroism					
Substance	λ_{max}	$\varepsilon \times 10^{-3}$ (cm^{-1} $mole^{-1}$)	λ_{max}	R_{max} $\times 10^{40}$ at 0°C	$\Delta\nu$ (cm^{-1})	λ_{min}	$R_{min} \times 10^{40}$ at 0°C	λ_0
At pH 7.5								
3′→5′								
Monomer 5′	272	9	272	11	—	—	—	—
Dimer	269	7.92	280	22	2600	—	—	—
Trimer	268	7.4	278	31	2900	—	—	—
Tetramer	268	7.25	278	34	3000	—	—	—
Pentamer	268.5	7.06	278	34	3000	—	—	—
Hexamer	268.5	6.82	—	—	—	—	—	—
Heptamer	268.5	6.63	276	34.5	3000	—	—	—
Octamer	268.5	6.34	276	34.5	3000	—	—	—
Nonamer	268.5	6.2	276	38	3000	—	—	—
Decamer	268.5	6.2	276	35	2900	—	—	—
Tridecamer	268	6.14	275	39.5	2900	—	—	—
Poly C	268	6.09	{276, 200}	57.2	3200	—	—	—

2′→5′								
Monomer 5′	272	8	272	11	—	—	—	—
Dimer	269.5	7.78	275	—	—	—	—	—
Pentamer	269	7.60	277	—	—	—	—	—
Hexamer	269	7.47	277	—	—	—	—	—
At pH 4.0								
3′→5′								
Monomer	280	—	—	—	—	—	—	—
Dimer	272.5	8.65	285	—	—	—	—	—
Hexamer	270.5	7.60	281	+27	—	257	—	—
Heptamer	272.5	8.0	285	+27	—	262.5	−4.2	270
Octamer	272.5	6.97	287	+27.5	—	264	−8.7	272.5
Nonamer	273	7.05	285.5	+31.2	—	263	−9.5	273
Decamer	273	6.99	286	+27.2	—	263	−7.8	272.5
Tridecamer	274	6.90	286	+30.6	—	264	−7.8	272.5
Poly C	275	6.5	287	+39.2	—	265.5	−13.7	274

[a] After Brahms, Maurizot, and Michelson, 1967.

[b] Symbols are as follows: for the absorption spectra λ_{max} is the wavelength of the maximum and ε is the molar extinction coefficient at the maximum. For the circular dichroism, λ_{max}, λ_{min}, λ_0 are the wavelengths of the maximum, minimum, and crossing point, respectively. R_{max} and R_{min} are the rotational strengths of the positive and negative bands, respectively, and $\Delta\nu$ is the half-width of the positive band. The circular dichroic spectra were measured at 0°C.

of Brahms, Maurizot, and Michelson (1967a) and Bush and Brahms (1967) papers can be summarized as follows:

(1) In the near UV region, the circular dichroism spectrum is composed of one main band of positive sign.

(2) The comparison of the maxima of the circular dichroism and absorption bands indicates that the λ_{max} of the monomer absorption is situated at a wavelength between λ_{max} for the circular dichroic spectra of CpC and oligomers and the absorption maxima of oligocytidylates. This strongly suggests the existence of exciton splitting. The higher wavelength component will have greater rotational strength, whereas the lower wavelength component will have higher oscillator strength (Table II).

One can conclude that CpC and cytidylates circular dichroism spectra indicate the complexity of different contributions: (1) exciton contributions, which lead to the conservation of overall rotational strength, and (2) other contributions not included in the exciton theory which give rise to the predominant nonconservative effects (Bush and Brahms, 1967; Brahms, Maurizot, and Michelson, 1967a).

E. Theoretical Analysis

Bush and Brahms (1967) have analyzed the nature of contributions which give rise to the nonconservative circular dichroic spectra. They consider the very general Tinoco equation (34), assuming that the effect of static field is negligible for polynucleotides and that only two terms contribute. The calculation of these two terms leads to the division of the summation over the excited states into two groups: (1) One should consider the interaction among transitions in the accessible region, the so-called near UV transitions. However, this will give rise to positive and negative values of rotational strengths, such as exciton terms, as shown before. (2) The interaction with "far UV" transitions gives rise to rotational strength in the "near UV." These transitions to higher energy states are not known in nucleotides, and are taken together using the Kirkwood polarizability approximation. This yields a contribution of a nonconservative character.

The following equation gives the rotational strength at ν_a arising from the interaction of the transition $0 \rightarrow a$ with all the far UV transitions above a certain energy ω which will be taken as about 50,000 cm^{-1} (or 200 nm)

$$R_{A\omega} = -2\sum_{i}\sum_{j\neq 1}\frac{1}{h(\nu_0^2-\nu_a^2)}\,\mathrm{Im}\left[\left(\frac{\pi i}{c}\,\nu_a\nu_0\boldsymbol{\mu}_{i0a}\cdot\mathbf{T}\cdot\boldsymbol{\alpha}_{\omega}\cdot\boldsymbol{\mu}_{i0a}\times\mathbf{R}_{ij} + \nu_0^2\boldsymbol{\mu}_{i0a}\cdot\mathbf{T}\cdot\boldsymbol{\alpha}_{\omega}\cdot\mathbf{m}_{ia0}\right)\right] \tag{61}$$

where $\boldsymbol{\alpha}_\omega$ is the polarizability tensor arising from all transition of energy greater than ω.

This equation yields a contribution of a nonconservative character. The first term is the Kirkwood polarizability term and depends on the polarizability differences in plane and out of plane. The second term contains the magnetic moment and the polarizability. Both terms depend on the direction of the electric dipole transition moment. This emphasizes the importance of the geometric factor.

More recently this problem of the origin of nonconservative and conservative spectra was examined by Tinoco (1968) in a particularly simple theory which predicts the shape of the CD curves of polynucleotides. The basic considerations are those of the exciton theory. In a polymer of N monomers, there will be N different singly excited polymer states with wave functions $|A_K\rangle$ and N monomer transitions occurring in the limited spectral region. Each such polymer state could also be in interaction with the remaining monomer excited states which are situated outside of the frequency range of interest, the far UV states, similarly to Bush and Brahms (1967).

Johnson and Tinoco (1969) express the N polymer states by

$$|K\rangle = |1K\rangle + |2K\rangle \tag{62}$$

where $|1K\rangle$ is $|A_K\rangle$ and $|2K\rangle$ is the mixing of $|1K\rangle$ with the remaining states. A modified Gaussian of a shape similar to that of the absorption band was chosen by Tinoco for the analysis of the shape of a circular dichroic band of a single electronic transition.

$$(\varepsilon_L - \varepsilon_R)_M(\nu) = \frac{2.458 \times 10^{38} \nu R_a \exp[-(\nu - \nu_a)^2/\Gamma^2]}{\Gamma} \tag{63}$$

where R_a is the rotational strength in cgs units. In a polymer, a single electronic transition of each monomer [Eq. (63)] will be combined into N polymer transitions.

$$(\varepsilon_L - \varepsilon_R)_p(\nu) = \frac{2.458 \times 10^{38}\, \nu}{\Gamma N} \sum_K^N R_K \exp[-(\nu - \nu_K)^2/\Gamma^2] \tag{64}$$

This can be simply written in the form of $\nu f(\nu - \nu_K)$ according to Johnson and Tinoco (1969).

$$(\varepsilon_L - \varepsilon_R)(\nu) = \nu \sum_K^N R_K f(\nu - \nu_K) \tag{65}$$

If the bands in the polymer are not too shifted relative to the monomer band,

Eq. (65) can be expanded in a Taylor series about $\bar{\nu}$, the average frequency. Only the first two terms in the series will be retained.

$$\frac{(\varepsilon_L - \varepsilon_R)(\nu)}{\nu} = \sum_{K=1}^{N} R_K f(\nu - \bar{\nu}) - \sum_{K=1}^{N} R_K \frac{\partial f(\nu - \bar{\nu})}{\partial \nu} (\nu_K - \bar{\nu}) + \cdots \tag{66}$$

This equation characterizes the shape of the CD curve of a polymer in which one term is proportional to the modified Gaussian and the second to its first derivative with respect to frequency.

Applying some approximation such as the Kirkwood polarizability for the expression of rotational strength owing to the mixing with high-energy states, and the monopole approximation to compute the interaction potential, V, Johnson and Tinoco derived the explicit expression for the circular dichroism curve.

$$\begin{aligned}(\varepsilon_L - \varepsilon_R)(\nu) = {} & \frac{\pi\nu}{c} f(\nu - \bar{\nu}) \sum_{\substack{i,j \\ i \neq j}}{}' \sum_{a} \nu_{i0a} \left[\sum_{s,t} \frac{\rho_s^{i0a}}{r_{st}^3} r_{st} \right] \cdot \boldsymbol{\alpha}_t^j \times \boldsymbol{\mu}_{i0a} \cdot \mathbf{R}_{it} \\ & + \frac{\pi\nu}{2c} \frac{\partial f(\nu - \bar{\nu})}{\partial \nu} \sum_{\substack{i,j \\ i \neq j}}{}' \sum_{a,b} \nu_{i0a} \left[\sum_{s,t} \frac{\rho_s^{i0a} \rho_t^{j0b}}{r_{st}} \right] \mathbf{R}_{ij} \cdot \boldsymbol{\mu}_{i0a} \times \boldsymbol{\mu}_{j0b}\end{aligned} \tag{67}$$

where ρ_s^{i0a} is the monopole charge of group s on monomer i due to the transition $0 \rightarrow a$; $\boldsymbol{\alpha}_t^j$ is the polarizability tensor of group t in the monomer j, and r_{st} is the magnitude of the distance between group s on i and group t on j.

In this equation, the first term represents the interaction of monomer transition $0 \rightarrow a$ on residue i with the other remaining states (far UV transitions) which, similarly to Bush and Brahms (1967), are expressed by a Kirkwood polarizability term. In calculating the polarizabilities, the contribution of near UV transitions was subtracted out. It is this term which will give rise to the nonconservative spectra depending on the geometry of the transition moment $0 \rightarrow a$ interacting with principal polarizabilities of neighboring polynucleotide bases.

The second term represents the first derivative of the absorption band $f(\nu - \bar{\nu})$ and involves the known dipole coupling term. This term will give rise to the conservative spectra, similar to the exciton effect of interaction among identical transitions, that is, to positive and negative circular dichroic bands.

Tinoco also illustrated how these two contributions change with the geometry of a helical polynucleotide. By fitting the observed circular dichroism spectrum, one can obtain the relative contribution of the modified Gaussian and its first derivative. The part with Gaussian shape will inform us about the interaction between different transitions, while the derivative will characterize the interaction among identical groups (see Fig. 9). By calculating R_K and $(\varepsilon_L - \varepsilon_R)$ from these explicit equations, Johnson and Tinoco (1969) and Tinoco (1968) were able to test a hypothetical structure against the measured parameters.

One can easily see from Fig. 9 that the shape of the circular dichroism and ORD curves represents the combination of two types of basic curves. Tinoco

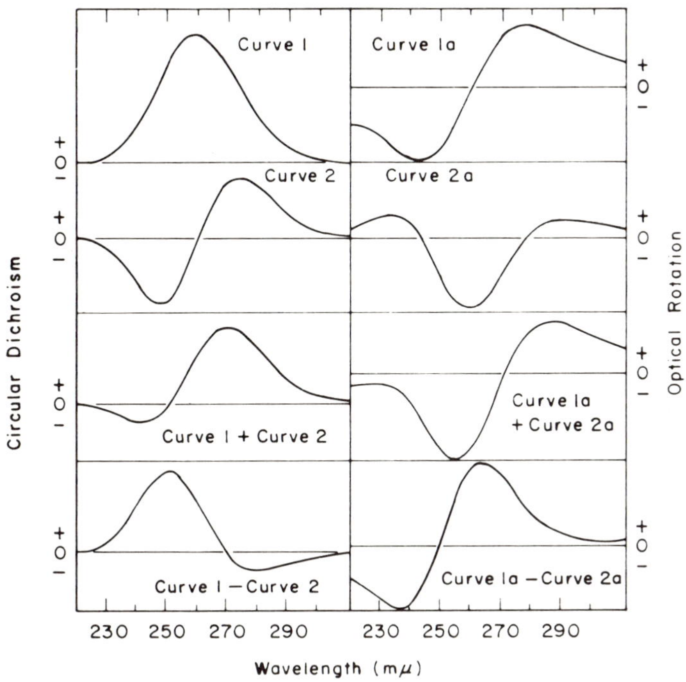

Fig. 9. The shape of circular dichroism spectra (left) and corresponding optical rotatory dispersion curves (right). The combination of the two top curves gives all others. (After Tinoco, 1968.)

also calculated the rotational strength and $(\varepsilon_L - \varepsilon_R)$ for a single-stranded polynucleotide for which the vertical distance of the bases is 3.4 Å, and the monomer has a single electric transition at 260 mμ. The conclusions reached from such approximate calculations show that the tilting of the transition moment to a plane perpendicular to the helix axis is important in order to obtain a contribution from both the conservative and nonconservative terms. Another important factor which influences the predicted shape of the spectra is the number of residues per turn. Tinoco found that the experimental parameters of circular dichroism spectra of single-stranded and double-stranded ordered polynucleotides will be useful for obtaining geometrical information even with approximated calculation.

F. Other Possible Contributions

In all the nearest neighbor (exciton) calculations of polynucleotides, only strong electron dipole allowed transitions were taken into consideration, which are perpendicularly polarized to the helical axis. However, one can expect on theoretical grounds, that a weak magnetic allowed transition of $n \rightarrow \pi^*$ origin will occur in purine and pyrimidine bases, as proposed by Kasha (1960) and Mason (1959). Some experimental evidence indicates that the strong $\pi \rightarrow \pi^*$ absorption band at 260 mμ of poly A may contain a hidden $n \rightarrow \pi^*$ transition occurring at the longer wavelength. The following observations support this possibility:

(1) In poly A at acid pH the positive circular dichroic band contains a shoulder, which according to Brahms (1964) allows the resolution into at least two components; the longer wavelength component is centered at about 280 mμ. (2) Bush and Scheraga (1969) and Adler, Grossman, and Fasman (1969) found that poly deoxy A (and not ApA or AMP) at neutral pH exhibits a weak circular dichroic band with a singularity at 278 mμ.

(3) Brahms, Sellini, and Brahms (1969) applied a high-resolution derivative spectroscopy method and have been able to resolve the main absorption band (at 260 mμ) of adenylates into three components. The two higher-energy components result from the exciton splitting of the $\pi \rightarrow \pi^*$ band, whereas the low-energy component, centered at about 280 mμ in poly A and at 278 mμ in poly dA, is probably of $n \rightarrow \pi^*$ origin. The presence of this band is also shown in the circular dichroism spectrum of ribo adenylates at neutral pH.

One can conclude that this $n \rightarrow \pi^*$ transition has to be taken into account in theoretical considerations of optical properties of polynucleotides.

IV. SEQUENCE CONTRIBUTION

The optical properties *a priori* are very sensitive to base sequence in an oligonucleotide. Using conventional difference spectra, Michelson (1963) has shown that one can distinguish two sequential isomers, ApC and CpA. However, much greater differences are obtained from the investigation of ORD and CD spectra. As shown in Figs. 3 and 4, where some examples of CD and ORD of dinucleotide isomers are shown, optical rotatory power can be used to determine the sequence in an oligonucleotide chain.

Cantor and Tinoco (1965) have developed a simple semiempirical method which is an attempt to predict the ORD spectra of single-stranded oligonucleotides. Their basic assumption is that mainly nearest neighbor interactions are contributing to the rotation. Thus, the rotations of any trinucleotide can be calculated knowing its constituent dinucleotides. For any dinucleotide phosphate (IpJ), Warshaw and Tinoco (1965) showed that the molar rotation can be written

$$[\phi'_{\mathrm{IJ}}(\lambda)] = N_{ij}(\lambda)+[\phi'_{\mathrm{I}}(\lambda)]+[\phi'_{\mathrm{J}}(\lambda)] \tag{68}$$

where N_{IJ} is the contribution of the nearest neighbor interaction to the rotation, $[\phi'_{\mathrm{IJ}}]$ is the molar rotation of the dinucleoside phosphate IpJ; $[\phi'_{\mathrm{I}}(\lambda)]$ and $[\phi'_{\mathrm{J}}(\lambda)]$ are the molar rotations of the mononucleosides I and pJ, which are usually very small compared with the dimer rotations. It is thus possible, according to the authors, to use the average of pJ and I. This separation of the molar rotation into three contributions may or may not have any physical meaning but mathematically is possible according to the authors.

Similarly for a trinucleotide, Cantor and Tinoco (1965) write

$$[\phi'_{\mathrm{IJK}}(\lambda)] = [\phi'_{\mathrm{I}}(\lambda)]+[\phi'_{\mathrm{J}}(\lambda)]+[\phi'_{\mathrm{K}}(\lambda)]+M_{\mathrm{IJ}}(\lambda)+M_{\mathrm{JK}}(\lambda)+M_{\mathrm{IK}}(\lambda) \tag{69}$$

where $[\phi'_{\mathrm{IJK}}(\lambda)]$ is the molar rotation of the trinucleotide, and $M_{\mathrm{IJ}}(\lambda)$ and $M_{\mathrm{JK}}(\lambda)$ are the nearest neighbor interactions in a trinucleotide. M_{IK}, the next nearest neighbor interaction is considered to be zero in this treatment and is not taken into consideration. If the bases in the dimer and in the trimer have the same relative conformation, one would expect that $M_{\mathrm{IJ}}(\lambda) = N_{\mathrm{IJ}}(\lambda)$ and $M_{\mathrm{JK}}(\lambda) = N(\lambda)$. Together with the equation for the dinucleotide (68), Eq. (69) becomes

$$[\phi'_{\mathrm{IJK}}(\lambda)] = [\phi'_{\mathrm{IJ}}(\lambda)]+[\phi'_{\mathrm{JK}}(\lambda)]-[\phi'_{\mathrm{J}}(\lambda)] \tag{70}$$

which, transformed into units of molar rotation per residue, gives

$$[\phi_{\mathrm{IJK}}(\lambda)] = \frac{2[\phi_{\mathrm{IJ}}(\lambda)]+2[\phi_{\mathrm{JK}}(\lambda)]-[\phi_{\mathrm{J}}(\lambda)]}{3} \tag{71}$$

This equation can be extended to larger oligomers and polymers; if the end effects are ignored, the molar rotation per residue of a single-strand polynucleotide is:

$$[\phi_{\mathrm{polynucl.}}(\lambda)] = \sum_{\mathrm{I}=1}^{4}\sum_{\mathrm{J}=1}^{4} 2\chi_{\mathrm{IJ}}[\phi_{\mathrm{IJ}}(\lambda)] - \sum_{\mathrm{I}=1}^{4}\chi_{\mathrm{I}}[\phi_{\mathrm{I}}(\lambda)] \tag{72}$$

where χ_{IJ} is the mole fraction of dinucleotide IpJ and χ_I is that of the monomer. (The mole fraction of any dimer is the number of times it appears in the polymer divided by the chain length.) Cantor, Jaskunas, and Tinoco (1966) assume a random sequence for an RNA of unknown sequence which gives the molar rotation per residue:

$$[\phi_{RNA}(\lambda)] = 2\sum_{I=1}^{4} \sum_{J=1}^{4} \chi_I \chi_J [\phi_{IJ}] - \sum_{I=1}^{4} \chi_I [\phi_I] \tag{73}$$

For a homopolymer, the equation is greatly simplified:

$$[\phi] = 2[\phi_{II}] - [\phi_I] \tag{74}$$

Using Eq. (71), Cantor and Tinoco have calculated the ORD curves for all 64 trinucleoside diphosphates containing the normal four bases, A, U, C, G. For nine trinucleoside diphosphates, the experimentally measured ORD spectra agreed very well with the calculated values. This was also confirmed by the measurements of circular dichroism spectra of trinucleotides. By calculating the circular dichroism curves by the method of Cantor and Tinoco and by using the values of dinucleotides, circular dichroism is in good agreement with experimental results (Aubertin, 1968; Brahms, Aubertin, Dirheimer, and Grunberg-Manago, 1969) whenever the trimer is in non-aggregated stacked conformation. This means also that the assumption of similarity of conformation between stacked bases in dimers and in trimers appears to be correct. Furthermore, the computed ORD data on 64 trimers should be very useful in the identification of oligonucleotides obtained in the course of sequence determination.

Cantor, Jaskunas, and Tinoco (1966) have also calculated ORD curves of the homopolyribonucleotides, poly A and poly C, at neutral pH. The authors noticed that the agreement in this case is only qualitative between experimental and calculated results, that is, in the general shape of the curves. But in all these polymers, the observed curves were slightly shifted to the blue from the predicted values from the dimers, and the experimentally measured spectra were more intense. The lack of quantitative agreement between experimental and calculated results for polymers may mean that some assumptions implicit in the calculations are invalid.

One can suppose that by calculating the optical activity in single-stranded polymers as the sum of dimer contributions taken pairwise, one is not taking into account the effect of chain-length dependence (see previous discussion). Furthermore, the polymers have electrostatic and geometric constraints which are absent in the dinucleotide. Other authors have investigated various other tri- and tetranucleotides by means of circular dichroism and optical rotatory dispersion (Zavelgilskij and Li, 1967; Inoue, Aoyagi, and Nokanishi,

1967; Brahms, Aubertin, Dirheimer, and Grunberg-Manago, 1969). While the agreement between calculated and observed spectra seems to be satisfactory for the single-stranded trimers, for the tetramers Inoue has already found a slight discrepancy. The results of Inoue and co-workers may indicate that better agreement can be obtained by calculating the data of the dimer from the experimental data on the tetramer (Inoue, Masuda, and Aoyagi, 1968).

The temperature may also have its effect; that is, the dimer and the polymer may not have the same degree of stacked bases at room temperature (see section V).

More recently is was shown that the presence of the 3′-terminal phosphate is reflected by a slight decrease of the intensity of the circular dichroic bands (Brahms, Maurizot, and Pilet, 1969; Bush and Scheraga, 1969). This can be easily taken into account when the results of dinucleotides rather than dinucleoside phosphates are used.

All these corrections will affect the intensity of the optically active bands. More difficult to predict is the contribution of guanine which, as shown recently, is rather complex (Brahms, Aubertin, Dirheimer, and Grunberg-Manago, 1969). First, it has a strong tendency for association. Second, in certain sequences that contain uridine, guanine seems to weaken stacking interactions. This peculiarity of guanine residues is, according to Brahms et al. (1969), to be related to the existence of *syn* and *anti* conformers and their possibility of interconversion. This may explain why GpApU is stacked while ApUpG is not, and its rotation is not much different from the sum of monomer components.

The most important discrepancies in this semiempirical method may arise from neglecting the contribution of next nearest neighbors. It was shown that the nearest neighbors interaction theory of Bradley, Tinoco, and Woody (1963) correctly describes the optical properties of certain oligonucleotides (see section III. A). This may not always be the case. Bradley, Lifson, and Hornig (1964) showed that this approximation may not be sufficient, and next nearest neighbors have to be considered. In general, the effect of $|j-i|$th neighbors is to add terms. Thus, the position of the perpendicularly polarized band is as follows, when the first three neighbors are taken into consideration:

$$\Delta\sigma_{\perp} = \sigma_N - \sigma_1 = 2V_{12}\cos 2\pi/p + 2V_{13}\cos 4\pi/p + 2V_{14}\cos 6\pi/p \qquad (75)$$

More recently, the Johnson and Tinoco new exciton calculation (see section III E) indicates that the nearest neighbor approximation is often not valid. For example, in double-stranded RNA, according to these authors, the second and the third neighbors together contribute more to the intensity of the circular dichroism curve than do the nearest neighbors.

In practice, despite the shortcomings of this semiempirical method, the calculated curves were in good agreement with experimental results of single-stranded RNA, such as tobacco mosaic virus RNA (Cantor et al., 1966; Bush and Scheraga, 1967) and ribosomal RNA (Bush and Scheraga, 1967; Cotter, McPhie, and Gratzer, 1967) at very low ionic strength when there is no base pairing. The discrepancies between calculated and observed curves appear important when RNA adopt double-stranded structure.

Effect of Base-Pairing

For the analysis of double-stranded structure, the situation is more difficult. The nearest neighbor interaction methods can be extended using an expression similar to Eq. (72):

$$[\phi] = 2 \sum \sum \chi_{\mathrm{IJ}}[\phi_{\mathrm{IJ}}^{D}] - \sum \chi_{\mathrm{I}}[\phi_{\mathrm{I}}^{D}] \tag{76}$$

where $[\phi_{\mathrm{IJ}}^{D}]$ is the molar rotation per residue of the double-stranded dimer, containing the sequence IpJ and its antiparallel complement. The summations are carried out over the base sequence of only one of the two strands. $[\phi_{\mathrm{I}}^{D}]$ is the molar rotation per residue of a single base pair.

In principle, the optical rotatory dispersion of an RNA containing the double-stranded structure may be calculated from the contribution of double-stranded poly A·poly U and poly G·poly C. But this is a very crude simplification which may be applied for fully base-paired structures. Cantor et al. (1966) recognize that other base pairs may be formed in the double-stranded regions of RNA, like G–G, which are not taken into account. Furthermore, it is reasonable to suppose that for short helices, the optical rotatory power will show a chain-length dependence. Again, one should emphasize that the effect of next nearest neighbors is not taken into consideration.

The calculations using Eq. (76) were performed for models of maximally double-stranded RNAs without including the effect of other conformations (Cantor et al., 1966). The predicted spectra for a double-stranded t-RNA indicate a blue shift (Cantor et al., 1966; Vournakis and Scheraga, 1966), which is qualitatively in agreement with experimental observations (see also Hashizume and Imahori, 1967). However, there is an ambiguity about the method of estimation of double-stranded regions. Gratzer and co-workers (Cotter, McPhie, and Gratzer, 1967) estimated the amount of base pairs in ribosomal RNA from thermal absorbancy spectra. They calculated the ORD spectra including both contributions: single- and double-stranded helical regions. The agreement with the experimentally obtained spectrum of ribosomal RNA at moderate ionic strength is quite satisfactory. It seems that at present, the nearest neighbor interaction method may be considered as qualitative.

V. INFLUENCE OF TEMPERATURE ON POLYNUCLEOTIDE CONFORMATION

The early studies of optical rotatory dispersion and of circular dichroism showed clearly that the manifestation of the Cotton effect is related to the presence of dissymmetrical "helical" conformation of polynucleotides. The intensity of optically active bands decreases and tends to disappear under experimental conditions where the helical conformation is unstable at higher temperature (Brahms, 1963; Sarkar and Yang, 1965a and b; Holcomb and Tinoco, 1965) and also in the presence of organic solvents (Fasman, Lindblow, and Grossman, 1964). Van Holde, Brahms, and Michelson (1965) investigated the unknown conformation of polynucleotides at neutral pH and attempted to characterize the nature of forces responsible for the maintenance of conformational stability. Since the rotational strength is related to the presence of a dissymmetrical stacked-base structure, it was used as a sensitive parameter to obtain thermodynamic parameters for the unstacking process. The experimental results indicate a gradual "melting" over a broad temperature region which suggests a noncooperative process. This allows one to apply a "two-state" model according to which, at low temperatures, the dimer is in a stacked form and at very high temperatures, unstacking occurs. At intermediate temperatures, one assumes the existence of an equilibrium $D_{\text{stacked}} \rightleftharpoons D_{\text{unst}}$. The rotational strength at any temperature can be considered as expressing the concentration of "stacked" molecules, and one can calculate an apparent equilibrium constant K:

$$K = \frac{C_{\text{u}}}{C_{\text{st}}} = \frac{R_0 - R_T}{R_T - R_n} \tag{77}$$

where R_T represents the rotational strength at a given temperature, and R_0 and R_n are the limiting values of R at low and high temperatures, respectively. The value of R_n at high temperatures approaches that of mononucleotide components. The changes of rotational strength with temperature for a series of adenylates are shown in Fig. 10. It can be seen that at low and high temperatures the curves for higher oligomers and the polymer present a plateau; extrapolation of the "plateau" of the sigmoidal "melting curve" yields the limiting values of R_0 and R_n at low and high temperatures. These values can also be obtained by making use of the inflection point, i.e., the midpoint of the transition $R_{\frac{1}{2}} = R_0/2$ (see Richards, 1968). Using these procedures, Van Holde et al. (1965) and Brahms et al. (1966) have calculated the equilibrium constants and constructed the Van't Hoff plot.

Figure 11 shows the Van't Hoff plot for the thermal denaturation process of a series of adenylates. The lines are parallel, which indicates that the standard state enthalpy changes, ΔH°, are essentially the same for all

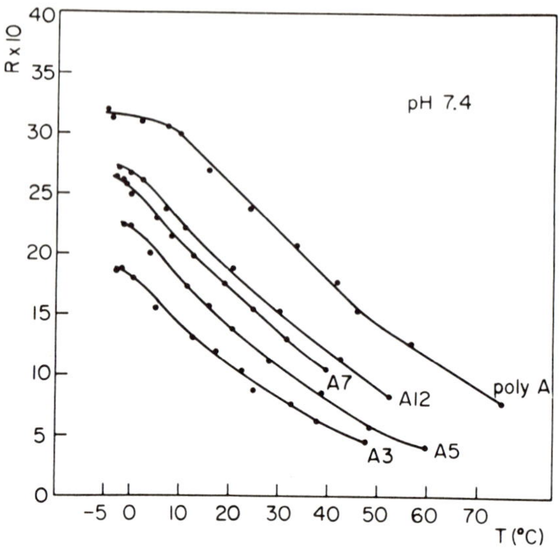

Fig. 10. Changes of rotational strength (R) of the positive CD band as a function of temperature for adenylate oligomers and poly A at pH 7.4. (After Brahms, Michelson, and Van Holde, 1966.)

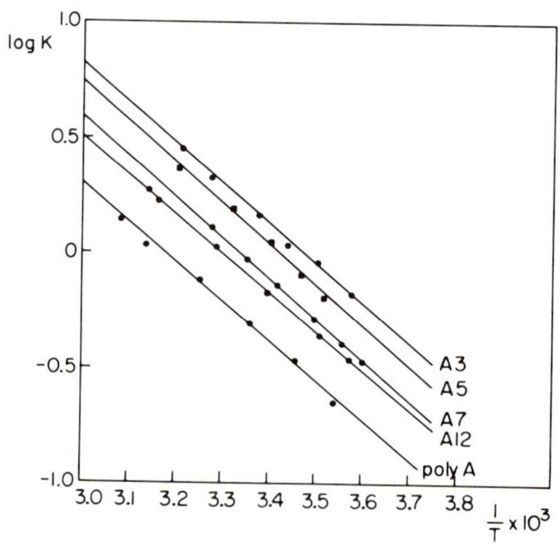

Fig. 11. A Van't Hoff plot of the thermal denaturation of adenylate oligomers and poly A at neutral pH. The data are from Fig. 10. (After Brahms, Michelson, and Van Holde, 1966.)

oligomers. Furthermore, since the thermodynamic parameters ΔH°, ΔF° for the unstacking process are not dependent upon chain length, it was concluded that the unstacking process is largely noncooperative (Brahms, Michelson, and Van Holde, 1966). Similar conclusions were reached from the studies of the temperature dependence of hypochromism by Leng and Felsenfeld (1966) and ORD changes by Poland, Vournakis, and Scheraga (1966) (see below). This means that the structure of poly A and adenylates at neutral pH must be represented as a collection of stacked dimers which may break independently.

Brahms, Maurizot, and Michelson (1967a) studied also circular dichroism temperature dependence of $3' \rightarrow 5'$ oligocytidylates at neutral pH and found similarly an essentially noncooperative behavior. Poland, Vournakis, and Scheraga (1966) applied statistical mechanical cooperative theory, using a multistate model to determine the thermodynamics of the neighbor–neighbor stacking interactions of oligo A's.

Experimentally these authors found that the effect of chain length on the ORD "melting curves" of oligo A's is small, in very good agreement with circular dichroism data of Brahms et al. (1966). The value of enthalpy changes of stacking is found to be -6.5 kcal/mole, again in good agreement with circular dichroism measurements. The conclusion of Poland, Vournakis, and Scheraga is that the effect of cooperation is so small that for most purposes it is adequate to assume that the successive bases can be stacked or unstacked independently, i.e., the state of one residue is essentially independent of the state of stacking of its neighbors.

Studies of conformational stability of simple dimers of various base composition were made in order to obtain a better understanding of RNA conformation in solution. Brahms, Maurizot, and Michelson (1967a) found that the "melting" of a series of $3' \rightarrow 5'$ dinucleoside phosphates is characterized by thermodynamic similarities; the Van't Hoff plots yield straight lines of parallel slope, and the values of ΔH° for the unstacking range from 6 to 8 kcal/mole (see Table III). The similarity of thermodynamic parameters suggests that the nature of stacking interaction is similar in dimers of different base composition. However, there are some differences in ΔF° (at 0°C) which indicate that the free energy changes for oligonucleotides containing uracil (or guanine) at neutral pH are certainly less in favor of stacking.

More recently Davis and Tinoco (1968) reexamined the thermodynamic parameters for the ORD unstacking process of dinucleotides using different solvents, i.e., 25% LiCl solution. Table III shows that the values of ΔH° obtained by Davis and Tinoco (1968) (column *b* in Table III) are in excellent agreement with that given by Brahms, Maurizot, and Michelson (1967a and b) (column *a*).

Glaubiger, Lloyd, and Tinoco (1968) analyzed the temperature dependence of optical rotatory dispersion and hypochromism of dinucleotides using the oscillating model. In this model, the bases of a dinucleotide are considered to be two circular disks above each other, connected by a spring. The two bases oscillate with respect to one another. At low temperatures, the oscillations would be negligible, and as the temperature increases the bases oscillate, remaining parallel with no solvent in between even at high temperature. The rotation of this torsional harmonic oscillator as a function of temperature will be given by

$$[\phi](T) = [\phi]_0 \, e^{-2kT/\kappa} \tag{78}$$

where $\kappa = \omega^2 I$ is the force constant of a classical torsional oscillator, and k is the Boltzmann's constant. $[\phi]_0$ is proportional to the low-temperature limit. Using several approximations and assumptions, Glaubiger et al. calculated the optical rotation at different temperatures. One supposes that the motion of the bases is constrained by a potential well of the form $\frac{1}{2}I\omega^2(\theta-\theta_0)^2$, where I is the reduced moment of inertia and ω is the natural frequency of the torsional harmonic oscillator. θ is the angle between the transition dipoles of base one μ_{10a} and base two μ_{20a}, θ_0 is the angle of minimum energy. The calculated results are compared with the experimental

TABLE III

Thermodynamic Parameters for Thermal Denaturation of $3' \rightarrow 5'$ Dinucleoside Phosphates (pH 7.5)

Substance	$\Delta H°$, kcal/mole		$\Delta S°$, e.u./mole		ΔF, kcal/mole
	a[a]	*b*[b]	*a*	*b*	*a*
CpA	7.0	7.3	24	27	+0.4
ApC	6.1	6.2	21	22	0.6
CpU	6.8	7.8	24	28	0.2
ApU	6.7	8.4	24	32	0.3
CpC	7.5	6.9	25	17	0.7
GpA	6.1	5.6	22	20	0.2
ApA	8.0	5.3	28	20	0.4
UpU		7.8		20	
UpC		6.2		22	
UpG		6.0		23	
CpG		4.8		17	
GpC		7.8		28	

[a] Data from Brahms, Maurizot, and Michelson (1967b).
[b] From Davis and Tinoco, 1968.

ORD results obtained by Davis and Tinoco (1968). At high temperature, the decrease of $[\phi]$ is exponential, which agrees well with the experimental results. However, at low temperatures, there is a disagreement with the experimental data. Similarly the model applied to the ApA absorbancy studies predicts that the hypochromism is almost independent of temperature, whereas experimentally the authors observed a 70% decrease between 0° and 100°C. This oscillating model fits the ORD data only partially as well as the two-state model and, according to the authors, does not predict the correct temperature dependence of hypochromism. The model predicts for single-stranded polynucleotides, even at high temperatures, significant stacking with no solvent between the bases. Thus it seems that the principal merit of the model is that it attracts attention to molecular oscillations, and it proposes an alternative to the two-state model. The comparison of the influence of temperature on the conformational stability of deoxyribo- and ribo-oligomers is considered in a following section.

VI. SINGLE- AND MULTIPLE-STRANDED HOMOPOLYNUCLEOTIDES

A. Poly A

The application of circular dichroism and ORD allowed us to make advances in the analysis of nucleic acid structure in solution. The models of structured polynucleotides were essentially based on X-ray diffraction results on fibers of DNA (Watson and Crick, 1953; Wilkins, 1952) and of poly A prepared from acid solution (Rich, Davies, Crick, and Watson, 1961). The concept of double or multichain helices stabilized by hydrogen bonding between the bases provided the basis of structural consideration. The conformation of single-stranded polynucleotides was previously not well understood.

More recently it has become evident that hydrogen bonding alone may not account for the stability of helical DNA and RNA structures (Geidushek and Herskovitz, 1961; Herskovitz, 1963; Levine, Gordon, and Jencks, 1963; Timasheff, Witz, and Luzzati, 1961, Luzzati et al., 1964). The possibility that stacking interactions between parallel bases may contribute an appreciable fraction to the helix stability was suggested also on the basis of theoretical considerations (DeVoe and Tinoco, 1961).

On the other hand, it was assumed that the polynucleotides that were not in double or multihelical conformation behaved as random coils, e.g., poly A, poly U, and poly C at neutral pH. At first the proposed models involved short intramolecular hydrogen-bonded regions of paired bases alternating with random coils (Fresco and Klemperer, 1959; Steiner and Beers, 1961).

In order to understand the structure of polynucleotides that are not in double-stranded form, model compounds and simple oligonucleotides were studied.

The first line of evidence of the ordered dissymmetrical structure of these polynucleotides was obtained from the studies of circular dichroism (Brahms, 1963, 1964) which indicated the presence of intense circular dichroic bands,

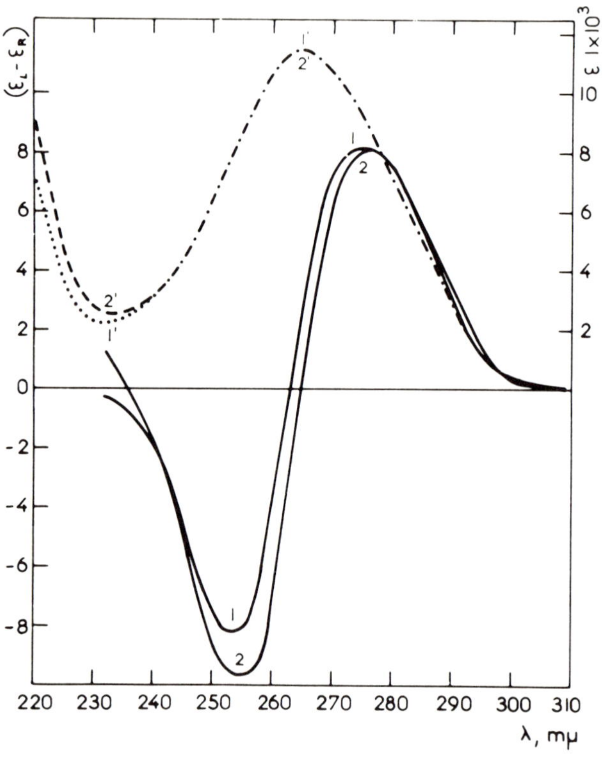

Fig. 12. Circular dichroic spectra of poly N^6-hydroxyethyl adenylic acid at pH 7.4 (curve 1) and 4.9 (curve 2) and at temperature about 4°C; the absorption spectra are also shown at pH 7.4 (curve 1′) and at pH 4.9 (curve 2′). (After Van Holde, Brahms, and Michelson, 1965.)

and of Cotton effects (Fasman, Lindblow, and Grossman, 1964; Holcomb and Tinoco, 1965; Sarkar and Yang, 1965a and b).

Second, Van Holde, Brahms, and Michelson (1965) showed that the polymer of N^6-hydroxyethyladenine, in which the 6-amino group is blocked and cannot form hydrogen-bonded double-stranded helices, exhibits similar CD spectra to that of ApA (at neutral and acid pH) and poly A (at neutral pH) (Figs. 1 and 5). Figure 12 shows that the circular dichroic spectrum of poly HEA is composed of a pair of bands of opposite sign, but approximately equal rotational strength. Their crossing point corresponds to the maximum of hydroxyethyladenine absorption. At acid pH, when poly A forms a double-stranded structure (Rich, Davies, Crick, and Watson, 1961; see below) the CD spectrum of poly HEA is unchanged (Van Holde et al., 1965). It was concluded that the dissymmetric conformation of poly HEA, ApA, and poly A at neutral pH must be common. This cannot be a hydrogen-bonded base-paired structure.

The third line of evidence demonstrating the inconsistency with a hydrogen-bonded base-paired structure was obtained from the studies of circular dichroism and ORD of ApA (Van Holde et al., 1965; Holcomb and Tinoco, 1965; Warshaw, Bush, and Tinoco, 1965) and of a series of adenylate oligomers (Brahms et al., 1966). The optical properties calculated for a single-stranded stacked model were in agreement with experimental results (see Section III). The circular dichroism of a series of adenylates indicates a progressive build-up of single-stranded helical structure (see Fig. 5).

Finally, the evidence that the conformation of poly A at neutral pH is different from double- or multistranded structures was obtained from the thermal denaturation process, which was not a cooperative phenomenon (see Section V; Brahms, Michelson, and Van Holde, 1966; Poland, Vournakis, and Scheraga, 1966; Leng and Felsenfeld, 1966).

More recently the circular dichroism measurements of the neutral form of poly A were extended to the spectral region below 220 mμ (Hashizume and Imahori, 1967; Wolfe, Oikawa, and Kay, 1969) which show the presence of a second pair of bands occurring below 220 mμ. From the linear dichroism measurements extended to the vacuum ultraviolet region, it is known that the direction of these bands are polarized in the plane of the base and probably can be correlated with the doubly degenerate E_{1u} band of benzene (Brahms et al., 1968).

The optical rotatory dispersion measurements of poly A at neutral pH were reported, describing the presence of multiple Cotton effects (Sarkar and Yang, 1965a and b; Lamborg et al., 1965) in the region of base absorption.

B. Double-Stranded Form

Brahms (1964) found that the structural transition of poly A from the neutral, single-stranded form to the acid form is reflected in the circular dichroism spectrum. From X-ray diffraction data on poly A fibers prepared from acid solution, it is known that its structure is of a double helix with parallel strand orientation (Rich et al., 1961). The circular dichroism spectrum of poly A at acid pH shows a blue shift over the entire wavelength region when compared to that of the neutral form (see Figs. 13 and 14). The rotational strength of the long wavelength positive band is two times greater in the acid form relative to that of the neutral form.

These spectral changes are not exhibited by small oligoadenylates at acid pH. Only starting from the heptamer does one observe changes of the shapes of the circular dichroic spectra and a sudden increase of the value of rotational strength. This is clearly demonstrated in Fig. 6 where the rotational strengths of oligo A are plotted. In contrast with the slow regular increase

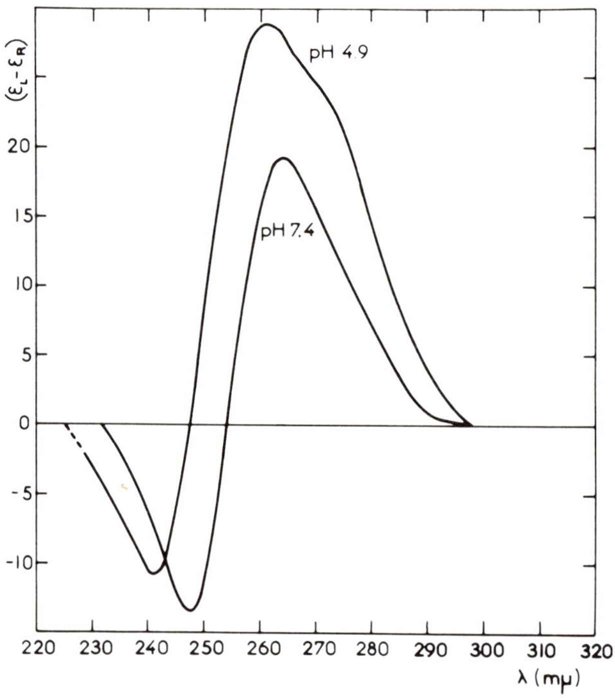

Fig. 13. Circular dichroism spectra of poly A at acid and neutral pH. (After Brahms, 1964.)

of R with N (degree of polymerization) observed in neutral solution from the dimer to the polymer, at acid pH, oligoadenylates exhibit a sudden increase in R at the level of the heptamer (Fig. 6). It is to be noticed that the circular dichroism spectrum of poly A at acid pH is characterized by the presence of a shoulder which can be considered as resulting from the contribution of a component centered at about 280 mμ. The tentative assignment of an $n \rightarrow \pi^*$ transition seems to find further support in the new high-resolution absorption spectra of poly A (Brahms, Sellini, and Brahms, 1969).

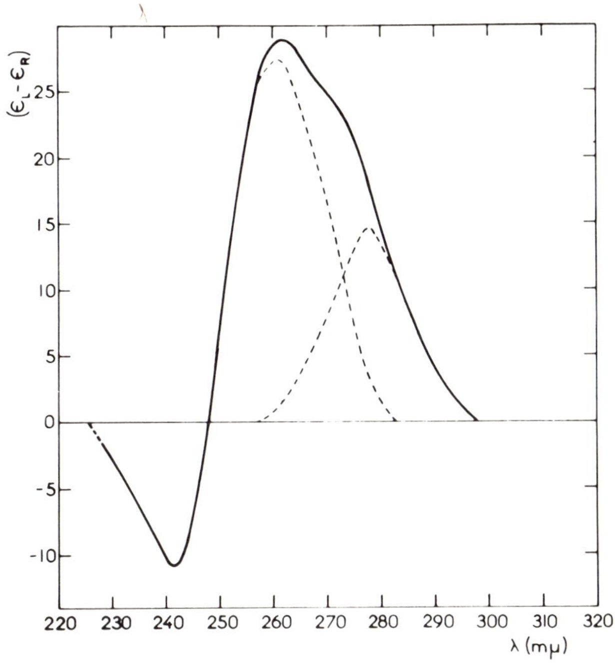

Fig. 14. Circular dichroism of poly A at pH 4.86 in 0.1 *M* NaCl, 0.04 *M* acetate buffer, 10^{-4} *M* EDTA measured at 22°C; dotted line corresponds to the tentative resolution of the positive band into two components. (After Brahms, 1964.)

C. Poly C

The single-stranded helical structure was suggested for the neutral form of poly C by Fasman et al. (1964). Fasman and co-workers were the first to propose that stacking interaction and not hydrogen bonding is responsible for maintaining the helical structure of poly C at neutral pH. This was demonstrated by ORD studies of poly C in the presence and in the absence of formaldehyde, a reagent which blocks the amino group of cytidine. Formylated poly C was shown to maintain a dissymmetrical structure. The Cotton effect of poly C and of the formaldehyde-treated polynucleotide remained essentially unchanged. Thus, the authors concluded that the hydrogen bonding of the amino group of poly C plays no role in stabilizing this helical structure. The important role of hydrophobic base–base interactions for the stability of the single-stranded helical structure was investigated by observing the decrease of the rotation and hypochromicity as a function of ethylene glycol concentration. The ORD measurements of the neutral form of poly

C were extended to 200 mμ (Ulbricht, Swan, and Michelson, 1966; Sarkar and Yang, 1965b; Cantor, Jaskunas, and Tinoco, 1965; Ts'o, Rapaport, and Bollum, 1966).

The existence of two different conformations of poly C in solution was shown by the analysis of circular dichroism spectra. At neutral or alkaline pH, poly C yields a characteristic nonconservative circular dichroic spectrum (Brahms, 1963; Brahms, Maurizot, and Michelson, 1967a), whereas at acid pH, where the structure of poly C fibers is known to be a double-stranded helix (Langridge and Rich, 1963), the CD spectrum is conservative, composed of a pair of bands of opposite sign (Fig. 8). This spectrum is red shifted, as is the entire absorption spectrum, and corresponds to the red shift of the cytidine spectrum upon protonation.

The interpretation of these spectra at acid and alkaline pH is facilitated by the investigations of the dimer, CpC, and oligocytidylates (Brahms, Maurizot, and Michelson, 1967a). In contrast to the polymer spectra, the general shape of CpC spectra is similar at acid and alkaline pH, i.e., nonconservative spectra with one main band are observed. There is no drastic change upon protonation except the above-mentioned red shift. Thus, the presence of a conservative or nonconservative spectrum is to be related with the double- or single-stranded conformation of poly C, respectively. The studies of a series of oligocytidylates at acid pH showed that the change from nonconservative to a conservative spectrum appears starting from the heptamer, similar to the formation of the double-stranded helices of oligoriboadenylics. Thus, poly C circular dichroism spectra provide a precise technique for detecting the conformational change from single- to double-stranded helical structures. It is to be mentioned that these striking differences between conservative and nonconservative spectra may not be readily seen from optical rotatory dispersion curves; thus, Sarkar and Yang (1965a and b) were not able to distinguish the effect of protonation of poly C from the conformational change. However, Thiéry (1968) showed that ORD curves of CpC can be transformed correctly by the application of Kronig–Kramers relations into CD spectra when the transitions occurring below 220 mμ are also taken into account.

In subsequent studies, the ORD measurements of poly C were extended to lower wavelengths (Sarkar and Yang, 1965a and b; Ts'o, Rapaport, and Bollum, 1966).

Poly I

Another example of a conformational transition from single- to multiple-stranded form, which is reflected in two types of circular dichroic spectra, is provided by poly I. Relatively little is known about the conformation of poly I in solution. X-ray diffraction data of poly I fibers were interpreted as indicating the presence of triple-stranded helix (Rich, 1958). The circular dichroism spectrum of poly I in high ionic strength solution (0.5 *M* NaCl) is essentially of conservative form, in the spectral region from 300 to 220 mμ (Brahms and Sadron, 1966). However, the shape is inverted; the longer wavelength band, centered at 274 mμ, is of negative sign, followed by a positive band centered at 250 mμ. At low ionic strength and at low temperature, poly I exhibits an essentially nonconservative spectrum, composed of one main band of positive sign centered at 249 mμ, and by a very weak negative component. Even if we take this weak negative bond into account, the observed CD spectrum is not conservative; that is, it does not obey the sum rule. Previously, poly I at low ionic strength was considered to be in random coil form. However, the presence of circular dichroic bands that do not change upon a further decrease of ionic strength allowed one to interpret this spectrum as characteristic of some elements of a single-stranded poly I conformation (Brahms and Sadron, 1966). Certainly further studies are necessary in order to obtain more precise correlation between the spectra and the conformation (Maurizot, Blicharski and Brahms, 1969). This is of importance since poly I is considered as an analog of poly G.

More recently the measurements of circular dichroism of poly I were extended to 190 mμ by Kay and co-workers (Wolfe, Oikawa, and Kay, 1969). The presence of intense bands in the region of 200 mμ was observed. The ORD investigations are essentially in agreement with these CD results (Sarkar and Yang, 1965b).

VII. POLYNUCLEOTIDE COMPLEXES

Several examples of double- or multiple-stranded complexes formed by interaction of different homopolynucleotides have been previously described (see excellent general reviews of Michelson et al., 1967; Felsenfeld and Miles, 1967). The application of circular dichroism or optical rotatory dispersion should provide a better understanding of their structure and of more complex nucleic acids. It is hoped that this will allow one to detect specific interactions of fundamental importance for structure and stability in solution.

One of the most studied is the complex between poly A and poly U. Their interaction leads to the formation of two-stranded (poly A·poly U) and three-stranded (poly A·2poly U) complexes at appropriate ionic strength and temperature conditions (Warner, 1957; Rich and Davies, 1956; Steiner and Beers, 1961; Fresco et al., 1962; Massoullié, 1966).

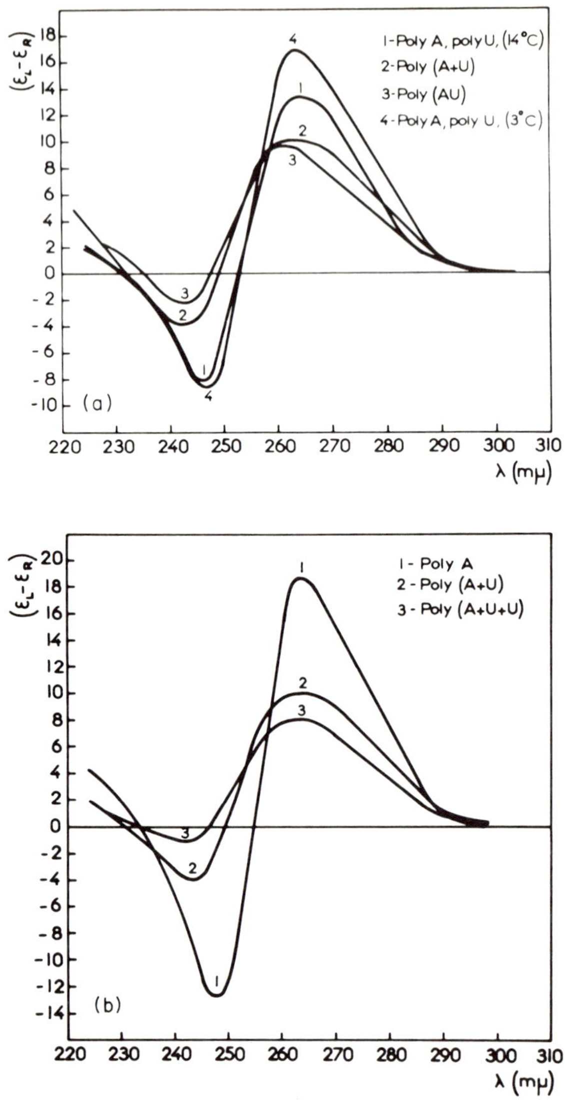

Fig. 15. (a) Circular dichroism spectrum of poly A: poly U complex (curve 2) in 0.005 *M* NaCl, 0.01 *M* tris, pH 7.4 at about 14°C. Curve 1 is the spectrum of the two components measured separately in the same conditions. Curve 4 is the spectrum of poly A and poly U measured separately: poly U in 0.005 *M* NaCl, 0.01 *M* $MgCl_2$, 0.01 *M* tris, pH 7.4 at about 2°C and poly A in 0.005 *M* NaCl, 0.01 *M* tris, pH 7.4 at 2°C. Curve 3 is the spectrum of poly (A+U) copolymer (1:1) in 0.1 *M* NaCl, 0.1 *M* tris, pH 7.4 at about 14°C. (After Brahms, 1965.) (b) Comparison of the circular dichroic spectra of poly A at pH 7.4 in 0.005 *M* NaCl–0.01 *M* tris at 14°C (curve 1) with poly (A+U), the two-stranded complex, in 0.005 *M* NaCl–0.01 *M* tris (pH 7.4) at 14°C (curve 2), and with poly (A+U+U), the three-stranded complex, in 0.02 *M* NaCl–0.01 *M* $MgCl_2$–0.01 *M* tris (pH 7.4) at 2 to 10°C (curve 3). (After Brahms, 1965.)

The circular dichroism spectrum of the double-stranded complex shows several interesting features when compared to that of the separated constituents (Brahms, 1965). The formation of the complex is reflected generally in the decrease of intensity of the circular dichroic bands and in the broadening of these bands as shown by the measurements of band width at one-half of the maximum. Furthermore, the maximum of the band is shifted toward the blue [see Fig. 15(a)]. This blue shift was also observed in the position of the ORD maxima of poly A·poly U (Sarkar and Yang, 1965a and b). These different parameters allow one to distinguish the formation of the complex from that of the separate components poly A and poly U. The copolymer of randomly distributed adenylic and uridylic acids (1:1), poly AU, exhibits a very similar spectrum when compared to that of the poly A·poly U double-stranded complex. The slight differences in the spectral characteristics, such as in the band width and in the intensity of the negative band [see Fig. 15(a)], are probably caused by a less-ordered structure of the copolymer when compared to that of poly A·poly U. In fact, the random copolymer undergoes a gradual transition (based on hypochromicity measurements) relatively broader than the poly A·poly U double-stranded complex. One can thus conclude that the formation of the double-stranded complex by hydrogen bonding between different nucleotides causes a decrease of the intensity of CD bands and of rotational strength, and a band-broadening blue shift of the position of both extrema. Furthermore, the splitting into positive and negative bands is much less pronounced in the spectra of the poly A·poly U double-stranded complex than in single-stranded poly A. In other words, the CD spectrum changes from conservative to a rather nonconservative type upon double-strand formation.

These changes are also characteristic and even more pronounced in the spectrum of the three-stranded poly A·2poly U (Brahms, 1965; Sarkar and Yang, 1965a and b). The circular dichroism spectrum of the three-stranded complex is characterized by the predominant positive broad band and can be considered of the nonconservative type. Again the melting curve, followed by changes either in rotation at the peak or by hypochromicity, shows a sharp transition (Sarkar and Yang, 1965a and b). These spectral characteristics of the three-stranded poly A·2poly U and the double-stranded complex poly A·poly U are compared in Fig. 15(b) with the essentially conservative spectrum of poly A at neutral pH. It seems that these changes are caused by the interaction of different monomeric units (Brahms, 1965). The decrease of the rotational strength upon the formation of the complexes is one of the important characteristics. In agreement with the exciton theory of Tinoco, Woody, and Bradley (1963), this may well occur from the mutual cancellation of the many positive and negative bands that result from the interaction between the bases on the same and adjacent strands. Johnson and Tinoco's (see previous section) recent calculations predict quite well the anomalously low intensity of the circular dichroic bands of nucleic acids and of polynucleotide complexes to be a natural consequence of the cancellation of rotational strengths of the numerous CD bands. However, the calculations for the alternating poly rAU:rAU copolymer yielded curves of too high intensity when compared with the experimental results.

The second important complex is between poly guanylic and poly cytidylic acids. Polymers of guanylic acid have been difficult to prepare and only recently became available (Thang, Graffé, and Grunberg-Manago, 1965). Ulbricht, Swan, and Michelson (1966) showed that at neutral pH, poly rG exhibits a rather complex ORD spectrum composed of two peaks and one trough (at 266, 243, and 222 mμ). The ORD curves survive complete protonation and change only little at pH 2. Thus, the polymer possesses a secondary structure with an extremely high stability. The alkaline dissociation is cooperative (Pochon and Michelson, 1965) and at pH 12 the polymer shows complete loss of secondary structure. At this pH, the ORD curve correspond to β mononucleotides (Ulbricht et al., 1966).

The difficulty in preparation of poly G·poly C complex probably causes some contradictory results. The complex between poly G and poly C was studied using two different preparations. Ulbricht, Swan, and Michelson (1966) studied poly G·poly C obtained by interaction of the homopolynucleotides, whereas Sarkar and Yang (1965a and b) used a polymer synthesized by RNA polymerase (from *Micrococcus lysodeikticus*) in the presence of poly C as template and GTP. Whereas some differences do exist in the position of the extrema, the general shape of the ORD curves is similar. The complex is characterized by a very marked thermal stability, as shown by very high T_m (Pochon and Michelson, 1965). Ulbricht, Swan, and Michelson (1966) showed that the complex is stable over a wide region of pH between 2.0 and 12.2. It is interesting that the corresponding polydeoxynucleotide complex, poly *d*G·poly *d*C, yielded circular dichroic curves (Brahms and Mommaerts, unpublished results) which are inverted by comparison with the majority of polynucleotides spectra (and DNA) and which have a negative band followed by a positive one. Also, the ORD curves are of unexpected shape, different from both poly rG·poly rC and DNA. The explanation of this inverted type of spectrum is not yet clear. One can consider that either the guanine electric dipole moment orientations are such as to yield an inverted type of band; this proposal is supported by calculations of G–G theoretical spectra (Bush, thesis, 1965) or nearest neighbor calculations (Aubertin, thesis, 1968). The other, more probable possibility is that multiple-strand associations, due to the known tendency of guanine for aggregation, are reflected in this inverted spectrum.

Further support for such interpretation of inverted circular dichroism spectra is probably provided by GMP ordered gels which, at low temperature, form a four-stranded helical structure (Gellert, Lipsett, and Davies, 1962). Again GMP gels yield (in 0.1 *M* KF and at pH 5) a circular dichroism spectrum that is inverted.

VIII. THE NOTION OF RANDOM COIL SINGLE-STRANDED POLYDEOXYNUCLEOTIDES

We consider now the extreme case of the random coil (i.e., disordered polynucleotide chain) optical activity. *A priori* one can assume that the disruption of the dissymmetrical helical conformation will be reflected in the disappearance of the Cotton effect, and the resulting spectrum of unstructured chain will be that of the sum of monomer constituents. However, it is exceedingly rare to observe a polynucleotide spectrum identical to that of isolated mononucleotides, even under the conditions favoring unstructured, random coil, e.g., at elevated temperatures, in organic solvents, or for some probably unstacked trinucleotides, containing uracil and guanine (see Brahms et al., 1969). Thus one is led to conclude first that a polynucleotide in a random coil form does not have a freely rotating backbone with bases assuming all possible orientations. Rather, the experimental results strongly suggest that only a *restricted* rotation about the five covalent bonds of the phosphodiester backbone occurs, corresponding to the conformations of minimum energy.

Second, the rotation about the glycosidic C—N bond of an isolated monomer and of the same residue in a polynucleotide chain may be slightly different. The majority of crystal structures of mononucleotides were found to be in *anti* conformations (Haschemeyer and Rich, 1967), which is also suggested by nuclear magnetic resonance studies in solution (Schweizer, Broom, Ts'o, and Hollis, 1968). However, the calculations of Haschemeyer and Rich (1967) have shown that for purines and particularly for guanosine there is only a small energy difference between the *anti* and *syn* conformations, which may render possible the existence of two isomers and their interconversion. This possibility of the existence of two conformers which differ by the rotation about the C—N bond may well explain the very weak negative CD band of purine monomer and stronger positive CD bands of pyrimidine monomers, in which rotation about the C—N bond is more restricted and the bases are not freely rotating.

The structure of single-stranded polydeoxyribonucleotide is unknown, and might provide some examples that are probably related to the notion of unstructured polymers. It is possible that the bases may even interact without formation of a dissymmetrical ordered structure.

The first indication of the great differences in structure between single-stranded polyribonucleotides and polydeoxynucleotides was obtained by comparing ribo and deoxy poly and oligocytidylates (Adler, Grossman, and Fasman, 1967), $2' \rightarrow 5'$ cytidylates with the neutral $3' \rightarrow 5'$ series (Brahms, Maurizot, and Michelson, 1967a), and different polydeoxyribonucleotides (Ts'o, Rapaport, and Bollum, 1966). Adler et al. (1967, 1968) found that the Cotton effects displayed by polydeoxy C and poly rC are of similar shape, and the position of the extrema is almost the same (Fig. 16). However,

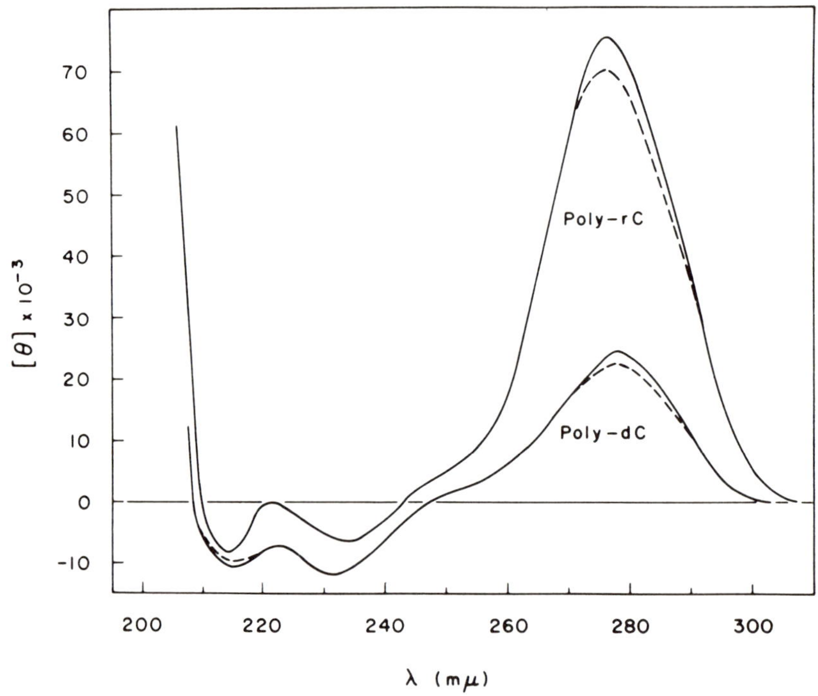

Fig. 16. Circular dichroism spectra of poly C (upper curves) and poly dC (lower curves). Effect of salt concentration at pH 10.0 2 M NaCl ———; 10^{-4} M salt — — —. (After Adler, Grossman, and Fasman, 1968.)

the intensity of the Cotton effect of the polydeoxyribonucleotide is greatly diminished when compared to its ribo analog. Adler, Grossman, and Fasman showed that a series of oligodeoxyribocytidylates at alkaline pH exhibit much lower rotatory power and hypochromicity than the corresponding nonprotonated ribooligomers.

Similarly, Brahms, Maurizot, and Michelson (1967a) showed that the $2' \rightarrow 5'$ oligocytidylates, in which the OH(2′) group is absent, exhibit circular dichroism spectroscopic characteristics which are different from both the natural 3′→ 5′ oligomers and the monomer. The positions of the CD extrema and intersections of 2′→ 5′ isomers are between those of the monomer and of the 3′→ 5′ natural isomers. Furthermore, the intensity of the circular dichroism band of 2′→ 5′ oligomers even at lowest temperature (−20°C) was less than that of the corresponding 3′→ 5′ derivatives by a factor of three, and there was no marked temperature dependence.

It was concluded both by Adler et al. and by Brahms et al. that the degree of dissymmetrical stacking is much lower in deoxyribonucleotides and in 2′ → 5′ ribonucleotides than in corresponding 3′ → 5′ ribooligonucleotides. This and studies from the same laboratories (Adler et al., 1968; Maurizot et al., 1969) indicate the important role of the OH(2′) for stabilizing the secondary structure of single-stranded ribooligonucleotides.

The structural differences between single stranded polyribonucleotides and polydeoxyribonucleotides seem to be related to the differences in their conformational stability. In order to reach a better understanding of the forces stabilizing both homologs, Maurizot, Brahms, and Eckstein (1969) studied dimer isomers containing both deoxyribose and ribose in different sequences: 3′ → 5′ rAp dA, dAprA and compared them with homodimers dApdA and rAprA. Studies of the circular dichroism as a function of temperature indicated that the compounds containing the 2′-hydroxyl group in the 3′ linked residue rAprA and rApdA yield similar sigmoidal "melting curves" typical of stacked dimers. In contrast compounds containing deoxyribose in the 3′ linked sugar, dAprA and dApdA, show a linear temperature dependence with very small slope. This comparison allows one to conclude that in a single stranded polynucleotide the presence of 2′-OH in the 3′ linked sugar is of importance for the stacked conformation.

The importance of the OH(2′) for the formation of the ordered structure was also shown by studying the interactions of the guanine monomers, 5′-GMP which form ordered helical aggregates, whereas the corresponding 2′-deoxy-GMP cannot (Brahms and Sadron, 1966). Ts'o et al. (1966) have made a comparative investigation of the ORD of poly dT with poly rU, poly dA with poly rA, and poly dC with poly rC. The poly dC ORD curve at 1°C was found to be quite similar in form to that of poly rC at high temperature (80°C), which implies, according to the authors, much less stacking interaction. Similarly, the ORD patterns of oligo deoxy T and of poly rU at low ionic strength and in the absence of magnesium are very similar; also, the absence of a temperature effect indicates very little stacking. In the region of 260 mμ, poly rA shows a strong Cotton effect, whereas according to Vournakis et al. (1967) the near UV Cotton effect of oligo dA is of very weak intensity and changes little with temperature. The interpretation of Ts'o et al. (1966) is that the structure of poly dA represents parallel stacked adenine, which cannot form a helical conformation. According to Ts'o, the angle of rotation of two neighboring bases is zero. This conclusion is different from that of Vournakis et al. (1967) and Bush and Scheraga (1969), who observed, in the region of 260 mμ, a Cotton effect and circular dichroism band of very weak intensity. This was confirmed by Adler, Grossman, and Fasman (1969) and Maurizot, Blicharski, and Brahms (1969). The intensity of this poly dA band represents only one-tenth of the poly rA positive band (Adler et al., 1969). However, the second absorption band of poly dA occurring in the region of 210–200 mμ, gives rise to intense circular dichroic bands as shown by Bush and Scheraga (1969) and Vournakis et al. (1967). These two optically active bands occurring at about 260 and 210 mμ have different relative intensities in poly dA and in poly rA. On this basis, Bush and Scheraga reject the hypothesis of Ts'o et al. of parallel stacking of the bases (with any increment angle α equal to zero or 180°), since the 211-mμ band should also be optically inactive, which is obviously not the case. It is reasonable to suppose that the tilting of the bases with regard to the helical axis would affect the 211- and 260-mμ bands differently since their polarization directions differ. It seems that the far UV Cotton effect centered in the region of 210–200 mμ is very sensitive to tilting and to conformational changes of nucleic acids (see Wolfe, Oikawa, and Kay, 1968; 1969). In poly dA at neutral pH, Bush and Scheraga (1969) observed a contribution of a distinct band at 280 mμ which they assign as arising from an $n \rightarrow \pi^*$ transition. The fact that the $n \rightarrow \pi^*$ band is not seen in the circular dichroism spectrum of the dimer but only in the polymer is ascribed to solvation differences of the adenine ring. Data obtained recently by Brahms, Sellini, and Brahms (1969) on highly resolved absorption spectra of deoxyadenylates allows one to detect in all oligomers the presence of a distinct band in this region of 280 mμ. This calculated spectrum for parallel stacked model is not confirmed by experimental results.

A series of deoxyadenylates $d(A)_2$, $d(A)_4$, and $d(A)_6$ was investigated at neutral pH by Vournakis, Poland, and Scheraga (1967). By following the rotation of the 210 mμ Cotton effect, Vournakis et al. observed an inverted chain-length dependence, that is, instead of an increase (as in rA), a decrease in rotation as a function of chain length. The authors postulated a model involving both attractive nearest neighbor and repulsive next nearest neighbor interactions that may involve the negative charges on phosphate groups. However, more recently, Adler, Grossman, and Fasman (1969) studied polydeoxyribo A as a function of ionic strength and did not observe any salt effect, which is difficult to reconcile with the proposed role of phosphate repulsion as responsible for this lowering of rotatory power as a function of chain length.

Adler, Grossman, and Fasman (1969) investigated the effect of dioxane on deoxy ApA and poly dA at neutral pH, and found that polydeoxy A rotation was nearly constant until a certain concentration of dioxane was reached, whereas deoxy ApA and similarly poly rA rotations were reduced by a small concentration of dioxane. Adler, Grossman, and Fasman consider that additional stability toward organic solvents (dioxane) may be caused by a distinctive base-stacking geometry.

Recent investigations of Maurizot, Blicharski, and Brahms (1969) on different oligodeoxyribonucleotides indicate that this inverted chain-length dependence (if any) is particular to oligo deoxy A at neutral pH, but cannot be considered as a general phenomenon for all deoxyribonucleotides. Oligo deoxy C and oligo deoxy T exhibit a small but progressive increase of rotational strength as a function of chain length. It is to be emphasized that this increase of R as a function of N is very weak, and the plateau is reached rapidly at the level of the pentamer. Furthermore the circular dichroism temperature dependence of different deoxyoligonucleotides, oligo T, oligo dI, oligo dC, never yields the form of a sigmoidal curve (see Fig. 11) characteristic of stacked adenylates. Finally the analysis of absorption and circular dichroism spectra of deoxyoligonucleotides does not provide indications of exciton type interactions between the bases. This is particularly clearly seen in the cases of oligo dC and oligo dT, where the form of the c.d. spectra ressembles that of the monomer (Maurizot, Blicharski, and Brahms, 1969; Brahms and Sellini, unpublished).

In conclusion, one can emphasize the absence of firm evidence indicating helical structure for polydeoxynucleotides in the single-stranded form. Our present results, based on high-resolution absorption spectra, may rather be interpreted in terms of less ordered form; restricted rotation about the backbone chain may cause some interactions and generate weak Cotton effects and weak circular dichroism bands which are different from that of the monomer (Brahms et al., 1969). These conclusions seem to find further support in recent X-ray diffraction analysis of nucleosides, mononucleotides, and polynucleotides by Arnott (1970) and Sundarlingam (1969). The comparison of the conformation of mononucleotides obtained from relatively precise single crystal X-ray diffraction analysis with that of low resolution analysis of polynucleotide and nucleic acid fibers showed that the conformational angles are restricted to narrow ranges. It is found that the backbone chain can generally adopt only a restricted number of possible conformations. According to Sundarlingam (1969) the values of torsional angles fall within $\pm 20°$ for the preferred backbone conformation in all the structures. Thus the interaction between the bases can be considered as a consequence of the restricted backbone conformation. However some exceptions were found such as in the DNA B and C form (Arnott and Hukins, 1969). According to Arnott and Hukins (1969) in less regular polynucleotides the angles ψ along P—O_4 and ϕ along P—O_1 are the possible sites of local variability to alternative ranges observed in monomers.

It appears that the great sensitivity of circular dichroism to conformational changes allows the detection of important differences in structures between single stranded stacked helical polyribonucleotides and unordered polydeoxynucleotides, where X-ray diffraction data are not available.

IX. NUCLEIC ACIDS AND DOUBLE-STRANDED POLYDEOXYNUCLEOTIDES

The measurements of circular dichroism of DNAs and RNAs led Brahms and Mommaerts to the following observations (Brahms, Mommaerts, and Sadron, 1964; Brahms and Mommaerts, 1964; Mommaerts, Brahms, Weil, and Ebel, 1964): The most striking difference is in the shape of the CD spectra of RNAs and DNAs. RNAs of various origins all exhibit CD spectra of very similar form. All show nonconservative spectra composed of one predominant positive band (Fig. 17) in the spectral region from 300 to 240 mμ, followed by a weak negative band. More recent measurements of different RNAs fully confirmed the nonconservative character of the spectra (Fig. 17c) (Sarkar, Wells, and Yang, 1967; Wolfe, Oikawa, and Kay, 1968; Samejima et al. 1968), and the presence of an exceedingly weak band at 300 mμ was observed. Even if one takes into account these two negative CD bands of relatively weak intensity at higher and lower wavelength, the sum of rotational strengths is not equal to zero.

In contrast, DNAs from different origins (bacterial and animal) yield circular dichroic spectra (Fig. 18*a*) which are indistinguishable from one another. All spectra were of conservative type composed of two bands of equal rotational strength, but of opposite sign, with an intersection point situated at about the wavelength of the absorption maximum. This shape of spectrum is in rather good agreement with the exciton theory developed by Tinoco (see section III. A). More recent measurements extended below 220 mμ show the presence of CD bands associated with the absorption band in the region of 200 mμ (E_1u). These profound differences between DNA and RNA circular dichroic spectra show the sensitivity of the method to conformation differences not detectable in conventional UV absorption spectra.

The second important observation concerns the intensity of these nucleic acid circular dichroic bands which are anomalously low, considerably smaller than those observed in homopolynucleotides. However, it was observed that RNAs exhibit a more intense circular dichroic band than do DNAs. This anomalously low magnitude of the nucleic acid CD band in the near UV seems to confirm again the exciton theory of Tinoco and co-workers (see Section III) of the circular dichroism bands arising from the interactions between bases oriented perpendicularly to the helical axis. The splitting will be reflected in the general shape of the circular dichroism curve of DNA, in which the crossing point corresponds approximately to ν_0 of the perpendicular band. A broad positive band on one side and a negative band on the other result from the overalapping of many transitions. However, in contrast to single-stranded homopolynucleotides, in which one type of interaction gives rise to intense circular dichroic bands, in DNA the various interactions may tend to cancel and only weak resulting bands are observed (see Fig. 18).

In order to explain the origin of differences between RNA and DNA, Brahms and Mommaerts studied the behavior of DNA in 80% alcohol (Fig. 18b). In this medium, DNA exhibits a circular dichroism spectrum with a predominant positive band. The rotational strength of this band is three times that of the positive band of native DNA. It was concluded that the CD spectrum of DNA in 80% alcohol "reflects the existence of a new conformation of DNA which can have the geometry of RNA. This may indicate the tilting of the bases with respect to the helical axis is similar to that found by analysis of X-ray diffraction of RNA and of the A form of DNA at low humidity." (Brahms, Mommaerts, and Sadron, 1964).

A similar kind of spectrum (Fig. 18a, curve 2), but less pronounced, with predominant positive band was observed for native DNA in saline solution at about 45°C (below the melting temperature). This can be considered as a transient appearance of an intermediate form, which occurs before the melting of DNA. Brahms et al. (1964) suggested that this intermediate form of DNA and the conformation of DNA in 80% alcohol may be similar to the A form of DNA and to RNA conformation. The characteristic feature of their structures is the titling of the bases with respect to the helical axis. Tinoco and co-workers recently applied the exciton theory by taking into account the interactions between different transitions, e.g., interaction with far UV transitions approximated by polarizabilities (see Section III. E). Tinoco's extensive calculation confirmed this hypothesis. When the strong electronic transitions are taken into account for DNA in the B form, Tinoco's calculations predict the conservative spectrum. The titling of the bases as in the DNA A form or in RNA is reflected in the contributions of polarizabilities, and the predicted spectrum is nonconservative. More recently, Tunis-Schneider and Maestre (private communication) measured the circular dichroism of DNA films at various degrees of relative humidity (r.h.). The authors found that circular dichroism spectra of hydrated DNA films (95% r.h. or higher) to be little different from that in solution (i.e., of conservative type). A transition was observed in the range of 95% to 81% r.h., and the CD spectrum became nonconservative. In the case of calf thymus DNA in 3% NaCl, the CD spectra were of nonconservative character (RNA-like), with a predominant positive band centered at about 260 mμ. This spectrum is considered as evidence that DNA is taking an A type geometry in good agreement with Brahms and Mommaerts. Under similar

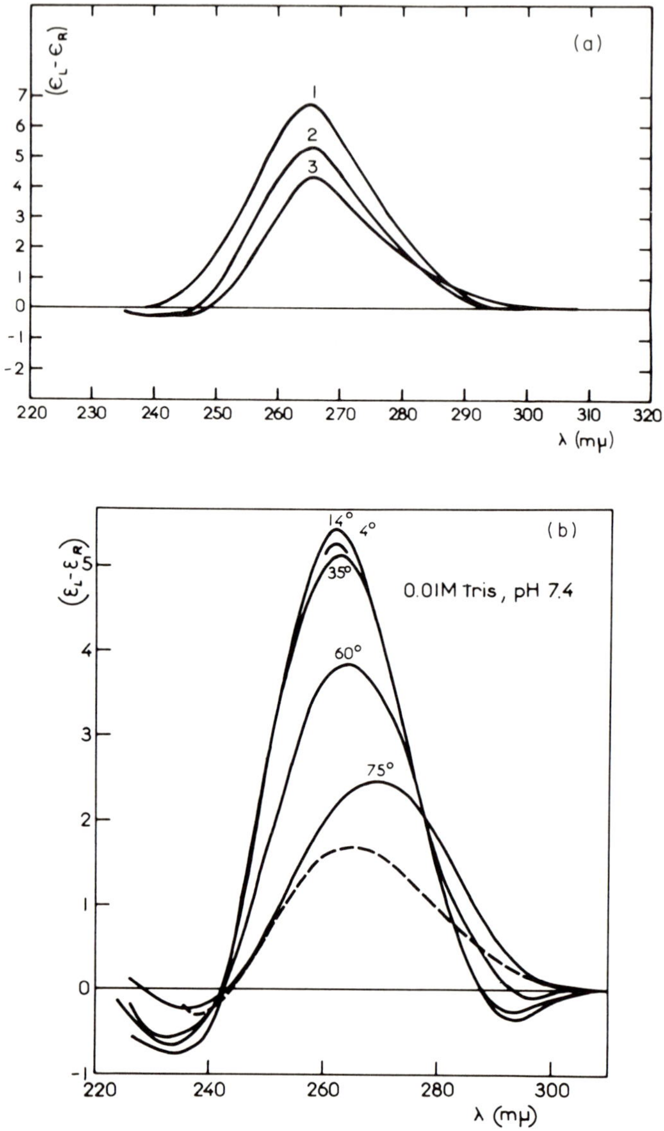

Fig. 17. (a) Circular dichroic spectra of RNA from tobacco mosaic virus in 0.15 *M* NaCl, 0.01 % tris buffer pH 7.4, 0.001 *M* EDTA at different temperatures: at 22°C (curve 1), at 45° (curve 2), at 60°C (curve 3). (After Brahms and Mommaerts, 1964.) (b) Circular dichroism of transfer RNA at low ionic strength (After Bakès, Dirheimer, and Brahms, 1968.)

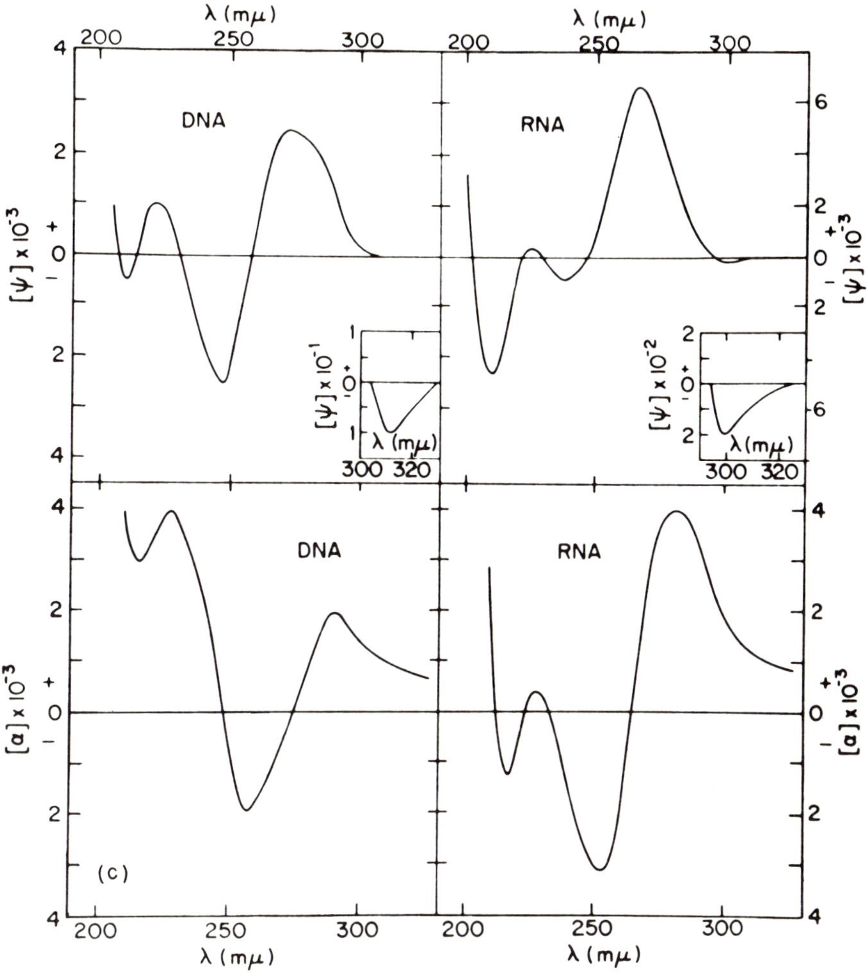

Fig. 17 *continued.*

(c) Comparison of CD (upper half) and ORD (lower half) spectra of salmon sperm DNA and ribosomal *E. coli* RNA. (After Sarkar, Wells, and Yang, 1967.)

conditions, *E. coli* DNA exhibits also a rather nonconservative spectum with a positive 260 mμ band but accompanied by a negative band in the region of 300–270 mμ. Both spectra are taken as reflecting the presence of A type geometry of DNA. The authors took precautions to avoid artifacts due to linear dichroism effects.

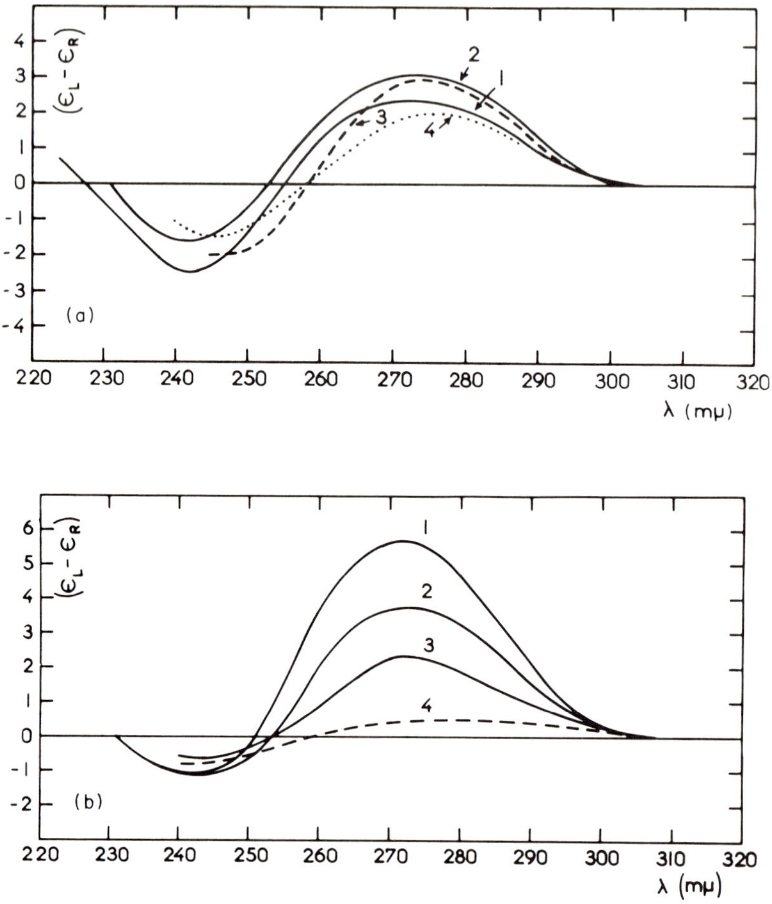

Fig. 18. (a) Circular dichroic spectra of DNA (calf thymus) in 0.01 *M* NaCl, 0.01 *M* tris buffer, 0.001 *M* EDTA measured at different temperatures: curve 1, native DNA at 20°C; curve 2 at 45°C; curve 4 at 80°C; curve 3 heated to the boiling temperature for 10 to 15 min followed by rapid cooling. (After Brahms and Mommaerts, 1964.) (b) Circular dichroism curves of DNA in 80% ethanol solution containing 0.001 *M* NaCl, 0.01 *M* tris buffer pH 7, 0.001 *M* EDTA at different temperatures: curve 1 at 22°C; curve 2 at 40°; curve 3 at 60°C; curve 4 at 80°C. (After Brahms and Mommaerts, 1964.)

Low molecular weight DNA prepared by sonication or by shearing was used, and the films were prepared by applying stirring motion in order to avoid orientation. In fact, whenever oriented macromolecular films are measured the apparent CD results from two contributions, one being the linear dichroism of the film, and the second the real circular dichroism.

In general, one may conclude that the differences in the circular dichroic spectrum between DNA and RNA depend primarily upon structural features. The differences in base composition and in sequences cannot lead to the observed spectral differences since one shape of the circular dichroic spectra is similar for all DNAs and another for all RNAs, regardless of origin.

Thus it seems possible to compare the CD spectra of polydeoxyribo and ribopolynucleotides in double-stranded ordered conformation in order to obtain information on their structure. Johnson and Tinoco (1969) calculated circular dichroism spectra of a series of polynucleotides in the conformation of DNA (B form) and RNA geometry with bases tilted with respect to the helical axis. Particularly interesting is the comparison between poly r(A,U) and poly dAT:dAT. (One can suppose that the presence of thymine instead of uracil will not influence the structure of the polymer.) The circular dichroic spectrum of poly r(A,U), the random copolymer, is essentially nonconservative as measured by Brahms (1965) [see Fig. 15(a)] and of similar shape to the calculated curve of Johnson and Tinoco (1969). The poly dAT:dAT circular dichroism experimental curve can be obtained, according to Johnson and Tinoco, by Kronig–Kramers transformation of the observed ORD curves (Samejima and Yang, 1965). More recently, Bernardi and Timasheff (1970) measured directly the circular dichroism of poly dAT:dAT. In both cases, the experimental curve is essentially conservative (see Fig. 19). Also conservative is the calculated curve of this alternating copolymer for the assumed geometry of the DNA B form. Since the structure of poly dAT:dAT fibers seems to be similar to the DNA B form (Davies and Baldwin, 1963), one can conclude that the polymer retains the same conformation in solution. This also means that the structural differences between these two ribo and deoxyribopolynucleotides reside primarily in the geometry: Double-stranded polyribonucleotides have a conformation similar to RNA, whereas polydeoxynucleotides are essentially similar to the DNA B form. Furthermore, Johnson and Tinoco calculated the circular dichroism spectrum of poly dAC:dGU and poly rAC:rGU, assuming again the DNA planar B form and the RNA conformation, respectively. The calculations indicate that poly rAC:rGU exhibits a non-conservative spectrum, whereas poly dAC:dGU is conservative, similar to the DNA curve. The calculated curves are in rather good agreement with their experimental results.

This apparent success of Tinoco's theory should not be overestimated. The authors show some results where the theory fails, e.g., for GC polymers. Also for poly dA:dT the experimentally obtained circular dichroism spectrum (Fig. 19) is complex (Bernardi and Timasheff, 1970; Brahms, Blicharski and Raquet, unpublished results) and does not agree with the calculated, conservative curve.

The origin of the weaknesses in the calculation is at present not known. This theory takes into account only $\pi \rightarrow \pi^*$ transitions of the purine and pyrimidine bases. However, it is not excluded that the $n \rightarrow \pi^*$ transition at about 280 mμ contributes weakly to the circular dichroic band (Bush and Scheraga, 1969; Brahms, Sellini, and Brahms, 1969). However, the merits of

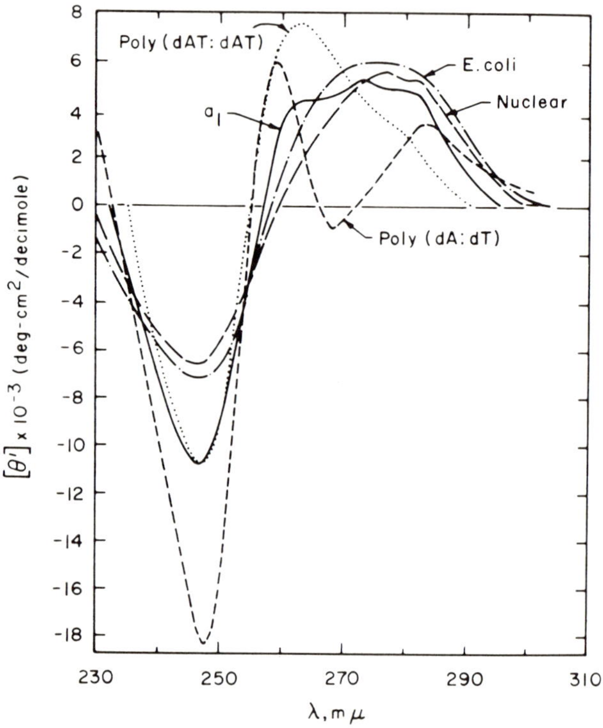

Fig. 19. Circular dichroism spectra of poly dAT: dAT, poly dA: dT and of the mitochondrial DNA from the "petite" mutant a_1, from *E. coli*, and from nuclear yeast DNA. (After Bernardi and Timasheff, 1970.)

the Tinoco theory are that it explains and predicts the nonconservative and conservative spectra of RNAs and DNAs and some other polynucleotides, mainly as a result of tilting of the bases. One can conclude that at present the theory of optical activity of polynucleotides seems to stand on a firm basis.

X. NUCLEIC ACID COMPLEXES: RIBOSOMES AND VIRUSES. FOLDING OF HELICES IN A TERTIARY STRUCTURE

Two different examples may illustrate the interaction of nucleic acids and proteins, which may provide the basis for the understanding of conformational changes that may possibly be due to the folding of helices in a tertiary structure.

McPhie and Gratzer (1966) and Blake and Peacocke (1965) studied ribosomes of different origins (yeast, *E. coli*, rabbit reticulocytes). The intact ribosomes yield ORD curves in the near UV region very similar to that of the free, extracted RNA. McPhie and Gratzer showed that dissociation of yeast ribosomes and their disorganization by high concentration of chelating agent essentially had no effect on their ORD results. The analysis of the ORD curves in the region of amide absorption indicate the presence of protein in partially α-helical form in the intact ribosome.

The RNA structure in ribosomes seems to be largely a base-paired helical conformation. Despite the ambiguity as to how the effect of pairing should be computed, two methods yielded similar results. Bush and Scheraga (1967) compared the ORD curves of RNA of *E. coli* ribosomes in salt-free solution, where the RNA is predominantly single stranded, with that in solutions of moderate or high salt concentration where it is base paired. The results indicate that RNA in *E. coli* ribosomes is in a largely base-paired conformation (see Section IV). Cotter, McPhie, and Gratzer (1967) estimated the proportion of paired bases to be 60% from thermal absorbancy difference spectra. This allowed them to calculate the ORD curves by taking into account single- and double-stranded contributions. The predicted curve is in rather good agreement with the experimental ORD spectrum of ribosomal RNA in salt solution (Fig. 20). One can thus conclude that the conformation of free RNA is similar to that of RNA in intact ribosomes, and has a predominantly double-stranded structure.

On the other hand, the conformation of ribosomal protein appears to depend on the interaction with RNA in the ribosomes. In the ribosomes, the protein was estimated to be partially (about 30%) in the α-helical conformation, but once removed, the soluble protein had no detectable α-helical content.

In contrast with ribosomes, the investigations of viruses have shown a completely different situation, i.e., the conformation of nucleic acids is different in the free state and inside the viruses. Maestre and Tinoco (1967) have measured the ORD of 16 viruses containing DNA and RNA. Despite the fact that each virus has a characteristic ORD spectrum, one can provide a classification into three essential classes of ORD spectra (Figs. 21 and 22): (1) The T-even bacteriophages have negative rotation above 250 mμ with a

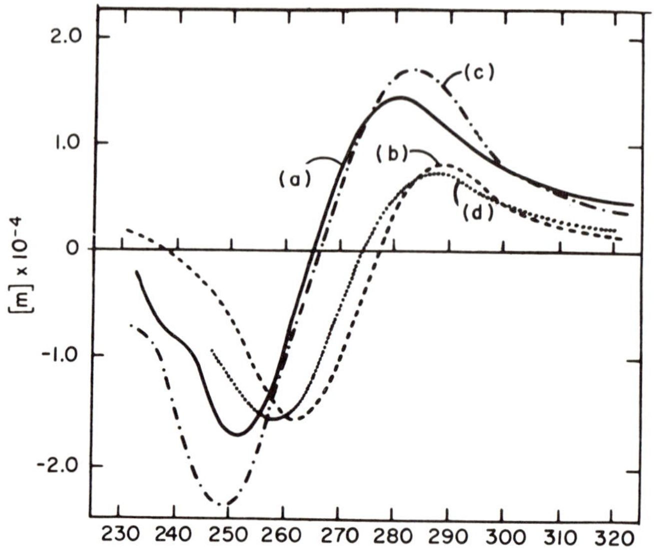

Fig. 20. Calculated and observed optical rotatory dispersion curves for yeast ribosoma RNA: (a) Observed curve for ribosomal RNA (ionic strength 0.1); (b) observed curve for ribosomal RNA in 10^{-4} *M* EDTA (no other salts present); (c) calculated ORD curve for 60% base pairing. The composition of the base-paired part was derived from thermal difference spectra. (d) Calculated ORD curve for a single-stranded chain with the base composition of ribosomal RNA, assuming a random sequence. (After Cotter, McPhie, and Gratzer, 1968.)

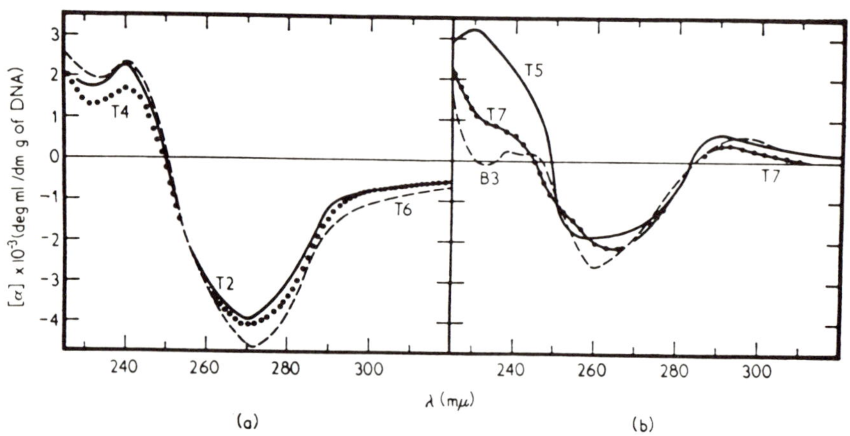

Fig. 21. Optical rotatory dispersion of (a) the T-even class of viruses and (b) the T-odd class. (After Maestre and Tinoco, 1967.)

minimum rotation at about 270 mμ (Fig. 21a); (2) The T-odd class has ORD similar to nucleic acids with a trough near 260 mμ and a peak near 290 mμ (Fig. 21b); (3) The third class shows a characteristic deep trough near 235 mμ in ORD caused by a high percentage of protein coat (Fig. 22). In this class are included different viruses: f2, MS2, and ϕX174. To this classification of Maestre and Tinoco, one can add that the ORD spectra of MS2 and of f2 are almost similar in the spectral region above 260 mμ, and are characterized by a relatively large positive rotation. It can be thus suggested

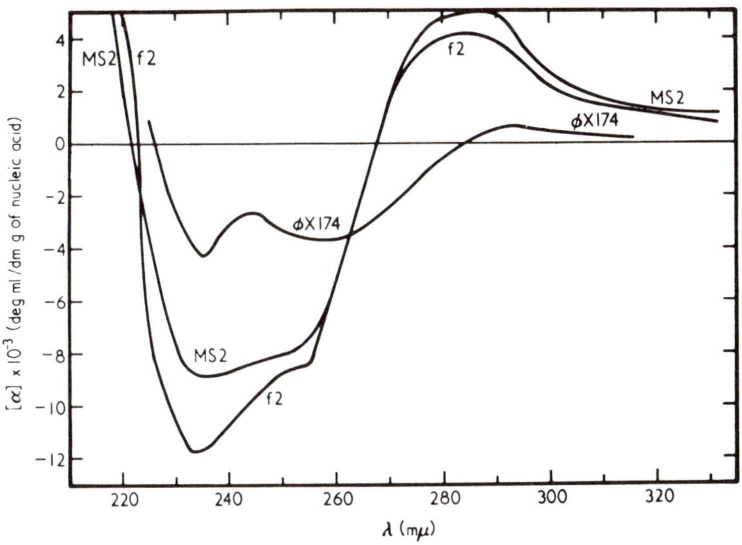

Fig. 22. Optical rotatory dispersion of viruses containing a large contribution from the protein coat: ϕX174, MS2, f2. (After Maestre and Tinoco, 1967.)

that structure of the DNA–RNA hybrid in f2 is similar to that of the double-stranded RNA in MS2, in agreement with the ORD results on nucleic acids isolated from these phages (Tunis and Hearst, 1968).

In order to interpret these spectra, Maestre and Tinoco compared ORD of osmotically shocked T2 phage with that of the intact phage. The ORD spectrum of osmotically shocked phage can be obtained by adding the ORD curves of the purified components: protein and DNA. In contrast, the spectrum of the entire T2 phage cannot be obtained from the sum of the isolated protein coat and nucleic acid. Thus the ORD contributions are nonadditive in this case, in contrast to ORD of ribosomes.

There is no quantitative explanation of these phenomena, but it is clear that the important additional contribution involves protein–nucleic acid interactions which cause some structural changes. The explanation of Maestre and Tinoco is that the folding of the DNA molecule in the head of the phage causes some alteration of the structure similar to dehydration effects. Maestre and Tinoco suggest that this coiling of the nucleic acid must involve a greater part of the DNA molecule and not only some selected points.

Another approach for studying the folding of polynucleotide helices was attempted in order to elucidate the three-dimensional structure of t-RNAs in solution. Figure 23 shows that the temperature dependence of circular dichroism of specific t-RNA (yeast phenylalanine) is not a simple sigmoidal curve like that of various oligonucleotides (see section V), but contains discontinuities at lower temperatures (Bakès, Dirheimer, and Brahms,

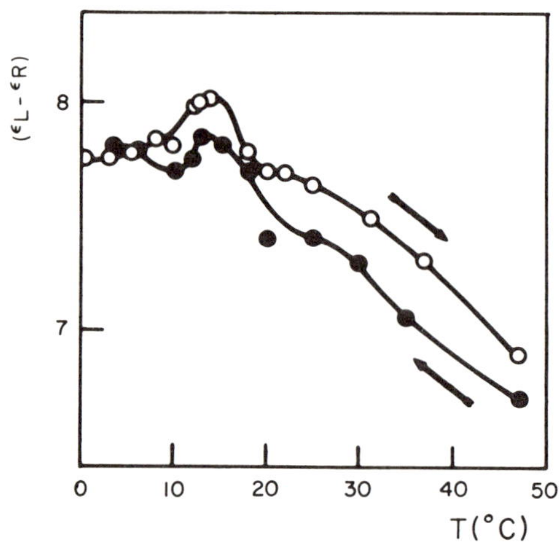

Fig. 23. Changes in circular dichroism as a function of temperature of phenylalanine t-RNA in tris 0.01 *M* pH 7.4, 0.0001 *M* EDTA (no other salts present). (After Bakès, Dirheimer, and Brahms, 1968. Unpublished results).

1969). This phenomenon is reversible and cannot correspond to changes in primary structure, e.g., to cleavage of covalent bonds, but may be interpreted as corresponding to changes in tertiary structure. The changes in circular dichroism are more accentuated at higher ionic strength and in the presence of magnesium.

It is probable that interaction of RNAs with proteins may involve similar changes in tertiary structure, as can be deduced from the results of Oriel and Koning (1968) on thermal denaturation of MS2 phage studied by ORD.

More recently, Davidson and Fasman (1969) demonstrated that the formation of a complex between poly L-lysine and single stranded poly A causes a drastic change of its ORD spectrum. The positive maximum at

280 mμ disappears and only a negative peak centered at 260 mμ is observed. Simultaneously a red shift of the UV absorption spectrum is observed.

Similar results have been reported for systems of poly-L-lysine and highly polymerized DNA (Cohen and Kidson, 1968; Shapiro, Leng, and Felsenfeld, 1969). Particularly remarkable are the CD curves of the poly-L-lysine–DNA (calf thymus) complex which indicate a broad negative band in the region of 300–240 mμ and centered at 269 mμ (Shapiro, Leng, and Felsenfeld, 1969). These authors suggest that this result can be interpreted on the basis of similarity with the CD spectum of DNA in 80% ethanol–water solution found by Brahms and Mommaerts (see Fig. 18b). Both spectra are of non-conservative character but of opposite sign. Recently Fasman et al. (1970) demonstrated similar spectra for f-1 histone:DNA complexes. Shapiro et al. suggest that base tilting, as in the A form of DNA proposed by Brahms and Mommaerts for the case of DNA in alcohol, can be invoked to explain the results of the poly lysine–DNA complex.

REFERENCES

Adler, A. J., L. Grossman, and G. D. Fasman (1967). *Proc. Natl. Acad. Sci., U.S.*, **57**, 423.

Adler, A. J., L. Grossman, and G. D. Fasman (1968). *Biochemistry*, **7**, 3836.

Adler, A. J., L. Grossman, and G. D. Fasman (1969). *Biochemistry*, **8**, 3846.

Arnott, S. (1970). *Prog. Bioph. Molec. Biol.* **21**, 265.

Arnott, S., and Hukins, D. W. L. (1969). *Nature* **224**, 886.

Aubertin, A. M. (1968). Thesis, University of Strasbourg.

Bakès J. (1969). Ph.D. thesis, University of Strasbourg.

Bakès, J., G. Dirheimer, and J. Brahms (1968). Unpublished results.

Bayley, P., E. B. Nielsen, and J. Schellman (1969). *J. Phys. Chem.*, **73**, 228.

Bernardi, G., and S. Timasheff (1970). *J. Mol. Biol.* **48**, 43.

Beychok, S. (1967), in *Poly-α-Amino Acids, Protein Models for Conformational Studies* (G. D. Fasman, ed.), Dekker, New York, p. 293.

Blake, A., and A. R. Peacocke (1965). *Biopolymers*, **4**, 1097.

Born, M. (1933). *Optik* (Springer Verlag, Berlin).

Bradley, D. F., S. Lifson, and B. Hornig (1964), in *Electronic Aspects of Biochemistry* (B. Pullman, ed.), Academic, New York, p. 77.

Bradley, D. F., I. Tinoco, Jr., and R. W. Woody (1963). *Biopolymers*, **1**, 239.

Brahms, J. (1963). *J. Am. Chem. Soc.*, **85**, 3298.

Brahms, J. (1964). *Nature*, **202**, 797.

Brahms, J. (1965). *J. Mol. Biol.*, **11**, 785.

Brahms, J. (1968). *J. Chim. Phys.*, **65**, 101.

Brahms, J., A. M. Aubertin, G. Dirheimer, M. Grunberg-Manago (1969). *Biochemistry*, **8**, 3269.

Brahms, J., J. C. Maurizot, and A. M. Michelson (1967a). *J. Mol. Biol.*, **25**, 481.

Brahms, J., J. C. Maurizot, and A. M. Michelson (1967b), *J. Mol. Biol.*, **25**, 465.

Brahms, J., A. M. Michelson, and K. E. Van Holde (1966). *J. Mol. Biol.*, **15**, 467.

Brahms, J., and W. F. H. M. Mommaerts (1964). *J. Mol. Biol.*, **10**, 73.

Brahms, J., W. F. H. M. Mommaerts, and C. Sadron (1964). *Roy. Acad. Sci.*, **258**, 2203.

Brahms, J., J. Pilet, H. Damany, and V. Chandrasekharan (1968). *Proc. Matl. Acad. Sci., U.S.*, **60**, 1130.

Brahms, J., and C. Sadron (1966). *Nature*, **212**, 1309.

Brahms, J., H. Sellini, and S. Brahms (1969). *III Intern. Biophys. Congr. Abstr.*, p. 171.

Bush, C. A. (1965). Thesis, University of California, Berkeley.

Bush, C. A., and J. Brahms (1967). *J. Chem. Phys.*, **46**, 79.

Bush, C. A., and H. A. Scheraga (1967). *Biochemistry*, **6**, 3036.

Bush, C. A., and H. A. Scheraga (1969). *Biopolymers*, **7**, 395.

Bush, C. A., and I. Tinoco, Jr. (1967). *J. Mol. Biol.*, **23**, 601.

Cantor, C. R., S. R. Jaskunas, and I. Tinoco, Jr. (1966). *J. Mol. Biol.*, **20**, 39.

Cantor, C. R., and I. Tinoco, Jr. (1965). *J. Mol. Biol.*, **13**, 65.

Cantor, C. R., and I. Tinoco, Jr. (1967). *Biopolymers*, **5**, 821.

Cohen, P. and Kidson, C. (1968). *J. Mol. Biol.*, **35**, 241.

Condon, E. V. (1937). *Rev. Mod. Phys.*, **9**, 432.

Condon, E. V., W. Altar, and H. Eyring (1937). *J. Chem. Phys.*, **5**, 753.

Cotter, R. I., P. McPhie, and W. B. Gratzer (1967). *Nature*, **216**, 864.

Davidson, B., and G. D. Fasman (1969). *Biochemistry*, **8**, 4116.

Davies, D. R., and R. L. Baldwin (1963). *J. Mol. Biol.*, **6**, 251.

Davis, R. C., and I. Tinoco, Jr. (1968). *Biopolymers*, **6**, 409.

Davydov, A. S. (1948). *J. Exptl. Theoret. Phys.*, **18**, 210.

de Malleman, R. (1924). *Theorie moleculaire de la polarisation rotatoire. Ann. Phys.* **2**, 137; **4**, 456.

DeVoe, H., and I. Tinoco, Jr. (1962). *J. Mol. Biol.*, **4**, 518.

Emeis, C. A., L. J. Oosterhoff, and G. de Vries (1967). *Proc. Roy. Soc.*, **A297**, 54.

Fasman, G. D., C. Lindblow, and L. Grossman (1964). *Biochemistry*, **3**, 1015.

Fasman, G. D., C. Lindblow, and E. Seaman (1965). *J. Mol. Biol.*, **12**, 630.

Fasman, G., B. Schaffhausen, L. Goldsmith, and A. Adler (1970). *Biochemistry* **9**, 2814.

Felsenfeld, G., D. R. Davies, and A. Rich (1957). *J. Am. Chem. Soc.*, **79**, 2023.

Felsenfeld, G., and H. T. Miles (1967). *Ann. Rev. Biochem.*, **36**, 407.

Fresco, J., and P. Doty (1957). *J. Am. Chem. Soc.*, **79**, 3928.

Fresco, J., and E. Klemperer (1959). *Ann. N. Y. Acad. Sci.*, **81**, 730.

Geidushek, E. P., and T. T. Herskovits (1961). *Arch. Biochem. Biophys.*, **94**, 99.

Gellert, M., M. N. Lipsett, and D. R. Davies (1962). *Proc. Natl. Acad. Sci., U.S.*, **48**, 2013.

Glaubiger, D., D. A. Lloyd, and I. Tinoco, Jr. (1968). *Biopolymers*, **6**, 409.

Gratzer, W. B., G. Holzwarth, and P. Doty (1961). *Proc. Natl. Acad. Sci., U.S.*, **47**, 1785.

Haschemeyer, A. E. V., and A. Rich (1967). *J. Mol. Biol.*, **27**, 369.

Hashizume, H., and K. Imahori (1967). *J. Biochem.* (Tokyo), **61**, 738.

Herskovits, T. T. (1963). *Biochemistry*, **2**, 335.

Holcomb, D. N., and I. Tinoco, Jr. (1965). *Biopolymers*, **3**, 121.

Inoue, Y., S. Aoyagi, and K. Nakanishi (1967). *J. Am. Chem. Soc.*, **89**, 5701.

Inoue, Y., M. Masuda, and S. Aoyagi (1968). *Biochem. Biophys. Res. Comm.*, **31**, 577.

Jaffé, J. H., H. Jaffé, and K. Rosenheck (1967). *Rev. Sci. Instr.*, **38**, 935.

Johnson, W. C., and I. Tinoco, Jr. (1969). *Biopolymers*, **7**, 727.

Kasha, M. (1960) in *Light and Life*, Johns Hopkins Univ. Press, Baltimore.

Kirkwood, J. G. (1937). *J. Chem. Phys.*, **5**, 479.

Kuhn, W. (1929). *Z. Physik Chem.*, **B4**, 14.

Kuhn, W., and M. Rometsch (1944). *Helv. Chim. Acta*, **27**, 1080.

Lamborg, M. R., and P. C. Zamecnik (1965). *Biochim. Biophys. Res. Commun.*, **20**, 328.

Lamborg, M. R., P. C. Zamecnik, T. K. Li, J. Kagi, and B. L. Vallee (1965). *Biochemistry*, **4**, 63.

Langridge, R., D. A. Marvin, W. E. Seeds, H. R. Wilson, W. C. Hooper, M. H. F. Wilkins, and L. D. Hamilton (1960). *J. Mol. Biol.*, **2,** 38.
Langridge, R., and A. Rich (1963). *Nature*, **198,** 725.
Leng, M., and G. V. Felsenfeld (1966). *J. Mol. Biol.*, **15,** 455.
Levine, L., J. A. Gordon, and W. P. Jencks (1963). *Biochemistry*, **2,** 168.
Litman, B. J., and J. A. Schellman (1965). *J. Phys. Chem.*, **69,** 978.
Luzzati, V., A. Mathis, F. Masson, and J. Witz (1964). *J. Mol. Biol.*, **10,** 28.
Maestre, M. F., and I. Tinoco, Jr. (1965). *J. Mol. Biol.*, **12,** 287.
Maestre, M. F., and I. Tinoco, Jr. (1967). *J. Mol. Biol.*, **23,** 323.
Mason, S. F. (1959). *J. Chem. Soc.*, 1240.
Massoulié, J. (1966). Thesis, University of Paris.
Maurizot, J. C., J. Blicharski, and J. Brahms (1969). *III Intern. Biophys. Congr. Abstr.*, p. 172.
Maurizot, J. C., J. Brahms, and F. Eckstein (1969). *Nature*, **222,** 559.
McPhie, P., and W. B. Gratzer (1966). *Biochemistry*, **5,** 1310.
Michelson, A. M. (1963). *The Chemistry of Nucleosides and Nucleotides*, Academic, New York.
Michelson, A. M., J. Massoulié, and W. Guschlbauer (1967). *Progr. Nucleic Acid Res. Mol. Biol.*, **6,** 83.
Moffitt, W. (1956). *J. Chem. Phys.*, **25,** 467.
Moffitt, W., and A. Moscowitz (1956). *J. Chem. Phys.*, **30,** 648.
Moffitt, W., R. B. Woodward, A. Moscowitz, W. Klyne, and C. Djerassi (1961). *J. Am. Chem. Soc.*, **83,** 4013.
Mommaerts, W. F. H. M., J. Brahms, J. H. Weil, and J. P. Ebel (1964). *Compt. Rend. Acad. Sci.*, **258,** 2203.
Moscowitz, A. (1960). *Optical Rotatory Dispersion* (C. Djerassi, ed.), McGraw-Hill, New York.
Moscowitz, A. (1962). In *Adv. in Chem. Phys.*, **4,** 67.
Moscowitz, A. (1961). *Tetrahedron*, **13,** 48.
Olins, D. E. (1969). *J. Mol. Biol.*, **43,** 439.
Oriel, P. J., and J. A. Koening (1968). *Arch. Biochem.*, **127,** 274.
Pochon, F., and A. M. Michelson (1965). *Proc. Natl. Acad. Sci., U.S.*, **53,** 1425.
Poland, D., J. N. Vournakis, and H. A. Scheraga (1966). *Biopolymers*, **4,** 223.
Rich, A. (1958). *Biochim. Biophys. Acta*, **29,** 502.
Rich, A., and D. R. Davies (1956). *J. Am. Chem. Soc.*, **78,** 3548.
Rich, A., D. R. Davies, F. H. C. Crick, and J. D. Watson (1961). *J. Mol. Biol.*, **3,** 71.
Richards, E. G. (1968). *Europ. J. Biochem.*, **6,** 88.
Rosenfeld, L. (1928). *Z. Physik*, **52,** 161.
Samejima, T., H. Hashizume, K. Imahori, J. Fujii, and K. Miura (1968). *J. Mol. Biol.*, **34,** 39.
Sarkar, P. K., and J. T. Yang (1965a). *J. Biol. Chem.*, **240,** 2088.
Sarkar, P. K., and J. T. Yang (1965b). *Biochemistry*, **4,** 1238.
Sarkar, P. K., B. Wells, and J. T. Yang (1967). *J. Mol. Biol.*, **25,** 563.
Schellman, J. A. (1966). *J. Chem. Phys.*, **44,** 55.
Schellman, J. A. (1968). *Accounts Chem. Res.*, **1,** 144.
Schellman, J. A., and E. B. Nielsen (1967). *The Conformation of Biopolymers* (G. N. Ramachandran, ed.), Academic, New York.
Schellman, J. A., and P. J. Oriel (1962). *J. Chem. Phys.*, **37,** 2114.
Schweizer, M. P., A. D. Broom, P. O. P. Ts'o, and D. P. Hollis (1968). *J. Am. Chem. Soc.*, **90,** 1042.
Shapiro, J. T., M. Leng, and G. Felsenfeld (1969). *Biochemistry*, **8,** 3219.
Simmons, N. S., and E. R. Blout (1960). *Biophys. J.*, **1,** 55.

Simmons, N. S., C. Cohen, A. Szent-Györgyi, D. B. Wetlaufer, and E. R. Blout (1961). *J. Am. Chem. Soc.*, **83,** 4766.

Steiner, R. F., and R. F. Beers (1961). *Polynucleotides*, Elsevier, Amsterdam.

Sundarolingam M. (1969). *Biopolymers*, **7,** 821.

Thang, M. N., M. Graffé, and M. Grunberg-Manago (1965). *Biochim. Biophys. Acta*, **108,** 125.

Thiéry, J. M. (1968). *J. Chim. Phys.*, **65,** 98.

Timasheff, S. N., J. Witz, and V. Luzzati (1961). *Biophys. J.*, **1,** 525.

Tinoco, I., Jr. (1962). *Adv. Chem. Phys.*, **4,** 133.

Tinoco, I., Jr. (1963). *Radiation Res.*, **20,** 133.

Tinoco, I., Jr. (1964). *J. Am. Chem. Soc.*, **86,** 297.

Tinoco, I., Jr. (1968). *J. Chim. Phys.*, **65,** 91.

Tinoco, I., Jr., R. W. Woody, and D. F. Bradley (1963). *J. Chem. Phys.*, **38,** 1317.

Ts'o, P. O. P., S. A. Rapaport, and F. J. Bollum (1966). *Biochemistry*, **5,** 4153.

Tunis, M. J. B., and J. E. Hearst (1968). *Biopolymers*, **6,** 1218.

Ulbricht, T. T. L. V., R. J. Swan, and A. M. Michelson (1966). *Chem. Commun.*, p. 63.

Van Holde, K. E., J. Brahms, and A. M. Michelson (1965). *J. Mol. Biol.*, **12,** 726.

Velluz, L., M. Legrand, and M. Grosjean, 1965. *Optical Circular Dichroism, Principles, Measurements and Applications*, Verlag Chemie, GMBH Weinheim, Academic, New York.

Vournakis, J. N., D. Poland, and H. A. Scheraga (1967). *Biopolymers*, **5,** 403.

Vournakis, J. N. and H. A. Scheraga (1966). *Biochemistry* **5,** 2997.

Warner, R. C. (1957). *J. Biol. Chem.*, **229,** 711.

Warshaw, M. M., C. A. Bush, and I. Tinoco, Jr. (1965). *Biochem. Biophys. Res. Commun.*, **18,** 633.

Warshaw, M. M., and I. Tinoco, Jr. (1965). *J. Mol. Biol.*, **13,** 54.

Warshaw, M. M., and I. Tinoco, Jr. (1966). *J. Mol. Biol.*, **20,** 29.

Watson, J. D. and F. H. C. Crick (1953). *Nature*, **171,** 737.

Wilkins, M. H. F., A. R. Stockes, and H. R. Wilson (1953). *Nature*, **171,** 738.

Witz, J., and V. Luzzati (1965). *J. Mol. Biol.*, **11,** 620.

Wolfe, F. H., K. Oikawa, and C. M. Kay (1968). *Biochemistry*, **7,** 3361.

Wolfe, F. H., K. Oikawa, and C. M. Kay (1969). *Can. J. Biochem.*, **47,** 637.

Woody, R. W., and I. Tinoco, Jr. (1967). *J. Chem. Phys.*, **46,** 4927.

Zavilgelskij, G. B., and L. Li (1967). *Mol. Biol.*, **1,** 323.

CHAPTER 4

BURIED AND EXPOSED GROUPS IN PROTEINS

Martin J. Kronman† *and Frederick M. Robbins*

PIONEERING RESEARCH LABORATORY
U.S. ARMY NATICK LABORATORIES
NATICK, MASSACHUSETTS

† Present address: Department of Biochemistry, Upstate Medical Center, State University of New York, Syracuse, New York.

I. INTRODUCTION

If one were to regard a globular protein molecule as a rigid impenetrable sphere, the idea of "buried" and "exposed" side chains and molecular "interior" and "surface" would be easily understood. Unfortunately, neither our knowledge of the solution properties of proteins nor the X-ray crystallographic structures that have been determined in the last few years will allow us to comfortably accept such a naive model. Rather than a rigid sphere, a globular protein would seem to be better represented as a "roundish crumpled bundle of wire." When confronted with questions such as whether a group is exposed, we are forced to counter with a question such as "Exposed to what?" or "Why do you want to know?"

The types of problems that fall into the "Why do you want to know?" category include those such as : "What amino acid side chains form the active site of an enzyme?"; "What steric limitations are there on approach of a low molecular weight or macromolecular weight substrate or co-factor to the active site?"; "What are the side chain–side chain and side chain–solvent

TABLE I

Amino Acid Residues "Involved" in the Activity of Various Enzymes

Residue	Enzyme	Reference
tyr	Fructodiphosphatase	*2*
	Pepsin	*3*
	Carboxypeptidase A	*4, 6*
	Carboxypeptidase B	*5*
trp	Lysozyme	*7*
his	Ribonuclease	*8*
	Yeast enolase	*9*
	Chymotrypsin	*10*
-cy SH	Glyceraldehyde phosphate Dehydrogenase	*11, 12*
	Phosphorylase	*13, 14, 15*
	Papain	*16*
	Aspartic aminotransferase	*17, 18, 18a*
lys	Ribonuclease	*19*
	Aldolase	*20*
met	Chymotrypsin	*21*
ser	Chymotrypsin	*22, 23*
	Trypsin	*24*
	Alkaline phosphatase	*25, 26*
	Phosphoglucomutase	*27*
	Thrombin	*28*

interactions which maintain the native conformation?" "What factors influence globular proteins to form higher molecular weight species (e.g., insulin, α-lactalbumin, chymotrypsin, enzymes with subunits, etc.)?" Indeed, as Fisher (*1*) has suggested, the principal determinant of whether a protein associates or polymerizes may well be its capacity for burying its nonpolar side chains in its monomeric state.

The range of amino acid side chains which have been implicated in the activities of enzymes is seen from the compilation in Table I. This is not intended to be a summary of all known cases nor does it represent information from critically selected sources. In some instances, e.g., glyceraldehyde phosphate dehydrogenase, the residue in question, has a primary role in the catalytic process. Other groups may be involved in binding substrate, e.g., ala 107, aspN 59, and trp 62 and 63 of lysozyme (*7*). Still others, by virtue of their proximity to "sensitive" areas of the substrate molecule, may produce electronic perturbations that contribute to the driving force of the catalysis, e.g., asp 52 and glu 35 of lysozyme (*7*). In other cases, the group may be involved in maintaining the appropriate conformation of the active site or of the overall protein structure. In any event, the possibility of interactions of the residue in question with solvent molecules or with small or macromolecular substrates in the medium would seem to be of prime importance. Thus, we might regard a group as buried if it were unavailable for such interactions, regardless of whether this were due to specific steric factors, to strong interactions with other structural moieties, or to actual insulation from contact with the medium.

While we may not be able to exactly define what we mean by "buried," "exposed," "interior," and "surface," we can give an operational meaning to these terms using techniques having as their basis interactions of the kind alluded to above for enzymes. Thus, virtually all of the techniques to be discussed make use of molecular probes in conjunction with some observational procedure. The ideal probe would be a small molecule or ion which, on interaction with a specific side chain, would "send back" a signal only if groups were freely accessible, but which would in no way perturb the molecular conformation of the protein. It is doubtful whether the latter requirement can ever be strictly met. However, as we shall see, in many instances such changes can be minimized or, perhaps more frankly stated, rendered undetectable by techniques available to us *at present*.

The amino side chains to be considered are summarized in Table II; the classification is that of Tanford (*29*). There are certain rules of thumb that tell us where to expect particular side chains to be located. Thermodynamic arguments have led to the view in recent years that, in general, polar residues should be on the surface of the protein molecule while nonpolar groups ought to be confined to the anhydrous interior.

TABLE II

Classification of Amino Acid Side Chains in Proteins

Group	Method for determining exposure
Charged Side Chains	
Arginine	Chemical techniques, potentiometric titration.
Lysine	
Histidine	
Glutamic and aspartic acids	
Hydrophobic side chains	
Tyrosine	Spectral and chemical techniques. Tyrosine evaluated also by potentiometric titration.
Tryptophan	
Phenylalanine	
Methionine	
Cysteine	Chemical techniques.
Cystine	
Proline	
Valine, leucine	No methods for individual amino acids. Hydrocarbon gas binding and fluorescent dye studies give information about clusters of hydrophobic groups.
Isoleucine	
Hydrocarbon portions of arginine, lysine, aspartic and glutamic acids	
Uncharged polar side chains	
Asparagine	
Glutamine	Chemical techniques
Serine	
Threonine	
Acidic and basic side chains at a pH at which they are uncharged	Potentiometric titration, chemical techniques

X-ray crystallographic studies of myoglobin (*30*) and of lysozyme (*7*) essentially support this point of view. It is quite clear, however, that the above generalization is not a hard and fast rule; the X-ray structure of lysozyme, for example, reveals a number of hydrophobic groups on the molecular surface including trp 62, trp 63, ileu 98, val 109, and trp 123. Likewise, solution techniques, including fluorescent dye binding and solvent perturbation, show that nonpolar groups need not be confined to the interior of the protein molecule. Instances of "burying" of polar groups could also be cited.

Ultimately, the overall energy of a given conformation will determine whether a group is in contact with the medium or buried in the interior. This in turn will be a complex function of the steric requirements for folding and the energetics of side chain–side chain and side chain–solvent interactions. The problem of folding a peptide chain has been considered from a number of points of view in Vol. 1 of this series (*31*) and will not be treated further here.

We restrict our consideration of side chain interactions to problems relating to the concept of exposure and burying.

Shown also in Table II are some of the methods used to determine their accessibility. These are discussed at length in subsequent sections of this chapter.

Chemical reactivity techniques have great advantage in that in many instances stable derivatives can be formed which later permit identification of the reactive side chain in the amino acid sequence. Alteration of the side chain of the protein, of course, entails the risk of disrupting the balance of forces stabilizing the native conformation either through introduction of a bulky group or by destroying local interactions of side chains. Subsequent reaction might then occur with normally buried groups. As we shall see, the specificity of reaction of a given reagent with the side chain of interest may not be as absolute as one might wish.

Potentiometric and spectrophotometric titrations, while relatively group specific for a given pH region, also entail risks of conformational alterations because of the necessity of titrating the protein to pH values quite removed from their "natural" state. "Pure" spectroscopic techniques, such as measurement of pH-, temperature-, or concentration-difference spectra and of absolute absorption and emission spectra, perhaps perturb the system being studied less than any technique that one could employ. Unfortunately, as we shall see, it is precisely these methods which give the most ambiguous answers about exposure when used by themselves. The use of a chemical additive in the solvent perturbation spectrophotometric procedure, while circumventing this type of ambiguity, also entails a risk of alteration of the molecular conformation, although perhaps less than might occur on chemical modification or potentiometric titration.

The use of molecular probes, in general, of necessity involves the risk of local perturbations of the molecular environment of the side chains in question. The allosteric character of enzymes testifies to the reality of such subtle conformational changes. However, gross changes in structure can be detected through the use of ancillary measurements or in many instances by critical examination of the exposure experiments themselves. We emphasize throughout this chapter the necessity of verifying that the measured exposure is that for the unperturbed protein molecule.

II. SPECTRAL TECHNIQUES

A. Introduction

Spectral techniques, because of their great sensitivity, relative simplicity of execution, and "innocuous" character with respect to potential alteration of molecular structure of proteins, have been widely used in studying changes in the environment of amino acid side chains. Infrared techniques (see Section III. D) have been used to some degree in such work. We, however, confine our discussion in this section to the ultraviolet region of the spectrum. There have been a number of general reviews of ultraviolet spectral properties of proteins; some of the more recent ones are by Beaven (*32*), Wetlaufer (*33*), and Weber and Teale (*34*).

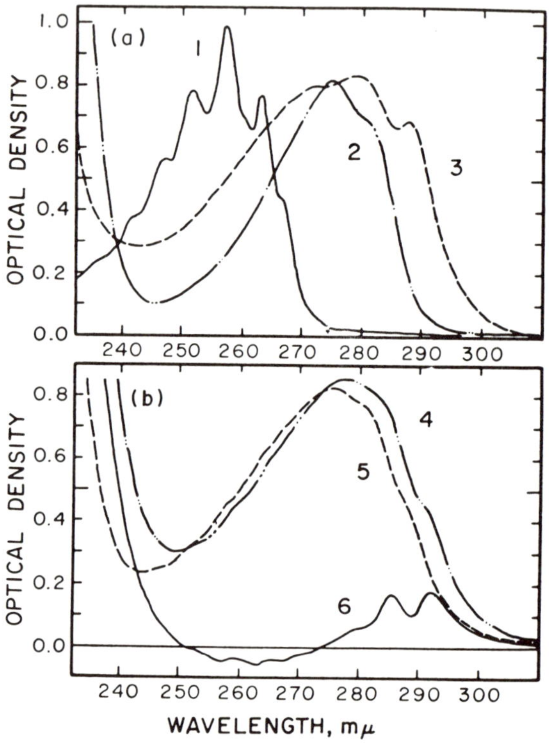

Fig. 1. (a) Ultraviolet absorption spectra of the aromatic amino acids: curve 1, 5×10^{-3} *M* phenylalanine; curve 2, 6.4×10^{-4} *M* tyrosine; and curve 3, 1.6×10^{-4} *M* tryptophan. (b) Comparison of the spectrum of 2×10^{-5} *M* pepsin (curve 4) with that of a mixture of the aromatic amino acids, curve 5, in the proportions in which they occur in pepsin. Curve 6, the difference spectrum obtained with the pepsin solution in the sample beam and the amino acid mixture in the reference beam [from Yanari and Bovey (*25*)].

The absorbing components of proteins are the aromatic amino acids, tyrosine, tryptophan, and phenylalanine; the sulfur-containing amino acids, cysteine, cystine, and methionine; histidine; and the peptide bond. Below 2300 Å all of these protein moieties have significant absorption, and study of a particular one is likely to be difficult except under rather special circumstances. For this reason, we do not, in general, consider spectral techniques involving this lower wavelength region. Absorption above 2300 Å is essentially from the aromatic amino acids. As Fig. 1(a) indicates, the molar absorbance from phenylalanine is relatively low as compared to tyrosine and tryptophan. This, together with the fact that proteins contain a significant number of tyrosine and tryptophan residues relative to phenylalanines, accounts for the fact that little use has been made of spectral techniques in studying the environment of the latter groups. Incorporation of the aromatic amino acid into a protein generally produces a long wavelength shift of the ultraviolet spectrum as illustrated by the observations of Yanari and Bovey (*35*) for pepsin and mixtures of the aromatic amino acids in the proportions in which they occur in this enzyme [Fig. 1(b)]. This type of spectral shift is conveniently determined as a difference spectrum shown in curve 6 of the figure. Comparable difference spectra but of opposite sign, i.e., blue shifts, are generally observed on denaturation of proteins. It has been common in these cases to conclude that such blue shifts arise from the unfolding of the protein molecule, which results in exposure to the polar solvent of chromophores that had been previously buried in nonpolar regions of the molecule. In many instances, such a conclusion may be qualitatively valid. Observation of a blue shift, however, is not conclusive evidence for enhanced exposure of groups since spectral shifts may have other origins, such as changes in the polarizability, polarity, and charge state of the immediate surroundings of the chromophore as well as its hydrogen-bonded state (*35,36*). Instances of spectral shifts that do not correspond to changes of exposure of chromophores are discussed subsequently in Part E of this section.

B. Solvent Perturbation

1. Introduction

As noted above, the spectral shift that protein undergoes with various treatments cannot usually be interpreted unambiguously in terms of specific molecular events. In order to circumvent this ambiguity, Herskovits and Laskowski (*86*) devised the solvent perturbation technique and have been instrumental in exploring its application to a number of systems (see Ref. *38a* and *59* for reviews of this work). The solvent perturbation technique restricts its goal to answering a rather limited question: namely, "Does the spectral shift involve an exposure of groups?" For example, if a change in pH of a protein solution were to produce a tyrosine spectral shift, the perturbation technique would concern itself with the degree of exposure of such groups at the initial pH and at the final pH, rather than with origin of the spectral shift observed.

The perturbation technique is based on the fact that the position of an absorption band of a chromophore depends upon the medium in which it is immersed. Thus, if one compares spectra of such a chromophore in a series of solvents, one will observe "red" (long wavelength) or "blue" (short wavelength) shifts of the spectrum depending on refractive index, dielectric constant, hydrogen bonding, or some other property of the solvent. For purposes of the solvent perturbation technique, the *origin* of the spectral shift need not concern us; we are only interested in its magnitude relative to that of a suitable standard.

The chromophores of primary interest in solvent perturbation are tyrosines and tryptophans; phenylalanine makes too small a contribution to the absolute spectrum to make it practicable in general to measure its exposure. A schematic view of a protein molecule containing only tyrosine as might be seen by solvent perturbation is given in Fig. 2. This model, as we shall see in

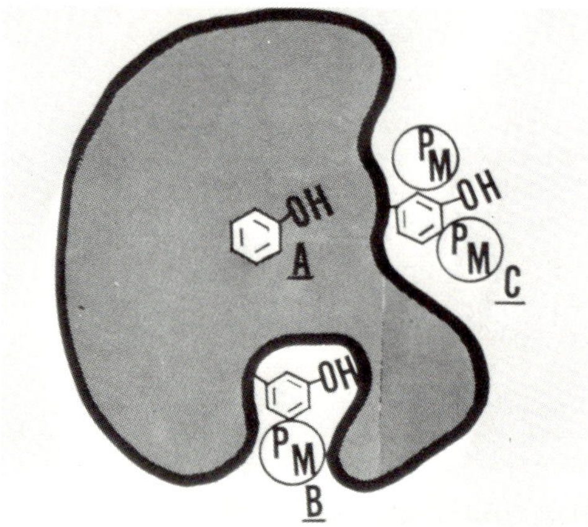

Fig. 2. Schematic view of a protein molecule illustrating interaction of a perturbant with an exposed group (C) and with one located in a crevice (B).

subsequent sections, grossly oversimplifies the meaning of exposure and burying of groups, even in terms of the solvent perturbation technique. Nonetheless, it is an instructive means of considering such chromophores. If the spectrum of such a protein in a purely aqueous medium was compared with that obtained in another solvent, the magnitude of the observed shift would depend upon the extent of contact of the tyrosine with the medium. Tyrosine A (Fig. 2), which we assume to be completely buried, would experience no spectral shift while tyrosine C, which is completely exposed, would experience the maximum possible shift. Tyrosine B, which we have depicted as lying in a "crevice like" fissure of the molecular surface of the protein molecule, would experience a partial shift depending upon the nature and molecular size of the perturbant solvent molecule, the size of the crevice, and the orientation of the chromophore within the crevice. We consider these determinants in greater detail subsequently. The spectral shift observed for such a protein molecule when compared with a suitable standard will yield an average degree of exposure reflecting complete burying, complete exposure, and intermediate states of the various chromophores within the protein molecule.

Perturbation spectra are determined most accurately and conveniently as difference spectra in a double beam recording spectrophotometer. The sample and reference cells contain solutions of the protein at fixed pH, ionic strength, and concentration. The solvent for one of these solutions, however, contains a perturbant such as glycerol or ethylene glycol at moderate concentrations (ca., 20%, v/v). As might be anticipated, the position of extrema of perturbation difference spectra are characteristic of the chromophore rather than of the perturbant. For example, in the perturbation spectra obtained for tyrosine and tryptophan using 20% ethylene glycol as perturbant (Fig. 3), the position

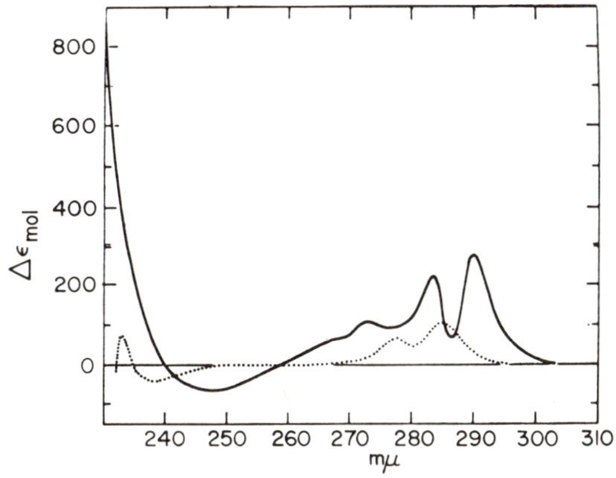

Fig. 3. Perturbation difference spectra of tyrosine and tryptophan caused by 20% ethylene glycol. Solid curve, tryptophan; dotted curve, tyrosine. These spectra were determined at a concentration of 8.4×10^{-4} *M* for tyrosine, 2.6×10^{-4} *M* for tryptophan, in neutral solution (*37*).

of the maxima at 285, 278, and 274, and 290, 283, and 273 mμ are characteristic, respectively, of tyrosine and tryptophan. As we shall see, however, the exact positions change somewhat on incorporation of the aromatic residue in the protein. The amplitude of these maxima in a protein will be determined by the degree of exposure of such groups.

2. Some Experimental Considerations

Because of the small optical densities encountered in perturbation difference spectra (as small as 0.02 absorbance units in some cases), experimental manipulations and spectral measurements must be made with extreme care.

Protocols for solution preparation have been given in detail by Herskovits and Laskowski (*38*) and are not elaborated on here.

Herskovits and Sorenson (*38b*) have recently critically examined the optimal spectrophotometer parameters for carrying out solvent perturbation measurements. Since the spectral shifts observed in solvent perturbation experiments are small, the protein concentrations employed should be as large as possible. Unfortunately, instrumental capabilities place an upper limit on this concentration. At very high optical densities where the stray light passed by the spectrophotometer monochromator becomes comparable to the unabsorbed incident light, spurious values of absorption may be observed. These stray light effects may be recognizable as distortion of the spectra or as deviation from Beer's law. This type of error, as well as that due to fluorescence of solutions, has been treated by Mehler (*39*), Fridovich et al. (*40*), Mehler et al. (*41*), and more recently by Mihalyi (*42*). The latter has shown that stray light in a single monochromator instrument like the Beckman DU spectrophotometer may be as much as two orders of magnitude larger than that in a double monochromator instrument. For this reason, it is best to employ double monochromator recording spectrophotometers such as the Cary 14 or 15 for solvent perturbation measurements. In principle, a double monochromator manual instrument such as the Zeiss PMQ II (MMR) should yield equally satisfactory results. However, the desirability, if not necessity, of scanning complete spectra over a somewhat extended wavelength range, renders a manual instrument of limited utility. The Cary Model 16 manual double monochromator spectrophotometer, on the other hand, may have great utility for solvent perturbation measurements because of the recently developed accessories which permit recording of spectra. The manufacturer's specifications for this instrument indicate a photometric accuracy ranging from 0.00024 (zero absorbance) to 0.005 (absorbance of 3) with the capability of full scale recording of any 0.1 optical density interval between zero and 1.9. To exploit this greater photometric capability of this instrument, it may be necessary to develop improved methods of solution preparation. It is yet to be seen whether this instrument will yield more precise perturbation data than the spectrophotometers currently being used.

Solvent perturbation measurements are most conveniently made in tandem cells as described by Herskovits and Laskowski (*38*), and Holmes and Kronman (*43*). The arrangement of solution and blanks are shown in Fig. 4. This type of cell configuration is employed to properly subtract the solvent blanks and to ensure that the optical paths for sample and reference beams are comparable. Some investigators have made separate optical density measurements for pairs of protein solutions and for pairs of blanks with the blank then being applied numerically. In our experience, however, this leads to less precise results. The difference in refractive index of solutions in the reference and sample beams, for example, may result in differences in illumination of the photodetector.

Laskowski and co-workers have employed a fused cylindrical tandem double cell (*38*). For reasons described elsewhere (*43*), we have frequently preferred to use a demountable tandem cell assembly, illustrated in Fig. 4(b).

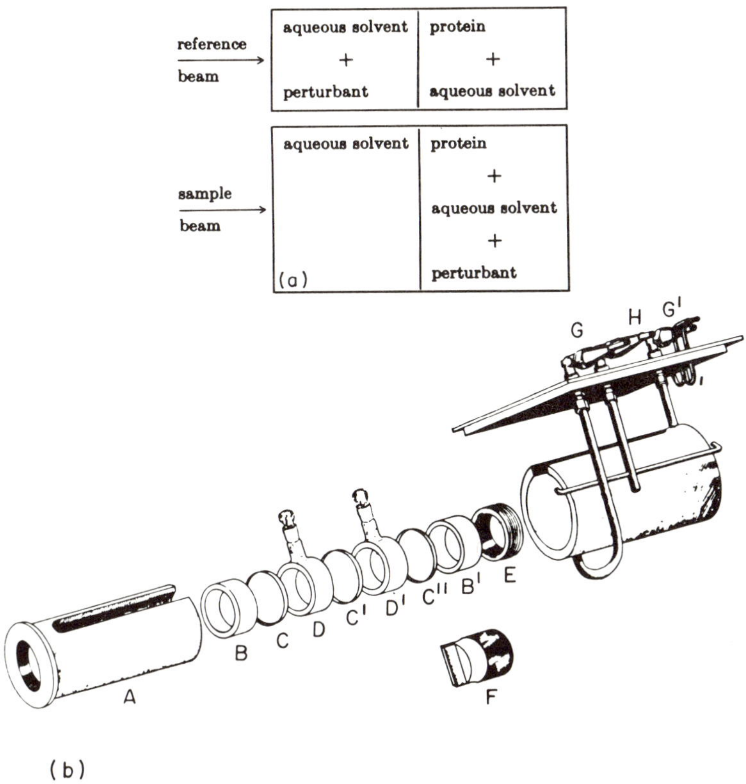

Fig. 4. (a) Arrangement of solutions in the spectrophotometer for a solvent perturbation measurement. (b) Exploded view of cell holder, cell, and cell thermostat: (A) cell holder, (B,B′) nylon spacers, (C,C′,C″) quartz windows, (D,D′) cell bodies, (E) tightening ring, (F) wrench for tightening E, (G,G′) water inlet and outlet, (H) inlet tube for dry gas, (I) openings for thermocouples.

This particular unit was made for a Cary Model 14 spectrophotometer, but slight modifications permit its use with the Cary Model 15 instrument whose reference compartment is placed underneath the sample compartment. A pair of these units are used, one in the reference compartment, the other in the sample compartment. This type of arrangement has proven of particular value where solvent perturbation measurements were carried out near 0°C where the temperature could be maintained to $\pm 0.1°C$. Yankeelov (*44*) has suggested the use of a double compartment square cell for difference spectra. This is a 1-cm square cell divided into two 0.5-cm compartments by fusing an internal window in the optical path. In our experience, this type of cell is not satisfactory for solvent perturbation measurements, particularly with solvents, such as ethylene glycol, that show a pronounced tendency to creep over the internal wall into the other compartment.

The low absorbance encountered in perturbation measurements makes it imperative to avoid uncompensated scattered light from protein solutions. Scattering by protein molecules in the sample and reference beams will generally cancel, unless, for example, improper solution preparation results in turbidity in *one* of the cells. Scattering can be recognized by scanning spectra at wavelengths above 300 mμ where absorption should be essentially absent. In principal, turbidity corrections can be made using a $1/\lambda^4$ relationship or an empirical turbidity correction. The magnitudes of perturbation difference spectra generally encountered makes this practice questionable in most cases.

3. Calculation of the Degree of Exposure of Chromophores

Since the amplitudes of the maxima in a perturbation spectrum will be a reflection of the degree of exposure of the constituent chromophores in a protein, selection of a suitable reference state for 100% exposure will permit quantitative evaluation of their fraction of the exposure, *R*. Amino acid esters in the appropriate molar ratios in which they occur in the protein, or the reduced denatured protein itself, may be used as a reference. The N-acetyl ethyl esters of tyrosine, tryptophan, and phenylalanine are to be preferred to the amino acids themselves since with the latter, spectral shifts may occur owing to changes in ionization of carboxyl and amino groups under the influence of the perturbant or of other additives such as urea. Secondly, the state of the chromophores in such a derivative more closely approximates that of the amino acid incorporated into a peptide chain. In the latter regard, the reduced denatured protein is probably a better reference than model compound mixtures. The relative merits of these two reference states are considered subsequently. In both cases, however, the following reduced difference spectral parameter has utility:

$$\Delta\varepsilon_\lambda/\varepsilon_{\lambda_{\max}} = \Delta D_\lambda/D_{\lambda_{\max}}$$

where ΔD_λ is the observed difference in optical density obtained at a peak of the perturbation spectrum, while $D_{\lambda_{\max}}$ is the optical density of the model compound mixture or reduced protein at the maximum of the absolute spectrum. This reduced parameter, $\Delta\varepsilon/\varepsilon$, is more useful than the molar quantities since neither the molecular weight, concentration, nor extinction coefficient of the protein need be known. The degree of exposure of chromophores is then defined as

$$R_M \equiv \frac{[\Delta\varepsilon_\lambda/\varepsilon_{\lambda\max}]_{\text{protein}}}{[\Delta\varepsilon_\lambda/\varepsilon_{\lambda\max}]_{\text{model compound}}} \tag{2}$$

$$R_P \equiv \frac{[\Delta\varepsilon_\lambda/\varepsilon_{\lambda\max}]_{\text{protein}}}{[\Delta\varepsilon_\lambda/\varepsilon_{\lambda\max}]_{\text{reduced protein}}} \tag{3}$$

Since the incorporation of a chromophoric residue in a polypeptide chain produces a slight shift in the spectrum toward longer wavelengths, the peaks of perturbation spectra for model compounds will not in general correspond to the wavelength observed with protein [see Fig. 1(b), for example]. Values of ΔD_λ employed in calculation of R, however, should be taken at the maxima. Representative values of $\Delta\varepsilon/\varepsilon$ for tyrosine and tryptophan are given in columns 3 and 4, Table III. The values for the former amino acid were

TABLE III

Characteristics of Perturbants[a]

Perturbant (1)	Effective diameter, Å (2)	$\Delta\varepsilon_{286-288}/\varepsilon_{279-278}$ (tyrosine) (3)	$\Delta\varepsilon_{290-293}/\varepsilon_{278-280}$ (tryptophan) (4)	Range effect (5)
90% Heavy water	2.0	−0.043[b]	−0.030[b]	Long
20% Dimethyl sulfoxide	4.0	0.158[d]	0.098[d]	Short
20% Ethylene glycol	4.4	0.067[b]	0.056[b]	Long
20% Glycerol	5.2	0.063[b]	0.052[b]	Long
20% Polyethylene glycol	9.2	0.145[c]	0.088[f]	Short
20% Sucrose	9.4	0.037[b]	0.036[b]	Long
20% Methanol	—	0.059[c]	0.038[f]	Short
20% Ribitol	—	0.036[e]	—	Long
20% Sorbitol	7.3	0.034[e]	—	Long
20% Erythritol	—	0.035[e]	—	Long

[a] Values obtained in aqueous medium at room temperature. Also see Appendix.
[b] Values obtained for N-acetyl tyrosine and tryptophan ethyl esters, respectively (*45*).
[c] Values obtained for model compound mixture for insulin (*46*).
[d] Values obtained for N-acetyl tyrosine and tryptophan ethyl esters, respectively (*38*).
[e] Values obtained from data given by Riddiford (*47*) for model compound mixtures for paramyosin.
[f] Values obtained for model compound mixtures for lysozyme (*48*).

obtained with amino acid esters and model compound mixtures for the tryptophan-free proteins, paramyosin and insulin. The values for tryptophan were obtained in a similar manner. Values of $\Delta\varepsilon/\varepsilon$ for tyrosine perturbation are uniformly higher than those observed for tryptophans. With the exception of heavy water, all of the perturbants produced a "red" shift, as might be anticipated on the basis of the increase in refractive index occurring on addition of the perturbant. Similarly, the "blue" shift observed with 90% heavy water is the consequence of a *decrease* in the refractive index as compared to that of ordinary water. The *magnitudes* of these $\Delta\varepsilon/\varepsilon$ values, as we shall see subsequently, are *not* solely a reflection of the refractive index or dielectric properties of the medium. In this regard, the rather large perturbations produced by polyethylene glycol and dimethyl sulfoxide are noteworthy (Table III). It should be emphasized that some of the values of $\Delta\varepsilon/\varepsilon$ shown in Table III are for model compound mixtures for particular proteins. While $\Delta\varepsilon_{290-293}$ may be little affected by the presence of tyrosine in a protein, variation of $\Delta\varepsilon_{278-280}$ with change in tyrosine/tryptophan ratios may produce significant changes in the parameter $\Delta\varepsilon/\varepsilon$. Similar conclusions can be drawn for values of $\Delta\varepsilon/\varepsilon$ for the tyrosine groups.

The values of $\Delta\varepsilon/\varepsilon$ given in Table III were obtained in dilute aqueous solutions of buffer or salt. These quantities, however, undergo quite significant changes in organic solvents or aqueous solutions containing high concentrations of urea. Williams et al. (*48*) have shown that $\Delta\varepsilon_{290-292}/\varepsilon_{280-282}$ (tryptophan) decreases by as much as 20% in going from water to 8 *M* urea (also, compare columns 2 and 3, Table III). These authors found also that tyrosine difference spectra are generally larger in 8 *M* urea than in water (see Appendix for $\Delta\varepsilon$ values in solutions of guanidine hydrochloride). Herskovits (*46*) has shown that solvent perturbation parameters are still larger in organic solvents such as chloroethanol or ethylene glycol. Parameters obtained with model compounds may also show a significant temperature dependence (*45,49*). It is thus of importance that values of $\Delta\varepsilon/\varepsilon$ for reference states be obtained under conditions comparable to those used for the protein itself.

A somewhat different method of measurement and calculation has been employed by Hamaguchi and co-workers (*50–55*) in their work on lysozyme. This procedure, which one might call the "slope method," involved measurement of the usual perturbation difference spectrum for both protein and model compound as a function of perturbant concentration. The number of groups exposed, N, was calculated as

$$N = \frac{[\Delta\varepsilon_\lambda/c]_{\text{protein}}}{[\Delta\varepsilon_\lambda/c]_{\text{model compound}}} \tag{4}$$

where $\Delta\varepsilon_\lambda/c$ represents the slope of the linear portions of plots of $\Delta\varepsilon_\lambda$ vs.

perturbant concentration, c. This procedure is essentially equivalent to comparison of the parameters $\Delta\varepsilon/\varepsilon$ as outlined above.

The reduced denatured protein may also be used as a reference for 100% exposure. Reduction of disulfide bonds may be accomplished using thioglycolic acid or by S-sulfonation according to the method of Swan (*56*). The S-sulfo protein may be considerably easier to work with since conversion of the sulfurs to S-sulfo groups is likely to yield a more soluble protein derivative than thioglycolate reduction. The thioglycolate (reduced) and S-sulfonated proteins are generally dissolved in concentrated urea or guanidine solutions to ensure absence of aggregation and to completely unfold the peptide chain. Detailed procedures for solvent perturbation measurements of reduced protein are given by Herskovits and Laskowski (*38*) and Williams et al. (*48*).

The requirement that reduced denatured protein be dissolved in high concentrations of urea necessitates a correction of the observed values of $(\Delta\varepsilon/\varepsilon)$ of the reduced denatured protein for the effect of urea on the chromophore if its value is to be used for 100% exposure of a protein in water (*48*). The correction is given by

$$\left(\frac{\Delta\varepsilon}{\varepsilon}\right)_{\text{corr}} = \left(\frac{\Delta\varepsilon}{\varepsilon}\right)_{\text{obs}} \left[\frac{(\Delta\varepsilon/\varepsilon)_{\text{model cmpd } H_2O}}{(\Delta\varepsilon/\varepsilon)_{\text{model cmpd urea}}}\right] \tag{5}$$

The magnitude of the correction can be seen from data obtained by Williams et al. (*48*) for lysozyme. The apparent degree of exposure, R_M, for reduced denatured protein was calculated using Eq. (2) for both the model compound mixture in water and in 8 *M* urea (columns 2 and 3, respectively, Table IV). The values of R_M obtained with the model compounds in 8 *M* urea are considerably closer to unity than those obtained with the water reference. The difference in R_M for the two methods of calculation is particularly striking for polyethylene glycol; in one case only 68% exposure was observed, while for the other, 98% or virtually complete exposure, was found.

Use of the reduced denatured protein as a reference for 100% exposure is very attractive since the molar ratios of tyrosine to tryptophan need not be known in such case. This may be particularly important with high molecular weight proteins where the relatively low accuracy of tryptophan analysis may lead to significant errors. Furthermore, tyrosines and tryptophans incorporated into a random peptide chain may be a more realistic reference than model compounds. On the other hand, solvent perturbation measurements with reduced denatured proteins in high concentrations of urea appear in general to be less accurate than those with model compounds. Secondly, the presence of certain peculiar sequences of amino acids in a protein may somewhat insulate tyrosines and tryptophans from the medium even in the reduced denatured state. No definitive answer seems possible at present as to which is the preferred reference state for 100% exposure. As suggested by Williams

TABLE IV

Comparison of Perturbation of Tryptophan in Model Compound Mixture with that of Reduced Denatured Lysozyme[a]

Perturbant (20%) (1)	$\Delta\varepsilon_{290-292}/\varepsilon_{280-282}$: Model compound mixture in water (2)	Model compound mixture in 8 *M* urea (3)	Reduced denatured lysozyme in 8 *M* urea (4)	R_M Reduced denatured lysozyme: Water[b] (5)	8 *M* urea[c] (6)
Sucrose	0.027	0.027	0.028	1.04	1.04
Glycerol	0.046	0.042	0.035	0.76	0.83
Ethylene glycol	0.049	0.042	0.037	0.76	0.88
Methanol	0.038	0.037	0.040	1.05	1.08
Dimethyl sulfoxide	0.088	0.081	0.075	0.85	0.93
Polyethylene glycol	0.088	0.066	0.060	0.68	0.91

[a] Data taken from Williams et al. (*48*).
[b] Calculated from data in columns 2 and 4.
[c] Calculated from data in columns 3 and 4.

et al. (*48*), it seems best at present to determine and tabulate exposures relative to both model compound mixtures and reduced denatured proteins.

The foregoing remarks concerning the use of reduced proteins or model compound mixtures pertain to proteins where the tryptophan–tyrosine ratio is either very high or very low. In such cases, perturbation spectra in the 290–294 mμ region will be largely dominated by tryptophan absorption and those in the 282–286 mμ region by tyrosine absorption. For intermediate tyrosine to tryptophan ratios, the tryptophans may make sufficiently large spectral contributions near 284 mμ to make an accurate determination of tyrosine exposure impossible. A high tyrosine-to-tryptophan ratio combined with a particularly low degree of exposure of tryptophans may likewise make it impossible to accurately determine the degree of exposure of the latter group. Alternately, calculations could be made using the following:

$$(\Delta\varepsilon_{282-286})_{\text{obs}} = n(\Delta\varepsilon_{282-286})_{\text{tyr}} + m(\Delta\varepsilon_{282-286})_{\text{tpt}} \tag{6}$$

$$(\Delta\varepsilon_{290-294})_{\text{obs}} = n(\Delta\varepsilon_{290-294})_{\text{tyr}} + m(\Delta\varepsilon_{290-294})_{\text{tpt}} \tag{7}$$

where n and m are the number of exposed tyrosines and tryptophans and the four extinction coefficients on the right-hand side of the equations are the corresponding intrinsic perturbation extinction coefficients for the tyrosine and tryptophan. Simultaneous solution of these equations would yield *n* and *m*. The procedure is less satisfactory than that outlined below since there may be uncertainty as to the appropriate wavelength of observation for the proteins because of poor resolution of the perturbation spectra and because the perturbation spectra of protein will be shifted relative to those of model compounds (see above).

A more satisfactory procedure involves a comparison of the complete experimental perturbation spectrum for the protein with those calculated for varying degrees of exposure of tyrosine and tryptophan from the perturbation spectra of the individual amino acid model compounds. Results obtained with yeast glyceraldehyde 3-phosphate dehydrogenase are given in Fig. 5 (*45*). The

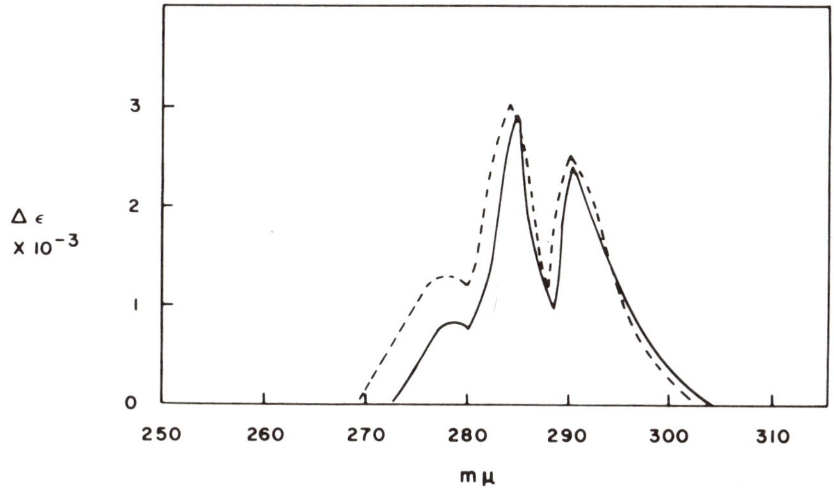

Fig. 5. Solvent perturbation spectrum of native glyceraldehyde-3-phosphate dehydrogenase. Temp., 0°C. Perturbant, 20% sucrose. ———— experimental spectrum. - - - - - - - - - -, spectrum calculated on the assumption of 20 exposed tyrosyl and 10 exposed tryptophyl groups [from Warrington and Kronman (*45*)].

dotted curve was calculated from perturbation spectra obtained in 20% sucrose with the N-acetyl ethyl esters of tyrosine and tryptophan. The molar difference extinction coefficients used in calculation of such curves are tabulated in the Appendix. [Similar numerical data have been tabulated recently by Herskovits and Sorenson (*38b*) for a somewhat larger group of perturbants in water and 8 *M* urea, however, only at 25°.] The experimental curve for the enzyme is given as a solid line. The agreement of the calculated and experimental curves is quite good and indicates that 10 of the 12 tryptophans and 30 of the 38 tyrosines of glyceraldehyde 3-phosphate dehydrogenase are exposed with respect to sucrose.

4. Properties of Perturbants

Substances used as perturbants should have certain obvious characteristics. They should be transparent in the wavelength region of interest (ca., 270–300 mμ), be sufficiently miscible with water to obtain relatively high concentrations, and be capable of producing relatively large spectral shifts. Most important, they should not induce structural changes in the protein at the perturbant concentration levels used. As we shall see, however, the magnitude of the spectral shift obtained with a perturbant is not unrelated to its potentially disruptive character.

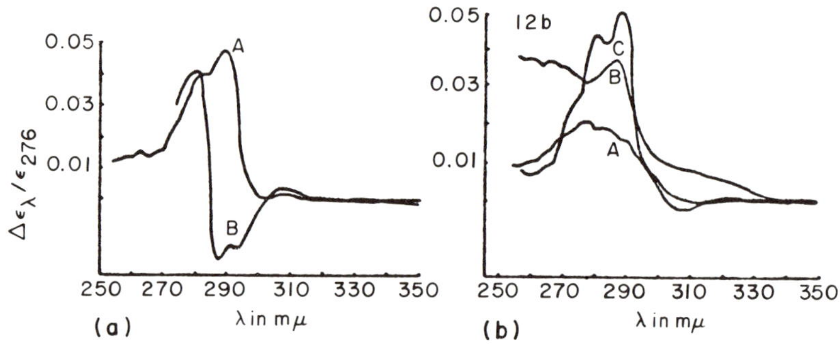

Fig. 6. Distortion of solvent perturbation difference spectra by the denaturing action of some solvents. (a) Difference spectra showing the cooperative denaturing effects of acid and 20% formamide. Curve A, difference spectrum obtained between BSA in water and in 20% formamide, pH 4.0, $\Gamma/2 = 0.25$. Curve B, difference spectrum obtained between bovine serum albumin (BSA) in acid and in 20% formamide, pH 1.5, $\Gamma/2 = 0.25$. (b) Distortion of and time effects on the difference spectra of BSA obtained with 20% methanol. Curve A, difference spectra obtained between BSA in water and in 20% methanol approximately 10 min after the addition of methanol, pH 1.5 $\Gamma/2 = 0.25$; curve B, the same solution as in curve A after 3 hr of standing at 25°; curve C, thioglycolic acid–reduced BSA in 8 *M* urea, pH 3.6, $\Gamma/2 = 0.2$. Perturbant, 20% methanol; protein concentration, 0.6%; temperature, 25 ± 1° [from Herskovits and Laskowski (*38*)].

Any method to be used for group counting must be evaluated in terms of its potential for *inducing* conformational changes in the protein. When critically examined, solvent perturbation spectra themselves may provide a clue as to the possible deleterious effect of a perturbant. When the perturbant drastically alters the protein structure, the corresponding spectrum will be radically distorted as illustrated in Fig. 6(a). Here, the large blue shift resulting from the denaturation of bovine serum albumin by 20% formamide cancels the perturbation red shift. Less extreme spectral distortion can also be used as a means of detecting induced conformational changes, as can time effects [Fig. 6(b)].

A further test of the "innocuous" character of a perturbant additive would be a linear relationship between $\Delta\varepsilon/\varepsilon$ and perturbant concentration. Such a check is inherent in the "slope method" of calculating exposure (see Section II. B. 3). Other pertinent tests would involve comparison of physical properties such as acid difference spectra or optical rotation dispersion of proteins in the presence and absence of additives. The perturbants that appear to be "safest" are the polyhydric compounds such as the sugars, glycerol, and ethylene glycol. Even with these compounds, however, one can cite instances

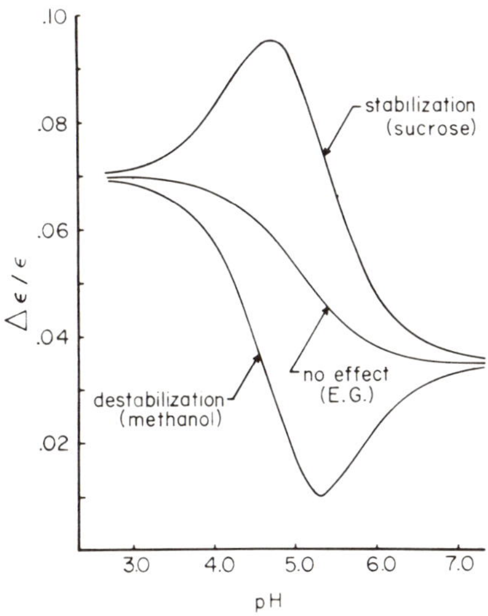

Fig. 7. Effect of a perturbant on a hypothetical protein undergoing a pH-dependent conformational change [from Laskowski (*59*)].

of induced molecular alteration. Herskovits and Laskowski (*57*), for example, have found that the perturbation spectra of ovomucoid are markedly distorted by 20% sucrose, and Green (*58*) has reported a similar effect of ethylene glycol on the spectrum of avidin. It is thus vital that each protein–perturbant system be critically examined to ensure the absence of induced conformational changes.

The effect of a perturbant in the transition region of a protein undergoing a conformational change is particularly noteworthy. Here, even so-called innocuous solvents may yield anomolous results. In such a transition region, where differences in the standard free energy for native and isomerized proteins are small, preferential interaction of the perturbant with one of the conformational isomers may shift the equilibrium in one direction or another. Such a situation is illustrated in Fig. 7, taken from a paper by Laskowski (*59*).

These curves were calculated for a system undergoing a pH-dependent transition resulting in 50% exposure of the chromophores. For a perturbant stabilizing the native protein, $\Delta\varepsilon/\varepsilon$ goes through a maximum near the midpoint of the transition. Since exposure of groups results in a "blue" shift, stabilization of the native state would result in a "red" shift which would add to the perturbation red shift, yielding a maximum value.

The minimum seen for a destabilizing perturbant can be accounted for in much the same way. Outside of the transition region, however, values of $\Delta\varepsilon/\varepsilon$ for nonstabilizing, stabilizing, and neutral perturbants converge. The above considerations apply to a conformational transition involving a spectral

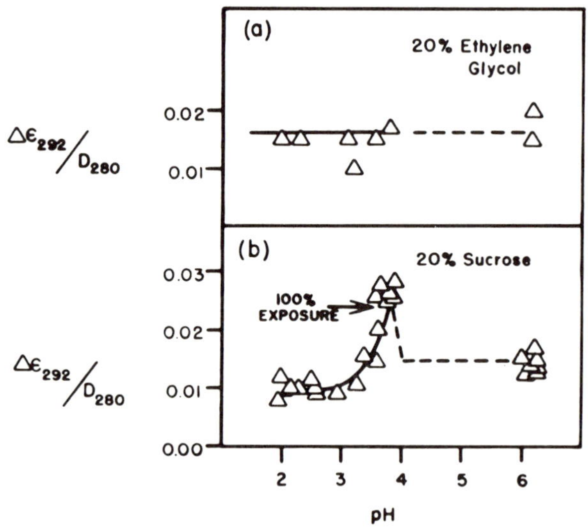

Fig. 8. Effect of perturbant on the parameter $\Delta\varepsilon_{293}/\varepsilon_{280}$ in the conformational transition region of α-lactalbumin [from Kronman and Holmes (*49*, *61*)].

change whether such a change results in exposure of groups or not. The effect of perturbant on the acid denaturation transition of α-lactalbumin is a good illustration of the latter. Between pH 3 and 4, this protein undergoes a blue shift of the ultraviolet absorption spectrum implicating tryptophan groups in the structural change (*60*) without, however, any increase in the exposure of these groups (*49*). Shown in Fig. 8 are the pH dependencies of $\Delta\varepsilon_{292}/\varepsilon_{280}$ for perturbation of α-lactalbumin by 20% ethylene glycol and 20% sucrose. In the case of sucrose, this parameter increases by almost a factor of 3 in the transition region (pH 3 to 4). Note particularly that the observed $\Delta\varepsilon_{293}/\varepsilon_{280}$ may be *larger* than that anticipated for complete exposure of tryptophans. By contrast, perturbation by ethylene glycol is relatively independent of pH over a wide range. It is quite apparent from these observations and the remarks of the preceding paragraph that sucrose stabilizes the native conformation of α-lactalbumin and that ethylene glycol is neutral in its effect. Measurements of acid difference spectra in the presence of perturbants support this contention (*49*). Anomalous behavior was also observed in the transition region for α-lactalbumin using 90% D_2O as perturbant. In this case, the usual *blue* perturbation shift (Table III) becomes inverted to yield a red shift, suggesting stabilization of the native form by D_2O. This conclusion would be in accord with the observations of other investigators (*62–65*) who found increased thermostability of ribonuclease, collagen, and a number of other proteins in D_2O as compared to H_2O.

If a particular chromophore in a protein is completely exposed such as tyrosine C in Fig. 1, use of perturbants of varying size would provide no information beyond that obtained with a single perturbant. If, however, a chromophore such as tyrosine B in Fig. 2 were to lie in a "crevice-like" fissure in the molecular surface, use of perturbants of different size may provide a means of estimating the critical dimension of the crevice. One assumes that a small perturbant in contact with a chromophore will give a perturbation spectral shift while none will result with a perturbant whose bulk will not permit contact with the chromophore. For lack of a better parameter one uses as a measure of perturbant size, the "effective diameter" of the molecule calculated from diffusion constants (*38*). Such an effective diameter, of course, does not allow for any specific mutual orientation of chromophore and perturbant molecule that might be required to give the spectral shift. This point is considered subsequently when we examine "range" effects. Shown in column 2 of Table III are the effective molecular diameters of commonly used perturbants, ranging from 2 Å (heavy water) to 9.4 Å (sucrose). Heavy water is a particularly useful perturbant because of its small size.

The utility of perturbants of different sizes is illustrated by studies with α-lactalbumin (*49*). At 25°, at pH 6, both sucrose (effective diameter, 9.4 Å) and heavy water (effective diameter, 2 Å) give perturbation shifts corresponding to two exposed tryptophans. These two groups must either lie in crevices having a critical dimension larger than 9.4 Å or must be freely accessible to the medium. As the temperature is lowered to 1°C, the sucrose and glycerol perturbation spectra for α-lactalbumin indicate zero exposure while the D_2O spectra still indicate two groups. The use of perturbants of different sizes in this system led to the conclusion that at low temperature, two of the tryptophans of α-lactalbumin lie in crevices having a critical dimension smaller than 5.2 Å (the effective diameter of glycerol) but larger than 2 Å. The characteristics of this conformational change are considered at greater length in a subsequent section. A similar phenomenon appears to occur with ovomucoid (*57*) where crevice contraction likewise does not exclude heavy water molecules from contact with the chromophores (*66*).

We have assumed heretofore in this discussion that perturbant-induced spectral shifts are the result of changes in nonspecific solvent properties such as refractive index and dielectric constant (see, for example, Ref. *35*). It can be shown, however, that in the case of some perturbant-additives shifts include spectral contributions owing to formation of molecular complexes. Such perturbants are generally those whose intrinsic spectral shifts are larger than might have been anticipated from the magnitude of their contribution

TABLE V

Effects of Various Perturbants on Difference Spectral Constants of N-acetyl-L-tyrosine ethyl ester and N-acetyl-O-methyl-L-tyrosine ethyl ester

Perturbant[a]	$\Delta\varepsilon_{286}$ mμ/ε_{275} mμ[b] R—  —OH	$\Delta\varepsilon_{284}$ mμ/ε_{274} mμ R— 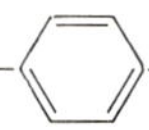 —O—CH_2	Ratio
20% Sucrose	0.040	0.041	0.98
20% Glycerol	0.060	0.051	1.17
20% Ethylene glycol	0.072	0.062	1.16
20% Polyethylene glycol	0.135	0.097	1.39
20% Dimethylsulfoxide	0.158	0.118	1.34
20% Methanol	0.056	0.035	1.60
20% Formamide	0.102	0.081	1.26
20% Dimethylformamide	0.152	0.103	1.48
20% Acetonitrile	0.117	0.094	1.24
5 *M* Potassium acetate[c]	0.195	0.083	2.35
5 *M* Acetic acid[d]	0.072	0.071	1.02

[a] The pH of the solutions was 5.4 to 6.1.
[b] With 1.5×10^{-3} *M* N-acetyl-L-tyrosine ethyl ester and 1.5×10^{-3} M N-acetyl-O-methyl-L-tyrosine ethyl ester.
[c] At pH 7.6, acetate solution contains 1% acetic acid.
[d] At pH 1.8.
[e] Taken from Ref. *38*.

to the refractive index of the medium. For example, 20% methanol and glycerol give roughly comparable perturbation shifts with tyrosine and tryptophan esters (Table III). The change in refractive index on addition of methanol to water, however, is only 5% of that of glycerol at comparable concentrations. It seems probable, therefore, that the perturbation shift observed with methanol must derive from processes other than nonspecific solvent effects. Similar conclusions can be drawn for the perturbants dimethylsulfoxide and polyethylene glycol (see Table III).

It is convenient to speak of perturbants as having "short-" and "long-" range effects. The former refers to perturbants that form molecular complexes with a chromophore and that thus must intimately contact a portion of the chromophore in order to achieve its intrinsic perturbation shift. Long-range perturbants conversely achieve their function through nonspecific solvent effects and hence should be effective at greater distances. Herskovits and Laskowski (*38*) have suggested a diagnostic criterion for distinguishing short- and long-range perturbants and have applied this directly to tyrosine as shown by the information in Table V taken from their studies. Values of $\Delta\varepsilon_{287}/\varepsilon_{278}$ are compared for a variety of perturbants for N-acetyl-L-tyrosine ethyl ester and N-acetyl-O-methyl-L-tyrosine ethyl ester. The decrease in $\Delta\varepsilon_{287}/\varepsilon_{278}$

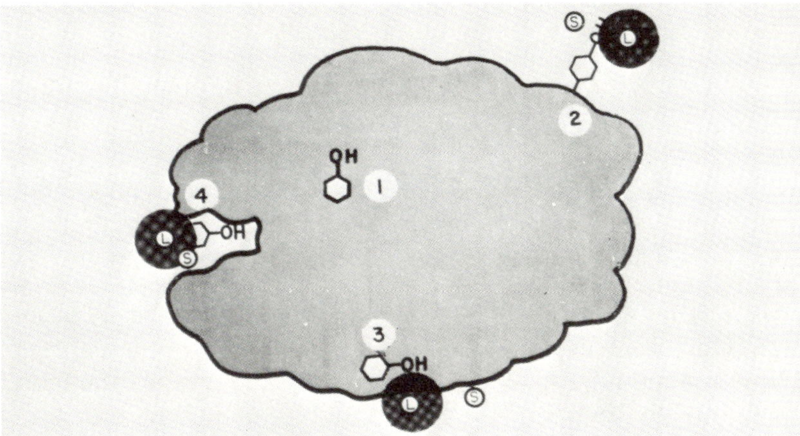

Fig. 9. Schematic view of a protein molecule illustrating the effect of short-range (Ⓢ) and long-range (Ⓛ) perturbants. The shading around the latter symbols designates the extended area over which these perturbants act (see text).

observed with some of the perturbants when the phenolic hydrogen is eliminated is indicative of hydrogen bonding between the tyrosine moiety and the perturbant. Large ratios are thus characteristic of short-range perturbants while long-range perturbants should have ratios closer to unity. On this basis we would tentatively classify dimethyl sulfoxide, polyethylene glycol, and methanol as short range, and glycerol and sucrose as long-range perturbants. Ethylene glycol appears to have an intermediate character. No evaluation of the range effect of perturbants has been made as yet with respect to tryptophan groups. It would appear from the relative magnitudes of $\Delta\varepsilon_{293}/\varepsilon_{278}$ for tryptophan (see Table III) that these perturbants have similar short-range effects as those exhibited with respect to tyrosyl groups.

Perturbants having different range effects might be expected to yield somewhat different apparent degrees of exposure for protein molecules having partially buried groups. This is illustrated in Fig. 9 for a hypothetical protein molecule containing four tyrosines. Ⓢ and Ⓛ represent perturbants of comparable molecular size but having short- and long-range effects, respectively. The shaded area around Ⓛ represents the range of effect of this perturbant. Tyrosine 1, which is completely buried, will not be seen by either short- or long-range perturbants, while tyrosine 2, which is completely exposed, will be seen by both. Tyrosine 3 which is depicted as lying just below the molecular surface, will experience a partial perturbation by the long-range solvent but none by the short-range perturbant since the requisite molecular complex cannot be formed. How deep must a chromophore be buried before it no longer experiences the effect of a long-range perturbant? Although no definitive answer can be given, it seems likely that this distance is of the order of 5 to 10 Å (*59*).

A short-range perturbant in the above case would give a lower apparent degree of exposure than the long-range one. Tyrosine 4, which lies in a crevice sufficiently large to admit both Ⓛ and Ⓢ, will experience nearly the full perturbation of the long-range solvent but for steric reasons might appear to be buried with respect to the short-range one.

As indicated above, short-range perturbants generally produce correspondingly large intrinsic spectral shifts, clearly a very desirable characteristic for a perturbant. Unfortunately, however, it is just those solvents which have the greatest potentiality for disruption of the protein molecule, since the mechanism by which such large shifts are produced may involve formation of new hydrogen bonds. If, for example, a chromophore were hydrogen bonded in the native protein structure and such bonding contributed to the stability of the molecule, the perturbation process *per se* would alter the structure. Thus, substances such as dimethyl-formamide and urea (*50,54*), while giving large intrinsic spectral shifts, would be dubious choices as perturbants.

5. Limitations of Solvent Perturbation

Solvent perturbation measurements cannot, as indicated earlier, be made in transition regions. The rather large red spectral shift occurring on tyrosine ionization (see Section II. C) would also restrict the measurements to pH values below 9 to 9.5. At higher pH, the relative absorbances of ionized tyrosine and tryptophan groups would preclude accurate group counting for the latter residues. Other than these restrictions, solvent perturbation probably has the widest pH range of application of any of the techniques used in determining the exposure of tyrosines and tryptophans.

A rather serious limitation on the applicability of the solvent perturbation technique in determining tyrosine exposure exists for proteins having high tryptophan-to-tyrosine molar ratios. In such cases, calculation of the degree of exposure from Eq. (2) using either a model compound mixture or a tyrosine standard alone may lead to large errors.

For a protein having a tyrosine-to-tryptophan molar ratio of 10:1, one can show that the measured value of R_{tyr} will have an uncertainty of about $\pm 20\%$, roughly comparable to experimental uncertainty. Thus, for proteins such as bovine serum albumin (molar ratio, 10.5:1) or ovomucoid (molar ratio, 12.4:1), the amplitude of the 282–286 mμ peak can be compared directly with that of a model compound mixture in order to determine R_{tyr}. On the other hand, for a tyrosine:tryptophan molar ratio of 2, an uncertainty of about $\pm 50\%$ in R_{tyr} will exist. Clearly, little information can be obtained about tyrosine exposure in proteins having low tryptophan:tyrosine molar ratios such as chymotrypsin (molar ratio, 0.6:1) or α-lactalbumin (molar ratio, 1:1) using the 282–286 mμ peak alone. Better estimates of tyrosine

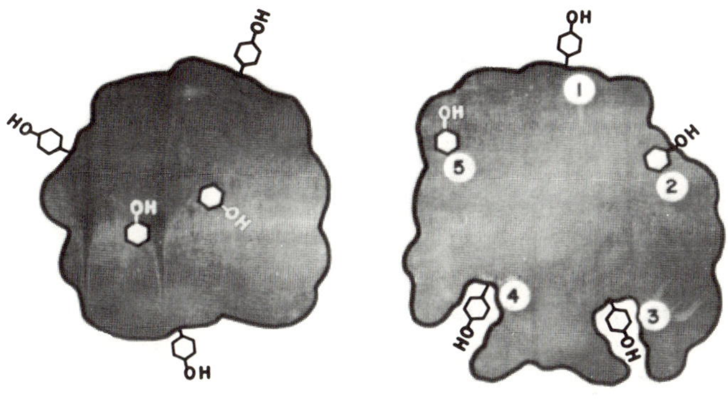

(a) (b)

Fig. 10. Schematic view of a protein molecule illustrating that the *apparent* exposure of three groups may be the result of complete exposure of three groups to perturbant (a) or exposure to varying degrees of more than three groups (b) (see text).

exposure can perhaps be made using the complete difference spectrum (see previous section) but even this method will yield relatively imprecise values due to the high absorbance of tryptophan as compared to tyrosine.

A more serious difficulty arises from the fact that solvent perturbation cannot distinguish a few completely exposed groups from a larger number of partially exposed ones. This is illustrated schematically in Fig. 10(a) where three of the five chromophores are depicted as completely accessible to solvent, while two are completely buried. In Fig. 10(b) four of the five groups are shown as partially exposed, two in crevices of varying size, and two either near the molecular surface or projecting somewhat through it. Use of a single perturbant could not distinguish between the two situations represented in

Figs. 10(a) and (b). Use of a variety of perturbants of different sizes and range effects (see previous section) may, however, in some cases permit a differentiation. The molecule represented in Fig. 10(a) should give essentially identical degrees of exposure with perturbants having different sizes and range effects, since the exposed chromophores are freely accessible to perturbants while the other two are quite well insulated from the medium. The molecule depicted in Fig. 10(b), on the other hand, should exhibit varying degrees of exposure with respect to perturbants. Chromophores 2 and 5 for example, behave differently from chromophore 1 with respect to short- and long-range perturbants. Similarly, chromophores 3 and 4 should exhibit differences in apparent exposure with respect to perturbants of different size. Native α-lactalbumin at 25° appears to correspond to the first case, i.e., distinct classes of buried and exposed groups (*49*). The variation in the apparent degree of exposure of tyrosines and tryptophans in L-glutamate dehydrogenase (*67*) with perturbants suggests that some groups are partially exposed for this enzyme. A similar conclusion might be made for lysozyme and α-chymotrypsin which likewise show variations in the apparent degree of exposure to various perturbants (*48*).

Williams and Laskowski (*68*) have proposed a method to resolve this ambiguity by specifically modifying the chromophore with a reagent which essentially abolishes its contribution to the solvent perturbation spectrum. If reaction with a few groups were to completely abolish the difference spectrum, it would indicate that these had been completely exposed. Partial exposure of a larger number of these groups would imply a gradual decrease in the amplitude of the difference spectrum as a larger number of groups were modified. The reagent employed was N-bromosuccinimide (NBS) (see also Section IV. C). In the case of chymotrypsinogen, Williams and Laskowski observed that oxidation of three tryptophans essentially eliminated the perturbation difference spectrum with no further change being evident upon reaction with the other four. From these observations, the authors concluded that chymotrypsinogen has three completely exposed tryptophans and four completely buried ones. On the other hand, with lysozyme, NBS oxidation of all six tryptophans was required to eliminate the perturbation spectrum, a more rapid decrease being evident in earlier phases of the oxidation. Williams and Laskowski concluded from these observations that all six tryptophans in lysozyme are partially exposed with an appreciable gradation in their exposure.

While we reserve consideration of the latter conclusion until later, some remarks should be made concerning the validity of the use of NBS oxidation in the manner indicated above. Oxidation of lysozyme by N-bromosuccinimide has been shown to involve attack of tyrosines and histidines in addition to oxidation of tryptophan (*69*) (see Section IV. C). These procedures are accompanied by conformational alterations. Furthermore, the observed spectral changes do not quantitatively reflect tryptophan destruction. It is thus questionable that NBS is the ideal reagent for use with solvent perturbation in the manner suggested by Williams and Laskowski.

6. Application of Solvent Perturbation

a. *Effect of Disulfide Bridges on Tyrosine and Tryptophan Exposure*

One of the more interesting differences between what solvent perturbation and other techniques "see" is illustrated by data obtained with proteins denatured with urea. Spectral titrations and chemical techniques generally

TABLE VI

Apparent Exposure of Tyrosines in Reduced and Denatured Proteins[a]

	Ribonuclease R_M[b]		Ovomucoid R_M[d]		Serum albumin R_M	
Perturbant[a]	Reduced	Denatured	Reduced	Denatured	Reduced	Denatured
Sucrose	0.88	—	1.00	—	—	—
Glycerol	0.87	—	1.10	0.95	1.05[c]	0.98[c]
Ethylene glycol	0.83	—	1.07	—	—	—
Methanol	0.86	—	1.29	—	—	—
Polyethylene glycol	0.82	—	0.74	—	—	—
Dimethyl sulfoxide	0.88	0.68	1.07	0.65	0.86[b]	0.60–0.68[b]

[a] Values calculated using Eq. (2).
[b] Reference: Model compound mixtures in 8 *M* urea (*46*).
[c] Reference: Model compound mixtures in 8 *M* urea (*57*).
[d] Reference: Model compound mixtures in aqueous medium (*38*).

indicate complete accessibility of tyrosine and tryptophan residues in the unfolded protein molecule (see Tables X, XIII, and XV). Solvent perturbation, by contrast, generally shows incomplete exposure of such groups in this state, particularly if the disulfide bridges remain intact. This conclusion is illustrated by the data summarized in Tables VI and VII for proteins denatured in 8 *M* urea, with and without reduction of disulfide bonds. In general, where the disulfide bonds were intact, apparent exposure ranged from 60 to 80% (Tables VI and VII).

TABLE VII

Apparent Exposure of Tryptophans in Reduced and Denatured Proteins[a]

Perturbant 20%	Lysozyme[b] R_M		α-Casein[c] R_M		α-Chymotrypsinogen[b] R_M		α-Lactalbumin[d] R_M	
	Reduced	Denatured	Reduced	Denatured	Reduced	Denatured	Reduced	Denatured
Sucrose	1.05	—		1.03	1.11	0.93	0.86	0.79
Glycerol	0.83	0.64		0.84	0.83	0.80	1.06	0.73
Ethylene glycol	0.88	0.67	No	0.89	0.81	0.83	1.08	0.88
Methanol	1.08	0.73	disulfide	0.91	—	—	—	—
Polyethylene glycol	0.91	0.62	bridges	0.81	0.72	0.77	—	—
Dimethyl sulfoxide	0.93	0.69		0.90	0.78	0.72	—	—

[a] Values of R_M calculated with respect to model compound mixtures in 8 *M* urea, using Eq. (2).
[b] Reference *48*.
[c] Reference *46*.
[d] Reference *49*.

With the exception of chymotrypsinogen, reduction of disulfide bridges resulted in increases in the apparent exposure of the order of 25%. The disulfide bridges themselves, or the bulky groups adjacent to the bridges, prevent free access of the perturbant to the chromophores. In some proteins, however, it is apparent that the disulfide bridges *per se* are not responsible for the incomplete exposure. For example, Herskovits (*46*) has found for β-lactoglobulin denatured in 2-chloroethanol or acidic methanol and for insulin denatured in 2-chloroethanol, acidic methanol, and acidic ethylene glycol, that R_M values (perturbant, dimethyl sulfoxide) are larger than those observed on denaturation in 8 *M* urea. Apparently, in these organic solvents, molecular conformations are possible which permit greater access to the chromophores in spite of the restrictions imposed by the disulfide bridges.

In the case of ovomucoid (Table VI), glycerol "sees" virtually all of the tyrosines irrespective of whether the disulfide bridges are intact, while for the less bulky perturbant, dimethyl sulfoxide, 35% of them remain buried after urea denaturation. The difference between glycerol and dimethyl sulfoxide is the result of the range effect of the two perturbants. Thus, with respect to dimethyl sulfoxide, solvent access to the tyrosyl groups appears to be somewhat restricted after urea denaturation.

Some tentative relationships can be pointed out for chymotrypsin, ribonuclease, and lysozyme for which amino acid sequences have been established. The latter two proteins show little change in R_M on denaturation in urea and in various organic solvents; indeed, in the case of lysozyme, R_M for tryptophan exposure is changed very little in 8 *M* urea compared to the native protein (*46*). Both of these proteins, however, show significant increases on cleavage of the disulfide bridges (Tables VI and VII). In the case of ribonuclease, three of the six tyrosyl side chains are in close proximity to cystines (*70*) while for lysozyme, a similar number of tryptophans are so situated (*71*).

Chymotrypsinogen is interesting in that denaturation in 8 *M* urea apparently results in as complete exposure of the tryptophans as is seen after reduction of disulfide bridges (Table VII). This observation is to be compared with the observations of Hartley (*72*) and Keil et al. (*73*) who found that the disulfide bridges are far removed from the tryptophyl groups.

Values of R_M obtained with reduced proteins in concentrated urea are in some instances somewhat less than unity (Tables VI and VII) even when the correct reference state is utilized. These observations suggest that even when the disulfide bridges are cleaved, exposure is incomplete, either because of some residual secondary and tertiary structure or because of shielding by bulky side chains adjacent to chromophores. It is difficult to say, however, how much significance should be ascribed to the small deviations of R_M from unity (ca., 0.1 to 0.15), particularly in view of the lack of correspondence of the perturbation spectra for reduced proteins and model compounds.

b. *Specific Proteins*

The Summary in Table VIII of the application of solvent perturbation to specific protein systems gives some idea of the scope of the use that has been made of this technique. The high points of some of these will be considered in this section.

Paramyosin is at present the only nonglobular protein which has been examined by the solvent perturbation technique. The maximum value of R_M for tyrosyl groups has been found by Riddiford (*47*) to be 0.64 (20% propylene glycol). This apparent burying of about 14 of the 38 tyrosyl residues (molecular weight, 220,000) is to be contrasted with the unequivocal completeness of ionization of tyrosyl groups at alkaline pH (*74*). The insulation of 14 tyrosines in paramyosin is somewhat puzzling in view of the asymmetric molecular shape of this protein. The 220,000 molecular weight unit has a length of 1330 Å (*75*); the molecular diameter has been inferred to be 20 Å (*75,76*). It is difficult to see how such a small internal space could accommodate such a relatively large side chain as the tyrosyl group. Secondly, it raises the question as to how deeply such a chromophore must be buried before it is unaffected by the external solvent environment. As we have indicated earlier, short-range perturbants require rather intimate contact with the chromophore. The incompleteness of exposure seen with paramyosin, however, was determined with relatively long-range perturbants (sucrose, glycerol, ethylene glycol, etc.) which might be expected to perturb tyrosines just below the molecular surface. Laskowski (*59*) has suggested that chromophores which lie 5 to 10 Å below the molecular surface should not be perturbed at all. It is thus rather difficult to imagine that a relatively bulky tyrosyl group could be accomodated into the 20 Å diameter paramyosin molecule and not be affected by the solvent environment. A more plausible but unproven explanation is that the observed values of R_M of 0.64 or less are the result of partial burying of considerably more than 14 tyrosines and that there are no completely buried groups. The fact that incomplete exposure of tyrosyl groups is observed even in urea-denatured proteins suggests that such shielding is possible. As Riddiford (*47*) has pointed out, the relative hydrophobic character of tyrosyl groups would lead one to expect them to be folded against the helical backbone. She has suggested that since the advance distance of one turn of the helix is of the same order of magnitude as the size of excluded solvents (ethylene glycol, 4.4 Å), the "crevices" may be identified with the advance distance. If this were indeed the case, one would anticipate that perturbation with heavy water would lead to values of R_M considerably closer to unity. These experiments, however, have yet to be done.

Weil et al. (*81*) have found that R_M for zinc insulin is lower than that of the zinc free hormone (0.46 and 0.67, respectively). They suggest that the apparent burying of tyrosyl groups is due to the conformational change on

TABLE VIII

Summary of Applications of Solvent Perturbation

Protein	Observations	Reference
Aldolase	Perturbation of the native and acid-denatured enzyme with ethylene glycol. Spectra are best fit on the assumption of exposure of one tryptophan and 14 tyrosines and 10 tryptophans and 42 tyrosines in the native and acid-denatured states, respectively. The latter correspond to the analytical values.	*37*
Avidin	Ethylene glycol perturbation of the avidin–biotin complex indicates virtually complete burying of the 14 tryptophans.	*58*
Bovine growth hormone	Exposure of tyrosine residues on urea denaturation is pH dependent.	*77*
Bovine and human serum albumin	The acid conformational change occurring at low pH involves an exposure of tyrosyl groups. Detailed description of solvent perturbation procedure.	*38*
α-Chymotrypsin	Perturbation of the cinnamoyl group indicates that burying of this chromophore occurs on formation of the enzyme substrate compound *trans*-cinnamoyl chymotrypsin.	*78*
Cytochrome c	Perturbation with 20% ethylene glycol showed all tryptophan groups and two of four tyrosines buried. Heme only partially exposed to medium.	*79*
Glutamic dehydrogenase	Tryptophan, tyrosine, and phenylalanine groups are buried to the extent of 58, 22, and 75% in the native state. The "crevice" sizes housing these residues are estimated.	*67*
Glyceraldehyde 3 phosphate dehydrogenase	Acid inactivation involves increased exposure of chromophoric residues.	*80*
Insulin	Zn-free insulin has one less buried tyrosine than Zn-insulin. Photo-oxidation has no effect on tyrosine exposure.	*81*
α-Lactalbumin, β-lactoglobulin, casein, ribonuclease, lysozyme, insulin, bovine serum albumin	The degree of exposure of tyrosines and tryptophans in aqueous medium are compared with those in 2-chloroethanol, acidic methanol, ethylene glycol, and 8 *M* urea. The differences are considered in terms of the ability of these proteins to unfold.	*46*

TABLE VIII (Continued)

Protein	Observations	Reference
α-Lactalbumin	Perturbants as large as sucrose or as small as heavy water reveal that in both native and acid denatured protein two out of the four tryptophan groups are buried. At pH 6 at 0° to 2°, two tryptophans are no longer accessible to large perturbants but are exposed with respect to heavy water.	*49*
Lysozyme, chymotrypsinogen	A combination of solvent perturbation and oxidation of tryptophans with N-bromosuccinimide was used to detect partially buried groups. Application to chymotrypsinogen showed that this protein has three completely exposed and four completely buried groups. All six tryptophans of lysozyme appear to be partially exposed. (See Section IV for a critique of use of NBS.)	*68*
Lysozyme, chymotrypsin, diisopropyl chymotrypsin, chymotrypsinogen	Exposure of tryptophans in lysozyme is independent of pH over the range 1.8 to 7 and is found to be 54–79%. Exposure of tryptophans in chymotrypsin and diisopropyl chymotrypsin are essentially identical (30–47%). Activation of the zymogen does not result in major changes in tryptophan exposure, although observations with short-range perturbants indicate a certain restriction of solvent access to tryptophans of chymotrypsin relative to chymotrypsinogen.	*48*
Lysozyme	Formation of the enzyme–substrate (glycol chiten) complex involves a burying of one to two tryptophan groups with respect to perturbation by polyethylene glycol and sucrose.	*82*
Lysozyme, ovomucoid, and bovine serum albumin	Perturbation of 1-dimethylamino-5-sulfonyl-naphthalene (DNS) conjugates of proteins. For lysozyme and ovomucoid the coupled DNS groups lie essentially on the surface (74 to 100% exposure) while for serum albumin the groups are more buried (32 to 50% exposure).	*83*
Lysozyme	Perturbation with dimethyl sulfoxide, dimethyl formamide, 2-chloroethanol, ethanol, dioxane, lithium chloride, lithium bromide, ethylene glycol, and sucrose indicate that about 67% of the tryptophans are exposed.	*50–55a*
Met myoglobin	Both tryptophyl groups appear to be partially exposed. NBS oxidation technique used.	*84*

TABLE VIII (Continued)

Protein	Observations	Reference
Ovomucoid	A limited conformational change occurs below pH 5 which involves a *decrease* in exposure with respect to larger perturbants. The low pH conformational change does not involve changes in exposure with respect to heavy water.	*57, 66*
Paramyosin	Exposure of 44–64% of tyrosines in the native state.	*47*
Pepsinogen	The degree of exposure of chromophoric residues on urea denaturation is pH dependent.	*85*
Ribonuclease	Tyrosine groups are about 50% exposed in the native state. Acid denaturation exposes an additional 30%.	*86*
Serum albumin	Binding of sodium dodecyl sulfate at pH 6.8 produces a conformational change resulting in increased exposure of tyrosyl groups.	*87*
Trypsin	Formation of the trypsin–pancreatic trypsin inhibitor complex involves only small changes in the exposure of chromophoric residues.	*88*
Pepsin	Two to three of the five tryptophan and 10–12 of the 17 tyrosyl groups exposed.	*38c*
Aldolase	Exposure of tyrosyl and tryptophyl groups with respect to ethylene glycol, glycerol, and polyethylene glycol in essential accord with observations of Donovan (*37*).	*38c*
Bovine serum albumin	One to two of the two tryptophans exposed in native and acid forms. Tyrosyl exposure in essential agreement with earlier study (*38*).	*38c*
Insulin	The degree of exposure of tyrosyl groups in Zn-free insulin is essentially uneffected by dimerization.	*81a*
β-Lactoglobulin	Tryptophan moieties of the bovine protein are located in portions of the molecule which have little to do with their association reactions or conformational transitions. About half of the tryptophans are exposed between pH 2 and 7.	*121b*
Ribonuclease	Measurements with the native protein with a number of perturbants are in good agreement with a model having two fully exposed tyrosines, two buried ones, and two partially exposed groups. Ribonuclease that has been acid- or urea-denatured can be refolded on anion binding to bury previously exposed tyrosyl groups.	*88a*

formation of the zinc complex. Although the degree of monomer association differs for the zinc and zinc-free proteins (*89–92*), the observed differences in exposure do not appear to depend upon the extent of dimer formation (*81a*).

Cross and Fisher's (*67*) observations with L-glutamate dehydrogenase (GDH) are a good illustration of the utility of heavy water as a perturbant. Sucrose, glucose, glycerol, and ethylene glycol gave values of R_M ranging from 0.24 to 0.36 for tyrosine and 0.36 and 0.44 for tryptophan. Perturbation by heavy water yielded values of 0.44 and 0.78, respectively. Thus, of the 353 tyrosine residues in GDH, about 200 are completely accessible, 28 lie in crevices having a critical dimension of 2.2 to 4.3 Å, and the remainder are exposed. Similarly, of the 54 tryptophans, about 12 are completely buried, 18 lie in crevices, and the remainder are exposed. The involvement of tryptophan in the active site of the enzyme (*93*) makes these groups of particular interest. The authors also found that R_M for phenylalanine ranges from 0.1 to 0.2. In view of the small contribution of this small side chain to the spectrum, however, the latter figure should probably be viewed with some caution. The authors conclude that since no spectral changes occur when the α form of GDH dissociates into three subunits, the exposed tyrosyl, tryptophyl, and phenylalanine groups cannot lie on the faces of the β subunits forming the α GDH molecule.

The trypsin inhibitor ovomucoid is unique in that the conformational change occurring below its isoelectric point involves burying of tyrosyl groups rather than exposure as is observed with serum albumin and ribonuclease. Herskovits and Laskowski (*57*) found that below pH 3.5, exposure of tyrosines with respect to five perturbants was relatively constant (0.3 to 0.35). Above this pH, however, marked differences were observed with different perturbants. Bulky perturbants such as glycerol and polyethylene glycol showed only small changes in R_M while with smaller ones such as ethylene glycol, dimethyl sulfoxide, and methanol, increases were observed. Herskovits and Laskowski conclude that some of the tyrosyl groups of ovomucoid lie in crevices which are capable of contraction at low pH. More recently, Donovan (*66*) has shown that this crevice contraction does not alter the accessibility of the chromophore with respect to heavy water.

c. *Perturbation of Chromophores Other Than the Aromatic Amino Acids*

In principle, solvent perturbation should have great utility in studying exposure of chromophores other than aromatic amino acids such as tyrosine, tryptophan, and phenylalanine. Comparatively little, however, has been done to date in this area.

Mercouroff and Hess (*78*) have used solvent perturbation in studying the environment of a cinnamoyl group attached to a serine residue at the active site of chymotrypsin. This cinnamoyl–chymotrypsin compound has been shown to be an intermediate in the α-chymotrypsin-catalyzed hydrolysis of cinnamoyl esters (*94*). They showed that the ethylene glycol perturbation spectrum of cinnamoyl–chymotrypsin is essentially the same as that of native chymotrypsin, corresponding to perturbation of tryptophan groups. Disruption of the native structure by peptic hydrolysis in addition to producing the anticipated increase in apparent tryptophan exposure, resulted in a solvent perturbation spectrum above 300 mμ, corresponding to the cinnamoyl group. The authors conclude that the cinnamoyl group ". . . becomes buried inside the protein molecule as a result of the conformational change which accompanies formation of the chymotrypsin substrate compound."

Stellwagen (*79*) has studied the crevice-like structure housing the heme in cytochrome c using the Soret absorption spectrum of the heme–hexadecapeptide from a chymotryptic digest of the protein as a reference for 100% exposure. The exposure of heme in the native protein was about 25 to 35% depending upon the perturbant employed. Replacement of one of the protein–heme coordinate bonds by cyanide resulted in increased exposure with respect to 20% ethylene glycol.

Wang (*83*) has carried out perturbation measurements on 1-dimethyl amino-5-sulfonyl naphthalene (DNS) groups coupled to lysozyme, ovomucoid, and bovine serum albumin. Since the DNS groups are hydrophobic in character, their apparent degree of exposure should be a measure of the density of nonpolar side chains in the region of potential coupling sites such as sulfhydryl, amino, and perhaps imidazole groups (*95*), as well as their ability to "enfold" the attached DNS groups. Wang found with dimethyl sulfoxide, polyethylene glycol, and ethylene glycol that the DNS groups for lysozyme and ovomucoid are almost completely exposed. On the other hand, as might be anticipated from the known configurational adaptability of serum albumin (*96*), DNS groups coupled to this protein are considerably buried (R_M, 0.32 to 0.5).

It has been suggested (*97,98*) that conformational changes involving certain areas of an enzyme molecule might be detected by the use of "reporter groups," i.e., chromophores coupled to side chains in the region of interest. Interpretation of spectral changes involving such chromophores are likely to suffer from the same ambiguities seen in interpreting similar changes for tyrosyl and tryptophyl groups. Solvent perturbation of such "reporter groups" may prove useful in resolving such ambiguities.

C. Spectrophotometric Determination of Tyrosine Ionization

1. Qualitative Aspects

As is well known, ionization of tyrosine to form the phenolate ion is accompanied by a long wavelength shift of the ultraviolet spectrum in the region above 220 mμ with a marked increase in the molar absorbance. With the exceptions cited below, we consider only those spectral changes occurring above 250 mμ. These form the basis for specifically determining the accessibility of tyrosyl groups to protons. We consider the more general case of titration of ionizable groups in Section III. B.

Spectral changes occurring on tyrosine ionization are conveniently determined as difference spectra. Ordinarily, ionized tyrosine exhibits a difference peak having a maximum close to 295 mμ. The relative constancy of the molar difference extinction coefficient, $\Delta\varepsilon$, for tyrosine ionization makes "group counting" of these residues possible in proteins. Assembled in Table IX are

TABLE IX

Difference Extinction Coefficients for Tyrosine Ionization

Substance	Moles tyr./ mole	$\Delta\varepsilon_{295}$ [a]	Reference
Tyrosine	1	2300	*99*
Actin	19–22	2520–2920	*100*
Conalbumin	18–19	2230–2350	*101*
α-Corticotropin (ovine)	2	2250	*102*
Ribonuclease	6	2630	*103*
Lysozyme	3	2230[b]	*104, 105*
Papain	17	2370	*106*
Bovine serum albumin	21	2430	*107*
α-Lactalbumin	4	2490[b]	*108*

[a] Includes only values where peak was at 295 mμ or where corrected spectra were obtained.

[b] Corrected for interference by tryptophan (see text).

values of $\Delta\varepsilon_{295}$ obtained for tyrosine- and tyrosine-containing proteins. With the exception of the high values for actin, which may be due to uncertainty in tyrosine content of this protein, $\Delta\varepsilon_{295}$ per group is close to 2300 with an uncertainty of the order of 10–15%. The approximate number of tyrosyl groups ionized at a given pH would thus be given by $n = \Delta\varepsilon_{obs}/2300$. More exactly, $\Delta\varepsilon$/group should be determined for a protein in question from the spectral change corresponding to complete ionization of all the tyrosyl groups. Where tyrosyl groups do not titrate freely, this value can often be obtained by denaturation at very high pH or through the use of a denaturing agent such as guanidine. In the latter case, some uncertainty may arise because of the effect of high concentrations of additive on the intrinsic tyrosyl difference extinction coefficient, e.g., Havsteen and Hess (*109*) found an increase in $\Delta\varepsilon$

for α-chymotrypsin on going from water to high concentrations of guanidine while Tachibana and Murachi (*110*) saw no change with stem bromalein under comparable conditions.

The accessibility of the tyrosines to protons *at the pH in question*, as well as the pK of these groups, will determine the magnitude of n. Unambiguous qualitative information about the extent of accessibility of protons to tyrosines can thus be obtained generally only at pH values well above the "normal" pK_{int} values of such groups (see Section III. B). As we shall see, however, in many cases conformational changes occurring in the pH region of tyrosine ionization preclude any determination of the degree of exposure of such groups in the *native* state.

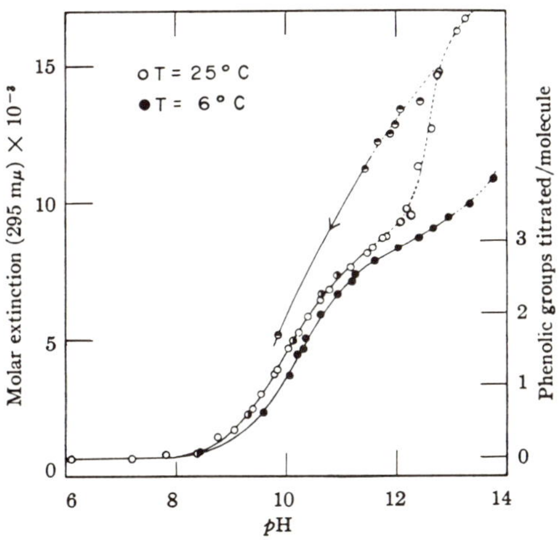

Fig. 11. Ionization of the phenolic groups of ribonuclease. The ionization is independent of time where solid lines have been drawn; it increases with time where dashed lines have been drawn: ◑, solutions at 25° reversed after exposure to pH 11.5; ◒, solutions at 25° reversed after exposure to pH 12.7 (*103*).

Time dependent spectral changes and a lack of correspondence of forward and reverse tyrosine ionization curves may also provide a clue as to the inaccessibility of such chromophores in the native state. This is seen in the titration studies of ribonuclease (*103*). Ionization of the six tyrosines in this protein occurs in two steps (Fig. 11). In the first of these, three groups ionize with a total change of $\Delta\varepsilon_{295}$ of 7900 (2630/group); titration is independent of time and reversible in this step (solid curves, Fig. 11). In the second step, the remaining three phenolic groups titrate, the process being time dependent and irreversible (dotted curves, Fig. 11). This clear-cut separation of the ionization into two steps while possible for ribonuclease, because of its rather high-iso-

TABLE X

Tyrosyl Ionization in Proteins

Protein	Total residues	Residues ionizing	Reference
Aldolase	44	11–13 (rapid) 31–33 (slow)	*111*
Bromelain, stem	19	9 (rapid) 10 (slow)	*110*
Carboxypeptidase A	19	7–8 (reversibly) 8–9 (irreversibly)	*112*
Carbonic anhydrase, human	7–8	4 (rapid) 4 (slow)	*113*
α-Chymotrypsin α-Chymotrypsinogen	4	2 (rapid) 1 (slow) 1 (un-ionized)	*109, 114, 115*
Conalbumin Conalbumin, iron complex	18–19	11 (rapid), 7 (slow) 5 (rapid)	*101*
Cytochrome c	4	4[a] *b* *c*	*116* *117* *79*
Hemoglobin, horse	12	8 (rapid) 4 (slow)	*118* *119*
Insulin	4	3 (rapid) 1 (slow)[d] 4[e]	 *120* *91*
β-Lactoglobulin	8	6[f] 8[g]	*121*
α-Lactalbumin	4	4[h]	*108*
Lactic dehydrogenase, chicken heart	30	10 (rapid) 20 (denatured)	*122*
Lysozyme	3	2 (rapid) 1 (slow)[i] 3	*120, 123* *104*
Myoglobin, sperm whale	3	2 (rapid) 1 (slow)[j]	*118*

TABLE X (Continued)

Myosin	19	17	
Tropomyosin	15	18	
Light meromyosin	9	8	*74*
Heavy meromyosin	21	20	
Paramyosin	18/100, 000 g	19	
Papain	17	11–12 (rapid) 5–6 (slow)	*106*
Pepsin	18	16 (rapid) 1 (slow)	*124*
Pepsinogen	18	15 (rapid) 2 (slow)	*124a*
Ribonuclease	6	3 (rapid) 3 (slow)	*103*
Serum albumin, bovine	18	19[h]	*107*
Trypsin	10	6 (rapid) 4 (slow)	*115, 125*
Trypsinogen	7–10	4 (rapid) 3–6 (slow)	*115*

[a] Horse heart protein. Titration curve shows a break at two groups.

[b] Horse heart protein. Titration curve analyzed into titration of four groups having values of pK_{app} of 10.05, 11.00, 12.35, and 13.10.

[c] Beef heart protein, curve shows break at two groups.

[d] Sample referred to as "crystalline beef insulin" rather than "Zn-insulin" or "Zn-free insulin."

[e] Amino and phenolate groups determined by potentiometric titration. Total number was equal to values obtained by amino acid analysis.

[f] Value probably low due to interference by tryptophan.

[g] Titration carried out in 1 *M* potassium bicarbonate, 10% dioxane.

[h] Linderstrom–Lang analysis indicates the occurrence of a conformational change in the region of titration.

[i] The differences in the observations of Tanford and Wagner (*104*) and Inada (*120*), and Tojo et al. (*123*) are quite puzzling. Tanford and Wagner do not report a time dependence to tyrosyl ionization in lysozyme as do the other investigators. Inada (*120*), on the other hand, has not indicated the occurrence of interference by tryptophyl spectral changes at the wavelength of observation, 295 mμ, in contrast with observations of Tanford and Wagner. Indeed, the value $\Delta\varepsilon_{295}$/group that he reports suggests that such interference is absent. The reason for these discrepancies is not at all clear. Since chemical reactivity studies (see Section IV.B.) indicate nonequivalence of the three tyrosines of lysozyme, it seems imperative that the titration of these groups be reinvestigated in a critical manner.

[j] As originally published, one of the rapidly titrating tyrosyl groups was believed to be abnormal. Re-evaluation of the data had led Hermans (*126*) to conclude that both rapidly titrating groups are "normal."

electric point and the relatively low value of pK_{int} for tyrosyl ionization, is not always as readily accomplished. Titration of the three "exposed" tyrosines of ribonuclease is virtually complete by pH 11.5. By contrast, with proteins having lower isoelectric points and higher pK_{int} values, the degree of ionization near pH 11.5 may be as low as 50%. Thus, estimation of exposure of tyrosines from the value of $\Delta\varepsilon_{295}$ at a break in the titration curve may lead to serious errors if the pH at which it occurs is relatively low. In such cases what is required in principle are "zero-time" titration curves extended to much higher values of pH. This approach, which would probably require rapid mixing techniques, has not been widely used.

Table X summarizes some of the available data on ionization of tyrosyl groups in proteins. No critical selection has been made. In some instances, numbers of "rapidly" and "slowly" ionizing groups were taken from inflections in $\Delta\varepsilon$ vs pH curves and must be questioned in quantitative terms (see subsequent section). "Rapid" may refer to values obtained as soon as possible after mixing while "slow" may be anywhere from 3 to 24 hr after mixing. In spite of the lack of quantitative significance to many of these values, the observation of time dependencies for tyrosyl ionizations for most of these proteins suggests that nonfreely ionizing tyrosyl groups in globular proteins may be more the rule rather than the exception. Bovine serum albumin, β-lactoglobulin, and α-lactalbumin are perhaps not exceptions to this rule since in all three cases complete tyrosyl ionization did not correspond to native protein. Of interest also is the fact that tyrosyl residues in the myosin group of proteins appear to be freely titratable, suggesting that the rodlike conformations of these proteins provide insufficient volume for insulation of the tyrosyl groups from the medium (see also solvent perturbation observations).

2. Interference by Other Chromophores

The foregoing remarks have been based on the assumption that the spectral changes observed at high pH between 290 and 300 mμ are due solely to tyrosine ionization. This, however, may not be true where the protein contains chromophoric prosthetic groups such as heme (see below) or where the tyrosine tryptophan ratio is particularly low. Such interferences have been noted with aldolase (*37*), lysozyme (*104*), trypsinogen (*125*), and α-lactalbumin (*108*). The observations made with the latter protein are a good illustration of the effect of a tryptophan spectral change on the tyrosine ionization spectrum. This protein has a tyrosine–tryptophan molar ratio of 1:1. The maximum value of $\Delta\varepsilon_{295}$ observed with α-lactalbumin corresponded to less than four tyrosines (based on a value of $\Delta\varepsilon$ per group of 2300) (*108*). Since the amino acid sequence (*127*) indicates four tyrosine residues, one might have concluded that one group is buried even above pH 13.† The complete difference spectrum however, showed a *maximum* at 298 mμ (dashed curve, Fig. 12) and a shoulder just above 290 mμ. Shown also for comparison is a "normal" tyrosine ionization difference spectrum as obtained with the tryptophan-free protein, ribonuclease. As anticipated in this case, the maximum occurs at 295 mμ and no shoulder is evident. The atypical tyrosine difference spectrum observed with α-lactalbumin is due to superposition of a tryptophan blue shift. The magnitude of the tryptophan shift observed at high pH appears to be comparable to that observed directly below pH 4 (*60*). Thus, use of a pH 2 reference (dot-dashed curve, Fig. 12) rather than the pH 6 reference,

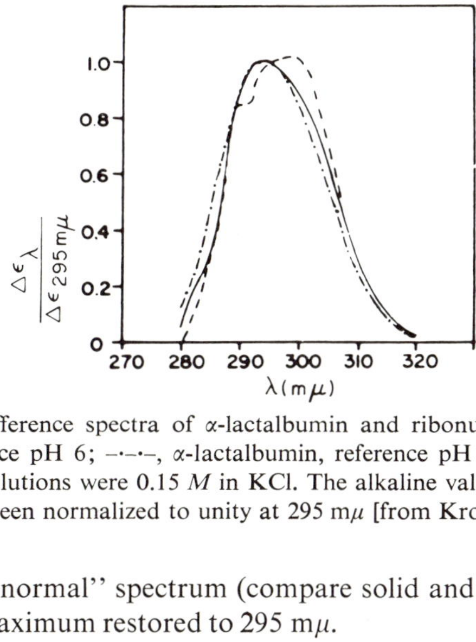

Fig. 12. Alkaline difference spectra of α-lactalbumin and ribonuclease at 25°; - - - -, α-lactalbumin, reference pH 6; -·-·-, α-lactalbumin, reference pH 2; ——, ribonuclease, reference pH 6. All solutions were 0.15 *M* in KCl. The alkaline values were about pH 12. All curves have been normalized to unity at 295 mμ [from Kronman et al. (*69*)].

results in a more "normal" spectrum (compare solid and dot-dashed curves, Fig. 12) with the maximum restored to 295 mμ.

The maximum value of $\Delta\varepsilon_{295}$ obtained for α-lactalbumin with pH 2 reference was 9940 corresponding to 4.3 groups, in good agreement with the analytical value. Similar use of an acid reference solution has been made by Donovan (*37*) for aldolase where the underlying blue shift in that system was due to exposure of tryptophan groups. Failure to use the appropriate reference in the case of aldolase results in apparent titration of seven too few tyrosines. For the reasons implied above, the detailed tyrosyl ionization curve of α-lactalbumin was obtained from data at 298 mμ (*108*). Smillie and Kay (*125*) have used a wavelength of 301 mμ in their study of trypsinogen for similar reasons. Another means of detecting interference from groups such as tryp-

† The molar extinction coefficients have been corrected to a molecular weight of 14.4×10^3 as obtained for the amino acid sequence. In our original paper we used a molecular weight of 15.5×10^3 and based our conclusions on the presence of five tyrosines.

tophan is to determine the pH dependence of the apparent degree of ionization using spectral data at a number of wavelengths. The degree of ionization of the tyrosyl groups of lysozyme (*104*) appears to be dependent on the wavelength of observation. In those experiments, the maximum value of $\Delta\varepsilon_{295}$ was 7500 or 2500 per group, somewhat higher than the "normal" value of 2300 but not outside of the range of values observed with proteins (see Table IX). Donovan et al. (*105*) have shown that the spectral anomaly observed with lysozyme is due to a pH-dependent tryptophan red shift; correction of the ionization spectrum for the change in tryptophan spectrum reconciles the data obtained for different wavelengths.

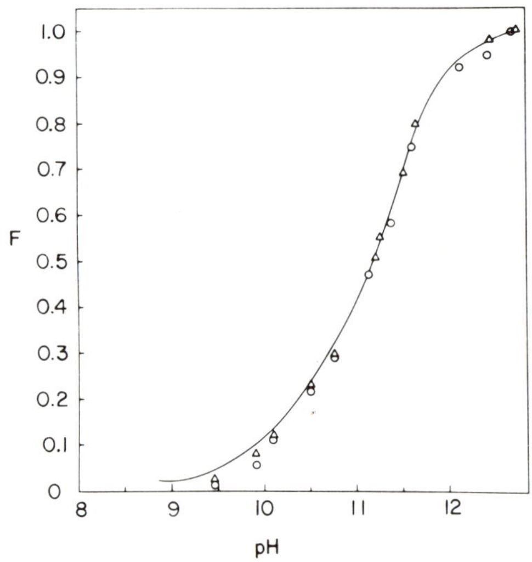

Fig. 13. Fraction of tyrosines, *F*, ionized in α-lactalbumin as a function of pH. —— calculated from spectral data at 298 mμ. Δ, calculated from 295 mμ data. O, calculated from 290 mμ data (*61*).

It should be emphasized, however, that the lack of a dependence of an apparent degree of tyrosyl ionization on the wavelength of observation is not *prima facie* evidence for the absence of tryptophan interference. For example, in the case of α-lactalbumin, in spite of the fact that a tryptophan shift gave anomalously low values of $\Delta\varepsilon_{295}$ at high pH (see above), pH dependence of the fractions of tyrosines ionized is relatively insensitive to the wavelength of observation (Fig. 13) (*61,108*). This apparent contradiction is the consequence of comparable pH dependencies to the tyrosine ionization and the tryptophan spectral shift. That this should indeed be the case can be demonstrated by simple algebraic manipulations:

$$(\Delta\varepsilon^{\lambda})_{\text{obs}} = F\Delta\varepsilon^{\lambda}_{\text{tyr}} + \gamma\Delta\varepsilon^{\lambda}_{\text{tpt}} \tag{8}$$

where F is the fraction of tyrosine ionized and γ is the fractional change in the tryptophan spectrum at the pH in question. The apparent degree of ionization is given by

$$F_{\text{app}} = \frac{\Delta\varepsilon^{\lambda}_{\text{obs}}}{[(\Delta\varepsilon^{\lambda})_{\text{obs}}]_{\text{tyr}}} \tag{9}$$

From Eqs. (8) and (9) we get

$$\frac{F_{\text{app}}}{F} = \frac{(\gamma/F)(\Delta\varepsilon^{\lambda}_{\text{tpt}} + \Delta\varepsilon^{\lambda}_{\text{tyr}})}{\Delta\varepsilon^{\lambda}_{\text{tpt}} + \Delta\varepsilon^{\lambda}_{\text{tyr}}} \tag{10}$$

Thus for $\gamma = F$, F_{app} and F are identically equal. For α-lactalbumin the pH dependencies of γ and F, as well as the magnitudes of $\Delta\varepsilon_{\text{tyr}}$ and $\Delta\varepsilon_{\text{tpt}}$, are such that within experimental error, F_{app}/F is unity (*108,128*). Thus, while an apparent wavelength dependency of tyrosine ionization may denote interference by tryptophan, the absence of such an anomaly does not indicate the absence of such interference. Inspection of difference spectra for distortion (see, for example, Fig. 12) and subsequent calculation of the extent of ionization from wavelengths above 298 mμ may be a more reliable procedure.

The presence of prosthetic groups in a protein may likewise complicate the use of ultraviolet spectra in assessing the ionization of tyrosyl groups. The presence of heme in proteins such as cytochrome (*79,116,117*), myoglobin (*118*), and hemoglobin (*118,119,129*) interferes with the 295-mμ difference peak; the 242–245 mμ peak however, has been used in such cases. The molar difference extinction coefficient for ionization of tyrosine has been reported to be 1.0×10^4 at 245 mμ (*118*) and 1.1×10^4 at 246 mμ (*130*). The values of $\Delta\varepsilon_{244-246}$ observed for tyrosines in proteins range from 0.97×10^4 (sperm whale myoglobin) (*118*) to 1.21×10^4 (horse and human hemoglobin) (*118*); the uncertainty in $\Delta\varepsilon_{245}$ thus appears to be of the same order of magnitude as that observed for $\Delta\varepsilon_{295}$. Since $\Delta\varepsilon_{245}$ is almost five times as great as $\Delta\varepsilon_{295}$, use of the lower wavelength peak would require lower concentrations of protein. This advantage is somewhat negated by the fact that measurements at the lower wavelengths are more subject to light scattering errors and interference by other chromophoric side chains.

Formation of the mercaptide ion at high pH is accompanied by an increase in absorption in the 230–250 mμ region (*33*). The molar difference extinction coefficient $\Delta\varepsilon_{243}$ for ionization of such sulfhydryl groups in aldolase has been reported to be of the order of 0.4×10^4 (*37*). Ionization of thiol groups should thus cause little difficulty except where the cysteine:tyrosine ratio is rather high. Cytochrome c, containing four tyrosyl and two cysteinyl groups, would be expected to present difficulties in this regard but is "well behaved" since the latter two side chains are covalently linked to the heme (*131,132*). The possible spectral contributions near 245 mμ due to alteration of tryptophan environment and changes in backbone conformation are somewhat difficult to assess. Glazer and Smith (*133*) have observed that the difference peak generated on denaturation of proteins containing or lacking tryptophan generally occur below 240 mμ. The magnitude of these changes, however, are sufficiently great that significant spectral contributions might still be made between 240 and 245 mμ.

3. A More Quantitative Approach

While it is sometimes possible to obtain qualitative information concerning the accessibility of tyrosyl residues by direct examination of the spectral titration curve (see above), valid quantitative information may be obtainable frequently only on more detailed analysis of the data. Such analysis as it pertains to the problem of group counting will be considered in Section III. B; we therefore anticipate some of this discussion as it pertains to titration of tyrosyl groups. Where we have no time-dependent changes in the pH region of the titration and the process is reversible, the titration should be describable to a first approximation by the Linderstrom–Lang equation,

$$\mathrm{pH} + \log\left(\frac{\Delta\varepsilon_{\max} - \Delta\varepsilon}{\Delta\varepsilon}\right) = \mathrm{p}K_{\mathrm{int}} - 0.868\, ZW \qquad (11)$$

$$\equiv \mathrm{p}K_{\mathrm{app}}$$

where Z is the net charge on the molecule, which in the absence of ion binding data is taken to be equal to the protonic charge. The latter is generally obtained by potentiometric titration. W, the electrostatic interaction factor, is given by

$$W = \frac{\varepsilon^2}{2DR\kappa T}\left[1 - \frac{\kappa R}{1+\kappa a}\right] \qquad (12)$$

where R is the molecular radius, a is the distance of closest approach of an ion, κ is the Debeye–Huckel parameter, and ε is the electronic charge.

It is evident from Eq. (11) that unless the pH of observation is well above that corresponding to $\mathrm{p}K_{\mathrm{app}}$, titration of tyrosyl groups will be incomplete. It seems trivial perhaps to emphasize this point except for the fact that one still encounters investigators who ascribe special significance to the fraction of ionization occurring at a pH at which a break in the curve is observed, without considering that ionization of that particular class of groups may be incomplete at this pH. It is quite apparent also from Eq. (11) that for proteins at pH values removed from their isoelectric point, $\mathrm{p}K_{\mathrm{app}}$ and $\mathrm{p}K_{\mathrm{int}}$ may be significantly different, e.g., for α-lactalbumin the complete spectrophotometric and potentiometric titration curves yield a value of $\mathrm{p}K_{\mathrm{int}}$ of 10.3 at 25° while $\mathrm{p}K_{\mathrm{app}}$ (the pH for 50% titration of tyrosyl groups) was more than one pH unit higher. These remarks are particularly important where attempts have been made to use the magnitude of the dissociation constant as a measure of shielding of tyrosyl groups from the medium (see Section III. B for a further discussion of this point). In this regard, one still encounters investigators who have concluded that tyrosyl groups in certain proteins are "abnormal" on the basis of the comparison of their pK_{app} value with pK_{int} for "normally" titrating tyrosyl groups. The relevancy of the electrostatic term, 0.868 ZW is obvious here.

The number of titratable tyrosyl groups of α-lactalbumin and serum albumin as we have indicated above, correspond to their analytical values (Table X). A detailed analysis of the titration data of these proteins however,

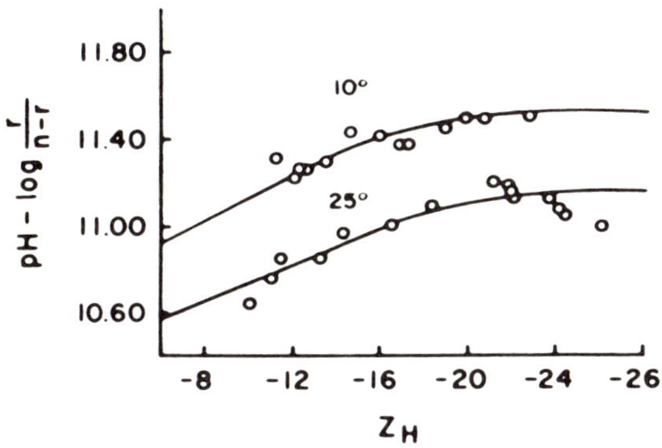

Fig. 14. Plot of the titration data of tyrosine groups in α-lactalbumin at 10 and 25°C according to eq. (11) [from Robbins et al. (*108*)].

reveals that at the pH at which ionization is complete, the protein is no longer in the native state. This kind of behavior is illustrated in Fig. 14, which is essentially a plot of Eq. (11) for the titration of tyrosyl groups in α-lact-albumin. Deviation from linearity is indicative of a change in pK_{int} or W or both, as the tyrosyl groups become titrated. Some measurements of the sedimentation coefficients of α-lactalbumin indicate that the molecule is more swollen at high pH than in the native state (*128*). It appears that the lack of linearity, at least in part, is due to a change in W. It seems likely therefore, that the complete accessibility of tyrosyl groups is not characteristic of native α-lactalbumin. This is borne out by chemical reactivity studies of the native protein (see Section IV. B).

The quantitative treatment implicit in Eq. (11) would be applicable to proteins whose spectral titration curves exhibited time dependencies if zero-time titration curves could be obtained. In some instances the time dependencies have been slow enough to measure zero-time curves by conventional manipulations. It may be, however, that detailed investigation of the titration behavior of many of the proteins of Table X will require rapid mixing techniques in conjunction with spectrophotometric and potentiometric observations.

D. Fluorescence Techniques

1. Introduction

Use of techniques based upon emission spectra offer advantages over those involving absorption spectra. Perturbation of chromophores by the protein fabric, while producing relatively small changes in absorption, tend to produce rather large changes in fluorescence. Thus, lower concentrations of protein can generally be employed in fluorescence measurements than for measurement of difference spectra, e.g., in studying environmental changes in tyrosine and tryptophan, absolute absorbances of 1.5 to 2 may be required to give changes of the order of 0.05 while fluorescence measurements can generally be made on solutions having absorbances of less than 0.1 with changes in emission intensity of the order of 50 to 100% not being uncommon. The low concentration levels used for fluorescence measurements may be particularly advantageous for relatively scarce proteins.

Measurement of ultraviolet fluorescence of the aromatic amino acids in proteins is potentially an extremely sensitive method for detecting and characterizing conformational changes in proteins including changes in the degree of exposure of such groups. However, perhaps even more than for absorption spectra, changes in fluorescence spectra have their origin in a very wide range of molecular events, thus making it extremely difficult to draw unambiguous conclusions from data in the absence of ancillary information. Nonetheless, because of their inherent advantages, some consideration will be given to fluorescence techniques.

Fluorescence measurements generate two parameters of interest to us, the quantum yield and wavelength distribution of the emitted light. The significance of these quantities for tryptophan and tyrosine fluorescence in proteins has been considered by a number of authors (*134–140*).

2. Tyrosine and Tryptophan Fluorescence in Proteins

The free amino acids, tyrosine and tryptophan, have emission maxima of 303 and 350 mμ, respectively, and quantum yields of 0.21 and 0.20 (*134*). Since the fluorescence of phenylalanine is not observed in proteins, it shall not be considered. Emission maxima for proteins such as insulin, zein, ribonuclease, and ovomucoid that contain tyrosines but not tryptophan, occur near 304 mμ (*135*), corresponding to the value observed with the free amino acid. The emission maxima for model substances such as poly L tyrosine and tyrosine derivatives show little dependence on solvent composition, which suggests that little information concerning the environment of tyrosines in proteins can be obtained from this spectral parameter. This is borne out by the fact that the positions of the emission maxima of tryptophan-free proteins are quite uniform and show no change after urea denaturation (*135*). On the the other hand, the quantum yield of tyrosine fluorescence in proteins such as those cited above, are considerably lower than 0.21, the value observed for the free amino acid (*61,134–136,141*).

No simple relationship appears to exist between the degree of exposure of tyrosine groups and quantum yield. For example, the exposure of tyrosyl groups in ribonuclease and insulin are roughly the same (solvent perturbation, Table VIII), but their quantum yields differ by more than a factor of two (*135*). Tyrosine quenching processes appear to be related, at least in part, to transfer of a proton from the excited phenolic group to a carboxylate or amino group (*142*). Thus, interactions of tyrosyl groups with other protein moieties may produce quenching and also may convey properties that we would construe as "burying."

The emission from proteins containing both tyrosine and tryptophan will generally show a single band corresponding to the fluorescence spectrum of the latter amino acid. Excitation at 280 mμ, however, where the fractional absorption of tyrosine is relatively high, may give rise to tyrosine fluorescence in addition to the dominant tryptophan band.

The position of the tryptophan emission band, in contrast with those of tyrosine, shows wide variation from protein to protein. Emission maxima range from 328 to 345 mμ (*135,61*) (see also Fig. 15). By contrast, the emission maxima for urea (*135*) or guanidine (*61*) denatured proteins all lie close to 350 mμ, the value observed with the free amino acid or with N-acetyl tryptophan ethyl ester (see Fig. 16, for example). These observations suggest that emission spectra for buried and exposed groups in proteins should differ appreciably. Indeed the effect of solvent on the emission spectra of indole and indole derivatives (*143*) suggests that tryptophans located in regions of low dielectric constant should emit at lower wavelengths than those in contact with water.

In spite of the implications of the above remarks, the magnitude of the emission maximum does not appear to be a reliable index of the degree of exposure of tryptophyl groups to the aqueous medium. For example, both aldolase and α-lactalbumin have emission maxima of about 328 mμ (compare curves *B* and *C*, Fig. 15). Yet, about 10% of the tryptophans of aldolase and 50% of those of α-lactalbumin are exposed (solvent perturbation, Table VIII). Similarly, although denaturation of lysozyme in 8 *M* urea does not involve an increase in exposure, at least as seen by solvent perturbation (Table VIII), it is accompanied by a 9-mμ red shift (*135*). This is not to say that differences

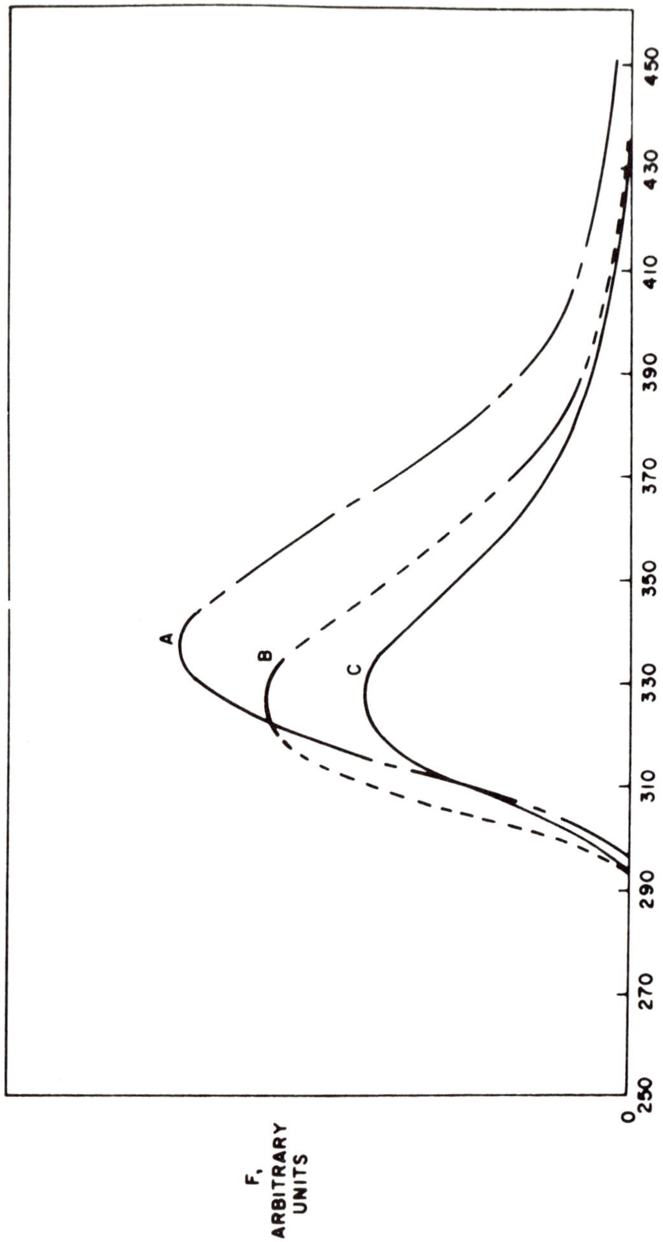

Fig. 15. Tryptophan emission spectra of native proteins. Solvent, pH 7.00, 0.05 *M* Tris; Temperature, 25.0°, curve A, egg-white lysozyme; B, muscle aldolase; C, α-lactalbumin. All spectra corrected for detector sensitivity. Excitation wavelength, 295 mμ. The ordinate is in arbitrary units; the integrated spectra, therefore, are not proportional to quantum yields [from Kronman and Holmes (*61*)].

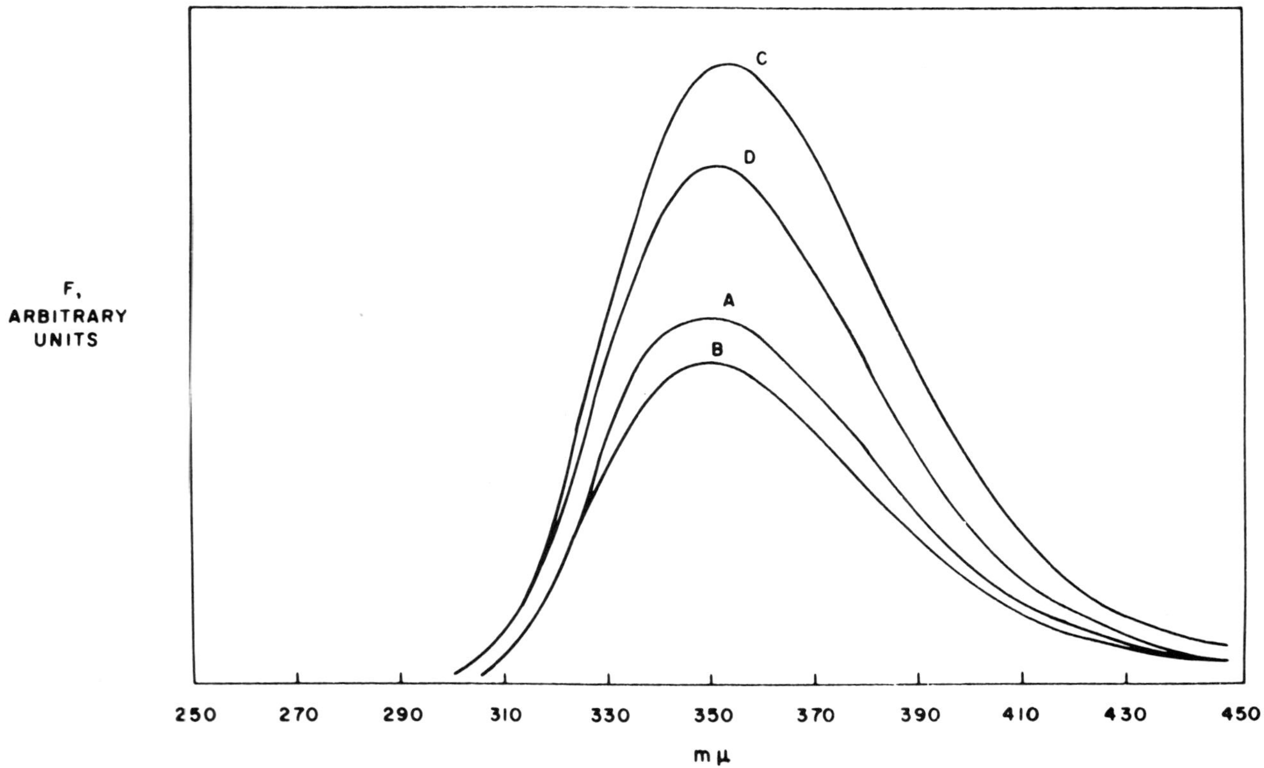

Fig. 16. Tryptophan emission spectra of guanidine-denatured proteins. Solvent, pH 7.0, 0.05 *M* Tris, 6.0 *M* guanidine hydrochloride; temp., 25.0°, Curve A, egg-white lysozyme; B, α-lactalbumin; C, tryptophan (absence of guanidine); D, muscle adolase. All spectra corrected for detector sensitivity. Excitation wavelength, 295 mμ. The ordinate is in arbitrary units such that the integrated spectra are *not* proportional to quantum yield [from Kronman and Holmes (*61*)].

in exposure *may* not be reflected in the position of the emission maximum. For example, Hyashi et al. (*82*) have found that one or two tryptophans of lysozyme become buried on binding of glycol chiton. Lehrer and Fasman (*144,145*) have shown that binding of a similar inhibitor (di- and tri-N-acetyl glucosamine) to lysozyme is indeed accompanied by a short wavelength shift of the emission maximum—the direction anticipated for transfer of a group from a polar to non-polar environment.

Recent work by Walker and co-workers (*146,147*) has considerably clarified the manner in which molecular environment influences the position of the emission maximum of tryptophan. They have shown that shifts in the fluorescence spectra of indole and related compounds, rather than being due to generalized solvent effects, have their origin in "exiplex" formation, i.e., formation of an excited state complex of indole and a polar molecule. In terms of such a mechanism, the position of the emission maximum for a given tryptophyl residue in a protein would depend upon its ability to form such an excited state complex. Exposed groups should be able to form those complexes with water molecules giving spectra with peaks close to 350 mμ, while groups buried in nonpolar regions of the protein molecule should be expected to have maxima close to 300 mμ. However, it also would appear that a tryptophyl group essentially insulated from the medium might also be capable of undergoing exiplex formation if its proximity to a polar group and freedom of rotation would permit the establishment of the requisite geometry for complex formation during the lifetime of the excited state (see Ref. *9* for a discussion of the application of this mechanism to a protein conformational change).

Thus, while transfer of a tryptophyl residue from a nonpolar region of the molecule to contact with water may well lead to a long wavelength shift of the emission spectrum, more subtle processes which might be considered to be changes in "exposure" may well lead to similar spectral shifts. An example of this is the low temperature behaviour of α-lactalbumin. When the crevices housing the two normally exposed tryptophans of this protein contract at low temperature, perturbants as large as sucrose and glycerol are completely excluded (*49*) and a short wavelength shift of the emission spectrum is observed (*139*). These observations taken by themselves would lead to the conclusion that the short wavelength shift of the emission spectrum was a result of change in the polarity of the surroundings of these two tryptophans. Perturbation measurements made with heavy water, however, demonstrated that the two tryptophans in the crevices are still accessible to water. The absence of a change in the absorption spectrum of α-lactalbumin during the crevice contraction process suggests also that no change in the polarity in the surroundings of the two tryptophans has occurred. A spectral shift accompanying this conformational change was explained in terms of inhibition of formation of an excited state complex of the tryptophyl group.

It appears with our present knowledge that little direct use can be made of the spectral distribution of tryptophan fluorescence in evaluating "exposure" of groups in the absence of ancillary measurements. The use of quantum yields for a similar purpose is, if anything, fraught with still greater ambiguities. Quantum yields for indole and tryptophan derivatives show a marked dependence on solvent polarity (*143*). The large differences in quantum yield for a variety of indole and tryptophan derivatives in the same solvent (*148,149*) makes it difficult to know what the appropriate model compounds should be for tryptophan fluorescence in protein. The problem is further complicated by the fact that the majority of experimental data available for tryptophan quantum yields in proteins at present has been obtained by excitation at 280 mμ where both tyrosine and tryptophan absorb. The significance in molecular terms of the yields thus obtained are somewhat in doubt because of the possibility of contributions from tyrosine fluorescence and from tryptophan fluorescence excited by energy transfer from tyrosine (*61*).

It is quite apparent that our understanding of the factors that influence fluorescence of aromatic residues in proteins is far too incomplete at present to make use of such parameters in determining the nature of the environment surrounding such residues in proteins. A more valid use at present seems to be a more restricted one of detecting conformational changes which may involve exposure and the evaluation of the thermodynamics of such processes. Characterization of the molecular events which produce these spectral changes is perhaps best left to other techniques.

3. Quenching or Exultation of Fluorescence from Aromatic Residues

The influence of chemical additives on the fluorescence of aromatic residues has been proposed as a means of estimating exposure of these groups in proteins. Steiner et al. (*138*) and Fratteli et al. (*85*) have shown that fluorescence of acetyl tryptophan amide (ATA) in water is enhanced by moderate concentrations of propylene glycol. The extent of enhancement of fluorescence in proteins relative to ATA has been suggested as a measure of the degree of exposure of tryptophans. The analogy with the solvent perturbation technique is obvious. In general, native proteins give less enhancement with propylene glycol than that observed with ATA; denaturation brings the observed values closer to those observed with the model compound. No quantitative estimation of exposure from such data has been attempted as yet however.

The observations of Shinitzky et al. (*150*) that compounds containing a protonated imidazole ring quench indole fluorescence in model compounds and in lysozyme suggests that such compounds might be useful in determining exposure of tryptophan by a quenching procedure. Since buried tryptophans and tryptophans in contact with an aqueous medium would be expected to have different emission maxima and different quantum yields (see above), one would anticipate that the degree of quenching of tryptophan fluorescence from a protein would depend upon the emission wavelength being monitored. Such quenching data might prove to be more useful than that obtained at a single wavelength.

The quenching of the tryptophan fluorescence of lysozyme by iodide ion (*151*) is a good illustration of the preferential quenching of exposed groups. In this system, quenching is accompanied by a short wavelength shift of the emission spectrum, the direction anticipated if the water-contacted indole rings were preferentially quenched. Quenching with N-acetyl glucosamine complexed lysozyme was markedly less than with free lysozyme. These observations are likewise consistent with the observation that complex formation buries tryptophan with respect to perturbing solvents (*82*). One of the attractive features of this quenching technique is the relatively low concentration of iodide required, 0.2 *M* or less.

Lehrer (*151*) has estimated the number of groups exposed in lysozyme from the quenching data using the relationship

$$\frac{N_{\text{exposed}}}{N_{\text{total}}} = \frac{\Delta Q}{Q_{\text{exp}}}$$

where ΔQ is the difference in yield on quenching and Q_{exp} is the quantum yield of an exposed tryptophan. N_{exposed} was calculated to be 3 ± 1 groups. The range given is a reflection of the uncertainty in a choice of Q_{exp}. The two extremes used by Lehrer, 0.06 and 0.12, corresponded, respectively, to values observed with tryptophan-containing peptides and with the polyamino acid.

We believe that a more realistic value of Q_{exp} would fall in the range 0.10 to 0.15; these are tryptophan yields obtained with a variety of guanidine-denatured proteins (*61*). The number of exposed groups in lysozyme, calculated on this basis from Lehrer's data, would be 2 ± 0.4. This is somewhat lower than the lower limits of values obtained from solvent perturbation (Table VIII).

A fundamental question, aside from the "numbers game" that we have been playing above, is whether one can speak of exposed tryptophans as having a unique quantum yield, an assumption implicit in Lehrer's calculations. As we have discussed above, tryptophans of α-lactalbumin, which remain exposed to water during the "crevice contraction," give rise to altered fluorescence properties during this process. One would anticipate that groups on the molecular surface of the protein molecule would be involved in interactions with other protein structural moieties which might give rise to quenching or exaltation of fluorescence. Indeed, the X-ray crystallographic structure of lysozyme (*152,153*) reveals that the active-site tryptophans, which are presumably exposed (see Section V), are involved in intramolecular interactions. Perturbations, probably induced by Asp 101 and Glu 35 markedly alter the chemical reactivity of trp 62 and 108 (see Section IV) and probably are also responsible for the spectral shifts observed by Donovan et al. (*105*).

Stryer (*154*) has observed a deuterium isotope effect on the fluorescence of indole, tryptophan, and tyrosine. The increase in yield appears to depend upon whether the excited species is involved in proton transfer reactions. This effect of heavy water on fluorescence, also observed with tyrosine- and tryptophan-containing proteins, suggests that heavy water might be useful as a fluorescence perturbant capable of determining whether a phenolic hydrogen or an indole N—H in a protein is in contact with the medium. This has not as yet been explored as a quantitative tool.

Chen and Cohen (*155*) have observed that the quenching of tyrosine fluorescence by phosphate ion is much greater in the free amino acid than for such residues incorporated into proteins. This decreased quenching of tyrosines in proteins appears to be an index of their inaccessibility to the medium. As in the case of the propylene glycol, no quantitative estimation of exposure has been attempted. Since the quenching appears to be due to HPO_4^{2-} ions *per se*, the pH range of potential application for such quenching phenomena in the determination of exposure would be rather limited.

A rather novel use of modifiers of aromatic amino acid fluorescence devised by Elkena (*156*) makes use of an additive which serves as an acceptor for transfer of the excited state of tryptophan. The method is based on a model which assumes that the tryptophans in a protein fall into two classes, low wavelength emitting (300–320 mμ) groups, assumed to be buried, and high wavelength emitting (340–355 mμ), assumed to be exposed (see above and Ref. *139*). Use of an acceptor molecule, e.g., salicylamide, whose absorption spectrum strongly overlaps the emission spectrum of the buried tryptophans, should produce preferential transfer of the excited state of these groups. Her preliminary measurements of lysozyme and salicylamide do indeed show a diminution of the short wavelength tryptophan fluorescence of this protein, and shift of the fluorescent peak to longer wavelength.

An important advantage of this "transfer method," as compared to the usual quenching procedures, is the low additive concentrations at which it is effective, e.g., with salicylamide, the concentration at which transfer quenching of the buried group appears to be complete is of the order of 10^{-4} M, while markedly higher concentrations of contact quenchers or exalters such as propylene glycol are required to effect significant changes. The influence of high concentrations of additives on the native protein structure may be of great significance.

4. Fluorescent "Probes"

Other than the specific spectral methods described for tyrosyl and tryptophyl groups, few, if any, methods are available for determining the extent of exposure of the *individual* nonpolar amino acid side chains, leucine, isoleucine, valine, etc. Fluorescent dye binding is a procedure which, although it does not yield quantitative information on the distribution of individual side chains, appears to provide an indication of the availability of hydrophobic regions in protein molecules. Other methods for studying nonpolar regions are discussed in Section III. A. The binding of fluorescent dyes to proteins is the subject of a recent review (*156a*).

Fluorescent compounds such as 1-anilino-8-naphthalene sulfonate (ANS, structure **I**) and 2-*p*-toluidinylnaphthalene-6-sulfonate (TNS, structure **II**) have extremely low quantum yields in water (ca. 0.004 and 0.0008, respectively) (*157,158*). In the presence of many proteins, the quantum yield of these dyes show dramatic increases with concomitant shift in fluorescent emission maxima, e.g., TNS in aqueous medium exhibits a 700-fold increase in quantum yield on binding to β-lactoglobulin together with a short wavelength shift of about 70 mμ (*157*).

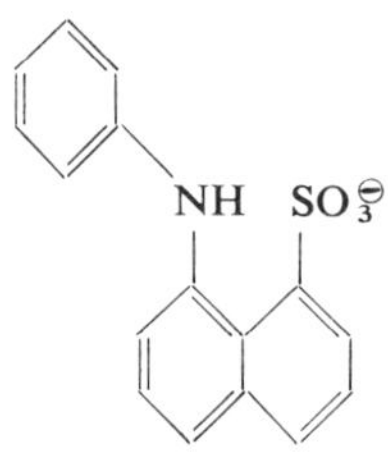

I

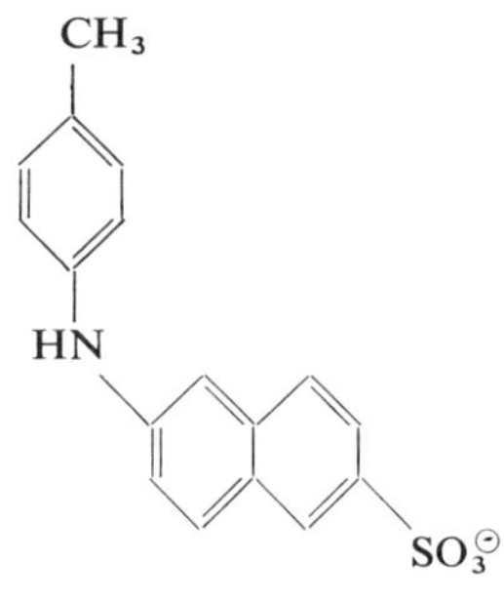

II

The dependence of ANS (*158*) and TNS (*157*) fluorescence on solvent composition suggests that interaction of these dyes with hydrophobic regions of the protein are responsible for the wavelength shift and increases in yields. The former property seems to be essentially determined by polarity of the medium surrounding the dye. The latter appears also to depend on the presence of the chemical groupings (*157*).

Proteins show considerable variation in their influence on the fluorescence parameters of both ANS and TNS. Bovine serum albumin, for example, produces a short wavelength shift of 54 mμ and a yield of 0.34 with respect to TNS, while chymotrypsinogen A exhibits a wavelength shift of 65 mμ but a yield of only 0.07 (*157*). Comparable observations have been made by Stryer (*158*) with respect to ANS. It is obvious that further work is required to distinguish low binding affinities of these substances for protein and an inherent low quantum yield of the bound dye. The potential utility of the fluorescence probe technique is illustrated by the following example.

Stryer (*158*) has employed ANS as a probe in determining nonpolar sites in apomyoglobin and apohemoglobin. Analysis of the fluorescence data indicates there is one site for binding of ANS per apomyoglobin molecule and one per quarter molecule of apohemoglobin. The fluorescence properties of ANS bound to the two proteins were quite similar (quantum yields of 0.92 and 0.98; emission maxima 457 and 454 mμ). The fact that myoglobin and hemoglobin do not bind ANS, together with the observation that hemin will displace bound ANS from apomyoglobin, is strong evidence that the single ANS site is the nonpolar crevice normally occupied by the heme.

TNS in the presence of β-lactoglobulin at pH 5 has a yield of 0.56 while a value of 0.36 was observed at pH 8 (*157*). While no unambiguous interpretation is possible in the absence of more detailed studies of the binding affinity of TNS, it may be that this change in yield is related to the conformational alteration that β-lactoglobulin undergoes near pH 7.5 (*121,157a*).

TNS in the presence of chymotrypsinogen shows a yield of 0.07 and a spectral shift of 69 mμ, while in the presence of chymotrypsin the yield was 0.18 and a shift of 54 mμ was observed (*157*). Although no tryptophan difference spectrum is generated on activation of chymotrypsinogen, solvent perturbation measurements indicate that a slight conformational change occurs which renders the tryptophyl groups of chymotrypsin less accessible to short-range perturbants (*48*). The change in fluorescence properties of TNS bound to chymotrypsinogen and chymotrypsin would appear to provide further evidence for a conformational alteration involving changes in exposure of nonpolar groups.

The conformational change that α-lactalbumin undergoes below pH 4 markedly increases its propensity for association and aggregation (*159,160*). We had attributed this behavior to the enhanced exposure of nonpolar side chains in the acid-denatured molecule. Exploratory studies of the interaction of α-lactalbumin with ANS appear to support this hypothesis (*161*).

The insert of Fig. 17 shows emission spectra of ANS in the presence of native (pH 6.1) and denatured (pH 2.2) α-lactalbumin. Fluorescence of ANS is enhanced in the presence of the protein (compare curves A and B with curve C), the emission maxima occurring in the range 462–467 mμ. The fluorescence intensity increases rather sharply below pH 4 (Fig. 17); the pH dependence is quite similar in form to that observed for the conformational transition, as measured by tryptophan absorption and emission spectra and by optical rotation dispersion (*60,139,162*). There is also the suggestion of a small short wavelength shift of the ANS emission spectrum on denaturation, which may not be evident from insert of Fig. 17.

The above observations could be ascribed to any of the following or combinations thereof: (a) A change in the number of ANS binding sites. (b) Alteration of the affinity of the site for ANS. (c) An increase in the quantum yield of the bound ANS. While we have not as yet carried out the experiments

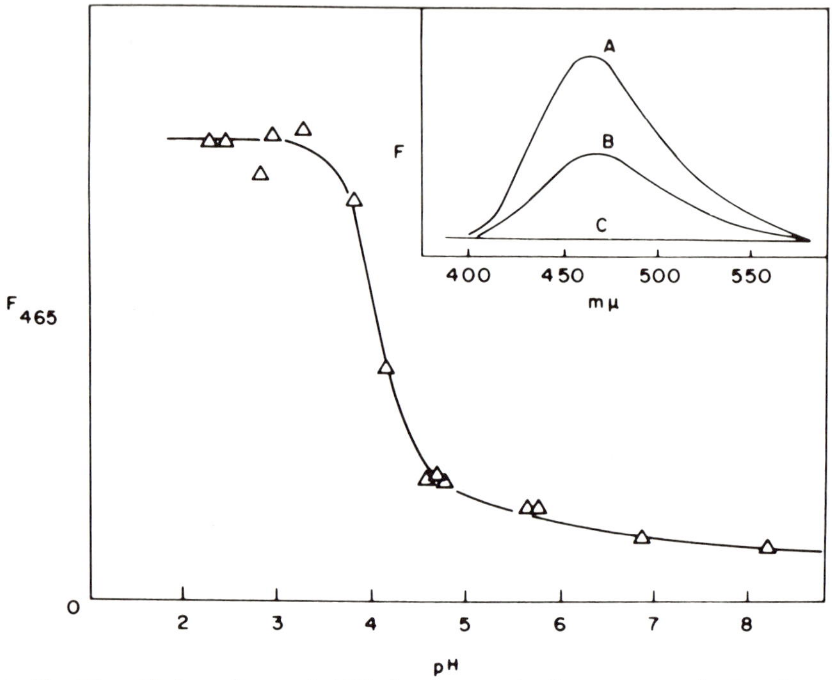

Fig. 17. The pH dependence of ANS fluorescence in the presence of α-lactalbumin. α-Lactalbumin concentration—5×10^{-6} *M*. ANS concentration—1.4×10^{-5} *M*. Insert: Emission spectra of ANS in presence and absence of protein. Curve A, α-lactalbumin, pH 2.2. Curve B, α-lactalbumin, pH 6.1. Curve C, absence of protein (*161*).

necessary to distinguish between these various possibilities, it is evident that the data reflect alterations of the hydrophobic character of the molecular surface on denaturation of α-lactalbumin. Similar observations have been made by Gally and Edelman (*163*) on thermal denaturation of the L chain of γ globulin.

Weber and Young (*164*) and Weber and Daniel (*165*) have employed ANS in a study of the subunit structure of bovine serum albumin and changes in the availability of nonpolar moieties during the acid conformational changes of this protein. Their observations are considered at greater length in Section III. A.

E. Spectral Changes and Exposure of Aromatic Side Chains

Spectral changes which occur during protein transconformation reactions have been commonly ascribed to unfolding of the molecule with subsequent exposure of the chromophoric group to the water of the medium. In view of the manifold ways by which spectra can become altered (*166–168*), such a *general* view must *a priori* be regarded as naive. While in many instances changes in absorption and emission spectra are paralleled by changes in exposure (as determined by one or more of the criteria discussed in this chapter), this correlation cannot be regarded as universal, even from an empirical point of view. The acid denaturation of α-lactalbumin, which has been cited above, represents an instance where shifts in absorption and emission spectra and changes in tryptophan quantum yield occur with no change in exposure, even with respect to perturbants as small as water molecules (see Section II. B). Similarly, changes in the absorption spectra of lysozyme and chymotrypsin are observed on acidification (*35*), but no change is observed with respect to a number of perturbants (*48*). A number of other examples could be cited where spectral changes occur without a parallel change in the exposure of these groups to water.

One can cite systems in which spectral changes occur but where the extent of the increase in exposure, if any, will depend upon how the determination is made. For example, treatment of lysozyme with urea at concentrations (ca. 8 M) too low to produce complete unfolding results in a long wavelength shift of the tryptophan emission maximum to a value comparable to that of free tryptophan in water (*134*). Yet, solvent perturbation indicates very little, if any, change in exposure of tryptophyl groups (compare Tables VII and VIII). By contrast, the reaction of the tryptophyl groups of lysozyme with 2-hydroxy-5-nitro benzyl bromide increases quite significantly after urea treatment (*169*) (see Section IV). The reason for the difference is not clear. Perhaps urea treatment has abolished a number of perturbing interactions of essentially exposed tryptophyl groups with other structural moieties. These interactions may inhibit reaction when lysozyme is in the native state. Finally, there are at least two instances known to us where spectral changes are observed during a conformational change with alteration of accessibility of the chromophore toward larger molecules of the medium, but where no change in exposure to water molecules is observed. Two exposed groups of α-lactalbumin become buried with respect to large perturbants when the temperature of a pH 6 solution is lowered from 25° to about 2° (*29*). Yet, no change is seen in perturbation with heavy water. This structural change is accompanied by a short wavelength shift of the tryptophan emission spectrum but with no change in absorption spectrum being detectable (*60*). A somewhat similar behavior is exhibited by the tyrosyl groups of ovomucoid. At acid pH, the tyrosyl groups of this protein are less exposed to large perturbants than near neutrality (*57*). However, as with the situation with α-lactalbumin, perturbation with heavy water is independent of pH (*66*). The acid conformational change of ovomucoid, a "crevice contraction," gives rise to a blue shift of the tyrosyl spectrum and an increase in the quantum yield for emission (*66*). It is noteworthy that for both α-lactalbumin and ovomucoid, spectral changes can occur without increased or decreased contact with water molecules. It may be that in both cases these spectral changes have little or nothing to do with the degree of exposure *per se*, but result from the alteration in perturbations of the chromophore by vicinal groups as contraction of the crevices occur.

It should be evident from the above remarks that spectral changes in themselves are not reliable indices of changes in exposure of tyrosyl and tryptophyl groups to water molecules in the medium. This is not to say that the environmental changes that these spectral alterations correspond to are unimportant in terms of the properties of the protein. It is apparent, however, that if one is concerned about a property of the protein molecule which depends upon the accessibility of such groups to the medium, such spectral measurements must be supplemented by other techniques such as solvent perturbation or chemical reactivity which provide more definitive information of this kind.

III. MISCELLANEOUS PHYSICAL METHODS

A. Binding of Ions and Neutral Molecules

1. Introduction

When one considers the broad range of structures available for use as molecular probes, one would anticipate that studies of binding phenomena on proteins would yield particularly valuable information about the degree of accessibility of a variety of amino acid side chains. Although extensive studies involving the binding of cations, anions, and neutral molecules have been carried out for more than 15 years (see Ref. *170* for a recent review of binding phenomena), we still know comparatively little about the sites at which such interactions occur. The latter remarks exclude, of course, proton binding studies to be discussed subsequently (see Section III. C and III. D) where the pH dependence of the titration of ionic groups in proteins and the titration behavior of model compounds allow us to identify the individual side chains involved. In this section we emphasize binding phenomena involving hydrophobic sites.

The binding site in a protein may well be a cluster of side chains rather than a single amino acid. The binding properties of cupric and zinc ions to metmyoglobin suggested that imidazole was a member of the chelating site (*171*). Application of the difference Fourier procedure of X-ray diffraction (see below) indicates that the initial cupric ion is bound to histidine A10 of myoglobin and probably sufficiently close to asparagine GH4 and lysine A14 as to permit bonding (*172*). Zinc was likewise found to bind to histidine GH1 and also close enough to the above cited asparagine and lysine to suggest bonding. The histidines in these sites appear to be exposed (chemical reactivity evidence and crystal structure, see Section V for more complete discussion of observations for myoglobin).

Binding of ions or small molecules in some instances induces structural alteration. Metmyoglobin undergoes structural change, for example, on binding of more than one to two cupric ions (*171*). Binding of hydrocarbons such as propane, butane, cyclopropane, and ethylene to human, bovine, and canine hemoglobin appears to result in dissociation of the tetramer on lowering of the ionic strength (*173*). Bovine serum albumin and β-lactoglobulin likewise exhibit conformational changes on binding of alkanes (*174*). It may thus be difficult in many cases to know if binding sites pre-existed in the uncomplex protein.

2. Binding to Hydrophobic Sites

As mentioned in the discussion on the binding of fluorescent dyes (Section II. D), no methods are available for determining the degree of exposure of individual nonpolar side chains with the exception of the aromatic amino acids. Binding studies, however, provide a means of obtaining information about clusters of nonpolar residues. Such side chains tend to be grouped together to minimize their contact with water (*175*), the majority forming relatively large nonpolar anhydrous regions in the interior of the molecule while the comparatively few groups located at the surface probably form clusters. Based on this physical picture, we can envision binding as occurring in two ways (*174*). (a) The bound nonpolar ligand could attach itself to the surface of a cluster located on the exterior of the protein molecule. Rearrangement of such a cluster might occur such that the probe molecule becomes "dissolved" in a nonpolar "droplet." Such a process would probably involve a minimal disruption of the native conformation and might not be detectable. The relaxation effect observed on the cooperative binding of ANS to serum albumin may be an instance of such a perturbation (*176*). (b) The penetration of the bound nonpolar molecule into the nonpolar region of the molecular interior, which might be expected to involve more drastic alteration of the protein conformation. The cooperative binding of detergents to serum albumin (see below) is probably an example of structurally disrupted binding (*177*). In practice, it may be quite difficult to distinguish between these two types of binding. One approach, however, might be to look for the large structural changes which would probably accompany the interior binding, but even this distinction may not be valid.

Kretsinger et al. (*178*) have made some rather extraordinary observations which have bearing on this problem of structural changes induced on binding of a ligand. HgI_3^- was found to be bound (X-ray difference Fourier method, see Section III. A. 5) to an "internal" site in sperm whale myoglobin. This site does not exist in a "preformed" state in the unliganded molecule; there is no cavity in this region large enough to accept even a water molecule. This ion is roughly disk shaped, has a Van der Waals thickness of more than 4 Å, and has a diameter of more than 8 Å. Yet, in spite of the fact that the static model does not reveal an open pathway from the molecular surface to the binding site, binding does occur, presumably as a result of dynamic fluctuations in the conformation. However, the X-ray observations show that the conformational changes which occur on binding must be rather local in nature. It would be interesting to know if complex formation leads to alteration of the macroscopic solution properties of myoglobin.

3. Binding of Hydrocarbon Gases

Measurements of the solubility of hydrocarbon gases such as propane, butane, and pentane in solutions of proteins indicate that such substances bind quite strongly to some proteins and rather weakly to others (*179–183*). For example, there appears to be a single site for binding of alkane to β-lactoglobulin (*183*). This site can apparently bind two molecules of butane, pentane, or xenon, or one of iodobutane. The magnitude of the thermodynamic binding parameters suggests to the authors that the site is a closed cleft near the molecular surface.

Although changes in the binding of hydrocarbon to β-lactoglobulin on dimerization or octomer formation (*184*,*185*) might have been anticipated, no differences in the binding parameters were observed (*182*,*183*). The authors conclude that the region of contact between monomer units and the dimer and octomer is not a large hydrophobic structure. While this conclusion cannot be disputed, alternatively it might be suggested that the binding of alkane inhibits formation of octomer and dimer. Pertinent to this hypothesis is the fact that butane and isobutane alter the optical rotation of β-lactoglobulin at pH 7.5 where it has been shown that the protein undergoes a reversible structural change (*121*,*157a*). Furthermore, Georges et al. (*186*) find that β-lactoglobulin dissociates into monomer in this pH range. One might also mention again the dissociating effects of alkane on hemoglobin, for instance, Taylor et al. (*173*).

Wishnia and Pinder (*182*) have found that pentane and xenon compete for binding to myoglobin, suggesting that a common site exists for these two ligands. Schoenborn et al. (*187*) have shown, using the difference Fourier technique (See Section III. A. 5), that the xenon binding site is located equidistant from the heme-linked histidine and the pyrole ring of the heme group. It is noteworthy that the xenon atom lies between a nonpolar region and one which is polar or charged. This site is in the same region as that found for binding of I_3^- and HgI_3^- (*178*) and for cyclopropane (*188*). It is apparent that the "site" is not specific for nonpolar ligands. Schoenborn et al. (*188*) and Kretsinger et al. (*178*) have suggested that the binding at this site depends upon interactions which are a function of the polarizability of the ligand. The above observations suggest that the use of nonpolar ligands to identify binding sites as hydrophobic regions may give rise to some ambiguity.

From the above remarks about the HgI_3^- binding site, it is apparent that: (a) the sites for binding of cyclopropane and presumably pentane to myoglobin are internal. Thus, binding of these substances does not reflect hydrophobic regions on the *surface* of the molecule. (b) Structural changes must occur in order for the ligands to *reach* the binding site. Thus, binding reflects the ability of the molecule to undergo such a change, as well as the ability to interact with the ligand. (c) Structural changes must occur in myoglobin to permit binding, since no preformed site exists.

4. Hydrophobic Sites in Serum Albumin

The binding characteristics of serum albumin have probably been more widely studied than that of any other protein. The observations made for a number of substances which appear to bind to hydrophobic sites on this protein are worth contrasting (*189*).

Foster (*189*) has proposed a model for the serum albumin molecule which provides a convenient frame of reference in considering interactions of nonpolar substances with hydrophobic regions of this molecule. This model (Fig. 18) is shown as consisting of four folded subunits formed from a continuous peptide chain. Hydrophobic regions are represented as holes in the N form (native closed) as well as the areas of subunit contact which become available in the F form. The N–F isomerization occurs at acid pH and involves changes in a variety of properties of serum albumin (see Ref. *189* for discussion of the N–F transition).

Alkyl benzenesulfonates at low concentrations bind rather strongly to 11 sites in serum albumin, while at higher detergent concentrations, two additional complexes are formed, having 38 and 76 sites, respectively (see Ref. *177* for the most recent observations of detergent binding to serum albumin). Formation of the higher complexes involves alteration of the conformation of the serum albumin molecule comparable to the N–F transition.

This change results in the destruction of the 11 strong binding sites. Foster (*189*) has suggested that the 11 strong sites exist in the native state ("holes," Fig. 18) while the larger number of sites come into existence only after the N–F transition, which can occur either on change of pH or on binding of detergent.

Reynolds et al. (*190*) have studied the binding of a series of alkyl sulfate (C_8 to C_{14}) and sulfonate (C_8 to C_{12}) anions. They find that the total number of strong binding sites is approximately 10 for all ligands. However, all of

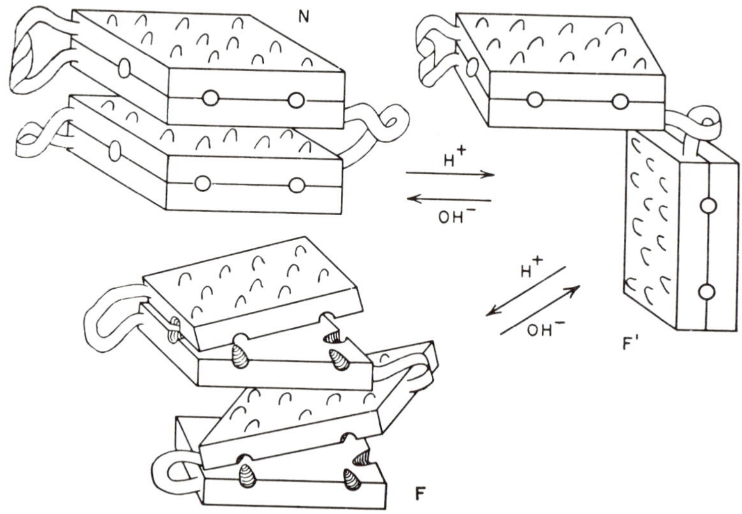

Fig. 18. A highly idealized model for the plasma albumin molecule to account for the N–F transformation and its relationship to the titration anomaly, cooperative detergent binding, and the altered solubility behavior of the low pH form. The model contains four folded amphipathic subunits, the hydrophobic surfaces being buried in the *N* form and exposed in the *F* form. Holes around the periphery of the molecule represent the 10 or 12 strong binding sites for detergent ions which are destroyed, upon isomerization, with the exposure of a large number of weaker sites [from Foster (*189*)].

these sites are not of equal affinity for detergent. The shorter chain anions (C_8, C_{10}) have about three to seven high affinity sites.

Teresi and Luck (*191*) also find two classes of binding sites for fatty acid anions and serum albumin, five strong sites, and about 30 weak ones. Wishnia and Pinder (*180,181*) have shown that at 1 atm butane at 25°, 5 moles of the hydrocarbon are bound per mole of protein although analysis of their binding data suggested the existence of more than 40 sites. The number of moles of butane bound was found by Wetlaufer and Lovrein (*174*) to be somewhat lower (ca. 2.1 to 2.7 groups). The extent of binding of pentane and butane to serum albumin decreases sharply with pH near pH 4 (the region of the

N-F transition) (*180,181*). This decrease in binding of alkane would appear to correspond to the destruction of the strong detergent binding sites. Binding of ANS (see Section II. D) to native serum albumin (pH 7) occurs at five sites (*176*) while at pH 2, the number of sites is considerably greater than five and the binding affinity is considerably lowered as compared to pH 7. The low pH site for ANS binding may correspond to the weak binding sites observed by Decker and Foster (*177*). There appears to be some uncertainty with respect to the number of sites available for nonpolar ligands. It is obvious on comparison of the observations made with the variety of substances cited above, that the apparent number of sites depend upon the chemical nature of the ligand. It would be of interest nonetheless to do competition experiments with various ligands to determine if the 10 to 11 detergent binding sites include those that bind other hydrophobic probes.

One might question the reality of the large number of weak binding sites for detergent. Wishnia and Pinder (*182,183*) have suggested that the very high degrees of binding observed with serum albumin might be due to formation of micelle-like structures with hydrophobic regions of the protein serving as nuclei. Reynolds et al. (*190*) have invoked a similar argument to explain the steplike increase in binding of the C_8 to C_{12} sulfonates and the C_8 and C_{10} sulfates to serum albumin since no conformational change could be detected at the corresponding detergent concentrations.

The effect of very short peptic digestion on serum albumin on the binding of ANS is also noteworthy. Weber and Young (*164*) have found that as digestion proceeds, both the number of sites and their affinity for ANS decreases. This short digestion appears to fragment the serum albumin molecule into subunits (see Fig. 18) (*176*). This decrease in binding appears to correspond to destruction of the strong binding sites (see Fig. 18) and is to be contrasted with the increase in ANS sites reported by Daniel and Weber (*176*) at low pH. This comparison suggests that a given site in the F form or in the expanded low pH form may be composed of nonpolar side chains located on connected subunits. Separation of such subunits by peptic digestion would thus destroy both strong and weak sites.

5. The Difference Fourier Method

If the structure of a protein molecule has been determined at the atomic level by X-ray diffraction, the atomic coordinates of the site of binding of a ligand or a chemically introduced group can be exactly specified by the difference Fourier method, provided that the derivative crystal is isomorphous with that of the free protein (*192*). In effect, one determines changes in the electron density map resulting from the introduction of new atoms or from the displacement of those of the protein molecule.

The difference Fourier method thus provides a means of considering the interactions of molecular probes with a protein molecule in exact three-dimensional structural terms. As such, this method is a link between techniques based upon solution properties of proteins and determination of absolute three-dimensional structures by X-ray crystallographic methods. Anticipating the discussion of Section V, it can be said that even with a molecular model constructed from high-resolution X-ray data, it may be quite difficult to assess the extent to which particular groups in a protein are accessible to the medium. The combination of X-ray and chemical techniques inherent in the difference Fourier method provides a means of overcoming this obstacle.

In chemical modification studies such as iodination it would, thus, be possible to identify the residue that has undergone reaction. However, unlike peptide analysis methods, which of course are used for identification of reactive residues, the difference Fourier technique is capable of delineating relatively subtle changes which may occur in the molecular conformation in the immediate vicinity of the modified group. The detection of more widespread structural changes is inherent in the X-ray method, i.e., radical alterations of the molecular conformation would likely lead to changes in crystal form such as occurs on oxygenation of hemoglobin (*197*).

Applications of this procedure are referred to where appropriate in this chapter. By way of summary, however, we have listed some of the more pertinent studies in Table XI.

TABLE XI

Studies of Ligand Binding and Chemical Reactivity for Proteins Using the Difference Fourier Technique

Protein	Observations	Reference
Myoglobin	Tyrosine H22 shown to be unreactive with respect to iodination.	*193*
	Internal site found for binding of I_3^-, HgI_3^-, xenon, and cyclopropane.	*178*
	Binding site determined for cupric and zinc ions.	*172*
Ribonuclease	Sites of iodination of tyrosines determined.	*194*
Lysozyme	Tryptophan 108 shown to be preferentially reactive to iodine.	*195*
	Sites of binding of N-acetyl glucosamine oligomer inhibitors determined.	*152*
	Site of binding of penicillin determined.	*196*

B. Proton Binding—Electrometric Titrations

1. General Considerations

In electrometric titration of proteins, the smallest of all probes is employed, the proton. Procedurally one compares the experimental titration curve with that anticipated from the known amino acid composition. The quantities that we derive from such a titration curve are: (a) The number of titratable groups for a given class, (b) the intrinsic dissociation constant or constants, pK_{int}, for such a group, and (c) the heats and entropies of dissociation of this class of group. Discussion of the analysis of hydrogen ion titration curves of proteins has been given in great detail by Tanford (*121*) and more recently

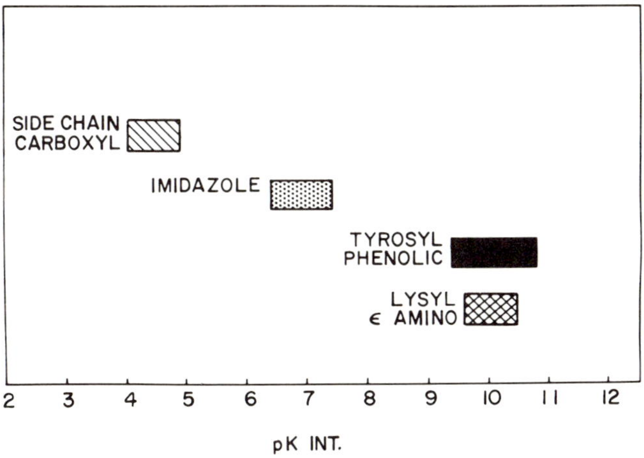

Fig. 19. Intrinsic dissociation constants of ionic side chains of proteins.

by Steinhardt and Beychok (*170*). We therefore only sketch a brief outline of titration phenomena and analysis, highlighting those features which bear on the question of exposure of groups.

The groups in question are side chain carboxyls of aspartic and glutamic acids, histidines, tyrosyl phenolics, and side chain amino groups of lysine. The range of pH over which these classes of groups must be titrated to protonate or deprotonate is seen from Fig. 19. The ranges for a given class of group reflect variations amongst proteins and, of course, represent average values of pK_{int} for the groups of that class. It is seen from this figure that a titration curve should be divisible roughly into three regions: Carboxyl groups in general should be completely protonated by pH 2 and essentially ionized at pH 6. The neutral pH region can be identified with titration of histidines and the α amino group of the N terminus. The alkaline region should correspond to titration of phenolic and side chain amino groups. The overlap of the latter two classes of groups can be resolved by spectrophotometric titration of the tyrosyl groups (see Section II. C). Quantitative analysis of the titration data for the carboxyl and imidazole groups may require some "computational manipulation" to allow for the small degree of overlap which may exist in some cases for these two classes of groups.

The most fundamental use of electrometric titration data is in counting titratable groups on the basis of the division cited in the above paragraph, i.e., "carboxyl region," "histidine region," and "tyrosyl-side chain amino region." Comparison of such figures with the amino acid composition should then be a measure of exposure with respect to protons. It should be noted however, that "abnormal" groups may have pK_{int} values quite different from the "normal" pK_{int} values and may thus not titrate in the anticipated pH region. For example, group counting in the titration of β-lactoglobulin (*121*) indicated two more imidazoles and two less carboxyls than shown by the amino acid analysis. It has been demonstrated however, that two carboxyls of β-lactoglobulin have a pK_{int} of 7.3 and hence titrate in the histidine region (*157a*).

2. Thermodynamic Analysis

If the titration of a particular class of ionizable group is reversible, it may be accorded thermodynamic analysis, the most widely used treatment being that of Linderstrom–Lang. The titration of a given type of group, *i*, would be described by

$$\log\left[\frac{x_i}{1-x_i}\right] = \mathrm{pH} - \mathrm{p}K_{int} + 0.868\,\bar{Z}w \qquad \text{(12a)}$$

$$= \mathrm{pH} - \mathrm{p}K_{app}$$

where x_i is the average degree of dissociation, $\bar{Z}$ is the average charge on the protein molecule, and w is defined by

$$w = \frac{\varepsilon^2}{2\,DRkt}\left(1 - \frac{\kappa R}{1+\kappa a}\right) \qquad \text{(12b)}$$

where ε is the unit of electrical charge, D the dielectric constant of the medium, k the Boltzman constant, T the absolute temperature, R the radius of a sphere representing the protein molecule, κ the Debye–Huckel parameter proportional to the square root of the ionic strength, and a the radial distance of closest approach of an average salt ion to the center of the protein molecule. A detailed discussion of the origin and limitations of this equation is given by Tanford (*121*).

We would expect that for a "well behaved" protein a plot of pH $-\log(x/1-x)$ vs $\bar{Z}$ would be linear and have a slope roughly corresponding to theory [Eqs. (12a) and (12b)]. The value of pK_{int} would then be derived from the intercept at $Z = 0$. Two types of anomalous curves which are of interest are given in Figs. 20 and 21. The Linderstrom–Lang plots at various ionic strengths for titration of carboxyl groups of lysozyme (Fig. 20) are steeper than Eqs. (12a) and (12b) would predict and do not come to common intercepts. This appears to be due to failure of an assumption implicit in the derivation of Eqs. 12, i.e., that all groups can be represented by the same

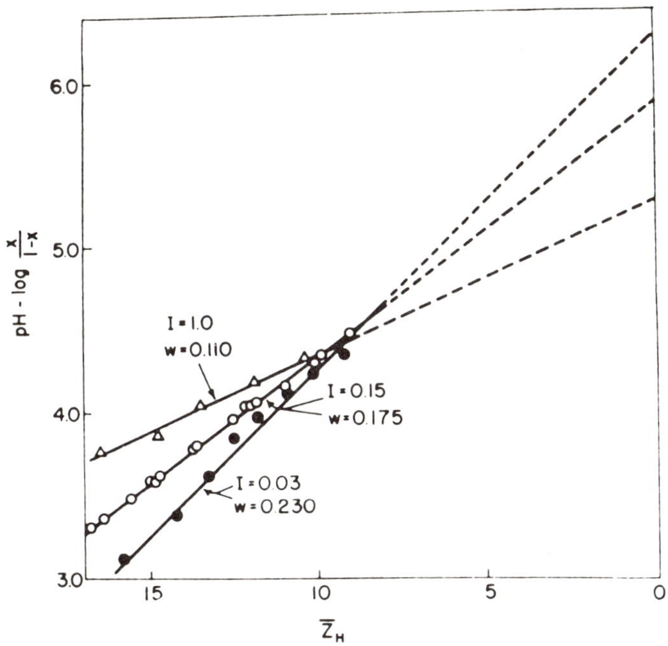

Fig. 20. Titration data for the carboxyl groups of lysozyme, at 25°, at three ionic strengths (*I*) plotted according to Eqs. (12) [from Tanford (*121*)].

pK_{int}. One would in general anticipate that the environment of different ionizable groups of a given type in a protein would be too different to allow representation of their titration by a single pK_{int}. However, with some notable exceptions, such as lysozyme (see below) the titration of carboxyl groups in proteins is generally rather well represented by a single p*K*. If however, certain intramolecular interactions were to occur between a relatively large number of carboxyl groups and other protein moieties, this might require more than one p*K* for representation of the titration data. This appears to be the case for lysozyme (*104*) and ribonuclease (*103*).

A second type of anomalous titration curve is illustrated in Fig. 21 (*108*). Here the Linderstrom–Lang plot for titration of α-lactalbumin shows marked curvature at high values of $\bar{Z}$. Reference to Eqs. (12a) and (12b) indicates that the nonlinearity could be due to changes in pK_{int} or in w, or both. Other physical measurements with α-lactalbumin (*60*,*139*,*162*) demonstrate the existence of a conformational change at a pH corresponding to the value at which the break in the curve occurs (Fig. 21). The curvature is most likely due to changes in w which reflect increases in molecular radius during the conformational change (*160*). Such anomalous plots have been observed with serum albumin (*198*) and other proteins.

It is apparent that any group-counting procedure involving titration into a pH region where denaturation or less drastic conformational changes occur need not refer to the accessibility of titratable groups in the *native* state of the

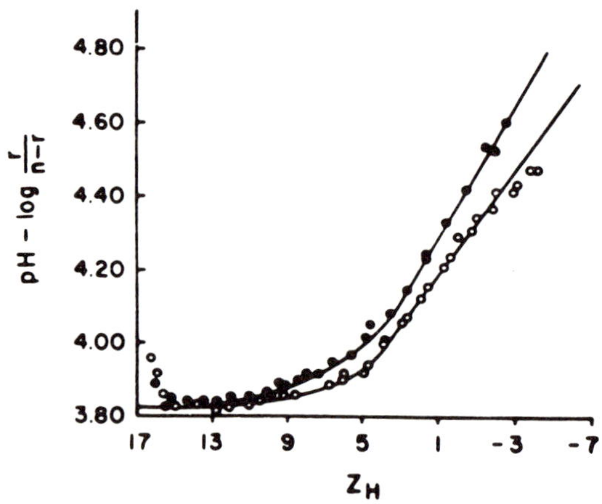

Fig. 21. Titration data for the carboxyl groups of α-lactalbumin plotted according to Eqs. (12) [from Robbins et al. (*108*)].

protein. This of course highlights a weakness of the titration procedure in determining burying of groups, i.e., the necessity in many instances of exposing the protein to extremes of pH where structural changes are likely to occur. In addition to the usual physical chemical means of testing for conformational changes, however, the proper analysis of titration data may also reveal such changes (see above).

Interpretation of "abnormally" high or low pK_{int} values as being indicative of burying of titratable groups is not uncommon. While there appear to be clear cut cases where pK_{int} differs sufficiently from that of a model compound or from "typical" values for a protein, e.g., the pK_{int} of 7.3 observed for the two carboxyls of β-lactoglobulin (see above), this facile interpretation must be viewed with some skepticism.

3. Burying of Titratable Groups

Fundamentally an acid–base titration curve of a protein provides a means of comparing the interaction of protons with an acceptor group in a protein with a comparable isolated group, e.g., a small molecule having the same functional group. A "significant" difference in the dissociation constant for the model compound and for the group in the protein can then be taken as a measurement of perturbing interactions of a titratable group with other structural moieties of the protein. Questions arise however, such as: What is a significant difference? What types of structural perturbations can alter the dissociation constant of an ionizable group and to what degree? Can ionizable groups be buried in both charged and uncharged states? What significance do such perturbations have for an operational classification of such groups as exposed or buried? Unfortunately, we cannot come up with entirely satisfactory or complete answers to any of these questions.

A priori, one would expect that the low dielectric constant within a protein molecule would preclude burying an ionizable group in its *charged* form. Indeed, the empirical evidence for burying of such groups in proteins is scanty. The X-ray crystallographic structure for lysozyme (*152*) suggests that the environment of asp 52 is such that it might be maintained in a charged state even at low pH; indeed, the pH dependence of binding of N-acetyl glucosamine oligomers indicates that lysozyme does have a titratable group with pK_{int} less than 2 (*199*). The titration curve of lysozyme suggests that three carboxylate groups may be buried but the uncertainty as to the validity of this curve (see Section V) renders any conclusion rather tenuous at present. Whether asp 52 could be considered "buried," however, is a moot point (to which we return in Section V). Another protein in which charged groups may be buried is native α-chymotrypsinogen, where three of the 13 lysyl groups do not appear to titrate, although the correct group count is obtained in a denaturing solvent (*114*). No explanation can be given at present for this observation. Perhaps this will be clarified when the X-ray crystallographic structure of this protein has been completed.

While the burying of single charged groups seems an improbable situation, such a process might occur for ion pairs and would be energetically a less costly process. Bovine serum albumin may well have such buried ion-pairs. Vijai and Foster (*198a*) have shown that about 40 carboxyl groups are stabilized as carboxylate ions in the native protein. The existence of a comparable number of unreactive amino groups (see Section IV. A. 1) suggested to these authors that they might be paired with carboxylate ions. They believe that these ion pairs lie in a hydrophilic cleft-like region of the molecule (see Fig. 18). Such a situation, however, does not appear to be common in proteins.

X-ray studies of the binding of HgI_3^- and I_3^- to sperm whale myoglobin indicates that burying of charged moieties can not be ruled out. Kretsinger et al. (*178*) have shown that these large anions, presumably together with appropriate gegenions, bind in a distinctly hydrophobic internal site of the protein.

Iron-complexed conalbumin is an interesting system in terms of pairing of charges of opposite sign. In this complex each of the two Fe^{3+} ions are coordinated to three phenolate ions (*200*). The latter groups in the complex are no longer available for titration (*101*). Such a group may well appear to be buried because of the strength of the complex and yet be on the surface of the molecule. It is by no means certain, however, that this is true of iron–conalbumin. On one hand, the reversal of complex formation with the iron complexing citrate ion suggests that the six phenolate ions are still on the surface. On the other hand, the dramatic increase in resistance of complexed conalbumin to denaturation and to proteolysis (*201–203*) suggest that structural alterations have occurred on iron binding which may well have buried the Fe^{3+}–phenolate locus. It should be added, however, that such a structural change would have to be rather "local" in character since large-scale conformational alterations do not seem to occur on complexation (*204*).

Anomalous pK intrinsic values may well occur where a titratable group on the protein surface is located in a region rich in charges of opposite sign. Under these conditions, the pK obtained by the Linderstrom–Lang treatment at $\bar{Z} = 0$ will not be the true intrinsic pK. If the titratable group were found in a region of high positive charge, the measured pK would appear to be low while for a group in a region of high negative charge, the opposite would be the case. Tanford (*121*) has suggested that the failure of the Linderstrom–Lang equation with the use of a single pK_{int} in describing the titration of lysozyme, ribonuclease, and chymotrypsinogen may be due to such unequal distribution of charges. It is rather difficult to estimate how large a difference in pK_{int} can result from this type of charge perturbation. In the case of ribonuclease cited above, Tanford and Hauenstein (*103*) found that the acid titration curve could be described with a single "reasonable" value of w [Eq. (12b)], if one assumed that five of the 10 carboxyl groups had a pK_{int} of 4.0 while the remainder had a value of 4.7. Hermans and Scheraga (*205*), on the other hand, showed that a fit to the ribonuclease titration curve could be obtained with eight groups of pK_{int} at 4.6, one with a value of 3.65, and another with a value of 2.5. One suspects that there is a certain amount of leeway in curve fitting of such potentiometric data. Thus, even if one were confident that the abnormal ribonuclease titration curves were the result of charge perturbation of carboxyl groups, the experimental data would not permit us to make reliable estimates of the magnitude of shifts in the dissociation constant.

Burying of groups in their un-ionized state is more plausible and indeed such behavior is known. As we have indicated above, two carboxyls of β-lactoglobulin have a pK_{int} of 7.3 and are assumed to exist in this state. Timasheff and Rupley (*211*) have used an infrared titration technique (see Section III. D) to show that a very weakly titrating group is found in lysozyme. We have already seen (Table X) that proteins having nontitratable un-ionized tyrosyl groups may be more the rule than the exception. This is probably no accident since tyrosyl groups are among the most hydrophobic (*29*) and are normally un-ionized at physiological pH. Un-ionized lysine side chains are also comparatively hydrophobic in character. The free energy change, however, required to maintain such a group in an un-ionized state near physiological pH is prohibitive. Un-ionized histidine groups are also hydrophobic but the magnitude of pK_{int} for these groups is comparable to physiological pH and the free energy required to maintain such a group in the uncharged form is small. Thus, buried un-ionized histidines should not be

TABLE XII

Histidine Titration in Native Proteins

Protein	Histidine groups, Analytical value	Histidine groups, no. titrated	Reference
Carbonic anhydrase B	11	4	*113*
Carbonic anhydrase C	12	5	
Conalbumin	13	13	*101*
Chymotrypsinogen	2	2	*114*
Hemoglobin	36	22	*206–208*
		12[b]	
Insulin, zinc-free		2	*91*
Insulin, zinc	2[b]	1[b]	
β-Lactoglobulin	4	6[a]	*121*
α-Lactalbumin	3	3	*108*
Lysozyme	1	1	*104*
Myoglobin	12	6	*209*
Paramyosin	24	24	*210*
Ribonuclease	4	4	*103*
Bovine serum albumin	17	17	*198*

[a] Groups counted as histidine are abnormal carboxyls.
[b] See Section V for a discussion of this difference.
[c] Based on a molecular weight of 5733.

uncommon (see Table XII). It is of particular interest that carbonic anhydrase, hemoglobin, and myoglobin all have buried histidines, since this offers great possibilities for ultimately determining the physical basis of burying of such groups. X-ray crystallographic structures for myoglobin and hemoglobin are now approaching a high degree of refinement and that for carbonic anhydrase will probably shortly be available at high resolution. We reserve our discussion as to the possible origin of burying of histidine groups until Section V where we contrast X-ray crystallographic observations with solution properties.

C. Hydrogen Exchange

1. Exchange of Labile Hydrogens

Except as noted below, the hydrogens exchanging instantaneously or relatively rapidly appear to be identified with the backbone rather than with side chains of proteins (see Ref. *212* for a review of hydrogen exchange in proteins). Since the general problem of protein conformation, including backbone folding, is somewhat outside the scope of this chapter, we only consider a few specialized aspects of this technique.

Wishnia and Saunders (*213*) have studied the exchange of hydrogens for deuterium in native and guanidinated ribonuclease using a nuclear magnetic resonance technique. The native protein was found to have 24 and the guanidinated derivative 54 very slowly exchanging NH hydrogens. The authors suggest that these hydrogens are attached to the guanidinium group of both proteins. Hvidt and Nielson (*212*), in commenting on this interpretation, point out two observations which militate against such a conclusion. First, the diffusion-controlled rate of reaction of hydroxyl ion with the guanidinium group and arginine is extremely high (*214*). Second, all of the labile hydrogen atoms in films of protamine (50% arginine) exchange quite rapidly (*215*). Unfortunately, arginine residues titrate at too high a pH to determine accessibility directly by potentiometric titration (see Section III. B). Thus, while no definitive answer can be given at present to the question of whether protons are freely accessible to arginine side chains, the distribution of ionic groups makes it seem likely that such arginines are freely accessible to the medium.

2. Tritium Labeling of Side Chains

A rather ingenious use of tritium exchange has been suggested by Saunders et al. (*216*) who have exposed chymotrypsinogen to a ^{60}Co gamma ray source in the presence of tritiated chloroform. They find extensive exchange of CH hydrogens for tritium without structural damage to the protein. They suggest that degradation studies could be used to identify which amino acid side-chains have incorporated tritium and which therefore must be at the molecular surface.

D. Infrared Techniques

As indicated in Section III. B, potentiometric titration of ionizable groups in a protein may lead to ambiguity where the regions of titration overlap. Spectrophotometric titration of tyrosyl groups (Section II. C) serves to resolve the overlap of tyrosyl and lysyl groups. The titration of carboxyl groups generally occurs in the pH region sufficiently removed from other side chains to avoid such ambiguity. Instances may arise, however, where burying of carboxyl-containing side chains may produce sufficiently large shifts in the dissociation constant to result in titration at unexpected pH values. Tanford et al. (*121*) for example, have found that two carboxyl groups of β-lactoglobulin are buried and have pK_{app} values of about 7.5. Susi et al. (*157a*) have shown that such groups could be detected through the use of infrared spectra of solutions of proteins in heavy water. The band resulting from the un-ionized carboxyl group was visible as a shoulder (ca. 1710 cm^{-1}) on the strong peptide band in agreement with the observation of Ehrlich and Sutherland (*217*). The amplitude of this shoulder showed a pH dependence which indicated two "abnormal" carboxyl groups titrating only above pH 7.5. These groups appear to be buried in the un-ionized form. Timasheff and Rupley (*211*) have applied this technique in studying an abnormal carboxyl group of lysozyme.

E. Nuclear Magnetic Resonance

Although there have been a number of NMR studies of proteins in solution in the last ten years, the ability to derive structural information from the spectra has been severely limited by the overlap of the chemical shifts of a very large number of structurally and environmentally nonequivalent protons. With the advent of spectrum accumulation and high-frequency NMR instrumentation, such limitations no longer appear to be absolute. The potentialities for obtaining structural information about *individual* residues of a protein warrant a brief account of some recent work, in spite of the fact that these data at present offer little that is concrete concerning exposure of groups.

The studies of McDonald and Phillips (*218*) with a 220 Mc/sec instrument dramatically illustrate the kind of resolution that can be obtained. The NMR spectra that they obtained with a number of proteins give well-resolved peaks that can be reasonably accounted for on a quantitative basis in terms of their amino acid composition. Spectra for native and denatured proteins showed marked differences in position and intensity distribution of resonances due to alterations in the mutual perturbation of side chains. Of particular interest are the so-called high-field satellite peaks observed with lysozyme which they attribute to the perturbing effects of the ring currents of aromatic residues on methyl and methylene protons. The X-ray crystallographic studies of lysozyme (*153*) have shown that the three active site tryptophans 62, 63, and 108 are sufficiently close to other side chains to allow such perturbations. A rather striking observation made by McDonald and Phillips (*218*) was the highly specific alteration of the high-field satellite region of the spectrum of lysozyme on binding of the inhibitor tri-N-acetyl glucosamine.

Another rather striking observation made with the 220 Mc/sec instrument was observed with lysozyme in H_2O rather than the usual D_2O (*219*). Under these conditions it has been possible to observe individual resonances from the indole N–H for five of the six tryptophans. These observations suggest that future refinements in the NMR technique may make it possible to obtain information about the environment of individual tryptophan groups in a protein molecule *in solution*.

Bradbury and Scheraga (*220*) had shown that the C(2) resonance of the histidine residues of ribonuclease shift down field on charging of the imidazole ring. Bradbury and Wilairit (*221*), exploiting this observation, were able to determine an approximate ionization constant for histidine titration in lysozyme from the pH dependence of the shift. Meadows et al. (*222*) have in a similar manner determined approximate ionization constants for *each* of the four hisitidine residues of ribonuclease and staphyloccal nuclease. Binding of calcium to staph nuclease, a process required for activation of the enzyme, was shown to shift one of the four C(2) resonances down field leaving the others essentially unchanged. Similarly, binding of nucleotide inhibitors to ribonuclease shifted and broadened no more than two of the four C(2) histidine resonances. This observation is of particular interest in view of the location of hist 12 and hist 119 at the active side of ribonuclease. These observations with histidine are quite important because one can now specifically "look at" such groups regardless of what pH region they titrate in (see Section III. B). More important, however, is the fact that it may now be possible to obtain information about the environment of individual histidine residues in a protein, and how this environment becomes altered in various states of protein.

The observations cited above, although very preliminary, demonstrate that NMR techniques, particularly those utilizing high-frequency instrumentation, provide a very powerful means of determining the environment of side chains in proteins. It may be that these techniques, used in conjunction with selective chemical modification or binding of various ligands will provide for proteins in solution the kind of information that heretofore has been available only from X-ray crystallographic studies of protein in the solid state.

F. The Spin-Labeling Technique

While chemical reactivity procedures (see Section IV) may make a distinction between completely unreactive and reactive amino acid side chains in a protein, it cannot readily distinguish between groups freely in contact with a medium and those which might lie in crevices and which therefore were less accessible. This problem is quite similar of course to the one encountered in solvent perturbation (see Section II. B) where one cannot distinguish a collection of completely buried and completely exposed groups from one containing a number of partially exposed groups. The spin-labeling technique appears to provide a means of obtaining information about the steric environment of chemically reactive groups. McConnell and co-workers (*223–226*) have shown that the hyperfine structure of the paramagnetic resonance spectrum of molecules containing the grouping:

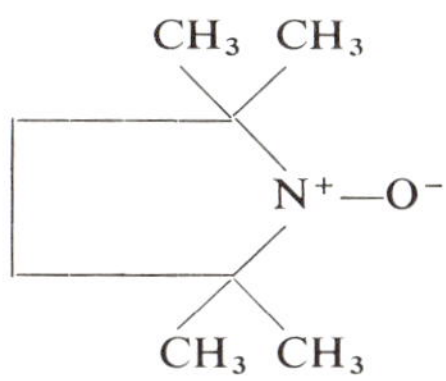

is quite sensitive to molecular motion. The group R can be so chosen as to permit reaction with specific functional groups in the protein. Griffith and McConnell (*223*) have used the spin-labeling compound N-(DL-2,2,5,5-tetramethyl pyrrolidinyl)-maleimide to show restricted freedom of rotation of SH groups in a number of enzymes. Boeyens and McConnell (*224*) have also used this compound and a mercurial spin label to show differences in freedom of rotation of the β93 SH group in hemoglobin and oxyhemoglobin. Berliner and McConnell (*225*) have used a spin-label substrate DL-2,2,5,5 tetramethyl-3-carboxyl pyrrolidone-*p*-nitrophenyl ester to label a serine at the active site of α-chymotrypsin. Analysis of the paramagnetic spectrum indicates that the spin-labeled acyl group is immobilized at the active site.

The spin-labeling technique appears to have a wide range of applications since choice of an appropriate R group should permit covalent coupling to particular amino acid side chains, or binding analogous with that observed with fluorescent dyes (Section II. D) or with co-factors such as NAD. It will be of interest to see how this technique is exploited in the future.

IV. CHEMICAL REACTIVITY OF AMINO ACID SIDE CHAINS

A. Introduction

1. Scope of the Evaluation

A more direct approach to the detection of buried or exposed groups in a protein is the determination of their reactivity with suitable chemical reagents. This review has been limited to methods for modifying tyrosine, tryptophan, and histidine residues since, from a pragmatic point of view, it is these groups which occur most frequently in a buried state. The reasons for this behavior has been alluded to elsewhere in this chapter. Furthermore, since the exposure of these side chains can be determined by spectral and potentiometric titration procedures, comparisons of these data can be made with those obtained by chemical means.

ε-Amino and carboxyl groups have generally been found to be on the surface of the protein molecule (see Section V) and hence may be regarded as "exposed," i.e., in contact with the medium. Nonetheless, strong interactions of such side chains with vicinal protein structural moieties may render them "abnormal" with respect to interactions with protons, e.g., lysozyme (*211*) or with chemical reagents, e.g., ribonuclease (*227,228*). Such "abnormality" may, from an operational point of view, be regarded as burying (see Section I).

The methods for determining reactivity of carboxyl groups in general have not proven too satisfactory, either because the conditions employed lead to denaturation of the protein or because the modifying reagent itself is too unstable in aqueous media. A recently reported method (*229*) employing water-soluble carbodiimides, may prove of value as a carboxyl reagent for proteins.

Picryl sulfonic acid has been used for studying the reactivity of ε amino groups (*230*, *231*). This reagent, which is water soluble, reacts specifically with ε amino groups under rather mild conditions; the degree of modification of such groups is determined spectrophotometrically in the visual absorption region. Goldfarb has studied the kinetics of reaction of picryl sulfonic acid with amino acids and peptides (*232*), and with human serum albumin (*233*). In the latter study he showed that of the 57 amino groups in this protein, only 19 were reactive. Analysis of the kinetics of reaction led to the conclusion of the nonequivalence of the groups. These observations are to be contrasted with the titration data (*234*) that indicate complete exposure of the ε amino residues. It should be noted, however, that titration of these groups occurs in a pH region in which a conformational change is known to occur.

Disulfide and sulfhydryl groups frequently occur in a buried state. Reactivity of these groups has been the subject of several reviews (*235–237*) and is not discussed here.

2. Reactivity and Exposure of Amino Acid Side Chains

In general, if a side chain is located on the surface of a protein molecule such that a potentially reactive group projects into the medium and if this group does not interact with other structural moieties, one would anticipate that the group should react with a suitable reagent at a rate comparable to that of a peptide or amino acid derivative having the same functional group. At the other extreme, if a side chain is located in the anhydrous interior of the protein molecule out of contact with the reagent, it obviously will be unreactive

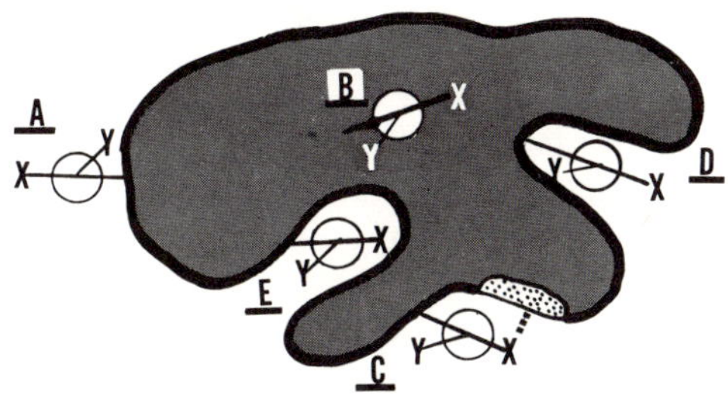

Fig. 22. Schematic view of a protein molecule having side chains, (X-ring-Y) capable of reacting with different reagents at position X and Y. The figure shows that specificity of reaction for X or for Y may determine if a group is counted as reactive or nonreactive (see text).

or buried (but see below for possible exceptions to this, however). The molecular states of amino acid side chains, however, do not fall that conveniently into these extreme cases.

A highly schematic diagram of a protein molecule with a side chain (X-ring-Y) capable of reaction at positions X and Y is shown in Fig. 22, corresponding, for example, to the phenolic OH and the benzene ring of a tyrosine residue. Residues A and B correspond to the two extremes cited above, i.e., a completely exposed and a completely buried side chain. Residue C, being on the surface of the protein molecule, might also be regarded as exposed. Strong interactions of X with other protein structural moieties, however, might markedly decrease its reactivity, e.g., hydrogen bonding of a phenolic OH might be expected to influence its rate of acylation (see Section V for specific examples). Conversely, strong interactions localized at X or at the entire side chain may *activate* the residue to reaction as compared to a low molecular weight model compound under comparable conditions (see reaction of trp 108 of lysozyme, for example, Section IV. C). Thus, the apparent exposure of a residue such as C may depend upon the reagent which is employed. Thus, residue C may be exposed with respect to a reagent specific for Y, but buried with respect to one specific for X, or conversely, with a change in the interactions at X and Y.

An amino acid side chain may likewise show differences in reactivity with respect to different reagents for purely steric reasons. Residues D and E (Fig. 22) are shown in crevice-like regions of the protein molecule. Although both residues may be in contact with solvent molecules, bulky reagents specific for X may "see" residue D as exposed and residue E as buried. The converse would be true for a reagent specific for Y. The reactivity of reagents specific for X and Y will depend, therefore, upon their bulk *per se*, since formation of the transition state complex of reagent and side chain functional group will require a specific geometry.

The above remarks have been restricted to a static view of the protein molecule. A more realistic picture probably involves a specifically folded structure in dynamic equilibrium with a variety of partially unfolded states. The binding of ligands to an *internal* site of myoglobin (see Sections III. A and V) suggests that such transitory processes may likewise be significant in determining reactivity of partially and perhaps even completely buried side chains.

3. Properties of Reagents Used for Determining Exposure of Side Chains

A chemical reagent suitable for studying reactivity or exposure in a protein would ideally have the following characteristics:

(a) It should be relatively stable and soluble in aqueous media over a somewhat wide pH range. The reaction rate should be sufficiently fast to avoid protracted experiments or use of very high concentrations of reagent.

(b) The specificity of reaction for a particular class of side chains should be high, preferably over a relatively wide pH range. The pH of optimal reaction should not lie at extremes of pH where the protein is apt to become denatured.

(c) The degree of modification of the side chain should be readily determined by a physical technique which can be used *directly* on the reaction mixture, e.g., by formation of a new chromophoric group having a characteristic absorption of light in the visible region of the spectrum.

(d) If direct determination of the degree of reaction cannot be made *in situ* because of interference by reagent or by reaction side products, these substances should be easily removable by procedures such as gel filtration or dialysis, and the modified side chain should be stable enough to permit the necessary purification.

(e) In many instances it will be useful if the chemical label on the side chain can be removed by gentle, but specific, means. This may be particularly important in assessing the effect of chemical modification on other molecular properties such as enzymatic activity. If, however, identification of the reactive residue is to be made in the amino acid sequence, the label should be quite resistant to treatment such as the acid or alkaline hydrolysis required prior to amino acid analysis.

Reagents actually used in determining exposure of amino acid side chains in general fall short of the above requirements. The advantages and disadvantages of the individual methods are discussed in detail in subsequent sections.

Ideally, studies of reactivity should involve measurement of the kinetics of the modification of a particular kind of side chain. Analysis of such data should then permit one to distinguish reactivities of individual or groups of side chains (see, for example, the reaction of picryl sulfonate with amino groups cited above). Still more ideally, one should measure the rates of modification of side chains at specific positions in the amino acid sequence. In practice, with very few exceptions, neither of these "ideal" approaches has been employed. This has been the case, in part, because the necessity in many procedures of removing interfering substances (see above) prior to quantitation precludes the detailed measurements necessary for a meaningful kinetic analysis. Furthermore, many of the reagents which are commonly employed are hydrolyzed at significant rates even under conditions for optimal reaction with side chains, thereby complicating any kinetic analysis of the reaction.

In practice, one commonly reacts the protein with a suitable concentration of reagent for a fixed time and then measures the extent of reaction either directly or after removal of interfering substances. Under these conditions the reagent may be essentially depleted by the primary reaction and by hydrolysis. Experimental conditions of temperature, pH, reaction time, reagent concentration (see below), etc., are established by reference to the reaction carried out with the free amino acid or a suitable derivative. Variation of the reagent concentration relative to the concentration of side chains may enable one to distinguish classes of reactive groups, i.e., a smooth reaction curve (groups reacted *versus* relative reagent concentration) might be indicative of a number of residues of comparable reactivity, while the presence of plateaus or inflections would suggest classes of groups having somewhat different reactivities. This method is described in greater detail when the individual procedures are considered. The presence of completely unreactive groups would be evident from the limiting value of reactive groups at high reagent concentration. The possibility of nonspecificity of reaction makes it particularly desirable that a complete amino acid analysis be made on the modified protein, especially when high concentrations of reagent have been employed.

Where the maximum number of groups reacting does not correspond to the analytical value for that particular amino acid residue, the reaction should be carried out on the protein which has been denatured in urea or guanidine and, in particularly questionable situations, on the reduced alkylated protein. Failure to achieve complete reaction in such cases may indicate an inaccuracy in the original analysis for that amino acid; analyses for tryptophan in proteins, for example, are usually subject to relatively large errors. It is possible, however, that a side chain will be unreactive because of its position in the amino acid sequence, rather than because of conformational determinants. We return to this point in subsequent sections.

Use of physical measurements to determine the effect of chemical modification on conformation would seem to be obligatory. Unfortunately few investigators have seen fit to carry out such measurements. These provide a means of determining if the reactivity of certain groups in a protein reflect the environment of such residues in the *native* protein. In addition (see Section IV. B), physicochemical measurements are useful in distinguishing uniquely reactive groups. Measurement of the inactivation of enzymes on chemical modification may serve the same purpose.

B. Tyrosine Modification

1. Introduction

Chemical modification procedures are of particular interest for tyrosyl groups because of the limitations in using solvent perturbation techniques with proteins having even moderately high tryptophan contents (see Section III. B). Chemical procedures also have the advantage of not requiring exposure of the protein to potentially denaturing conditions, i.e., high pH, as is the case with spectral titrations.

In a sense, the chemical reactivity techniques for tyrosine are *potentially* more discriminating than either solvent perturbation or spectral titration procedures. With the availability of a wide range of reagents, it should be possible, for example, to discriminate between the availability for reaction of the phenolic OH and the benzene ring of a tyrosyl group (see also Fig. 22). In this sense the parallel use of N-acetyl imidazole (OH reagent) and tetranitromethane (ring reagent) should, in theory, provide more detailed information about the accessibility of a tyrosyl group than would a purely physical method. Unfortunately, our limited understanding at present of the manner in which these reactions proceed, e.g., the requisite geometry for the activated complex of reactant and reagent, restricts exploitation of this approach.

a. *Acetylimidazole*

The usefulness of acetylimidazole (NAI) for modifying reactive tyrosines in proteins was first demonstrated by Simpson et al. (*4*) with carboxypeptidase

A, then extended to other proteins. The conditions of acetylation are quite mild with tyrosine side chains being converted to O-acetyl tyrosine derivatives. The reaction can be readily reversed by deacetylation of the modified tyrosines with aqueous hydroxylamine. Amino groups react appreciably with some proteins; aliphatic hydroxyl and sulfydryl groups appear to be indifferent in their reactivity.

Reaction of NAI with proteins is carried out at constant pH in aqueous buffer by use of a pH stat, employing a suitable molar excess of reagent to protein (*238*). The pH of reaction is generally near 7.5, where the reagent has its maximum hydrolytic stability and free tyrosines are readily acetylated.

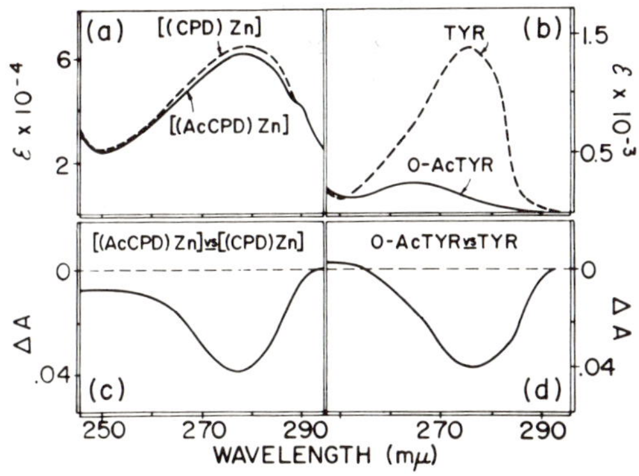

Fig. 23. Absorption spectra of carboxypeptidase and Ac$_I$-carboxypeptidase (a), and tyrosine (Tyr)* and N,O-diacetyltyrosine (O-AcTyr) (b). Difference spectra of Ac$_I$- and native carboxypeptidase (c), and tyrosine* and N,O-diacetyltyrosine (d). The protein concentration in (a) and the amino acid concentrations in (b) were identical for each pair. The concentration in (c) was 7.5×10^{-6} *M* and in D was 3.2×10^{-5} *M*. Cary recording spectrophotometer Model 15, 1-cm cell. All spectra were obtained in 0.01 *M* tris–1 *M* NaCl buffer at pH 7.5, 23° (*4*).

The reaction has been carried out as low as pH 5.5 with some proteins; however, under these conditions the hydrolysis of NAI by water is quite rapid and the reagent concentration must be increased over that used at pH 7.5 to obtain maximum reactivity with tyrosine side chains. High concentrations of nucleophilic buffer ions such as Tris should be avoided during the reaction step and subsequent handling of the acetylated protein to avoid deacetylating the modified tyrosine side chains. Upon completion of the reaction, excess NAI and hydrolytic side products are removed by gel filtration or dialysis since they interfere with quantitation of the degree of modification.

The degree of modification can be determined by the decrease in absorbance of the modified protein in the region of 278 to 280 mμ as a result of formation of O-acetyl tyrosine side chains (Fig. 23). Thus, by comparing the absorbance of a solution of the modified protein with the same solution treated with 0.1 hydroxylamine both at pH 7.5, the number of acetylated tyrosines per mole of protein can be determined. In the calculations, the difference molar extinction coefficient, $\Delta\varepsilon_{278}$ mμ, 1160, is used for obtaining the number of acetylated tyrosines; the protein concentration is determined from the absorbance of the hydroxylamine deacetylated protein at 280 mμ. Corroboration of the number of reactive tyrosines can be obtained by use of the colorimetric ferric hydroxamate procedure employing hydroxylamine at pH 7.5 (*239*). The colorimetric hydroxamate method may be particularly useful in proteins having high tryptophan contents. The relatively high "background" absorption in these cases may make accurate determination of the extent of reaction difficult by the direct spectral procedure. Furthermore, acetylation of tyrosine residues in some instances results in tryptophan spectral shifts which may obscure the changes due to the primary reaction (see below for α-lactalbumin); in such instances the hydroxamate procedure may be more reliable. By carrying out the colorimetric procedure at high pH, the number of acetylated aliphatic hydroxyl groups can be determined by difference.

Acetylated amino groups cannot be deacetylated by hydroxylamine under these conditions, but may be determined by the reduction in ninhydrin color value compared to the unmodified protein (*238*) or by comparison of the reactivity of the acetylated and unmodified protein with 2,4-dinitro fluorobenzene (*240*).

The facile deacetylation of acetylated tyrosines in the presence of hydroxylamine is a particularly useful property which can be employed to detect reversible and irreversible conformational changes occurring in a protein as a result of modification. Thus, by comparing the properties of the modified protein before and after deacetylation with the native protein by such conformation sensitive techniques as optical rotatory dispersion, circular dichroism, etc., the effect of modification can be assessed. This technique has been particularly effective in elucidating the role of reactive tyrosines in enzymes (see below). Facile deacetylation, however, is also a defect of this reagent, since unlike CyF and TNM, it cannot be used for location of reactive groups in the amino acid sequence.

b. *Cyanuric Fluoride*

Kurihara et al. (*241*) introduced cyanuric fluoride (CyF) for modification of tyrosine and demonstrated its capability for detecting reacting tyrosine side chains in proteins. CyF is an active acylating agent which attacks the hydroxyl group of tyrosine; the reaction product, however, has not yet been characterized. While other amino acid side chains can react with CyF, they do not appear to interfere with subsequent quantitation of its reaction with tyrosine in proteins (Fig. 24), although they may produce secondary effects through alteration of the balance of forces stabilizing the native conformation. One of the principal advantages to the use of CyF for detecting reactive tyrosines in a protein is the wide pH range (9 to 12.5) over which the reaction can be carried out. This enables one to classify the tyrosine side chains according to the pH dependency of their reactivity and to further correlate this reactivity with pH-dependent conformational changes in a protein. The

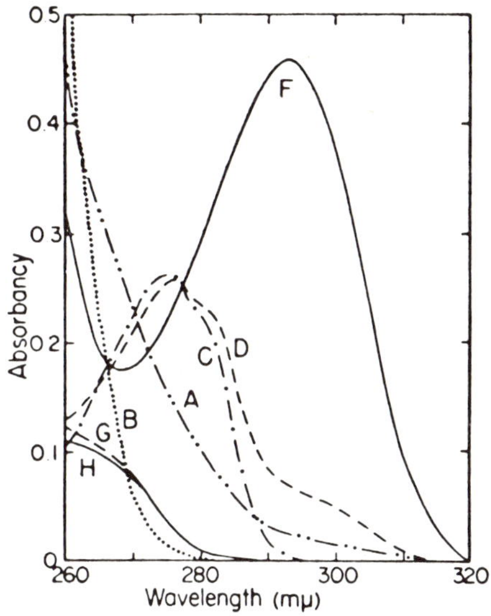

Fig. 24. Absorption spectra of CyF, cyanuric acid, tyrosine, and the reaction products of CyF and tyrosine: 2.22×10^{-2} *M* CyF in dioxane (A): 2.22×10^{-2} *M* cyanuric acid in 10% dioxane at pH 9.1 (B); 0.196 m*M* tyrosine in water at pH 7.0 (C); 9.1 (D); 12.6 (F); 0.196 m*M* tyrosine plus 2.22×10^{-2} *M* CyF in 10% dioxane at pH 9.1 (G); 12.6 (H) after 1.5-hr reaction (*241*).

necessity of using a 10% dioxane solution to solubilize CyF (see below) may prove to be a disadvantage in some protein systems where the conformational stability might be low. While dioxane probably produces gross structural changes only at concentrations exceeding 20%, the possibility of inducing more subtle changes at lower concentrations should not be ignored. The effect of dioxane on the protein in question should certainly be determined by ancillary measurements.

The reaction of CyF with proteins is carried out in aqueous solution containing 10% dioxane to keep the reagent soluble. The number of reactive tyrosines can be determined directly on the reaction mixture, since CyF, which has an appreciable absorption in the spectral region employed for quantitation, hydrolyzes during the reaction to cyanuric acid which does not interfere (Fig. 24). To quantitate the reaction, separate aliquots of the reaction mixture are adjusted to pH 7 and to pH 12 to 13, and the number of reactive tyrosines determined by difference spectroscopy. Since reaction of CyF with tyrosine prevents its ionization at alkaline pH (Fig. 24), the difference in absorbance between the two solutions at 295 mμ is a direct measure of the number of ionized and, therefore, unreactive tyrosine side chains in the protein. Where the unreacted tyrosines do not ionize readily at high pH, it may be necessary to accelerate the process by denaturation with urea (*241*) or guanidine (*242*).

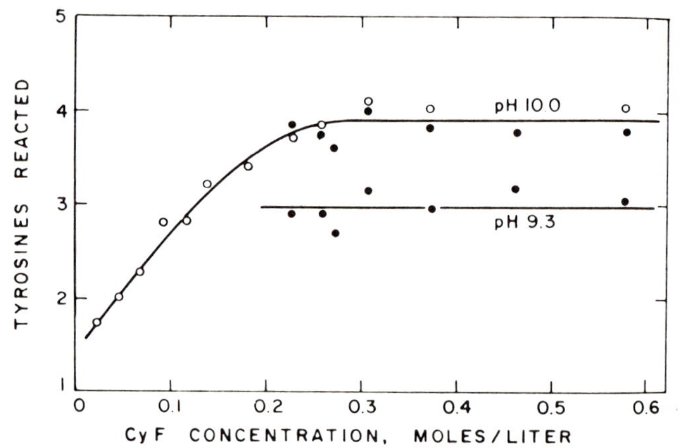

Fig. 25. Dependence of α-lactalbumin tyrosine reactivity on CyF concentration. (○) pH 7 standard. (●) pH 13 standard [from Gorbunoff (*121a*)].

In proteins having high tryptophan contents, interference due to perturbation of these residues may occur and may be quite serious with the CyF method, since the spectral change measured is for *residual* tyrosine groups and occurs fairly close to a maximum in the tryptophan difference spectrum. As in measurement of spectral titration curves, such errors can be minimized by carrying out the spectral measurements at somewhat longer wavelengths.

The reactivity is determined at a given pH employing increasing molar ratios of CyF to protein until a plateau is reached when the number of reactive tyrosines is plotted against the molar ratio of CyF to protein. This procedure is then repeated at various pH values to establish the pH dependency of tyrosine reactivity of the protein (Fig. 25).

c. *Tetranitromethane*

Tetranitromethane (TNM) was first used for modification of proteins by Wormall (*243*). Herriot (*3*) also recommended it as a nitrating agent which could be used for proteins under mild conditions. Recently, Sokolovsky et al. (*244*) have shown that TNM can be used for the specific nitration of reactive tyrosines of proteins in aqueous solution. The product of the reaction with tyrosine is 3-nitrotyrosine [Eq. (13a)]. The only other amino acid that is reactive is cysteine which is oxidized to cystine [Eq. (13b), below].

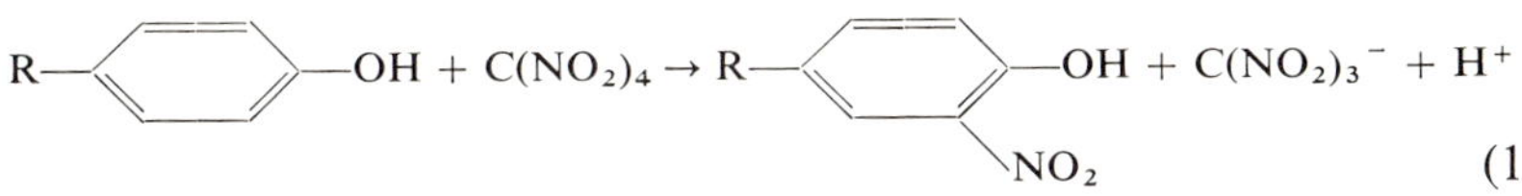

(13a)

$$RSH + C(NO_2)_4 \rightarrow RS\ NO_2 + C(NO_2)_3^- + H^+$$
$$R{-}S\ NO_2 + RSH \rightarrow R{-}S{-}S{-}R + NO_2 + H^+ \qquad (13b)$$

The rate of reaction of TNM with N-acetyltyrosine increases by more than a factor of ten over the pH range 7 to 11. The choice of pH 8 as the "optimum" value for reactions with proteins was made to avoid alkaline denaturation of proteins at higher pH and decomposition of reagent. It seems likely, however, that the useful range could be extended to higher and perhaps somewhat to lower pH values, thereby making it possible to classify reactive groups on the basis of the pH dependence of reactivity, as has been done with CyF (see below).

Nitration of proteins with TNM is carried out by adding TNM dissolved in 95% ethanol to an aqueous solution of the protein that is either buffered or maintained at the desired pH by a pH stat. After completion of reaction, the mixture is subjected to gel filtration or dialysis to remove nitroformate ion which interferes with subsequent quantitation because of high molar absorbance at 350 mμ.

Nitration of tyrosine to 3-nitrotyrosine results in an increase in molar absorption at 275 mμ and the appearance of a new absorption band, with a maximum at 428 mμ (ε_{428}, 4100) at alkaline pH. The absorption spectrum of 3-nitrotyrosine is pH dependent (Fig. 26), such that at acid pH the maximum absorption of the visible band is shifted to 360 mμ with an isobestic point at 381 mμ (ε_{381}, 2200). These absorption properties are used for quantitation of the degree of nitration. Thus, by measurement of the absorption at the isobestic point, the degree of nitration is obtained directly, independent of the pH of the measurement. The degree of nitration can also be determined by amino acid analysis since 3-nitrotyrosine withstands acid hydrolysis and can be separated by the usual ion exchange methods.

The pH-dependent absorption spectra of nitrated proteins should be useful in obtaining information on the microscopic environment of modified tyrosines by following changes in the spectral properties as revealed by perturbation spectra or similar methods. This has been demonstrated with nitro-carboxypeptidase A (*245*), where binding of the competitive inhibitor, β-phenylpropionate, causes a shift in p*K* of the 3-nitrotyrosyl residue from 6.3 to 7.2, indicative of changes in the environment of the chromophore.

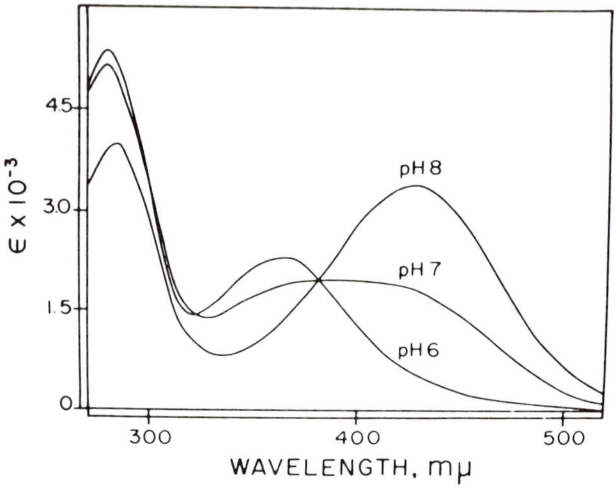

Fig. 26. Effect of pH on the absorption spectra of N-acetyl-3-nitrotyrosine in 0.05 *M* tris (*244*).

d. *Iodination*

Iodination of proteins leads to substitution of tyrosine and histidine as well as oxidative reaction with tryptophan, sulfur-containing amino acids, serine, and threonine (*246,247*). The oxidative reactions can be minimized by avoiding a large excess of iodine (*248*) and by working at low temperatures and alkaline pH (9.0 to 9.5) where the rate of substitution is increased. The iodinating agent is KI_3.

The iodination of tyrosine residues in proteins proceeds via the monoiodo derivative, which can be further iodinated but at a decreased rate to the diiodo derivative. Iodination shifts the p*K* value of the phenolic group of tyrosine to lower pH values; the acidic and basic iodinated forms have absorption maxima at longer wavelengths than unreacted tyrosine. The shift in absorption maxima of tyrosine at alkaline pH from 293 to 305 mμ (monoiodo) or 311 mμ (diiodo) can be used as a measure of the extent of formation of these products (*249*). The total number of unreactive tyrosines can also be determined by amino acid analysis.

The difference in iodine bound by the protein and that consumed in the reaction, as determined by titration with a suitable reducing agent such as thiosulfate, reflects the extent of oxidative side reactions.

2. Application to Specific Systems

a. *General Remarks*

In the following paragraphs, we compare the extent of exposure of tyrosyl groups in proteins with respect to the four reagents, NAI, CyF, TNM, and I. It should be noted that a critical comparison is impossible at present because of a paucity of reactivity data for all of the reagents except NAI (see Table XIII). Furthermore, the NAI data, since they were obtained generally at single concentrations of reagent, are useful only for rough "classification" purposes. Only by using a wide range of reagent concentration would one be able to distinguish very reactive groups from moderately reactive ones and unreactive tyrosines from slowly reacting ones. Furthermore, only with few exceptions has there been any determination whether reaction results in conformational change with concommitant exposure of formerly buried groups.

Where possible, comparisons will be made with spectral titration data (Table X) and with observations made by solvent perturbation (Table VIII). Agreement, or lack of agreement, of the chemical reactivity observations and those made by spectral titration should be considered in terms of the following limitations: (a) In a number of instances titration of tyrosyl groups occurs in a region of conformational change. Thus the titration results do not reflect the molecular properties of the native state. (b) For systems in which time-dependent ionizations are observed, the number of groups classified as "freely titrating" may be rather uncertain.

The above remarks may explain why, where differences exist, the chemical results are frequently lower than those obtained by titration. It is also possible that such differences may reflect the greater steric requirements of the transition state for the organic reagents with tyrosine as compared to hydrogen ion. For some proteins where the observations are rather sketchy, we present the observations in Table XIII without further comment. Lysozyme and myoglobin are discussed in Section V in order to contrast observations of solution properties with information derived from X-ray crystallographic structures.

b. *α-Lactalbumin*

The original amino acid analysis of α-lactalbumin showed five tyrosines for a molecular weight of 15,500 (*263*); the recently published amino acid sequence, as well as a more recent amino acid analysis (*264*), however, show only four groups for a molecular weight of 14,500 (*127*). The spectral titration curve for α-lactalbumin, making allowance for the lower molecular weight shows 4.3 groups in good agreement with the analytical value (see Table X). Analysis of the curve by the Linderstrom–Lang procedure (see Section II. C) shows, however, as do other physicochemical measurements, that a conformational change has occurred in the pH region of titration.

These observations suggest that the complete exposure of tyrosines of α-lactalbumin, seen at high pH, does not reflect behavior of the native state. This is supported by chemical reactivity data (Table XIII).

Three tyrosyl groups react with CyF below the conformational transition region, i.e., at pH 9.3, while at higher pH a fourth group reacts. Four groups were shown to ionize instantaneously at high pH in the CyF reaction solvent, dioxane, 1 *M* bicarbonate. It was concluded that the fifth group based on the original amino acid analysis was both unreactive and nonionizable. No attempt was ever made, however, to establish its presence after denaturation with guanidine or urea. As indicated above, the amino acid sequence, as well as a more recent amino acid analysis, now obviates any conclusions about a fifth tyrosyl group.

Observations made with NAI would seem to be more discriminating than those with CyF. At low levels of reagent (Fig. 27) less than three groups react. It is only with a threefold increase in NAI concentration that the third and fourth groups react. Reaction of these higher reagent levels, however, may not reflect accessibility of tyrosyls in the native state (see below). The increased reactivity of tyrosyl groups of α-lactalbumin after urea denaturation (Fig. 27) suggests that the low reactivity of two of the tyrosines is determined by the manner in which the native molecule is folded. The failure to achieve complete reaction of 8 *M* urea may result from increased lability of the acyl group in urea solutions.

The data shown in Fig. 27 were obtained, with the exceptions indicated, by the colorimetric hydroxamate procedure; observations made by the spectral technique (see above) were systematically higher than those obtained by the hydroxamate method due to an underlying tryptophan spectral shift. The molecular changes producing the shifts in absorption spectra were evident as changes in quantum yield and spectral shifts for the tryptophan emission spectrum (Figs. 28 and 29). At very low levels of NAI, an increase in yield (Fig. 28) is observed that can be reversed by deacylating with hydroxylamine. The occurrence of a sequence tyr 103, trp 104 in α-lactalbumin (*127*) indicates that the reversible perturbation of a tryptophan at low reagent concentration may be the result of acylation of tyr 103. The sequence homology of α-lactalbumin and lysozyme (*127*), as well as the X-ray crystallographic structure of the latter protein (*152*, *153*) suggests that tyr 103 of α-lactalbumin should be on the surface of the molecule. This identification, however, is presently without proof. Nonetheless, these data illustrate the value of ancillary physical measurements in demonstrating differential reactivity.

Above a molar excess of NAI of about 40:1, the fluorescence properties do not return on deacylation to those of the native protein (Figs. 28 and 29), demonstrating that reaction of the third and fourth tyrosines (compare Figs. 27–29) takes place in a molecule which is not structurally identical with

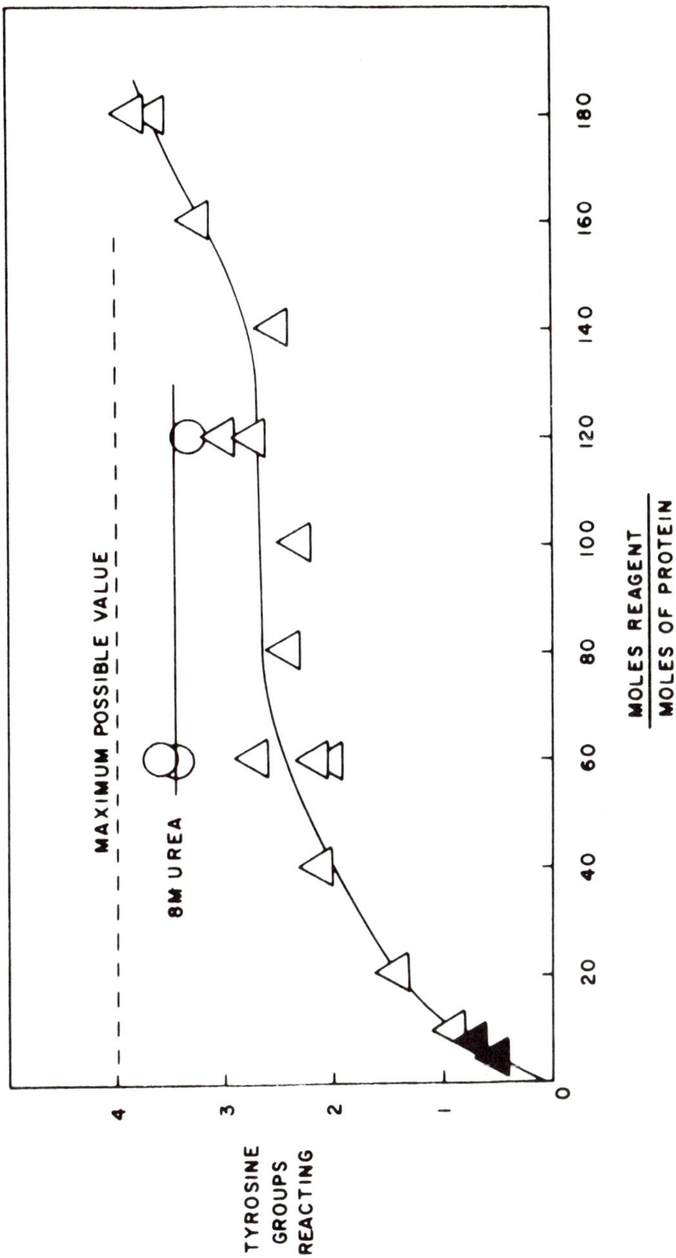

Fig. 27. Effect of reagent concentration on acylation of the tyrosyl groups of α-lactal-bumin. Filled and unfilled symbols correspond, respectively, to data obtained by the spectral and hydroxamate procedures (see text) [from Robbins et al. (*256*)].

TABLE XIII

Reactive Tyrosyl Groups in Proteins

Protein	Total residues	Reactive groups[a]				Literature references			
		NAI	CyF	TNM	I	NAI	CyF	TNM	I
Aldolase, rabbit muscle	44	10	—	—	—	*238*	—	—	—
		25				*230*			
Bromelein (stem)	19	—	7–8+5–7	—	—	—	*110*	—	—
Carboxypeptidase A	19	5.9	—	7.1	5–6	*4, 238*	—	*244*	*250*
		19[b]				*238*			
α-Chymotrypsin	4	2	2	—	—	*238*	*242*	—	—
		4[b]				*238*			
DIP deriv.		—	1	—	—	—	*242*	—	—
Conalbumin	18–19	10–11	—	—	—	*251*	—		—
Fe-complex		5	—	—	—	*238*	—	—	—
Cytochrome, c, Fe^{3+}	4	2, 4[b]	—	—	2[g]	*252*	—	—	*253*
Fe^{2+}		0.4–0.5	—	—	—	*252*	—	—	—
Hemoglobin	12	7.6	—	—	—	*238*	—	—	—
Fructose diphosphate dehydrogenase	39	3+3+4	—	—	—	*2, 254*	—	—	—
Insulin	4	4.1[c]	2+1+1[d]	2[c]	4[d]	*238*	*241*	*244*	*255*
		4[b c]				*238*	—	—	—
α-Lactalbumin	4	2+1+1	3+1	—	—	*256*	*121a*	—	—
		4[b]				*256*	—	—	—

β-Lactoglobulin, bovine A	8	4+2 (slow)	4+2[h]	—	—	*121b*	*121a*	—	—
goat	8	4+2 (slow)	4+2[i]	—	—	*121b*	*121b*	—	—
Lysozyme	3	2, 3[b]	2	2.8	1, 1, 3[b]	*257*	*241*	*244*	*258, 259*
Myoglobin	3	0.8	—	—	2	*238*	—	—	*193*
Ovomucoid	5–6	5.3 0.7–3+ 3 (slow)	2+1+1	—	—	*238,* *121c*	*121c*	—	—
Ovalbumin	9–10	1.5 2+6 (slow)	2+1+1+1	5.6	—	*238* *121d*	*121d*	*244*	—
Pepsin	16	9.2	—	—	—	*240*	—	—	—
Pepsinogen	17	8–10 13.5 (denat.)	7+2+2 +2+1	—	—	*240* *121c*	*121c*	—	—
Ribonuclease	6	3, 6[b]	1+1+1	3.1	3[e] 2+1+1[f]	*238* —	*121a* —	*244* —	*194* *260*
Serum albumin, bovine	18	4	—	—	—	*238*	—	—	—
Soybean trypsin inhibitor	4	2+1 (slow)	1+1	—	—	*121c*	*121c*	—	—
Trypsin	10	6.7, 10[b]	—	—	7–9	*238* *261*	—	—	*262*

[a] NAI, N-acetyl imidazole; Cyf, cyanuric fluoride; TNM, tetranitromethane; I, iodine.
[b] Denatured state.
[c] "Bovine insulin."
[d] Zn insulin.
[e] pH 6.7.
[f] pH 9.
[g] Spectral results based on assumption of formation of diiodo tyrosine (pH 9.5).
[h] Four groups react at pH 9.3; two additional on reaction at 10.8 where protein is irreversibly denatured.
[i] Four groups react at pH 9.3; two additional on reaction at 10.0. Protein is irreversibly denatured above pH 8.7.

native α-lactalbumin. The fact that reversal does not occur after treatment with hydroxylamine makes it probable that acylation of amino groups is responsible for the observed changes. These observations emphasize the need for monitoring chemical reactivity studies with physical measurements to identify conformational changes where they may occur.

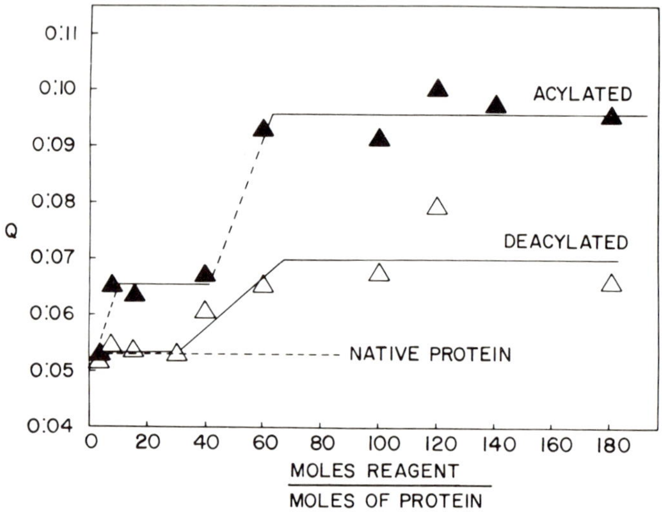

Fig. 28. Effect of acetyl imidizole concentration on the quantum yield of tryptophan fluorescence of α-lactalbumin. Filled symbols, acylated protein. Unfilled symbols, deacylated protein [from Robbins et al. (*256*)].

c. *Ribonuclease*

Observations of exposure of tyrosyl groups of ribonuclease are in amazingly good agreement (see Table XIII). CyF shows three reactive tyrosines, each having a distinctive reactivity. NAI likewise indicates three reactive groups; the high level of reagent required suggests, however, that some of these residues are less reactive. Acylation of the three reactive tyrosyl groups of ribonuclease had no significant effect on the optical rotation dispersion either in the side chain Cotton effect region or at 228 mμ, proving that no structural changes have occurred on reaction (*265*). Studies of the rates of iodination indicate three reactive residues, 73, 76, and 115. Of the three "unreactive" residues, 92, 97, and 25, one, tyr 92, becomes reactive at pH 9. These observations are in good agreement with the spectral data (Table X) which show three exposed groups. The solvent perturbation observations (*88a*) by contrast appear to be best explained on the basis of two completely exposed groups and two partially exposed ones. The physical basis for the burying of such residues is examined in Section V.

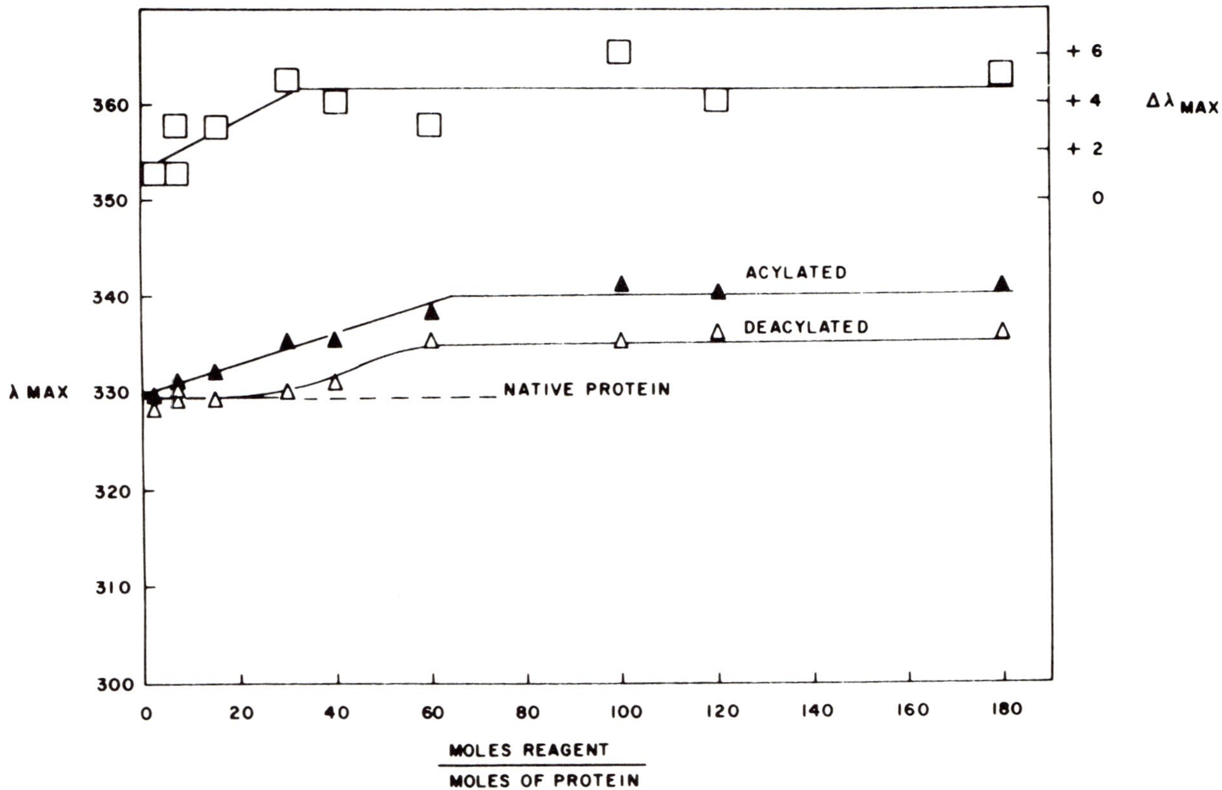

Fig. 29. Effect of acetyl imidizole concentration on the emission maximum of the tryptophan fluorescence spectrum of α-lactalbumin. Spectra corrected for detector sensitivity. ▲, acylated protein. △, deacylated protein [from Robbins et al. (*256*)].

d. *Insulin*

The reactivity of tyrosyl groups in insulin appears to depend upon the reagent employed (Table XIII). Thus, CyF is able to distinguish differences in reactivity, while the other phenolic OH reagent, NAI, shows reaction with all four groups. One nucleophilic reagent, TNM, "sees" only two reactive groups, while another, iodine, reacts with all four.

Studies of insulin seem to indicate that the four tyrosines are not equally accessible to the medium. Spectral titration data indicates one slowly titrating group (Table X) (see footnote to Table X, however) and solvent perturbation indicates exposure of two or three tyrosines for zinc and zinc-free protein, respectively (Table VIII).

There is indication that removal of the zinc alters the exposure of tyrosyl groups. In addition to the solvent perturbation observations cited above, two other studies provide evidence for this conclusion. Cory and Frieden (*266*) have shown that while all four tyrosines of zinc-insulin are oxidized by tyrosinase at nearly the same rate, removal of the zinc results in a marked increase in the rate of oxidation of one of these. Similarly Covelli and Wolff (*255*) have found that although all four tyrosines of Zn-insulin and Zn-free insulin are iodinated, removal of the zinc alters the *rate* at which some of the tyrosines react.

The studies of Aoyama et al. (*267*) indicate which groups react with CyF. Peptide analysis of CyF-treated insulin demonstrated that the reactive tyrosines were A19 and B16. A "buried" group, which could be normalized by tryptic digestion, was identified as B26 while the other "buried" group was A14. This is consistent with observations made by Friedman et al. (*260*). Aoyama et al. propose that residue B26 is unreactive because of hydrogen bonding to Arg B22, while A14 is unreactive and ionizes slowly because of hydrogen bonding to Glu B13. Since the three-dimensional structure of this protein is at present unknown, these conclusions must be regarded as tentative.

e. *Cytochrome c*

Two tyrosyl groups of cytochrome c (ferri) are acylated by NAI. The spectral change on iodination is consistent with two reactive groups (see footnote to Table XIII). This agrees well with solvent perturbation data (Table VIII), which also indicate two groups exposed in the native state. The spectral titration curve, while not readily analyzed in an unambiguous manner, is likewise consistent with two accessible groups (see Table X and pertinent footnotes).

Reduction of the protein to the ferro form results in a marked diminution of the reactivity of tyrosyl groups (Table XIII). The spectral titration curve, as might be anticipated, is shifted to higher pH on reduction (*253*), although the magnitude of the change in terms of groups titrating is difficult to ascertain. Reduction of cytochrome c is known to cause extensive changes in its properties, including its molecular conformation (*268*).

f. *β-Lactoglobulin*

The observations made with CyF and NAI are in close correspondence for both the bovine A and goat proteins. In the case of the former, reaction of NAI with four residues occurs with no change in molecular conformation while further acylation results in significant changes in the circular dichroic spectrum (*121b*). While the *number* of reactive tyrosines appear to be the same for both the bovine and goat proteins, the *rate* of acylation is much slower in latter case.

g. *Bovine Serum Albumin*

Four of the 18 tyrosines of bovine serum albumin react with NAI (Table XIII). Observations with solvent perturbation lead to comparable values (Table VIII); bulky perturbants "see" five to six groups, while small ones indicate about nine residues. The titration curve by contrast (Table X) shows ionization of all 18 residues, but the Linderstrom–Lang analysis of the data (Section III. C, Table X) reveals conformational change in the titration region.

h. *Carboxypeptidase A*

In view of the imminent availability of a high-resolution X-ray crystallographic structure for this enzyme, the observations made with this protein are worth examining. NAI, TNM, and I (Table XIII) indicate about six to seven reactive groups out of a total of 19. Spectral titration shows that seven to eight residues ionize readily (Table X). The involvement of tyrosyl groups in the active site of carboxypeptidase A is considered below.

j. *Enzymatic Properties and Tyrosine Reactivity*

One can frequently take advantage of the catalytic properties of an enzyme in making distinctions between its tyrosyl residues. In the cases cited below, the involvement of such groups in the enzymatic process is briefly considered.

NAI is particularly useful in this regard because of the mild conditions employed for acylation and deacylation. With carboxypeptidase A (*4*), as acylation takes place, peptidase activity decreases to zero while a sevenfold increase in esterase activity occurs. These changes parallel the acylation of two tyrosines. When the enzyme is acylated in the presence of the competitive inhibitor β-phenylpropionate, only two tyrosines are acylated compared to six in the unprotected enzyme. Treatment of the unprotected enzyme with aqueous hydroxylamine shows that two of the six modified tyrosines are deacetylated at a much faster rate than the others, and parallel a concommitant regain of the original enzymatic activity. Of the remaining modified tyrosines, two are deacetylated at a slower rate identical with the rate of deacetylation of the two modified tyrosines in the protected enzyme. Modification of carboxypeptidase A by various other reagents has shown a similar effect, and also indicates one or two of the exposed tyrosines are involved in the active site of the enzyme. Thus, a single tyrosine can be modified by nitration with TNM (*245*), or by coupling with 5-diazo-1H-tetrazole (*271*) using a limited molar excess of either reagent. Both modifications increase the esterase activity of the enzyme; however, only nitration significantly decreases peptidase activity. These results suggest that the two reagents each modify a different tyrosine and that the groups may have different functional roles in the enzymatic activity.

Another interesting example of the use of NAI in discerning a functional role for tyrosines in an enzyme is with fructose diphosphatase (*2,254*), where reactive tyrosines are involved in the active site as well as the allosteric site. Reaction with NAI shows that 10 of the 39 tyrosines in the enzyme are acetylated with essentially complete loss of catalytic activity. When acetylation is limited to two or three tyrosines, the modified enzyme retains all properties of the native enzyme. By comparison, when acetylation is increased to six tyrosines, full catalytic activity is retained, but the enzyme is completely desensitized to allosteric inhibition by adenosine monophosphate. All other properties of this derivative are identical to the native enzyme and it is still capable of being inhibited by an excess of the substrate, fructose-1,6-diphosphate. Acetylation in the presence of substrate leads to modification of only six tyrosines, and protects the enzyme from inactivation but it is not sensitive to allosteric inhibition by adenosine monophosphate. By contrast, acetylation in the presence of adenosine monophosphate leads also to acetylation of six tyrosines with no loss in allosteric inhibition, but with a large loss in enzymatic activity. With all of these derivatives, the original properties of the enzyme were restored by deacetylation with hydroxylamine. Similar results have also been obtained by iodination of the tyrosines of fructose diphosphatase (*269*).

In a similar study with aldolase (*270*), reaction with NAI also leads to changes in the catalytic properties of the enzyme. These changes are associated with acetylation of 25 of the 40 tyrosines and appears to affect primarily the site binding the 6 phosphate group of the substrate, fructose-1,6-diphosphate and secondarily the ability of the Schiff base intermediate to dissociate from the enzyme. The original properties are regained upon deacetylation of the modified enzyme.

In the case of trypsin, acetylation with NAI in aqueous buffer leads to reaction of 6.7 of the 10 tyrosines with no loss in activity. In 8 *M* urea, all 10 tyrosines are acetylated with complete loss of activity. Deacetylation of the completely acetylated enzyme restores full activity. Since unmodified trypsin can be reversibly denatured under the same conditions, loss of activity upon acetylation in 8 *M* urea must be associated with three buried tyrosines which are apparently essential for stabilizing the active conformation of the enzyme.

C. Tryptophan

1. Introduction

Aside from their inherent utility in determining exposure of tryptophans in proteins, chemical reactivity procedures effectively supplement solvent perturbation (see Section II. B). As we have pointed out earlier, solvent perturbation is limited in that it cannot distinguish full exposure of a few groups from partial exposure of a larger number. Chemical reactivity procedures may permit resolution of this ambiguity. These methods also have the advantage of being able to label an exposed group in the amino acid sequence, thereby providing a means ultimately of assessing the nature of its environment.

N-bromosuccinimide (NBS) has been widely used for estimating accessible tryptophans in proteins. As we consider subsequently, there is considerable indication that this reagent is far from ideal for this purpose (see Section IV. A for a discussion of the requirements for reagents used for group modification in proteins). While the reaction of ozone with the tryptophans of proteins appears to be rather specific (*272*), the use of anhydrous formic acid as a solvent precludes the use of this method in determining accessibility of tryptophyl groups in *native* proteins. Two relatively new procedures involving reaction with 2-hydroxy 5-nitrobenzyl bromide and with hydrogen peroxide appear to provide more reliable, less ambiguous routes for the determination of tryptophan reactivity. These, together with reactions involving NBS, are discussed below.

a. *N-Bromosuccinimide*

Oxidation of proteins by N-bromosuccinimide (NBS) at acid pH (3 to 4) has been used extensively for detecting reactive tryptophan side chains. The oxidation is not limited to tryptophan, however. Other amino acids including tyrosine, histidine, methionine, cysteine, cystine, and lysine can be oxidized (*273*), but this process has been generally assumed to be minimal by comparison with the oxidation of tryptophan if a low molar ratio of NBS to protein was maintained. The details of the method have been thoroughly reviewed by Witkop (*273*). This section is therefore limited to some general comments on the use of NBS for oxidation of proteins and to a brief description of recent work in which the oxidation has been extended to higher pH values.

NBS oxidation of tryptophan side chains in the pH range 3 to 4 leads to formation of an iminolactone which can be further oxidized to a dioxindole with concommitant hydrolytic cleavage of the polypeptide chain of the protein [Eq. (14), see Appendix, p. 407]. By contrast, Green and Witkop (*274*) have shown that the oxidation of tryptophan can also be carried out in the pH range 5 to 7 with the formation of an oxindole which does not lead to scission of the polypeptide chain [Eq. (14)]. This observation suggests that these conditions, rather than oxidation at lower pH, may be preferable for detecting exposed tryptophan side chains in native proteins since the possibility of exposing buried tryptophans through cleavage of the polypeptide chain is eliminated (see below, however).

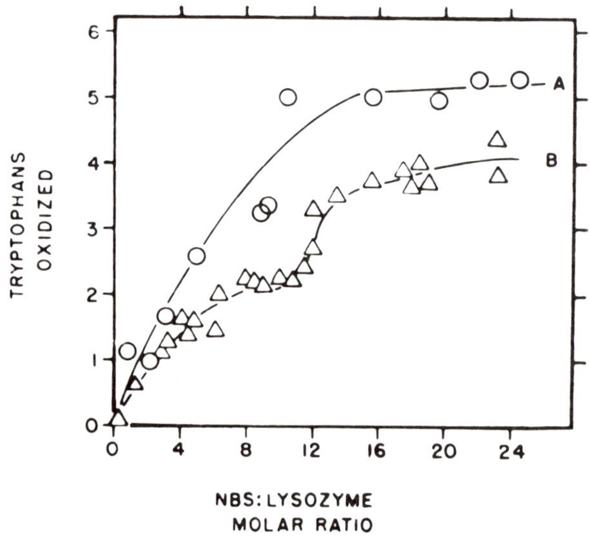

Fig. 30. Extent of oxidation of tryptophan residues of lysozyme as a function of N-bromosuccinimide (NBS)/lysozyme molar ratio. Curve A, reaction in pH 4.00, 0.1 *M* acetate buffer. Curve B, reaction in pH 5.52, 0.02 *M* acetate buffer. The number of tryptophan groups oxidized was determined by the spectral procedure [from Kronman et al. (*69*)].

Recent studies (*274,275*) on the oxidation of reactive tryptophan in proteins by NBS indicates an apparent marked pH dependency of their accessibility. The extent of tryptophan oxidation was determined using the spectral titration method (*273*), i.e., through measurement of the decrease in absorbance of the solution of oxidized protein at 278 mμ. The empirical factor 1.31 (*273*) relating the decrease in absorbance to the degree of oxidation of tryptophan was found to be invariant over the pH range 3 to 7. This work showed that with the majority of proteins the reactivity of tryptophan to NBS is greatest in the pH range 3 to 4, and decreases substantially as the pH is raised toward pH 7. Since most of the proteins that exhibited this behavior had isoelectric points above pH 7, it was suggested that the data might reflect conformational changes induced by the increasing positive charge with decreasing pH (*274*). However, as the authors themselves point out, none of the proteins, with the exception of myoglobin, are known to undergo conformational change in the pH region in question.

An alternative explanation, which was not considered however, is the possibility that at pH 4, cleavage of peptide bonds occurs [see Eq. (14)], liberating previously buried tryptophans which then undergo reaction. Lower reactivities might thus be expected to be observed at higher pH where cleavage of the peptide bond does not occur [Eq. (14)]. We have found, however, that the latter explanation is not sufficient to explain the anomalous behavior of

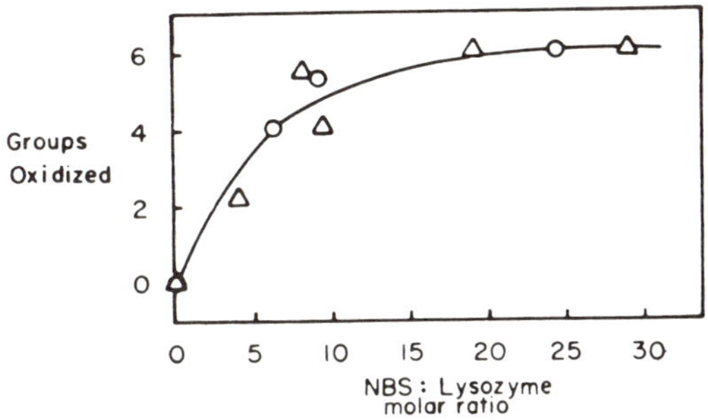

Fig. 31. Extent of oxidation of tryptophan groups of lysozyme as a function of N-bromosuccinimide (NBS)/lysozyme molar ratio. Values determined by amino acid analysis after alkaline hydrolysis. △, reaction carried out in pH 5.52, 0.02 *M* acetate buffer, ○, reaction carried out in pH 4.00, 0.1 *M* acetate buffer [from Kronman et al. (*69*)].

NBS. Our observations (*69*) with the lysozyme–NBS system which support this contention are discussed below.

In this study we examined the reactivity of the tryptophans of lysozyme with NBS at pH 4 and 5.5. Spectral observations of reactivity of tryptophans at pH 5.5 suggest that two of the four reacting groups (see Fig. 30) are more reactive. The difference in reactivity at pH 4 and at pH 5.5 is in accord with the observations of other investigators (*274*, *276*). Amino acid analysis of the tryptophan content of alkaline hydrolysates of NBS-oxidized lysozyme showed, however, that the conclusions drawn from the spectral analysis are in error; the number of groups reacting was essentially independent of pH (Fig. 31). Furthermore, the absence of a plateau in Fig. 31 suggests that the differential reactivity observed by the spectral procedure is an artifact (compare Figs. 30 and 31.)† The spectral analyses for tryptophan oxidation in

† We do not wish to imply that the tryptophans of lysozyme are equally reactive. Indeed, peptide analysis of lysozyme treated at pH 4.5 with very low levels of reagent (NBS: lysozyme molar ratios of about 1:1) indicates the preferential reactivity of trp 62 (*82*). It should be noted however, that the concentrations of NBS employed were less than one-tenth of those used in our studies.

lysozyme were without exception lower than those obtained by amino acid analysis (compare columns 3 and 4, Table XIV). We believe that this difference is the result of a systematic error in the spectral procedure due to oxidation of tyrosyl groups (see below). Oxidation of tyrosines in proteins is known to produce an *increase* in absorbance (*277*). Such an increase would tend to make the apparent spectral changes due to tryptophan oxidation too low. We believe, furthermore, that the apparent pH dependence of the NBS reaction with proteins is illusory and likewise the consequence of the error

TABLE XIV

Comparison of Oxidation of Tryptophan by Spectral Procedure and by Amino Acid Analysis

pH (1)	N-Bromosuccinimide/lysozyme molar ratio (2)	Tryptophans oxidized: Spectral (3)	Tryptophans oxidized: Amino acid analysis (4)	Tryptophans oxidized: Ratio[a] (5)
4.0	6.1:1	1.9	4.1	0.46
4.0	9.2:1	3.2	5.3	0.60
4.0	24.5:1	5.3	6.0	0.88
5.5	4.0:1	1.7	2.2	0.77
5.5	6.1:1	2.2	4.1	0.54
5.5	8.2:1	2.8	5.5	0.51
5.5	9.6:1	2.6	4.0	0.65
5.5	19:1	4.3	6.0	0.71
5.5	29:1	4.5	6.0	0.58

[a] Ratio of number of groups oxidized, determined spectrally to those obtained by amino acid analysis (*69*).

inherent in the spectral procedure (see Ref. *69* for a more complete discussion of this point).

Our studies with lysczyme also demonstrate that the assumption usually made of specific reaction of NBS with tryptophans is in error. Amino acid analysis revealed that significant destruction of tyrosine and histidine occurred at relatively moderate concentrations of NBS, i.e., reagent concentrations at which the spectral observations indicated incomplete reaction with tryptophan (compare Figs. 30 and 31). The occurrence of an NBS-induced conformational change at pH 4 (*276*) and pH 5.5 (*69*) makes it difficult to know if tyrosines and histidines react when the protein is in the native state or whether, on the contrary, oxidation of these groups can occur only after the molecule has become disrupted.

It should be noted that the high molar ratios of NBS to lysozyme required for oxidation of tryptophan at pH 5.5 in themselves are indicative of consumption of the reagent in side reactions such as oxidation of tyrosyl and histidyl groups, e.g., although the spectral procedure indicates a requirement of at least 5 moles/mole of tryptophan oxidized, amino acid analysis shows only about 2 moles/mole (compare Figs. 30 and 31). The latter value is comparable in magnitude with that required for oxidation of carbobenzoxytryptophan (*278*).

The above remarks illustrate the kinds of pitfalls that may be encountered with the use of NBS. It is clearly impossible to categorically state that the anomalous behavior of NBS on reaction with lysozyme will be the rule with all proteins. Unfortunately, there appear to have been few instances where investigators using NBS have monitored their reactions by amino acid analysis or by physicochemical techniques. One example known to us, other than lysozyme, is the oxidation of horse heart cytochrome c by NBS (*279*). Reaction of the single tryptophan in this system is accompanied by changes in viscosity, phenolic ionization, visible spectrum, and alteration of both methionine residues. Thus this protein, like lysozyme, undergoes a significant structural change on reaction with NBS. It is impossible, therefore, to assess the *quantitative* validity of most of the NBS data in terms of exposure of tryptophans in *native* proteins. There are, however, systems in which *changes* in NBS reactivity occur on interaction of the protein with another substance. These changes probably do reflect altered exposure of tryptophans, although such observations should be regarded as purely qualitative.

b. *2-Hydroxy-5-Nitrobenzyl Bromide*

Koshland and co-workers (*169,280,281*) introduced 2-hydroxy-5-nitrobenzyl bromide (HNBB) for the modification of tryptophan residues in proteins. The reagent has a high degree of specificity for tryptophan at acid pH, although the sulfhydryl group of cysteine does react at a slower rate. At alkaline pH, cysteine and tyrosine react at about the same rate as tryptophan. In contrast with other alkylating agents, no reaction occurs with histidine or lysine. Methionine apparently forms an unstable sulfonium salt which rapidly decomposes to the original methionine with formation of the HNB carbonium ion. HNBB can be reacted with proteins in aqueous buffer or 8 *M* urea, but solutions of guanidine hydrochloride should be avoided, since a reaction of unknown character occurs which leads to substitution of residues other than tryptophan, cysteine, and tyrosine (*169*). The product of the reaction of HNBB with tryptophan has not been identified; it was thought to be, however, a monosubstituted derivative (*281*). More recently, however, evidence has been presented which suggests that a di-substituted derivative may form at molar ratios of HNBB to tryptophan in excess of 20:1 (*169*).

The reaction with HNBB is generally carried out by addition of the reagent dissolved in acetone to a buffered solution of the protein at the desired pH, so that the final concentration of organic solvent is 5% by volume. Alternatively, the solid reagent can be added to a rapidly stirred solution of the protein.

Removal of the hydrolytic side product, HNB alcohol, must be carefully carried out since its spectral properties are nearly identical with those of the groups coupled to the protein and would interfere with subsequent quantitation of the extent of reaction by spectral analysis (see below). Gel filtration of

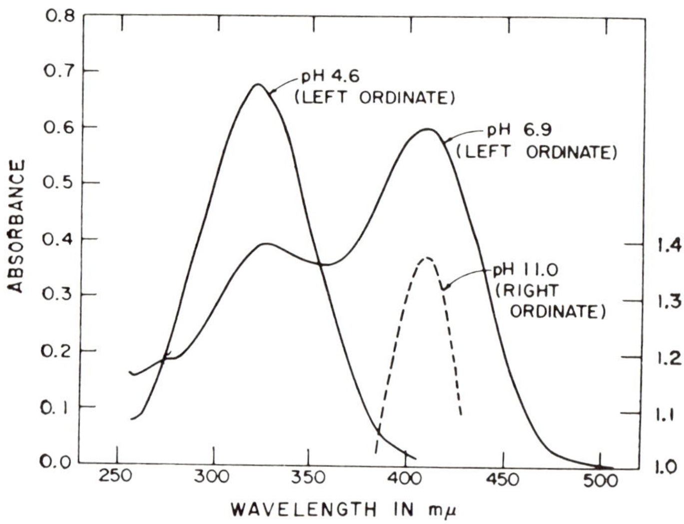

Fig. 32. Absorption spectra of 2-hydroxy-5-nitrobenzyl alcohol at various pH values (*280*.)

the reaction mixture may be effective in many instances. Where the alcohol is tenaciously absorbed to the protein, however, it may prove necessary to carry out the separation initially on a gel column equilibriated with 8 *M* urea followed by dialysis against water or buffer to remove the last traces of the alcohol.

The spectrum of the HNB moiety lies at longer wavelengths than the region of absorption of proteins (Fig. 32). Its pH dependence, furthermore, permits one to shift the spectrum into the visible region if necessary. This pH dependence might also serve as a sensitive index of changes in the environment of modified tryptophans. The number of HNB groups coupled to the protein molecule is generally determined in 0.1 *N* NaOH from the absorbancy of these groups at 410 mμ (ε_{410}, 18,000) (*169*) after removal of the alcohol side product. The protein concentration can be obtained from either the dry weight of a salt-free lyophilized sample or by amino acid analysis.

The possibility of forming di-substitution products of tryptophan suggests that the visible spectral characteristics of the HNB group should not be used as the sole criterion for group counting in native proteins, since even at relatively low molar ratios of HNBB, di-substitution of highly reactive tryptophans may occur. This possibility is less when one is interested in determining the total tryptophan content of a protein, since the reaction can be carried out under denaturing conditions on the reduced and alkylated protein, where all tryptophans are probably similar in reactivity (*169*). However, when HNBB is used for detection of reactive tryptophans in native proteins, where a large proportion of them may be partially buried, the high concentrations of reagent required to ensure complete reaction may favor di-substitution with subsequent errors in estimation of the number of exposed groups.

It would seem mandatory that the modification of tryptophan by HNBB be followed by some independent method capable of detecting unreacted tryptophan in the modified protein. This can be done by carrying out an alkaline hydrolysis of the modified protein and detecting the unreacted tryptophan by direct amino acid analysis. In this regard HNB tryptophan does not interfere in the standard ion exchange chromatography employed for amino acid analysis. It is also possible to employ the Spies and Chambers method (*281a*) for estimating unreacted tryptophan in HNB modified proteins as suggested by Dasgupta et al. (*283*).

HNBB has not yet been used extensively for detecting reactive tryptophan in native proteins. It should be quite useful because of its relatively high specificity for tryptophan and the wide pH range (pH 2–7) over which it can be employed.

c. *Alkaline Hydrogen Peroxide-Dioxane*

Hachimori et al. (*282*) have investigated various organic peroxides and hydrogen peroxide as reagents for the detection of exposed tryptophan residues in proteins. The most suitable reagent found was hydrogen peroxide in pH 8.1–9.4, 0.5 *M* bicarbonate buffer containing 10% dioxane. Since the reagent shows no absorption in the tryptophan region, the reaction can be followed spectrally by measurement of the decrease in absorbance at 282 mμ (Fig. 33) using a difference molar extinction of 3.49×10^3 for the oxidation of tryptophan. The rate of oxidation is dependent on pH, dioxane concentration (Figs. 34 and 35), and type of buffer ion. With no dioxane present, the rate of oxidation with hydrogen peroxide to determine an end point, but is too slow in the presence of 10% dioxane, the rate is maximal. A constant maximal rate is obtained above 0.45 *M* bicarbonate at pH 8.4. In phosphate buffers practically no change in absorbance occurs at 282 mμ. These observations indicate that dioxane and bicarbonate are involved in the oxidation, probably through formation of an organic peroxide with a greater oxidizing power than hydrogen peroxide.

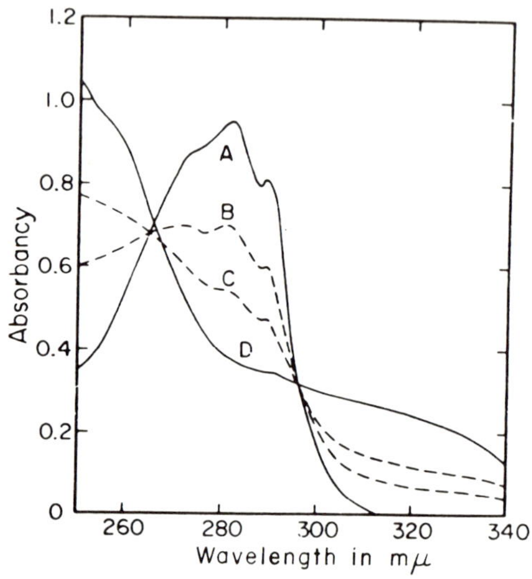

Fig. 33. The spectral change on the oxidation of 0.166 m*M* N-acetyl-L-tryptophan by hydrogen peroxide in 0.5 *M*, pH 8.4 bicarbonate buffer containing 10% dioxane; curve A, the spectrum before oxidation; curves B, C, and D, spectra after treatment with 1.50, 2.50, and 20.0 m*M* hydrogen peroxide, respectively, for 0.5 hr (*282*).

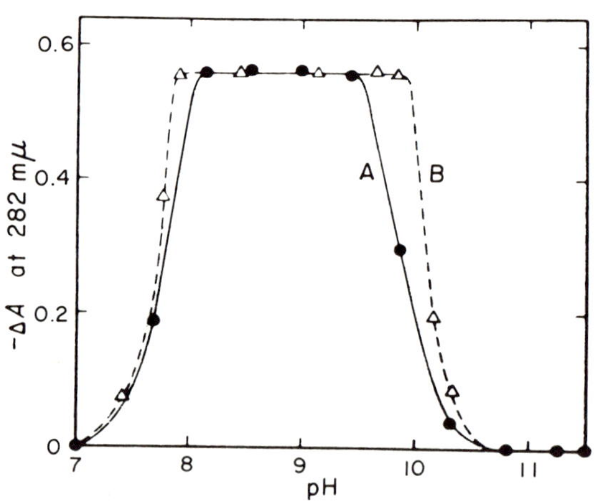

Fig. 34. The effect of pH on the degree of oxidation of 0.166 m*M* N-acetyl-L-tryptophan by 1.67 (curve A) and 16.7 (curve B) m*M* H_2O_2 in carbonate–bicarbonate buffers, as estimated from the lowering of absorbancy at 282 mμ (*282*).

The effect of the reagent on amino acids other than tryptophan appears to be minimal, at least, as judged from spectral changes. The spectrum of phenylalanine was unaffected, while that of tyrosine showed a fractional increase of absorbancy of only 3% at 275 mμ, and less than 5% at longer wavelengths. Free histidine appeared to be oxidized as indicated by changes in absorption in the region of 230 mμ. However, in the proteins studied, no evidence was found for oxidation of this amino acid. Other amino acids showed no changes in absorption near 282 mμ when treated with the reagent.

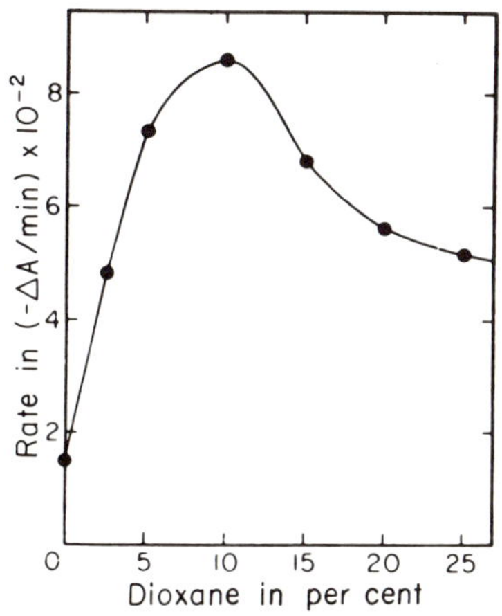

Fig. 35. The effect of dioxane concentration on the rate (ΔA at 282 mμ per min) of oxidation of 0.166 m*M* N-acetyl-L-tryptophan with 1.0 m*M* H_2O_2 in 0.5 *M* bicarbonate buffer (pH 8.4) (*282*).

This reagent is convenient for determining the reactivity of tryptophan residues in proteins. However, the precautions pointed out in the use of NBS (see above) also apply to this technique and the results obtained by spectral measurement should be verified by amino acid analysis and other techniques if possible. The fact that a substantially higher concentration of reagent is required for oxidation of tryptophans in proteins, relative to the model compound N-acetyl tryptophan, suggests that oxidation of other residues may occur, which could interfere with interpretations based on spectral measurements.

2. Application to Specific Systems

Information about the reactivity of tryptophans in proteins is summarized in Table XV. Anticipating our discussion of individual proteins, it can be

TABLE XV

Chemical Reactivity of Tryptophans in Proteins

Protein	Total residues	Method[a] NBS[b]	H_2O_2	HNBB	Reference NBS	H_2O_2	HNBB
Biotin	14	12	—	—	58	—	—
avidin complex		0	—	—	58	—	—
Chymotrypsin	7–8	7, 7.5	5+1	7–8 } [d]	274, 275	282	284
		7[c]	7[c]	1 } 8.1[c]	274	282	284, 169
DIP derivative		—	4+2	—	—	242	—
BGEE derivative		—	5+1	—	—	242	—
Chymotrypsinogen	7–8	8, 7	5+1+1	—	275, 68	282	—
Cytochrome c, horse heart	1	1	—	—	252, 279	—	—
Cytochrome c, tuna	2	2	—	—	252	—	—
Lysozyme	6	6	5+1	1.5	69, 274	282	283
				5.6[c]	—	—	169
				1–6[e]	—	—	285
Metmyoglobin	2	2[f]	—	—	84	—	—
Pepsin	6	3	—	—	274	—	—
Pepsinogen	6	1.5	—	—	274	—	—
Trypsinogen	4	4	2+1 4[c]	—	286, 275	282	—
Trypsin	4	3.7	2+1	4.2[c]	286, 275	282	169
Trypsin inhibitor, soybean	3	1, 1.4	2	3[c]	274, 287	287	287
Trypsin	4	3[g]	—	—	262	—	—
complex with soybean inhibitor		2[f,g]					
complex with lima bean inhibitor		1[f,g]	—	—	262	—	—
complex with pancreatic inhibitor		2[f,g]	—	—	262	—	—
complex with ovomucoid		2[f,g]	—	—	262	—	—

[a] NBS, N-bromosuccinimide; HNBB, 2-hydroxy-5-nitrobenzyl bromide.

[b] Unless otherwise indicated, reaction carried out between pH 3 and 4.

[c] Denatured protein.

[d] The higher value purported to be monomer, the lower dimer, see text.

[e] Value dependent upon concentration of chloroethanol in the medium, see text.

[f] Reaction carried out at pH 5.0 to 5.2.

[g] Value for trypsin soybean inhibitor complex obtained with the assumption that none of the three tryptophan groups of the latter protein react with NBS. None of the other inhibitors contain tryptophan.

stated that the use of chemical reactivity as a means of determining exposure of tryptophans is far from satisfactory at present. The quantitative aspects of these observations should be taken with some reservation for a variety of reasons: (a) In a number of instances no attempt was made apparently to determine reactivity as a function of reagent concentration. Thus, one cannot be certain to have obtained maximal reaction. Admittedly, such a procedure might result in side reactions or in multisubstitution. (b) Problems arising from the use of NBS have not been previously appreciated and taken into consideration (see above). (c) Most investigators have neglected to determine what effect chemical modification of tryptophans might have on other side chains in the protein in question and on the conformation of this protein. We believe, therefore, that the quantitative aspects of the results of these chemical studies, particularly the importance of differences observed by various procedures, should not be accorded great significance at present.

a. Lysozyme

Chemical reactivity data for the tryptophans of lysozyme are too fragmentary and uncertain to make internal comparisons particularly worthwhile. Likewise, a meaningful comparison of chemical and solvent perturbation data seems impossible at present. Some of the reactivity data, however, do highlight the potential difficulties one faces in formulating a straightforward interpretation of chemical data for tryptophan in terms of exposure.

X-ray crystallographic studies (see Section V for a more complete discussion) indicate that trp 62, 63, 108, and 123 lie at the molecular surface, although inspection of the molecular model strongly suggests that all of these are not equivalent in terms of accessibility to the medium. Reaction of lysozyme with very low concentrations of NBS (see also above) results in oxidation of trp 62 and virtually complete loss of enzymatic activity (*82*). This is not unexpected in view of the rather exposed character of this group inferred from the molecular model. By contrast however, iodine at pH 5.5 rapidly and preferentially oxidizes trp 108 (*195*,*288*,*289*) which, like trp 62 and 63 lies in the active site cleft of the enzyme (*152*). Identification of the modified residue has been made by peptide analysis (*289*) and by the difference Fourier technique (*195*) (see also Section III). Only at higher concentrations of iodine was it possible to oxidize trp 62 in addition to 108 (*195*). The difference in reactivity of 62 and 108 toward iodine is rather unexpected in light of their relative reactivity toward NBS, as well as what appear to be significant differences in exposure to the medium as judged from the molecular model (see Section V).

Hartdegen and Rupley (*289*) suggest that the enhanced reactivity of trp 108 is the result of hydrogen bonding from its indole nitrogen to a main chain carboxyl with the 2-carbon being brought into close contact with the carboxyl group of glu 35. The resulting increase in electron density of the indole ring, it is stated, should activate it to electrophilic attack by iodine. It is difficult to see, however, why this same situation should not produce preferential attack of trp 108 by the equally electrophilic bromonium ions present in solutions of NBS.

It is interesting that ozonation of lysozyme in anhydrous formic acid results in conversion of trp 108 and 111 to N-formyl kynurenine side chains with no loss in activity (activity determined after a 1:500 dilution with water) (*288a*). The retention of activity after modification of trp 108 is to be contrasted with the observation of Hartdegen and Rupley that activity is lost after oxidation of the same residue. Blake had shown by the difference Fourier method that in addition to oxidation of trp 108, treatment with iodine resulted in the displaçement of glu 35 from its normal position, which in itself might alter or abolish the enzymatic activity (see Section V). This ambiguity makes it difficult to say whether interaction of a substrate with lysozyme would depend upon the accessibility of trp 108.

The reaction of trp 111 cited above perhaps is not unusual even though this group appears to be quite buried (X-ray crystallographic model). Anhydrous formic acid might be expected to drastically alter the structure, thereby exposing trp 111. Since only two of the six tryptophans of lysozyme react under these conditions, the protein molecule may not be in a disordered state but rather is folded into a new conformation where previously exposed groups are now buried. Such refolding in nonaqueous solvents has been observed with some proteins (*289a,290*). This behavior should be borne in mind where reactions are carried out in nonaqueous solvents, and reaffirms the necessity for proving that the exposure measured refers to that of the native molecule.

Oxidation of the tryptophans of lysozyme by hydrogen peroxide demonstrates two classes of reactive groups; five groups which react initially and one group which reacts at higher reagent concentration (*282*). Five exposed groups seems too high in view of the X-ray (see Section V) and solvent perturbation observations (see Table VIII). The fact that all six groups ultimately react suggests that conformational changes may have been induced by the modification. Unfortunately, this information was not obtained.

In view of the X-ray and solvent perturbation observations, the value of 0.5 to 1.5 groups (*283,285*) obtained by reaction of HNBB with native lysozyme seems too low. All six groups are reactive in 6.3 *M* urea (*169*), even though the molecule is not completely disrupted under these conditions (*291*). Since the study with native proteins was with a rather limited range of reagent concentration, the low value may reflect reaction of a particularly reactive group. It is clear that more detailed work is required with this system.

b. ***Avidin***

The biotin binding site of avidin is known to contain tryptophan groups (*58*), which are clearly altered during formation of the avidin–biotin complex (Table XV). In the absence of biotin about 12 of the 14 groups are apparently rapidly oxidized at relatively moderate NBS concentrations (NBS:tryptophan molar ratio about 2.5) (*58*). The complex, on the other hand, shows virtually no reaction except after protracted exposure to rather high levels of reagent. It is noteworthy that the biotin binding capacity of avidin is virtually completely abolished after treatment of the protein at reagent-to-tryptophan molar ratios of only 1, where many of the effects cited for lysozyme (see above) are probably absent. At this level of reagent, about six tryptophans appear to be oxidized.

The altered reactivity of tryptophans in the biotin–avidin complex suggests that these groups are now shielded from the medium. This conclusion is consistent with spectral observations of: (a) a red shift of the absorption spectrum (*58*); (b) decreased perturbation of the tryptophan spectrum by ethylene glycol (*58*). These data unfortunately do not lend themselves to calculation of exposure in the usual way because of spectral distortions; and (c) a short wavelength shift of the tryptophan emission spectrum with apparent preferential quenching of long wavelength emitting groups (exposed groups) (see Section II. D) (*291a*).

c. ***Chymotrypsin–Chymotrypsinogen***

The tryptophans of chymotrypsinogen appear to be completely reactive to hydrogen peroxide. The reaction occurs in three steps; five groups react initially, while at higher concentrations two additional plateaus are observed, each corresponding to reaction of a single group (*282*). Chymotrypsin by contrast shows only six reactive groups, five of which react as a class at lower concentration of reagent (*282*). These observations differ significantly from solvent perturbation data (Table VIII) which show 2.7 to 3.4 groups to be exposed in chymotrypsinogen and 2.1 to 3.3 in chymotrypsin. The small differences in the exposure of the enzyme and its zymogen are in accord with those observed on chemical reaction and suggest a "tightening" of the structure on activation. In the absence of information concerning the effect of the chemical modification on the molecular conformation, it is difficult to know what significance should be ascribed to the observation of the rather high degree of reactivity of the tryptophans of chymotrypsin and chymotrypsinogen.

Formation of DIP chymotrypsin seems to decrease reactivity to hydrogen peroxide relative to chymotrypsin (Table XV). Solvent perturbation measurements are in accord with these observations (*48*) although the differences observed may be marginal.

Oza and Martin (*284*) have made some rather unusual observations in studying the pH dependence of the reaction of HNBB with chymotrypsin. Protein so modified in the pH range 2 to 5.6 consisted of two fractions, an inactive one having seven to eight modified tryptophans and an active fraction having a single modified group. The distribution of these fractions depended upon the pH of reaction; the fully modified protein was formed almost exclusively at pH 2 with the other one being preferentially formed at pH 3.5. The similarity of the pH dependence for reaction and for the sedimentation properties of the enzyme suggested to the authors that the fully modified fraction might be monomeric chymotrypsin, while the singly reacted fraction might be dimer.

If the relative amounts of reaction products were to reflect the distribution of monomer and dimer at a given pH, as indicated by Oza and Martin, it would be necessary that the rate of reaction of HNBB with tryptophan and the rate of hydrolysis of the reagent be quite short as compared to the rates of monomer–dimer interconversions. Since the lifetime of the reagent in solution according to Horton and Koshland (*281*) is less than 30 sec, this may indeed be the case.

The implication of the above observation is that tryptophan groups become buried at monomer–monomer interfaces on formation of dimer. Solvent perturbation measurements (*48*) do not reveal a similar pH dependence of tryptophan exposure, perhaps because these measurements were carried out at a much lower protein concentration than those employed in the chemical study of Oza and Martin. It would be of interest to make solvent perturbation measurements at comparable concentrations, making use of very short path length cells. Alternatively, the measurement of *concentration* difference spectra as outlined by Fisher and Cross (*292*) might reveal if tryptophan residues are buried on dimerization.

d. *Trypsin–Trypsinogen*

Reaction of hydrogen peroxide with both trypsin and trypsinogen show that two of the four tryptophans are relatively reactive, while a third is somewhat less so. Thus, unlike chymotrypsin and chymotrypsinogen, activation of the zymogen does not alter the environment of tryptophyl groups.

Data obtained with NBS are somewhat ambiguous; two separate studies from the same laboratory present contradictory observations; Spande et al. (*275*) conclude that tryptophans in trypsin are oxidized at the same rate or slightly faster than those in trypsinogen. Viswantha et al. (*286*) claim just the reverse.

While it is not known whether activation of trypsinogen alters the environment of tryptophan groups, it would appear that such groups lie on or near the active site region. Although the numerical values may have limited quantitative significance, it is apparent that formation of trypsin inhibitor complexes (Table XV) decreases the reactivity of tryptophans in trypsin to NBS. The apparent number of groups which become unreactive is one to two. The spectral changes observed on formation of the complex are consistent with burying of tryptophans (*287*).

Soybean trypsin inhibitor appears to have one to two reactive groups (Table XV) with respect to NBS and hydrogen peroxide reactions. Solvent perturbation indicates two exposed groups (*287*).

D. Histidine

1. Introduction

Chemical modification of histidine residues is particularly useful in determining exposure of these groups in proteins, since their titration is frequently subject to uncertainty due to overlap with the titration of carboxyl and α-amino groups (see Section III. B). Furthermore, the usual spectral techniques of analysis are not applicable because of interference from other chromophores in the protein (see Section III).

The methods most frequently employed for detecting reactive histidine in proteins have been photo-oxidation, and alkylation with activated halogen-containing compounds. A technique less frequently employed is based on the catalysis of hydrolysis of *p*-nitrophenylacetate by uncharged imidazole groups.

A recently proposed colorimetric method for detecting histidines based on coupling with diazonium-1-H-tetrazole (*293*) has been shown to give ambiguous results due to strong interference from coupling with tyrosine (*294*) and is not discussed. A brief outline of each method is presented below.

a. *Photo-oxidation*

Exposure of protein to light in the presence of oxygen and a suitable photosensitizer such as methylene blue (*295*) or rose bengal (*9*), may effect oxidation of certain side chains. The steps involved in this process probably involve a binding of the dye to the protein molecule in the region of the side chain to be modified, followed by reaction of the bound dye with oxygen to form a labile peroxide which in turn performs the oxidation (*296*). Cystine, tryptophan, tyrosine, and methionine may all be oxidized in addition to histidine (*295,297*). Oxidation of free histidine in the presence of methylene blue (*295*) or of proflavine (*297*) falls off markedly with decrease in pH, such that by pH 4 the rate of reaction approaches zero. For this reason photo-oxidation has generally been carried out near neutrality where the rate of oxidation of histidine is of a reasonable magnitude.

Oxidation of other amino acids is likewise pH dependent, but appears to depend also on the nature of the dye used as a photosensitizer. For example, the rates of oxidation of histidine and tyrosine are about equal at pH 7 in the presence of methylene blue (*295*). By contrast, the oxidation of histidine in the presence of proflavine is perhaps as much as 50 times faster than that of tyrosine (*297*). It is possible, therefore, that the specificity of reaction could be improved by selection of a suitable sensitizer and appropriate experimental conditions. To our knowledge no such systematic studies have been made in this direction.

The presumed requirement of binding of dye to protein prior to photo-oxidation (see above) should provide another means of controlling the specificity of reaction. Thus, Brake and Wold (*298*) found that photo-oxidation of yeast enolase with the cationic dye methylene blue led to reaction at a number of side chains, as well as to alteration of the molecular conformation (changes in hydrodynamic properties). Westhead (*9*) chose to use the anionic dye, rose bengal, as a sensitizer on the basis of the observation that the active site of enolase binds anionic substrates. The rate of inactivation observed with rose bengal was about 500 times greater than with methylene blue. Furthermore, oxidation in the presence of this dye resulted in no alteration of hydrodynamic properties of enolase in contrast with methylene blue.

It is apparent that insufficient study has been given to the factors which control photo-oxidation of amino acid side chains. The few observations cited above, however, suggest that such knowledge might provide a means of controlling the specificity of photo-oxidation. The extent of modification of photo-oxidized groups is determined by amino acid analysis of the oxidized proteins. It has been shown by Ray and Koshland (*299*) that the extent of modification of methionine cannot be determined by the usual method employing acid hydrolysis, since methionine sulfoxide, the major product of its oxidation, reverts to methionine. In order to determine the extent of its oxidation, an alkaline hydrolysis must be employed prior to amino acid analysis.

In spite of potential problems with the specificity of photo-oxidation, it has proven to be valuable in identifying reactive histidines in the active site of enzymes. Kinetic analysis has been used by Ray and Koshland (*299*) in correlating the rate of inactivation of the enzyme with the rate of photo-oxidation of a single amino acid or several amino acids. This kind of analysis proved to be quite useful in demonstrating the role of a single histidine and methionine residue in the active site of chymotrypsin (see below).

b. *Alkylation*

Alkylation of proteins using various reactive halogen compounds such as halo aliphatic acids and amides, or 2,4-dinitrofluorobenzene when carried out at pH values near neutrality, is fairly specific for histidine. Methionine and cysteine groups can also react, while tyrosine and amino groups generally react only at more alkaline pH.

Reaction with histidine occurs at nitrogen 1 or 3 of the imidazole ring to produce the monosubstituted derivative (**I** and **II**). Di-substituted derivatives (**III**) can also be formed by reaction at both nitrogens.

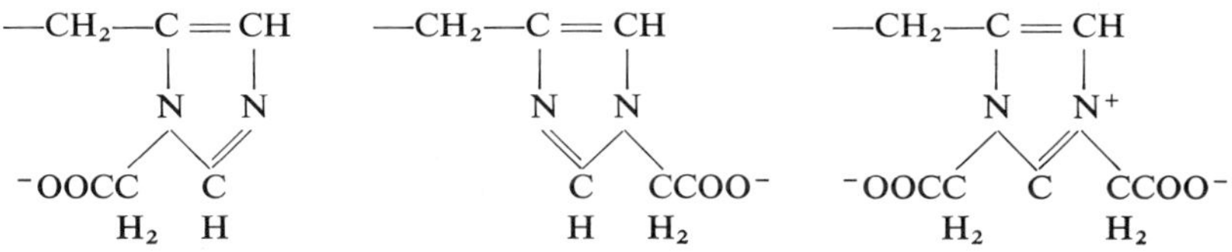

I **II** **III**

Experiments with α-N-acetyl-L-histidine and iodo acetate (*300*) indicate that positions 1 and 3 are not equally reactive. On steric grounds it would appear that reaction at position 3 should go more readily than at position 1. This distinction may be of some importance in distinguishing accessibility of the two histidine nitrogens in a protein. Alkylation occurs with the basic form of histidine and should be favored by pH values well above its p*K* (ca. 7). The pH actually employed, however, is somewhat restricted by the need to avoid reactions with other side chains (see above). The mild conditions for reaction and the relatively high specificity for histidine appear to lead to minimal changes in the conformation of native proteins.

c. *Hydrolysis of p-Nitrophenylacetate (NPA)*

This method is based on the fact that histidine in its basic form is capable of catalyzing the reaction (*301*):

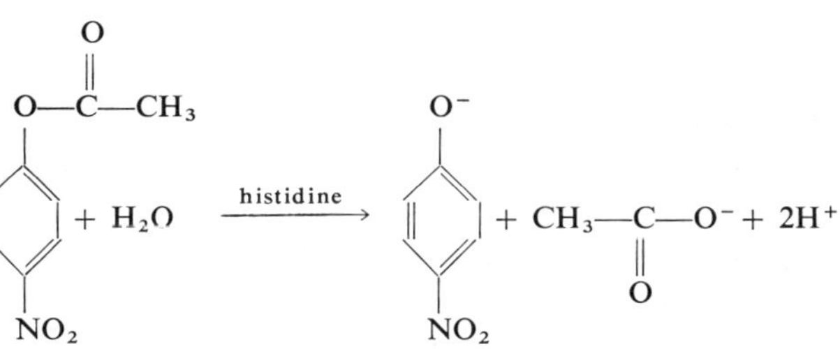

(15)

Since the intrinsic rate constant for this reaction is independent of pH, the apparent rate constant reflects the concentration of unprotonated imidazole groups at the pH in question. This in turn will be determined by the dissociation constant and the number of histidines which are available for titration.

The rate of hydrolysis can be determined from: (a) The rate of appearance of nitrophenolate ion as measured by the spectral change at 400 mμ; (b) the rate of disappearance of NPA as measured by the spectral change at 273 mμ; or (c) the rate of liberation of protons as determined by the consumption of alkali in a solution maintained at constant pH in a pH stat (*209*).

The major disadvantage of this procedure is that amino groups in their basic form will also catalyze the hydrolysis of NPA. Since the rate constants for hydrolysis of NPA by histidine and by α and ε amino groups differ significantly, such interference should be recognizable (see below for myoglobin). Furthermore, it may be possible in favorable instances to avoid the pH region in which amino groups titrate.

Another potential source of error exists with enzymes capable of undergoing acyl transfer reactions. In these systems an acyl group will be transferred from NPA to a side chain at the active site, e.g., a thiol group in glyceraldehyde phosphate dehydrogenase (*315*) and a serine in chymotrypsin (*316*). The reaction in such a system will reflect steps involving acyl transfer to enzyme followed by slow deacylation. Measurement of the kinetics, particularly that of alkali uptake, therefore, should enable a distinction to be made between catalytic hydrolysis by histidine and such transfer processes.

At the present time this method has been applied to only two proteins, myoglobin and soybean trypsin inhibitor, although it should be quite useful in detecting exposed histidine residues in other proteins.

2. Application to Specific Systems

Observations made for histidine reactivity in proteins are summarized in Table XVI. References corresponding to the systems cited below have generally been omitted from the text but are included in the table.

a. *Myoglobin*

Of the 12 histidines in sperm whale myoglobin, seven to eight are specifically alkylated by bromoacetate at pH 7. Heat denaturation results in reaction of an additional residue. Denaturation of myoglobin in 8 *M* urea now permits alkylation of all but one group. Parenthetically, it should be stated that the methionines of myoglobin are unreactive to bromoacetate in native or urea-denatured myoglobin. By contrast, reaction does occur at pH 5 or 11 after thermal denaturation. The reason for this difference is not apparent. Since this lack of reactivity cannot be ascribed to conformational determinants, it must be the result of a unique sequence of amino acids.

Observations made of NPA hydrolysis in the presence of myoglobin indicate six titratable histidines. The type of data obtained is illustrated in Fig. 36, a plot of the second-order rate constant versus $\bar{\nu}_H$, the number of protons bound per molecule of isoionic myoglobin. Note that the rate of hydrolysis falls to zero at $\bar{\nu}_H$ equal to about 6, i.e., after six histidines have been protonated. The deviation of points from the line at low values of $\bar{\nu}_H$ are the

TABLE XVI

Reactivity of Histidines in Proteins

		Chemical reaction			References		
Protein	Total histidines	H⁺ titration	Alkylation	Photo-oxidation	H⁺ titration	Alkylation	Photo-oxidation
Zinc insulin	2	1	1[a]	2	*91*	*255*	*81*
Zinc-free insulin	2	2	2[a]	2	*91*	*255*	*81*
Trypsin	3	3	1[b], 1[c]	—	*302*	*303, 304*	—
Chymotrypsin	2	2	1[d], 1[e]	1+1	*109*	*305–307*	*299*
Ribonuclease	4	4	1+1[f]	2+1	*103*	*227*	*308*
Horse heart cytochrome c, Fe^{3+}	2	1	1[g], 2[h]	—	*309, 310*	*311*	—
Horse heart cytochrome c, Fe^{2+}	2	—	1[g]	—	—	*311*	—
Hemoglobin	38	22[f]	22–23[g,l]		*312, 208*	*312*	—
		16[j], 26[k]	31[h,l]	—	*206–208a*		
Myoglobin, sperm whale	12	6, 6[i]	7–8[g]	—	*209*	*313, 314*	—
Soybean trypsin inhibitor	2	2[i]	—	—	*287*	—	—

[a] Iodination.
[b] Iodoacetamide.
[c] 1-Chloro-3-tosylamido-7-amino-2-heptanone.
[d] L-1-tosylamido-2-phenylethyl chloromethyl ketone.
[e] 2,4-Dinitro fluorobenzene.
[f] Iodoacetate.
[g] Bromoacetate.
[h] Bromoacetate in 8 *M* urea.
[i] *p*-Nitrophenylacetate.
[j] Human protein.
[k] Horse protein, see Section V for explanation of difference.
[l] Bovine protein.

result of titration of amino groups which now also catalyze the hydrolysis. The data illustrate that interference by such groups is readily recognized, at least in this system.

The titration results (Table X) and the NPA hydrolysis results cited in the paragraph above are in good agreement, but are in sharp contrast with observations of bromoacetate reactivity (see above). This difference is probably not unexpected in view of the irreversible character of the alkylation, i.e., dynamic structural changes of the myoglobin molecule might permit periodic short-term "exposure" of a potentially reactive group which might otherwise be inaccessible to protons. Such an alteration appears to occur on binding of HgI_3^- to the internal site of myoglobin (*178*). A more detailed discussion of the chemical reactivity data is given in Section V.

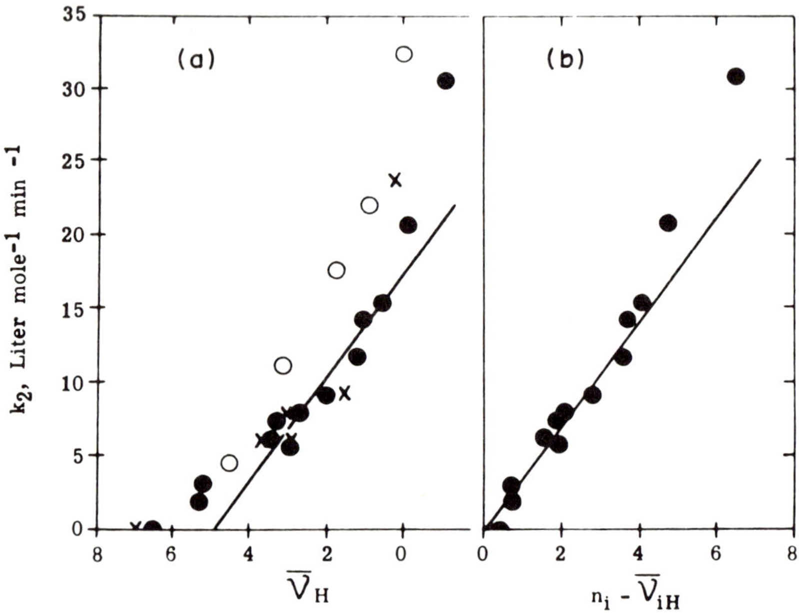

Fig. 36. Dependence of k_2 on (a) $\bar{\nu}_H$, and (b) $n_i - \bar{\nu}_{iH}$: ●, pH-stat method, $\mu = 0.06$; ×, pH-stat method, $\mu = 0.16$; ○, spectrophotometric method, $\mu = 0.09$ [from Breslow and Gurd (*209*)].

b. *Hemoglobin*

Reaction of native bovine hemoglobin with bromoacetate at pH 9 results in modification of 22 to 23 of the 38 histidines. Reaction in 2 to 8 *M* urea leads to modification of an additional three to eight residues.

A large decrease in absorption in the Soret region occurs as a result of the modification of both the native and denatured protein. With native protein, it could not be determined whether this was due to modification of heme-linked histidines or the result of oxidation of the heme iron by bromine produced during the reaction. It has been reported, however, that heme-linked residues F8 and E7 are not modified by bromoacetate (*317*). Possibly extensive reaction occurs with lysine side chains at pH 9 which might have led to conformational and spectral changes in the protein. No attempt was made, however, to determine the extent of modification of these residues although tyrosine groups were found to be unreactive.

A comparison of the chemical data and hydrogen ion titration data is made in Section V.

c. *Horse Heart Cytochrome c*

Reaction of bromoacetate with horse heart ferri- or ferrocytochrome c at neutral pH results in modification of only one of the three histidines. This causes only minor changes in the visible absorption spectra of the proteins. In 8 *M* urea, all histidines are modified and marked spectral changes occur.

Horinishi et al. (*318*) have reacted cytochrome c with diazonium-1-H-tetrazole and have identified the reactive residue as Hist 33. It is interesting that this residue is not necessarily conserved in cytochromes from various species, making it unlikely that it is heme bound (*268*). There is some evidence that the heme in cytochrome c is linked to the protein through at least one histidine group (*268*). The other two histidines, 18 and 26, are conserved in cytochromes from various species and would be likely candidates to be heme linked and hence unreactive.

d. *Insulin*

Photo-oxidation of zinc and zinc-free insulin resulted in oxidation of the two histidines (based on a molecular weight of 5733) as long as the temperature of the reaction was kept below 10°. Reaction at higher temperature appeared to involve tyrosyl groups. The apparent rate of reaction of the histidine residues was faster for the zinc-free enzyme. The kinetics of histidine oxidation was faster for the zinc-free protein, suggesting that the two groups were equivalent. Unfortunately, no comparable experiment was carried out for the zinc protein, so that the photo-oxidation data do not permit a decision as to whether binding of zinc reduces the reactivity of one or both histidines.

While no changes in the tyrosyl solvent perturbation spectra were observed on photo-oxidation of the two proteins, changes of the order of 40% were observed in the optical rotation dispersion parameter, a_0, suggesting that changes in conformation may have occurred on reaction; the parameter, b_0, by contrast remained unchanged.

Iodination of the histidines of insulin likewise suggest a difference for zinc and zinc-free insulin (*255*). In the former case, only one of the two groups was reactive, while for the latter, reaction was complete.

These chemical data indicate that formation of the zinc–insulin complex reduces the accessibility of histidines to the reagent. The fact that photo-oxidation of free histidine is unaffected by formation of a zinc complex indicates that complex formation *per se* is not responsible for the decreased reactivity. The iodination studies and the titration curve (Table X) indicate that one of the two groups is rendered less accessible. This burying of a histidyl group is probably the result of a complex formation occurring on binding of zinc; this is reflected as a decrease in tyrosyl exposure (see Section II. C for a discussion of these data) and a small change in the optical rotation dispersion parameter, b_0 (*81*).

e. ***Ribonuclease***

The reactivity of histidines in ribonuclease has been of particular interest because of the implication of these residues in its enzymatic activity. Photooxidation leads to inactivation of the enzyme with his 105 and 119 being rapidly oxidized at roughly the same rate and his 12 reacting somewhat more slowly (*319*).

Ribonuclease is likewise inactivated by the alkylating agents iodoacetate (*8*) and bromoacetate (*320,321*). The pH dependence of *inactivation* is somewhat unusual, i.e., the rate is most rapid at pH 5.5 to 6, decreases at slightly lower pH and increases again at pH 2.8. Furthermore, the rate of inactivation at pH 8.5 and 10.5 is *lower* than at 6. It is puzzling that although alkylation of histidine proceeded rather rapidly at pH 5.5, denaturation of the enzyme in 8 *M* urea reduced the alkylation of histidines to virtually zero. The observations made at pH 5.5 are unusual from two points of view: The optimal pH for alkylation of histidines should be higher than 5.5 since reaction takes place with the basic form of imidazole. Secondly, treatment with urea should increase the rate of reaction since the potentially hindered groups would be rendered more accessible to the reagent.

A thorough study of the alkylation of ribonuclease with iodoacetate at pH 5.5 (*227,228*) has shown that two distinct monoalkylated histidine derivatives are formed in the ratio of 8:1. The major product is inactive enzyme containing the 1-alkylated derivative of histidine 119. The minor product has less than 7% of the activity of native ribonuclease and has histidine 12 modified as the 3-alkylated derivative. These derivatives are formed at the same specific rate and are alternative products, with reaction at one of the histidines, preventing reaction at the other. The reaction is inhibited by 3′-cytidylic acid, an enzymic product and competitive inhibitor of the enzyme, by cupric ion which is also an inhibitor, and by divalent anions such as phosphate or sulfate, suggesting that histidine 12 and 119 are about 5 Å apart in the native protein.

The alkylation results suggest that due to the folding of the native protein, histidine 119 is abnormally reactive to iodoacetate at pH 5.5 and the methionines abnormally unreactive, probably due to being buried in the molecule. The alkylation reaction apparently takes place in two steps, with the positively charged form of one histidine electrostatically binding iodoacetate and then positioning it for attack on the basic form of the other histidines. Destruction of the native conformation eliminates this unique situation and prevents modification of the histidine residues. Since no studies appear to have been made of the alkylation of histidines in denatured ribonuclease at pH values at which optimal reaction should occur with these groups, it is impossible to tell if ribonuclease, like myoglobin and hemoglobin, contain residues whose position in the amino acid sequence renders them unreactive.

The difference in rates of inactivation at various pH values referred to earlier in this section reflect reaction with different amino acid side chains. As shown above, inactivation at pH 5.5 can be identified with carboxymethylation of his 119. Inactivation at acid pH is associated with modification of one or more methionines and at alkaline pH with reaction with one or two ε amino groups.

f. *Trypsin*

One of the histidines in trypsin is alkylated by the specific reagent 1-chloro-3-tosylamido-7 amino-2-heptanone (the chloromethyl ketone derived from N-α-tosyl-L-lysine) with loss of activity. This histidine has been identified as residue 46 (*322*). This residue is thought to be at the active site of the enzyme. Since denatured trypsin or trypsinogen do not react with the reagent, the unique geometry of the active site is apparently required for reaction. The N-α-tosyl lysine side chain confers substrate specificity on this reagent, which allows it to bind at the active site of the enzyme, while the chloromethyl ketone group enables it to function as an alkylating agent. An analogous compound derived from phenylalanine reacts similarly with chymotrypsin (see section on chymotrypsin).

Similar results have been obtained with trypsin, using alkyl guanidines and iodoacetamide (*303,304*). In this case the alkyl guanidine apparently serves the same purpose as the N-α-tosyl lysine side chain, and iodoacetamide functions as the chloromethyl ketone grouping. This work sheds some light on the mechanism of reaction of the N-α-tosyl-L-lysine chloromethyl ketone. In this work it was found that binding of methyl guanidine increased the esterolytic activity of trypsin by a factor of seven, while butyl guanidine inhibited the activity. When the enzyme is alkylated with iodoacetamide in the presence of the alkyl guanidines, one histidine is modified, and the rate of modification and inactivation is found to be six times faster with the methyl compound and 2.5 times slower with the butyl compound than reaction in their absence.

The fact that the rate of reaction of iodoacetamide with histidine parallels the observed effect of the alkyl guanidines on the esterolytic activity suggests that these compounds are bound at the specificity determining site of the enzyme, which in turn activates a histidine in the catalytic site. Butyl guanidine is thought to be inhibitory due to a steric effect of the large butyl group, which decreases access to the activated histidine.

g. *Chymotrypsin*

Chymotrypsin is inactivated by photo-oxidation in the presence of methylene blue. The rate of inactivation can be correlated with the rapid oxidation of a single histidine and a single methionine, suggesting their presence as fully exposed groups in the active site of the enzyme. The other histidine and methionine residues in the portein are oxidized at a much lower rate and may be partially exposed in the enzyme.

L-1-tosyl amido-2-phenylethyl chloromethyl ketone, the chloromethyl ketone of N-tosyl-L-phenylalanine is a specific alkylating agent for chymotrypsin and reacts with a single histidine (Table XV). The modification leads to inactivation of the enzyme, suggesting that histidine is in the active site. By analogy with the reaction of the N-α-tosyl lysine chloromethyl ketone with trypsin, reaction occurs only with the native protein; DIP chymotrypsin and chymotrypsinogen do not react, suggesting again that this reagent requires the geometry of the active site for reaction to occur. In this regard it has been found (*323,324*) that the reactive histidine is attached to a cystine residue in the B chain which greatly restricts its movement. The other histidine also occurs in the B chain and is held in the same general area due to an intrachain disulfide link, and may play an auxiliary role in the catalytic mechanism in a manner similar to one of the histidines of ribonuclease.

Some evidence that the two histidines of chymotrypsin may be involved in the active site comes from alkylation of the enzyme with 2,4-dinitrofluorobenzene. When chymotrypsin reacts with this reagent at a pH value close to neutrality (bicarbonate buffer) (*306*), only one histidine is reactive, and the modified derivative has a decreased binding capacity for substrate but the same maximum velocity. At pH 10.7, however, a single histidine is modified and the enzyme is activated. Although in the latter case inactivation could not be positively correlated with modification of histidine since an equivalent amount of reaction occurs at an unknown group, it suggests that under different conditions different histidines are modified by 2,4-dinitrofluorobenzene. At pH 10.7, modification apparently affects only the histidine at the catalytic site (i.e., that modified by the chloromethyl ketone) while at the lower pH, the other histidine, which is apparently involved in the substrate binding site, is modified.

V. SOME CRITICAL COMPARISONS—X-RAY CRYSTALLOGRAPHIC STRUCTURES AND SOLUTION PROPERTIES

A. Introduction

The recent advent of X-ray crystallographic structures of proteins may be regarded as "the moment of truth" for protein chemists, since we now have the opportunity (or obligation) to compare our interpretations or guesses with what is now relatively detailed information about the internal geometry of these molecules. It should be emphasized, however, that many of the conclusions drawn from the molecular models are quite tentative, since the manner in which such models should be used in predicting solution properties is in many instances far from clear. For example, at present it is quite difficult to tell from the models, except in some clear cut instances, to what degree solvent or particular reagents or perturbants can penetrate certain regions of the molecule. Furthermore, it should be emphasized that the resolution achieved in many instances is insufficient to establish the exact position in space of the side chains. The latter *caveat* is particularly important in postulating specific side chain–side chain and side chain–main chain interactions when steric relationships and distances seem appropriate.

In the following sections we attempt to contrast the solution properties of ribonuclease, lysozyme, myoglobin, and hemoglobin, as these pertain to exposure of side chains, with information gleaned from the X-ray crystallographic models. Conclusions drawn from the models by the X-ray crystallographers have been carefully identified. In a number of instances these conclusions have been supplemented by those of one of the authors (M.J.K.) from his inspection of the suitable models. These have been identified as "observations, M.J.K." If these latter conclusions prove to be wide of the mark, the responsibility should not be laid at the door of the original investigators who so kindly made their models available for inspection.

B. Ribonuclease

1. Introduction

Preliminary structures have been published by Wykoff et al. (*194*) using ribonuclease S at a resolution of 3.5 Å, and by Kartha et al. (*325*) (also see Ref. *326*) using ribonuclease crystallized from 55% 2-methyl-2,4-pentanediol at a resolution of 2 Å. The solvent used in the latter study should be borne in mind since the three-dimensional structure may not correspond in all its detail to that of the protein in a purely aqueous medium.

2. Tyrosyl and Carboxyl Groups

Both groups of investigators find that of the six tyrosines in ribonuclease three, 76, 93, and 115, appear to be on the surface of the molecule with their phenolic OH groups in contact with the medium. The most buried group appears to be tyr 97 which is located well below the molecular surface. Its location, however, is such that it could not be involved in the kind of interaction with asp 83 postulated by Scheraga and co-workers (see Ref. *327*). Tyr 25 appears to lie somewhat below the molecular surface with its phenolic

OH not too accessible to the medium. This residue lies close to asp 14, consistent with the interaction postulated by Scheraga and co-workers, but the refinement of the structure does not permit any definitive conclusion as to interaction of these side chains. Tyr 92 lies close to asp 38, as predicted by Scheraga and co-workers, but in this case the geometry does not seem entirely appropriate for interaction (*326*). Tyr 92 seems to be somewhat buried, but its phenolic group may project into the medium (*327*).

The number of exposed groups, as derived from the titration curve (Table X) and from chemical reactivity studies (Table XIII), correspond rather well to what is seen from the crystallographic structure. The groups that one would judge from the model as exposed, tyr 76, 93, and 115, are the residues which are iodinated in solution (*328*). Likewise, iodination of crystalline ribonuclease S with subsequent X-ray analysis showed that tyr 73 and 115 are reactive (*194*). Work in progress (*326*) on iodination and acylation of crystalline ribonuclease in the solvent system employed by Kartha et al. (*325*) should provide further means of relating activity of tyrosines to structural information derived from X-ray crystallographic studies.

Although two of the three unreactive carboxyl groups (asp 14, asp 38, asp 83) (*329,330*) lie in proximity to the tyrosine groups they are purported to interact with (see above), the X-ray structure provides at present no help in understanding how the environment of these groups, or for that matter any of the carboxyl groups, alters their chemical reactivity or their abnormal titration properties (see Section III. C). Perhaps such information will be forthcoming when molecular models become available based on X-ray data obtained at higher resolution.

3. Histidines

According to Wykoff et al. (*194*) "his 105 and his 119 are freely accessible, his 12 is partly buried and his 48 is rather inaccessible." The photo-oxidation observations of Kenkare and Richards (*319*) are entirely consistent with conclusions about exposure of histidines. By contrast, the hydrogen ion titration curve (*103*) gives no evidence of abnormal histidine residues, nor does the NMR analysis of the titration (*222*) (see Section III. E) of the four individual groups. In the latter study, the range of pK observed was 5.6 to 6.1, somewhat lower than the range of what would be considered "normal" (see Fig. 18). It should be noted, however, that these are apparent dissociation constants. This difference in the photo-oxidation and titration results may be a reflection of the differences in size of the two molecular probes which were employed.

The X-ray crystallographic model of ribonuclease provides ample confirmation of the hypothesis of Crestfield et al. (*227,228*) in explaining the alkylation of its histidines (see Section IV. D). His 12 and 119 are in close proximity with the nucleotide binding site and sufficiently close to make it unlikely that even a water molecule would fit between them. In this regard it should be mentioned, but perhaps not taken too seriously, that the pK values observed for the histidines of ribonuclease by the NMR technique (*222*) are such that a significant fraction of both basic and protonated groups could exist at pH 5.5, as required by the above cited mechanism. The apparent lack of reactivity of his 105 toward alkylation is perhaps unexpected in view of the above cited conclusion of Wykoff et al. about its exposure. The fact, however, that it was found to be unreactive at pH 5.5 in contrast with the high reactivity of 119, may reflect the abnormally *high* reactivity of the latter residue, rather than an abnormally low reactivity of the former one. The absence of alkylation studies of histidines in ribonuclease at higher pH in water or in urea solutions does not permit an unequivocal answer as to the reactivity of his 105, which the X-ray structure suggests is exposed. However, a residue may be unreactive by virtue of its position in the amino acid sequence (see Section IV).

C. Myoglobin

1. Tryptophan Groups

Solvent perturbation measurements (*84*) of native sperm whale metmyoglobin suggest that both tryptophans are on the surface of the molecule, i.e., accessible to heavy water molecules. The fact that larger perturbants "see" only 0.8 to 1.2 groups suggests that one or more of these lie in crevice-like regions of the molecule. The X-ray evidence (*30*) indicates that both tryptophans are at the surface of the molecule although the NH group of one of these is bonded to a glutamine residue.

2. Tyrosyl Groups

Myoglobin contains three tyrosyl groups, G4, H22, and HC3 (Kendrew notation, Ref. *331*). Spectrophotometric titration (245 mμ) of the horse and sperm whale proteins shows two readily titratable groups and one which does not ionize even above pH 12 (*118*). Reaction with acetyl imidazole at a single reagent level (molar ratio, 60:1) shows but one reactive group (*238*). Differences in the rates of iodination of the three residues also suggest non-equivalence of these groups. Tyr H22 is much less reactive than either of the other tyrosines (*332*).

Kretsinger (*193*) has shown by X-ray crystallographic techniques that tyrosines G4 and HC3 form monoiodo derivatives while H22 remains unreactive. Iodination appears to occur more readily at G4, the ratio of iodination of G4 to HC3 remaining constant at 2:1 over a sixfold range of tri-iodide concentration. This difference in the reactivity of the two groups may relate to the observation of a single tyrosine reactive with acetyl imidazole (see above paragraph).

It would appear that the diiodo derivatives of the two reactive tyrosines are not formed since one of the two ε carbon atoms (the notation is that of Edsall et al., Ref. *333*) in each residue is buried (*193*). It is also of interest to note that for reaction to occur at G4, the aromatic ring must rotate about the C_β—C_β bond. It is apparent that in this case the "local" structural change has no far reaching effects in terms of the overall molecular conformation. It also explains, perhaps, why the diiodo derivative begins to be formed at the highest concentration of tri-iodide employed.

Both the titration curve and chemical reactivity data appear to be consistent with their molecular environments, as seen from the crystallographic models. Reference to these models indicates that G4 and HC3 are on the surface of the molecule with their phenolic OH groups projecting into the medium, while H22 is somewhat below the surface with the phenolic group hydrogen bonded to ileu FG5 (observations, M.J.K.). Kretsinger (*193*) states that "tyrosine H22 is deeply buried in the interior of the molecule as to be completely inaccessible to the aqueous environment."

3. Histidyl Groups

Analysis of the potentiometric titration curve and of the kinetics of the catalytic hydrolysis of *p*-nitro phenyl acetate are both consistent with the presence of six reactive groups out of a total of 12 (*209*). On the other hand, treatment with bromoacetate shows seven to eight reactive histidines (*313,314*). Unfortunately, we cannot tell if groups which titrate readily or react with *p*-nitro phenyl acetate correspond to those which react with bromoacetate (see Section IV. D). Identification of the sequence positions of the histidine residues which undergo the latter reaction permits comparison of these observations with the group environments deduced from the X-ray crystallographic structure (see Table XVII).

Two of the four unreactive histidines, F8 and FG3, are linked to the heme group. Of these, FG3 appears to be on the surface of the molecule but is linked to a heme propionate group. The unreactive character of F8 seems to be reasonably ascribed to its heme-linked character since it is reactive in apomyoglobin (*334*). The latter explanation, however, should be accepted with reservation since there is reason to believe that removal of heme *per se* may result in conformational changes in regions of the molecule removed from the heme binding locus (*334*). It is even more difficult to relate the non-reactivity of FG3 to its bonded character (see Table XVII). This residue is unlike the others in the myoglobin molecule in that its lack of reactivity may result from its proximity to certain groups in the amino acid sequence (see Section IV. D).

The fact that histidines A10, G14, and GH1 complex readily with metals and metal derivatives without undergoing structural changes (difference Fourier criteria, see Section III. A) suggests that these groups are not in contact with the medium and should react readily with bromoacetate, as

TABLE XVII

Histidine Residues in Myoglobin

Residue No.	Reactivity with bromoacetate[a,b]	Observations from X-ray crystallographic model
A10	R	Site of binding of Cu^{2+} [d]
B5	R	Near his GH1. Internally bonded[a,b]
C1	U	Close to potentially coordinating groups Glu C3, Phe G7, and Glu G10
CD6	R	On surface
E7	(R ?)	Distal heme linked through water molecule[c]
EF4	R	Exposed plane of imidazole ring projects from molecular surface[g]
EF5	U	Nitrogen N(3) hydrogen bonded to Asp H17. Nitrogen N(1) appears to be bonded to a water molecule[b]
F8	U	Heme linked[c]
FG3	U	On surface but coordinated to propionate side chain of heme[g]
G14	(R)	On surface, does not appear to be coordinated to other groups. Site of binding of *p*-chloro mercuri-benzene sulfonate[f]
G17	(R)	Appears to be on surface and uncoordinated to other groups[g]
GH1	R	Site of binding of aurichloride,[e] mercury diamine,[e] and Zn^{2+} [d]

[a] References *313* and *314*.

[b] Values in parenthesis were done by indirect methods. No change in Soret band was seen on alkylation of myoglobin. Hence, the purported reactivity of the heme-linked residue E7 is somewhat in doubt. The homologous heme-linked histidines in hemoglobin were found to be unreactive (*317*).

[c] Reference *209*.

[d] Reference *172*.

[e] Reference *30*.

[f] Reference *336*.

[g] Observations, M.J.K.

indeed is the case (Table XVII). Similarly, we might anticipate that the doubly bonded residue, EF5, might be unreactive with respect to bromoacetate. The fact that residues EF4 and CD6 form dicarboxymethyl derivatives (*313,314*) suggests that these histidines are freely accessible to the medium. Since his GH1 appears to form only the 1-carboxymethyl derivative (*313,314*), the 3 nitrogen of this residue may be somewhat hindered. The above *ad hoc* reasoning is not particularly convincing, it must be admitted. We can only conclude at present that the bases for the nonreactivity of the histidines of myoglobin is far from clear. It is hoped that more satisfactory explanations will be forthcoming when crystallographic structure becomes available for myoglobin at still higher levels of resolution.

D. Hemoglobin

1. Introduction

The structural information presented here is largely derived from a hemoglobin model proposed by Perutz (*335*) as formulated from a combination of the three-dimensional Fourier synthesis of horse oxy-hemoglobin at 5.5 Å resolution and that of sperm whale myoglobin at 1.4 Å resolution. The information that is available for comparisons is not as complete as one might wish. Nonetheless, consideration of the available data in terms of the hemoglobin model is instructive in our attempts to come to grips with what is meant by a buried group.

Of particular interest in the hemoglobin structure is the central cavity passing through the molecule along its axis of dyad symmetry. This cavity has a length of about 25 Å and a width of from 5 to 10 Å, and appears to be filled with water molecules. Ionizible groups in the central cavity include carboxyls asp H9α*, Glu G3β, lysine G6α, hist G10α, and tyrosine C1β. From the viewpoint of accessibility to protons, these groups should be seen as exposed. The same could not be said as confidently for solvent perturbation or for chemical reactivity measurements.

One would anticipate that the solvent perturbation spectrum of the tyrosine in the central cavity would decrease markedly as the perturbant size began to approach the size of the opening of the central cavity. Unfortunately, no test of this prediction will probably ever be possible, since the tryptophan: tyrosine ratio in hemoglobin is relatively high and the tyrosine in question is only one of a total of six.

Chemical reactivity of tyr C1β and hist G10α should also clearly depend upon reagent size as would those of the two serines and two threonines which are also located in the central cavity. As yet no such studies have been reported in which reactive residues have been identified.

2. Tyrosines

Both human and horse hemoglobin contain a total of 12 tyrosyl groups; the distribution of these in the two α and the two β chains is shown in Table XVIII. Eight of these groups appear to ionize "normally" while four titrate slowly or not at all (*118*,*119*,*129*). Nagel et al. (*129*) have carried out spectral titrations of human hemoglobin variants, A, F, and H, which differ in the kinds of chain per molecule or in the deletion of tyrosyl groups. They conclude from their studies that β35 is the abnormal residue in the β chain and that the abnormal α residue is either α24 or α42. The reader is referred to the original paper for details of the analysis leading to this conclusion.

TABLE XVIII

Tyrosine Residues in Horse and Human Hemoglobin[a]

α Chain	β Chain
B5 (24)	C1 (35)
C7 (42)	H8* (180)
H23* (140)	H23* (145)

[a] Letter notation that of Kendrew (*331*) as modified by Perutz (*335*). Numbers in parenthesis refer to residue number in sequence.

Examination of the hemoglobin model indicates that residue β35, which is considered to be abnormal, lies in the central cavity. As indicated above, this residue should be accessible to protons since the cavity is almost certainly filled with water. Its failure to titrate, however, may be the consequence of hydrogen bonding which may exist with the carboxyl group of asp H9α*. Another hydrogen-bonded tyrosine in the Perutz model may be C7α (α42), probably linked to the carbonyl of leu FG3β. Thus, of the two possible abnormal residues proposed by Nagel et al., α24 and α42, the latter seems the more likely choice. Direct verification of this conclusion is certainly required.

It is interesting that tyr H23 (H22 in Kendrew's notation) which is abnormal in myoglobin (see Section V. C) is not so in either the α or β chain of hemoglobin (Table XVIII). It may be that the hydrogen-bonded contact of this residue with FG5 in myoglobin cannot be made in hemoglobin. This difference between the two proteins is worth mentioning as a cautionary note in any attempts to draw conclusions about the environment of side chains in hemoglobin from the more detailed information available for the myoglobin structure.

3. Histidines

The sequence positions of the 38 histidine residues per horse hemoglobin molecule are given in Table XIX. The number of abnormally titrating residues according to Steinhardt and co-workers (*206,207*) is 22 per molecule. Tanford and Nozaki (*208*), on the other hand, find about 12 abnormal residues. The difference between these observations is apparently a consequence of the method of analysis of the data (*208*) and represent, respectively, the maximum and minimum number of abnormal histidines.

TABLE XIX

Histidine Residues in Horse Hemoglobin[a]

α Chain	β Chain
B1 (asp)	E7 (his)
CD3 (arg)	E13 (thr)
D6 (lys)	E20 (lys)
E7 (his)	EF1 (lys)
EF1 (lys)	F8 (his)
F8 (his)	FG4 (lys)
FG1 (thr)	G19 (arg)
G10 (glu)	H21* (ala)
G19 (arg)	H24* (lys)
H5* (gln)	

[a] Numbering of residues is that of Perutz (*335*). Residues in parenthesis are those in myoglobin at comparable positions.

Reaction of bovine hemoglobin with bromoacetate shows about 15 unreactive groups (*312*) which corresponds more closely to the titration value obtained by Tanford and Nozaki. As yet, there has been no identification of the reactive residues with the exception of those indicated below.

Eight unreactive residues can be accounted for. Hist E7α, F8α, E7β, and F8β are heme linked and by analogy with myoglobin would not be expected to titrate or react with bromoacetate. These eight residues in hemoglobin indeed have been shown to be unreactive with bromoacetate (*317*).

Perutz's model indicates that hist H21β* and H24β* are hydrogen bonded to AsN H17β*, forming a bridge between two β chains. This might account for four additional abnormal histidines.

Actual burying of histidines in the interior of the hemoglobin molecule seems improbable, since, as Tanford and Nozaki (*208*) have pointed out, virtually all of its non-heme bonded residues correspond to ionic or polar residues in myoglobin (see Table XIX). Since the overall features of the folding of the individual chains of hemoglobin are similar to those of myoglobin, it might be anticipated that the histidines of hemoglobin would be exterior groups.

The possibility should not be ignored that the lack of reactivity of some of the histidines of hemoglobin may be a reflection of features of the amino acid sequence rather than of the conformation in the native state (for example, residue FG3 of myoglobin, see Section V. C). This finds some support in the observation of Nakatani (*312*) that even with urea-denatured hemoglobin, 7 to 13 histidine residues remained unreactive to bromoacetate.

E. Lysozyme

1. Tryptophyl Groups

The six tryptophans of lysozyme lie at sequence positions 28, 62, 63, 108, 111, and 123 (*71,337*). Of these, four groups, 62, 63, 108, and 123 lie on the molecular surface (*152,153*) although inspection of the model (observations, M.J.K.) suggests that these groups are not equivalent. For example, NH groups of tryptophan 62, 63, and 123 are not hydrogen bonded to other structural moieties (observations, M.J.K.) which is not unexpected since bonding between two of these groups and carbohydrate residues occur on binding of substrate (*152*). If short-range perturbants (see Section II. B) interact with the tryptophan NH, we would anticipate that solvent perturbation with short-range perturbants would "see" a minimum of three groups (62, 63, and 123). Perturbation with methanol, dimethyl sulfoxide, and polyethylene glycol, classified as short-range perturbants, showed three to four exposed groups (Table VIII).

Residues 28 and 111 appear to be buried (*152,153*) with their NH groups hydrogen bonded to main chain carbonyl groups (observations, M.J.K.). These tryptophans would probably appear buried with respect to both short- and long-range perturbants. Tryptophan 108, which lies in the active site cleft (*152*), appears to be somewhat accessible to solvent but less so than 62 and 63 (observation, M.J.K.). Thus, we would expect that solvent perturbation with long-range perturbants would see more than three, but less than six groups. Measurements with ethylene glycol, glycerol, and sucrose show from four to five groups (Table VIII). It would thus appear that observations made by solvent perturbation are qualitatively in agreement with the features of the X-ray structure.

Observations of chemical reactivity of tryptophans in lysozyme have been too fragmentary and of too uncertain a quality to warrant discussion in terms of all six residues. A limited comparison has already been made in Section IV. C.

2. Tyrosyl Groups

The three tyrosyl groups lie at positions 20, 23, and 54 (*71,337*). The X-ray structure indicates that all three groups are on the surface of the molecule with the phenolic group of residue 53 hydrogen bonded to ser 50 or to aspN 74. The other two groups appear to be freely accessible and are not sufficiently close to potential donors or acceptors to form hydrogen bonds.

In spite of the ambiguities surrounding the titration curve of lysozyme, one of the three tyrosyl groups is probably less free to titrate than the other two (see Section II. C). Reactivity with respect to cyanuric fluoride (*241*) and acetyl imidazole (*257*) likewise shows one unreactive group. It is tempting to identify this unique residue as the one involved in hydrogen bonding, residue 53. Such an assignment should be regarded as quite tentative, however, and must await direct confirmation by chemical and peptide analysis techniques.

3. Carboxyl Groups

In 1962, Tanford (*121*), on examination of the observations that had been made of the titration of lysozyme, concluded that "it would appear that the problem of the carboxyl groups of lysozyme merits further investigation." Since the above was written, little has been published which clarifies the following observations: (a) Although the amino acid sequence of the protein shows 11 carboxyl groups (*71,337*), titration in a denaturing solvent reveals 11–13.5 of these (*121*). (b) Three carboxyl groups titrating in a denaturing solvent appear to be in the carboxylate form in the native protein. This conclusion, however, is inconsistent with the maximum positive charge obtained experimentally (*104*). (c) The carboxyl titration region cannot be described by the Linderstrom–Lang equation with a single pK_{int} (see Section III. *B* and Fig. 20). Although these observations have still not been explained, additional evidence exists for "abnormal" carboxyl groups in lysozyme.

Infrared spectral measurements of the titration of carboxyl groups in lysozyme reveals the presence of a single group with a p*K* of about 6.5 (*211*) (see Section III. D). The pH dependence of the binding of N-acetyl glucosamine oligomers leads to the conclusion that two of the carboxyls at the binding site have values of p*K* close to 2 and 6.5, respectively (*338*). These values are well outside of the "normal" range of titration of such groups (see Fig. 18).

The structure of lysozyme derived from X-ray crystallographic studies (*152,153*) reveals that asp 52, 101, 103, and glu 35 are located in the active site cleft. Glu 35 lies in a predominantly nonpolar environment and might thus be expected to have an abnormally high pK_{int}. The geometry of binding of monomeric and trimeric N-acetyl glucosamine makes it likely that glu 35 is the carboxyl having a p*K* of 6.5. Asp 52 is involved in a complex system of hydrogen bonds and would thus perhaps be maintained in ionized form. This would probably correspond to the group with a p*K* of 2, except that asp 52 is somewhat removed from the site of binding of trimer which is quite sensitive to the presence of a group with such a low p*K*. This has yet to be clarified.

Although glu 35 and asp 52 might be expected to be abnormal in their titration behavior, one is reluctant to classify them as "buried" groups. First of all, the structure indicates that these two residues are on the surface of the molecule. Secondly, it has been proposed that they are involved in the actual catalytic process and hence must be capable of relatively intimate contact with bound substrate, i.e., *exposed.* This situation illustrates the ambiguity that may result when "abnormal" groups are classified as buried with all the connotations attached to this term.

Note Added in Proof

The literature surveyed in preparation of this chapter was essentially completed through 1967 with the exception of certain specific papers published at a later date which were available to us in manuscript form. Subsequent to preparation of this chapter, a number of pertinent papers were published (*339–347*).

Detailed solvent perturbation study of ribonuclease indicates a gradation in the exposure of tyrosyl groups which is in accord with the X-ray crystallographic structures (*339*). Hayashi et al. (*340*) have demonstrated that tyr 53 of lysozyme is resistant to iodination as predicted from its hydrogen-bonded character (see section IV E.2). The acid–base titration of lysozyme has been reinvestigated (see Section V E.3). Of the three carboxyls, three remain ionized even below pH 2 (*341*). The titration curve for the eight groups can be fit if we assume six groups have a pK_{int} of 4.3, one group with a pK_{int} of 3.5, and another with a value of 6.0. Dopheide and Jones (*342*) have found that four of the six tryptophan residues of pepsin react with 2-hydroxy-5-nitrobenzyl bromide. This is somewhat higher than the 2–3 groups found by solvent perturbation (see Table VIII). Menendez et al. (*343a,b*) have in an extensive solvent perturbation study of native and modified insulin found that the protein exhibits one exposed, two buried, and one partially buried tryosyl groups (compare with Tables VIII, X, and XIII). A number of papers have appeared which describe the use of NMR techniques in studying the environment of amino acid side chains in proteins (*344a–e*) (see Section III E). From a study of the reaction of 2-hydroxy-5-nitrobenzyl bromide with model compounds, Spande et al. (*345*) suggest that this reagent converts the tryptophan side chains of peptides and proteins to tetracyclic indoline derivatives. In addition to nitrating tyrosine, tetranitromethane has been found to nitrate tryptophan in staphylococcal nuclease (*346*) and papain (*347*).

Appendix

MOLAR DIFFERENCE EXTINCTION COEFFICIENTS FOR SOLVENT PERTURBATION†

TABLE A-I

N-Acetyl Tyrosine Ethyl Ester, Solvent, pH 7.0, 0.05 *M* Tris

	$\Delta\varepsilon$							
	86% D_2O		20% sucrose		20% ethylene glycol		20% glycerol	
Wavelength mμ (1)	2° (2)	25° (3)	2° (4)	25° (5)	2° (6)	25° (7)	2° (8)	25° (9)
300	0	0	0	0	0	0	0	0
299	0	0	0	0	0	0	0	0
298	0	0	0	0	0	0	0	0
297	0	0	0	0	0	0	0	0
296	0	0	0	0	0	0	0	0
295	−1.31	−1.28	0	0	0	0	0	0
294	−2.62	−2.57	0	0	1.68	2.68	3.33	4.02
293	−6.56	−6.43	0	0	4.20	4.69	4.99	6.70
292	−7.87	−7.71	2.62	1.34	8.40	8.71	8.31	12.06
291	−9.18	−8.99	6.56	3.69	11.76	13.40	11.64	16.08
290	−13.12	−12.85	13.12	9.24	19.31	21.44	18.29	22.78
289	−19.68	−19.28	19.68	18.48	31.91	33.51	29.93	34.18
288	−27.55	−26.99	27.55	33.26	51.22	50.93	45.73	49.59
287	−40.67	−39.84	40.67	40.65	71.37	70.36	65.68	67.01
286	−55.10	−53.98	49.86	49.88	89.01	86.44	81.48	82.42
285	−62.98	−61.69	59.04	54.52	100.76	99.84	93.12	93.14
284	−65.60	−64.26	59.70	55.19	101.60	100.52	93.53	93.81
283	−59.04	−57.83	47.23	44.34	82.29	84.43	78.98	82.42
282	−43.30	−42.42	36.74	34.18	56.26	63.66	58.20	60.31
281	−22.30	−21.84	26.24	24.02	34.43	41.55	38.24	46.91
280	−14.43	−14.13	24.92	21.25	26.87	37.53	30.76	41.55
279	−19.68	−19.28	31.49	27.71	36.11	45.57	37.41	49.59
278	−26.90	−26.35	36.74	31.41	51.22	57.63	47.39	58.97
277	−35.42	−34.70	36.74	30.48	52.90	57.63	49.88	58.97
276	−31.49	−30.85	34.11	23.09	42.82	50.93	41.57	49.59
275	−22.30	−21.84	23.62	18.48	26.87	36.86	27.44	37.53
274	−11.80	−11.56	13.12	7.39	13.44	24.12	16.21	25.46
273	−5.25	−5.14	7.87	6.03	3.36	10.72	9.15	17.42
272	0	0	3.94	2.68	0	6.70	2.49	10.72
271	0	0	0	1.34	−0.84	4.02	0	6.70
270	0	0	0	0	−1.68	0	0	4.02

† Unpublished data of Warrington and Kronman.

TABLE A-2

N-Acetyl Tryptophan Ethyl Ester, Solvent, pH 7.0, 0.05 *M* Tris

Wavelength	Δε							
	86% D_2O		20% sucrose		20% ethylene glycol		20% glycerol	
mμ	2°	25°	2°	25°	2°	25°	2°	25°
(1)	(2)	(3)	(4)	(5)	(6)	(7)	(8)	(9)
300	−66.27	−82.83	29.95	50.40	29.96	33.59	47.84	42.00
299	−78.70	−95.26	41.93	63.00	38.95	50.40	59.92	54.60
298	−84.91	−103.55	58.40	75.59	56.93	63.00	77.91	73.49
297	−93.19	−111.83	71.87	83.99	74.91	83.99	104.87	83.99
296	−113.90	−120.12	89.84	96.59	98.87	109.19	131.83	117.51
295	−124.26	−128.40	107.81	119.69	133.33	136.49	161.79	138.59
294	−132.54	−140.82	125.78	142.79	167.79	172.19	194.76	167.99
293	−144.97	−149.11	161.72	165.89	215.72	214.18	239.70	209.99
292	−163.61	−161.53	197.65	184.79	278.65	266.68	293.63	251.98
291	−167.75	−165.68	215.62	199.49	329.59	308.68	341.577	289.78
290	−163.61	−157.39	197.65	188.99	320.60	310.78	338.5	262.48
289	−136.68	−107.69	119.79	130.19	230.71	214.18	242.69	163.79
288	−31.06	−20.71	38.94	69.30	86.89	104.98	89.89	54.60
287	−37.28	−16.57	14.97	42.00	23.97	42.00	17.98	8.40
286	−24.85	−12.43	44.92	69.30	59.92	77.69	56.93	54.60
285	−31.06	−28.99	92.84	100.79	134.84	155.39	131.83	117.59
284	−66.27	−62.13	122.79	134.39	194.76	201.59	197.76	153.29
283	−82.84	−66.27	119.79	134.39	208.24	209.99	212.73	163.79
282	−51.77	−41.42	98.83	109.61	182.77	188.99	179.77	134.39
281	−24.85	−12.43	65.88	88.19	134.84	144.89	170.79	94.49
280	0	−16.57	44.92	79.79	103.36	125.99	101.87	75.59
279	−4.14	−20.71	34.44	67.20	89.89	107.09	77.91	56.70
278	−16.57	−28.99	23.96	63.00	74.91	94.49	59.92	42.00
277	−41.42	−41.42	13.48	46.20	53.93	83.99	47.94	31.50
276	−41.42	−43.49	14.97	46.20	56.92	83.99	35.95	29.40
275	−33.14	−41.42	23.96	56.70	65.92	90.29	44.95	39.90
274	−16.57	−41.42	31.44	63.00	83.90	96.59	62.92	42.00
273	−12.43	−41.42	29.95	63.00	77.91	96.59	62.92	39.90
272	−28.99	−43.49	13.48	46.20	64.42	83.99	56.93	29.40
271	−41.42	−45.56	0	33.60	49.44	67.20	41.94	12.60
270	−53.84	−66.27	0	27.30	29.96	54.60	23.97	0

TABLE A-3

N-Acetyl Tyrosine Ethyl Ester, Solvent, pH 7.0, 0.05 *M* Tris, 6 *M* Guanidine Hydrochloride

	Δε					
	25°			2°		
Wavelength mμ (1)	20% sucrose (2)	20% glycerol (3)	20% ethylene glycol (4)	20% sucrose (5)	20% glycerol (6)	20% ethylene glycol (7)
300	0	0	0	0	0	0
299	0	0	0	0	0	0
298	0	0	0	0	0	0
297	0	0	0	0	0	0
296	0	0	0	0	0	0
295	0	0	0	0	0	0
294	0	2.7	2.7	1.3	1.3	2.7
293	1.3	5.3	5.3	4.0	2.7	4.0
292	4.0	6.7	8.0	5.3	4.0	6.6
291	5.3	13.4	10.7	5.3	8.0	10.7
290	8.0	18.7	18.7	9.3	15.9	17.3
289	12.0	25.3	34.6	10.7	27.9	34.7
288	16.0	42.7	53.4	21.3	42.5	50.6
287	24.0	57.3	69.3	26.6	59.9	70.6
286	30.6	67.9	84.0	32.3	73.2	85.1
285	31.0	72.0	85.3	37.2	78.5	90.5
284	29.3	69.3	82.6	32.3	77.2	86.5
283	26.6	58.6	72.0	30.6	65.2	73.2
282	21.3	48.0	58.6	25.2	50.6	57.2
281	19.3	38.7	48.0	21.3	38.6	44.0
280	24.0	44.0	52.0	22.7	38.6	46.6
279	25.3	53.4	61.3	26.6	53.2	59.9
278	26.6	57.3	69.3	29.3	59.9	70.6
277	25.3	57.3	66.6	27.9	58.6	69.2
276	19.9	48.0	57.3	25.2	51.9	61.2
275	17.4	38.7	48.0	20.0	38.6	47.9
274	14.7	32.0	38.7	14.6	27.9	35.9
273	12.0	26.6	34.6	13.3	26.9	30.6
272	12.0	26.6	30.6	13.3	18.6	29.3

TABLE A-4

N-Acetyl Tryptophan Ethyl Ester, Solvent, pH 7.0, 0.05 *M* Tris, 6 *M* Guanidine Hydrochloride

	Δε					
	25°			2°		
Wavelength mμ (1)	20% sucrose (2)	20% glycerol (3)	20% ethylene glycol (4)	20% sucrose (5)	20% glycerol (6)	20% ethylene glycol (7)
300	0	0	0	0	0	0
299	5.2	5.2	0	5.2	5.2	5.2
298	15.7	20.8	5.2	15.6	10.4	10.4
297	20.8	41.8	10.4	20.8	15.6	20.8
296	26.1	52.2	20.8	20.8	20.8	31.2
295	41.8	57.4	31.3	26.0	26.0	46.7
294	57.4	62.7	41.8	36.4	41.6	57.1
293	73.1	83.5	62.7	51.9	57.1	72.7
292	78.3	114.9	114.9	71.5	88.3	129.9
291	67.9	130.5	146.2	62.3	117.3	161.0
290	57.4	94.0	125.3	41.6	119.1	187.0
289	36.5	67.9	109.6	20.8	77.9	114.3
288	15.7	26.1	67.9	0	26.0	57.1
287	20.8	10.4	47.0	10.4	10.4	46.7
286	31.3	20.8	62.7	26.0	36.4	57.1
285	36.5	36.5	67.9	41.6	57.1	77.9
284	41.8	36.5	88.8	41.6	62.3	98.7
283	36.5	47.0	104.4	39.0	62.3	103.9
282	20.8	41.8	88.8	15.6	51.9	93.5
281	15.7	26.1	73.1	7.8	46.7	72.7
280	10.4	20.8	57.4	5.2	36.4	67.5
279	5.2	15.7	41.8	10.4	20.8	72.7
278	5.2	10.4	41.8	10.4	10.4	51.9
277	10.4	5.2	26.1	10.4	10.4	41.6
276	15.9	5.2	20.8	20.8	20.8	51.9
275	26.4	15.7	31.3	23.4	26.0	51.9
274	31.3	10.4	41.8	23.4	20.8	41.6
273	31.3	10.4	47.0	15.6	15.6	41.6

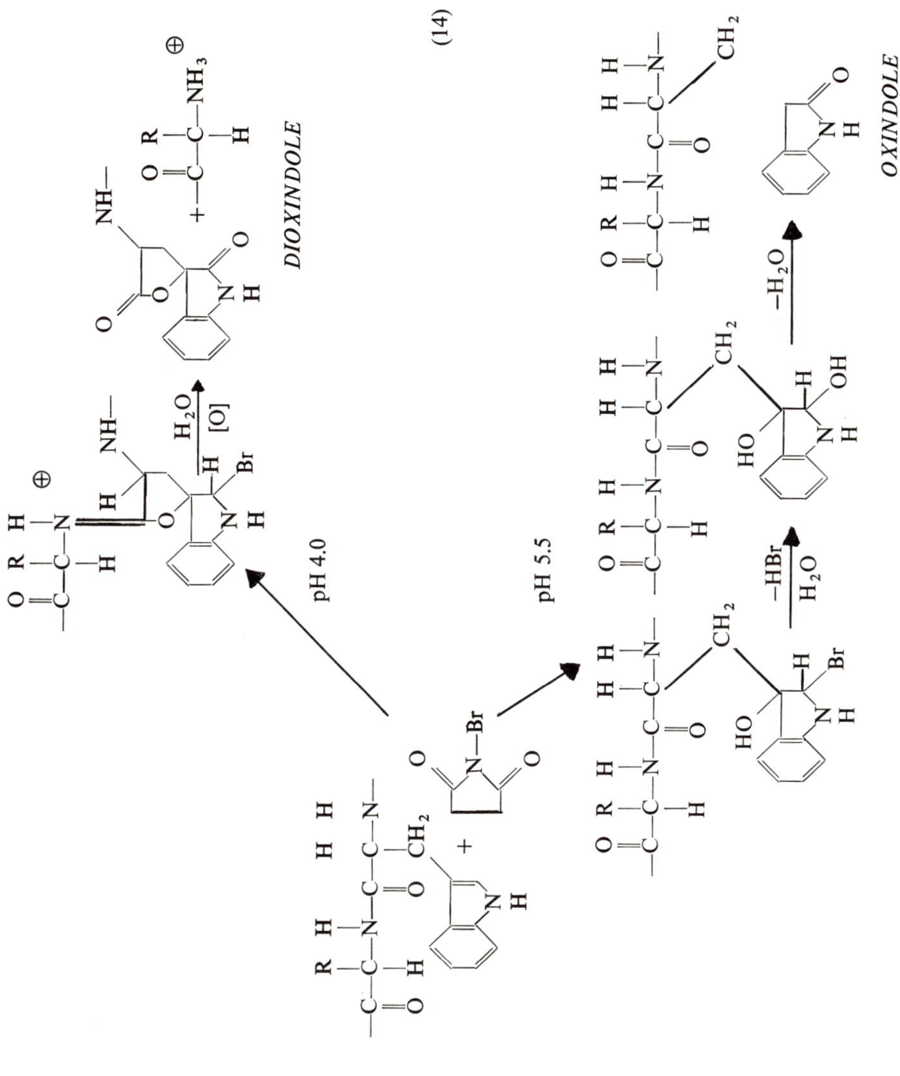

(14)

(See p. 368)

ACKNOWLEDGMENTS

We wish to thank a number of our colleagues who provided us with manuscripts of papers prior to publication and who have permitted us to quote from their unpublished observations. We are also pleased to acknowledge the comments of Dr. W. Blatt and Dr. G. Mandels on the manuscript and the assistance of Mr. R. E. Andreotti and Miss V. Monahan in its preparation.

REFERENCES

1. H. F. Fisher, *Proc. Natl. Acad. Sci. U.S.*, **51,** 1285 (1964).
2. S. Pontremoli, E. Grazi, and A. Accorsi, *Biochemistry*, **5,** 3072 (1966).
3. R. M. Herriot, *Advan. Protein Chem.*, **3,** 170 (1947).
4. R. T. Simpson, J. F. Riordan, and B. L. Vallee, *Biochemistry*, **2,** 616 (1963).
5. T. H. Plummer, Jr., and W. B. Lawson, *J. Biol. Chem.*, **241,** 1648 (1966).
6. B. L. Vallee, J. F. Riordan, and R. T. Simpson, *Biochemistry*, **2,** 616 (1963).
7. D. C. Phillips, *Proc. Natl. Acad. Sci. U.S.*, **57,** 484 (1967).
8. H. G. Gundlach, W. H. Stein, and S. Moore, *J. Biol. Chem.*, **234,** 1754 (1959).
9. E. W. Westhead, *Biochemistry*, **4,** 2139 (1965).
10. L. Weil, S. James, and A. R. Buchert, *Arch. Biochem. Biophys.*, **46,** 266 (1953).
11. I. Krimsky and E. Racker, *Science*, **122,** 319 (1955).
12. J. I. Harris, B. P. Meriwether, and J. H. Park, *Nature*, **198,** 154 (1963).
13. N. B. Madson, *J. Biol. Chem.*, **223,** 1067 (1956).
14. N. B. Madson and C. F. Cori, *J. Biol. Chem.*, **223,** 1055 (1956).
15. N. B. Madson and F. R. N. Gurd, *J. Biol. Chem.*, **223,** 1075 (1956).
16. E. L. Smith, *J. Biol. Chem.*, **233,** 1392 (1958).
17. P. L. Polianovski, *Biochimia*, **27,** 734 (1962).
18. C. Turano, P. Giartosio, and P. Fasella, *Arch. Biochem. Biophys.*, **104,** 524 (1964).
18a. C. Turano, A. Giartosio, and F. Riva, *Enzymologia*, **25,** 196 (1963).
19. C. H. W. Hirs, M. Halman, and J. H. Kycia, *Biological Structure and Function*, Vol. 1, Academic, New York, 1962, p. 41.
20. J. C. Speck, P. T. Rowley, and B. L. Horecker, *J. Am. Chem. Soc.*, **85,** 1012 (1963).
21. D. F. Koshland, Jr., D. H. Strumeyer, and J. Ray, Jr., *Brookhaven Symp. Biol.*, **15,** 101 (1962).
22. N. K. Schaffer, L. Simet, S. Harshman, R. Engle, and W. Drisko, *J. Biol. Chem.*, **225,** 197 (1957).
23. F. Turba and H. G. Grundlach, *Biochem. J.*, **327,** 196 (1955).
24. G. H. Dixon, D. L. Kaufman, and H. Neurath, *J. Biol. Chem.*, **233,** 1373 (1958).
25. J. H. Schwartz, A. M. Crestfield, and F. Lippman, *Proc. Natl. Acad. Sci. U.S.*, **49,** 722 (1963).
26. C. Milstein, *Biochim. Biophys. Acta*, **67,** 171 (1963).
27. C. Milstein and F. Sanger, *Biochem. J.*, **79,** 456 (1961).
28. J. A. Gladner and K. Laki, *J. Am. Chem. Soc.*, **80,** 1263 (1958).
29. C. Tanford, *J. Am. Chem. Soc.*, **84,** 4240 (1962).
30. J. C. Kendrew, *Brookhaven Symp. Biol.*, **15,** 216 (1962).
31. G. D. Fasman (Ed.), *Poly α-Amino Acids*, Vol. 1, Dekker, New York, 1967.
32. G. H. Beaven, *Advan. Spectry.*, **2,** 331 (1961).
33. D. B. Wetlaufer, *Advan. Protein Chem.*, **17,** 304 (1962).
34. G. Weber and F. W. J. Teale in *The Proteins* (H. Neurath, ed.), Vol. III, Academic, New York, 1965, p. 445.

35. S. Yanari and F. A. Bovey, *J. Biol. Chem.*, **235,** 2818 (1960).
36. S. J. Leach and H. A. Scheraga, *J. Am. Chem. Soc.*, **82,** 4790 (1960).
37. J. W. Donovan, *Biochemistry*, **3,** 67 (1964).
38. T. T. Herskovits and M. Laskowski, Jr., *J. Biol. Chem.*, **237,** 2481 (1962).
38a. T. T. Herskovits, *Methods in Enzymology*, **11,** 748 (1967).
38b. T. T. Herskovits and Sr. M. Sorenson, *Biochemistry*, **7,** 2523 (1968).
38c. T. T. Herskovits and Sr. M. Sorenson, *Biochemistry*, **7,** 2533 (1968).
39. A. H. Mehler, *Science*, **120,** 1043 (1954).
40. I. Fridovich, W. Farkas, G. W. Schwert, Jr., and P. Handler, *Science*, **125,** 1141 (1957).
41. A. H. Mehler, B. Bloom, M. E. Ahrendt, and D. W. Stetten, *Science*, **126,** 1285 (1957).
42. E. Mihalyi, *Arch. Biochem. Biophys.*, **110,** 325 (1965).
43. L. G. Holmes and M. J. Kronman, *Anal. Biochem.*, **7,** 124 (1964).
44. J. A. Yankeelov, *Anal. Biochem.*, **7,** 342 (1964).
45. T. L. Warrington and M. J. Kronman, unpublished experiments.
46. T. T. Herskovits, *J. Biol. Chem.*, **240,** 628 (1965).
47. L. Riddiford, *J. Biol. Chem.*, **241,** 2792 (1966).
48. E. J. Williams, T. T. Herskovits, and M. Laskowski, Jr., *J. Biol. Chem.*, **240,** 3574 (1965).
49. M. J. Kronman and L. G. Holmes, *Biochemistry*, **4,** 526 (1965).
50. K. Hamaguchi and A. Kurono, *J. Biochem.* (*Tokyo*), **54,** 111 (1963).
51. K. Hamaguchi, A. Kurono, and S. Goto, *J. Biochem.* (*Tokyo*), **54,** 259 (1963).
52. K. Hamaguchi and A. Kurono, *J. Biochem.* (*Tokyo*), **54,** 497 (1963)
53. K. Hamaguchi, L. Hayashi, T. Imoto, and M. Funatsu, *J. Biochem.* (*Tokyo*), **55,** 24 (1964).
54. K. Hamaguchi, *J. Biochem.* (*Tokyo*), **55,** 333 (1964).
55. K. Hamaguchi, *J. Biochem.* (*Tokyo*), **56,** 441 (1964).
55a. A. Kurono and K. Hamaguchi, *J. Biochem.* (*Tokyo*), **56,** 432 (1964).
56. J. M. Swan, *Australian J. Chem.*, **14,** 69 (1961).
57. T. T. Herskovitz and M. Laskowski, Jr., *J. Biol. Chem.*, **237,** 3418 (1962).
58. N. M. Green, *Biochem. J.*, **89,** 599 (1963).
59. M. Laskowski, Jr., *Federation Proc.*, **25,** 20 (1966).
60. M. J. Kronman, L. Cerankowski, and L. G. Holmes, *Biochemistry*, **4,** 518 (1965).
61. M. J. Kronman and L. G. Holmes, unpublished experiments.
62. R. H. Maybury and J. J. Katz, *Nature*, **177,** 629 (1956).
63. O. Shugar and E. Gajewska, *J. Polymer Sci.*, **31,** 281 (1958).
64. M. Feughelman, A. R. Haly, and T. W. Mitchell, *Textile Res. J.*, **29,** 564 (1959).
65. D. S. Berns, *Biochemistry*, **2,** 1377 (1963).
66. J. W. Donovan, *Biochemistry*, **6,** 3918 (1967).
67. D. G. Cross and H. F. Fisher, *Biochemistry*, **5,** 880 (1966).
68. E. J. Williams and M. Laskowski, Jr., *J. Biol. Chem.*, **240,** 3580 (1965).
69. M. J. Kronman, F. M. Robbins, and R. E. Andreotti, *Biochem. Biophys. Acta*, **147,** 462 (1967).
70. D. G. Smith, W. H. Stein, and S. Moore, *J. Biol. Chem.*, **238,** 227 (1963).
71. R. E. Canfield and A. K. Liu, *J. Biol. Chem.*, **240,** 1997 (1965).
72. B. S. Hartley, *Nature*, **201,** 1284 (1964).
73. B. Keil, A. Prusik, and F. Sorm, *Biochim. Biophys. Acta*, **78,** 559 (1963).
74. S. Lowey, *J. Biol. Chem.*, **240,** 2421 (1965).
75. S. Lowey, J. Kucera, and A. Holtzer, *J. Mol. Biol.*, **7,** 234 (1963).
76. S. Lowey and C. Cohen, *J. Mol. Biol.*, **4,** 293 (1962).
77. H. G. Burger, H. Edelhoch, and P. G. Condliffe, *J. Biol. Chem.*, **241,** 449 (1966).

78. J. Mercouroff and G. P. Hess, *Biochem. Biophys. Res. Commun.*, **11**, 283 (1963).
79. E. Stellwagen, *Biochemistry*, **3**, 919 (1964).
80. S. Libor, P. Elodi, and Z. Nagy, *Biochim. Biophys. Acta*, **110**, 484 (1965).
81. L. Weil, T. S. Seibles, and T. T. Herskovits, *Arch. Biochem. Biophys.*, **111**, 308 (1965).
81a. C. J. Menendez and T. T. Herskovits, Abstracts, 156th National Meeting, American Chemical Society, Atlantic City, N.J., Sept., 1968.
82. K. Hayashi, T. Imoto, G. Funatsu, and M. Funatsu, *J. Biochem. (Tokyo)*, **58**, 227 (1965).
83. C. C. Wang, Ph.D. thesis, Purdue University (1963).
84. E. J. Williams, *Arch. Biochem. Biophys.*, **115**, 21 (1966).
85. V. Frattali, R. F. Steiner, and H. Edelhoch, *J. Biol. Chem.*, **240**, 112 (1965).
86. T. T. Herskovits and M. Laskowski, Jr., *J. Biol. Chem.*, **235**, PC 56 (1960).
87. W. J. Leonard and J. F. Foster, *J. Biol. Chem.*, **236**, PC 73 (1961).
88. H. Edelhoch and R. F. Steiner, *J. Biol. Chem.*, **240**, 2877 (1965).
88a. T. T. Herskovits and M. Laskowski, Jr., *J. Biol. Chem.*, **243**, 2123 (1968).
89. J. M. Creeth, *Biochem. J.*, **53**, 41 (1953).
90. J. L. Oncley and E. Ellenbogen, *J. Phys. Chem.*, **56**, 85 (1952).
91. C. Tanford and J. Epstein, *J. Am. Chem. Soc.*, **76**, 2163, 2170 (1954).
92. K. Marcker, *Acta Chem. Scand.*, **14**, 194 (1960).
93. H. F. Fisher and D. G. Cross, *Biochem. Biophys. Res. Commun.*, **20**, 120 (1965).
94. M. L. Bender, G. R. Schonbaum, and B. Zerner, *J. Am. Chem. Soc.*, **84**, 2540 (1962).
95. V. Massey and B. Hartley, *Biochim. Biophys. Acta*, **21**, 58 (1956).
96. F. Karush, *J. Phys. Chem.*, **56**, 70 (1952).
97. M. Burr and D. E. Koshland, Jr., *Proc. Natl. Acad. Sci. U.S.*, **52**, 1017 (1964).
98. M. E. Kirtley and D. E. Koshland, Jr., *Biochem. Biophys. Res. Commun.*, **23**, 810 (1966).
99. J. L. Crammer and A. Neuberger, *Biochem. J.*, **37**, 302 (1943).
100. K. Mihashi and T. Ooi, *Biochemistry*, **4**, 805 (1965).
101. A. Wishnia, I. Weber, and R. C. Warner, *J. Am. Chem. Soc.*, **83**, 2071 (1961).
102. J. Leonis and C. H. Li, *J. Am. Chem. Soc.*, **81**, 415 (1959).
103. C. Tanford and J. D. Hauenstein, *J. Am. Chem. Soc.*, **78**, 5287 (1956).
104. C. Tanford and M. L. Wagner, *J. Am. Chem. Soc.*, **76**, 3331 (1954).
105. J. W. Donovan, M. Laskowski, Jr., and H. A. Scheraga, *J. Am. Chem. Soc.*, **83**, 2686 (1961).
106. A. N. Glazer and E. L. Smith, *J. Biol. Chem.*, **236**, 2948 (1961).
107. C. Tanford and G. L. Roberts, *J. Am. Chem. Soc.*, **74**, 2509 (1952).
108. F. M. Robbins, R. E. Andreotti, L. G. Holmes, and M. J. Kronman, *Biochim. Biophys. Acta*, **133**, 33 (1967).
109. B. H. Havsteen and G. P. Hess, *J. Am. Chem. Soc.*, **84**, 448 (1961).
110. A. Tachibana and T. Murachi, *Biochemistry*, **5**, 2756 (1966).
111. L. F. Hass and M. S. Lewis, *Biochemistry*, **6**, 1368 (1964).
112. R. T. Simpson, J. F. Riordan, and B. L. Vallee, *Biochemistry*, **2**, 616 (1963).
113. L. Riddiford, R. H. Stellwagen, S. Mehta, and J. T. Edsall, *J. Biol. Chem.*, **240**, 3305 (1965).
114. P. E. Wilcox, *Federation Proc.*, **16**, 270 (1957).
115. Y. Inada, M. Kamata, A. Matsushima, and K. Shibata *Biochim. Biophys. Acta*, **81**, 323 (1964).
116. T. Flatmark, *Acta. Chem. Scand.*, **18**, 1796 (1964).
117. J. A. Rupley, *Biochemistry*, **3**, 1648 (1964).
118. J. Hermans, Jr., *Biochemistry*, **1**, 193 (1962).

119. J. Hermans, Jr., *Biochemistry*, **2,** 453 (1963).
120. Y. Inada, *J. Biochem.* (*Tokyo*), **49,** 217 (1961).
121. C. Tanford, *Advan. Protein Chem.*, **17,** 70 (1962).
121a. M. Gorbunoff, *Biochemistry*, **6,** 1606 (1967).
121b. R. Townend, T. T. Herskovits, S. N. Timasheff and M. J. Gorbunoff, *Arch. Biochem. Biophys.*, **129,** 567 (1969).
121c. M. J. Gorbunoff, *Biochemistry*, **7,** 2547 (1968).
121d. M. J. Gorbunoff, *Biochemistry*, **8,** 2591 (1969).
122. G. DiSabata, *Biochemistry*, **4,** 2288 (1965).
123. T. Tojo, K. Hamaguchi, M. Imanishi, and T. Amano, *J. Biochem.* (*Tokyo*), **60,** 538 (1966).
124. Y. Inada, A. Matsushima, M. Kamata, and K. Shibata, *Arch. Biochem. Biophys.*, **106,** 326 (1964).
124a. G. E. Perlmann, *J. Biol. Chem.*, **239,** 3762 (1964).
125. L. B. Smillie and C. M. Kay, *J. Biol. Chem.*, **236,** 112 (1961).
126. J. Hermans, Jr., personal communication.
127. K. Brew, T. C. Vanaman, and R. L. Hill, *J. Biol. Chem.*, **242,** 3747 (1967).
128. M. J. Kronman, L. G. Holmes, and F. M. Robbins, *Biochem. Biophys. Acta*, **133,** 46 (1967).
129. R. L. Nagel, H. M. Ranney, and L. L. Kucinskis, *Biochemistry*, **5,** 1934 (1966).
130. D. Eisenberg and J. T. Edsall, *Science*, **142,** 50 (1963).
131. H. Theorell, *Biochem. J.*, **298,** 242 (1938).
132. H. Theorell, *Enzymologia*, **6,** 88 (1939).
133. A. N. Glazer and E. L. Smith, *J. Biol. Chem.*, **236,** 2942 (1961).
134. F. W. J. Teale and G. Weber, *Biochem. J.*, **65,** 476 (1957).
135. F. W. J. Teale, *Biochem. J.*, **76,** 381 (1960).
136. J. A. Gally and G. M. Edelman, *Biochem. Biophys. Acta*, **60,** 499 (1962).
137. J. A. Gally and G. M. Edelman, *Biopolymer Symp.*, **1,** 367 (1961).
138. R. F. Steiner, R. E. Lippoldt, H. Edelhoch, and V. Frattali, *Biopolymer Symp.*, **1,** 355 (1964).
139. M. J. Kronman, *Biochim. Biophys. Acta*, **133,** 19 (1967).
140. R. F. Chen in *Fluorescence* (G. G. Guilbault, ed.), Dekker, New York, 1967.
141. G. Weber and L. B. Young, *J. Biol. Chem.*, **239,** 1424 (1964).
142. G. Weber and K. Rosenheck, *Biopolymer Symp.*, **1,** 333 (1964).
143. B. L. Van Durren, *J. Org. Chem.*, **26,** 2954 (1961).
144. S. S. Lehrer and G. D. Fasman, *Biochem. Biophys. Res. Commun.*, **23,** 133 (1966).
145. S. S. Lehrer and G. D. Fasman, *J. Biol. Chem.*, **242,** 4644 (1967).
146. M. S. Walker, T. W. Bednar, and R. Lumry, *J. Chem. Phys.*, **47,** 1020 (1967).
147. M. S. Walker, T. W. Bednar, and R. Lumry, *J. Chem. Phys.*, **45,** 3455 (1966).
148. R. W. Cowgill, *Arch. Biochem. Biophys.*, **104,** 84 (1965).
149. R. W. Cowgill, *Biochim. Biophys. Acta*, **75,** 272 (1963).
150. M. Shinitzky, E. Katchalski, E. Grisaro, and J. Sharon, *Arch. Biochem. Biophys.*, **115,** 232 (1966).
151. S. S. Lehrer, *Biochem. Biophys. Res. Commun.*, **29,** 767 (1967).
152. C. C. F. Blake, L. N. Johnson, G. A. Mair, A. C. T. North, D. C. Phillips. and U. R. Sarma, *Proc. Roy. Soc.*, **167B,** 378 (1967).
153. C. C. F. Blake, G. A. Mair, A. C. T. North, D. C. Phillips, and U. R. Sarma, *Proc. Roy. Soc.*, **167B,** 365 (1967).
154. L. Stryer, *J. Am. Chem. Soc.*, **88,** 5708 (1966).
155. R. F. Chen and P. F. Cohen, *Arch. Biochem. Biophys.*, **114,** 515 (1966).

156. J. Elkena, *J. Phys. Chem.*, **72,** 3654 (1968).
156a. G. M. Edelman and W. O. McClure, *Accounts Chem. Res.*, **1,** 65 (1968).
157. W. O. McClure and G. M. Edelman, *Biochemistry*, **5,** 1908 (1966).
157a. H. Susi, T. Zell, and S. N. Timasheff, *Arch. Biochem. Biophys.*, **85,** 437 (1959).
158. L. Stryer, *J. Mol. Biol.*, **13,** 482 (1965).
159. M. J. Kronman and R. E. Andreotti, *Biochemistry*, **3,** 1145 (1964).
160. M. J. Kronman, R. E. Andreotti, and R. Vitols, *Biochemistry* **3,** 1152 (1964).
161. M. J. Kronman, unpublished experiments.
162. M. J. Kronman, R. Blum, and L. G. Holmes, *Biochemistry*, **5,** 1970 (1966).
163. J. A. Gally and G. M. Edelman, *Biochim. Biophys. Acta*, **94,** 175 (1965).
164. G. Weber and L. B. Young, *J. Biol. Chem.*, **239,** 1415 (1964).
165. G. Weber and E. Daniel, *Biochemistry*, **5,** 1900 (1966).
166. E. G. McRae, *J. Phys. Chem.*, **61,** 562 (1957).
167. H. C. Longuet-Higgins and J. A. Pople, *J. Chem. Phys.*, **27,** 192 (1957).
168. G. J. Brealey and M. Kasha, *J. Am. Chem. Soc.*, **77,** 4462 (1955).
169. T. E. Barman and D. E. Koshland, Jr., *J. Biol. Chem.*, **242,** 5771 (1967).
170. J. Steinhardt and S. Beychek in *The Proteins* (H. Neurath, ed.), Vol. II, Academic, New York, 1964, p. 139.
171. E. Breslow and F. R. N. Gurd, *J. Biol. Chem.*, **238,** 1332 (1963).
172. L. J. Banazak, H. C. Watson, and J. C. Kendrew, *J. Mol. Biol.*, **12,** 130 (1965).
173. J. F. Taylor, D. P. Pugh, and B. F. Van Osdol, *Federation Proc.*, **25,** 928 (1966).
174. D. B. Wetlaufer and R. Lovrien, *J. Biol. Chem.*, **239,** 596 (1964).
175. W. Kauzmann, *Advan. in Protein Chem.*, **14,** 1 (1959).
176. E. Daniel and G. Weber, *Biochemistry*, **6,** 1893 (1966).
177. R. V. Decker and J. F. Foster, *Biochemistry*, **5,** 1242 (1960).
178. R. H. Kretsinger, H. C. Watson, and J. C. Kendrew, *J. Mol. Biol.*, **31,** 305 (1968).
179. A. Wishnia, *Proc. Natl. Acad. Sci., U.S.*, **48,** 2200 (1962).
180. A. Wishnia and T. W. Pinder, Jr., *Biochemistry*, **3,** 1377 (1964).
181. A. Wishnia and T. W. Pinder, Jr., *Federation Proc.*, **23,** 160 (1964).
182. A. Wishnia and T. W. Pinder, Jr., Abstracts, Biophysical Society Meeting, Boston, Mass., 1966.
183. A. Wishnia and T. W. Pinder, Jr., *Biochemistry*, **5,** 1534 (1966).
184. S. N. Timasheff and R. Townend, *J. Am. Chem. Soc.*, **83,** 470 (1961).
185. S. N. Timasheff and R. Townend, *J. Am. Chem. Soc.*, **83,** 464 (1961).
186. C. Georges, S. Guinard, and J. Tonnelat, Abstracts of the International Biophysics Congress, Stockholm, August, 1961.
187. B. P. Schoenborn, H. C. Watson and J. C. Kendrew, *Nature*, **207,** 28 (1965).
188. B. P. Schoenborn, *Nature*, **214,** 1120 (1967).
189. J. F. Foster in *The Plasma Proteins* (F. W. Putnam, ed.), Academic, New York, 1960, Chap. 6.
190. J. A. Reynolds, S. Herbert, H. Polet, and J. Steinhardt, *Biochemistry*, **6,** 937 (1967).
191. J. D. Teresi and J. M. Luck, *J. Biol. Chem.*, **194,** 823 (1952).
192. L. Stryer, J. C. Kendrew, and H. C. Watson, *J. Mol. Biol.*, **8,** 96 (1964).
193. R. H. Kretsinger, *J. Mol. Biol.*, **31,** 315 (1968).
194. H. W. Wyckoff, K. D. Hardman, N. M. Allewell, T. Inagami, L. N. Johnson, and F. M. Richards, *J. Biol. Chem.*, **242,** 3984 (1967).
195. C. C. F. Blake, *Proc. Roy. Soc.*, **167B,** 435 (1967).
196. L. N. Johnson, *Proc. Roy. Soc.*, **167B,** 439 (1967).
197. M. F. Perutz, W. Bolton, R. Diamond, H. Muirhead, and H. C. Watson, *Nature*, **203,** 687 (1964).

198. C. Tanford, S. A. Swanson, and W. S. Shore, *J. Am. Chem. Soc.*, **77,** 6414 (1955).
198a. K. K. Vijai and J. F. Foster, *Biochemistry*, **6,** 1152 (1967).
199. J. A. Rupley, L. Butler, M. Gerring, F. J. Hartdegen, and R. Pecoraro, *Proc. Natl. Acad. Sci. U.S.*, **57,** 1088 (1967).
200. R. C. Warner and J. Weber, *J. Am. Chem. Soc.*, **75,** 5094 (1953).
201. A. N. Glazer and H. W. McKenzie, *Biochim. Biophys. Acta*, **71,** 109 (1963).
202. P. R. Azarl and P. E. Feeney, *J. Biol. Chem.*, **232,** 293 (1958).
203. P. R. Azarl and P. E. Feeney, *Arch. Biochem. Biophys.*, **92,** 44 (1961).
204. M. J. Kronman and A. J. Budreau, unpublished experiments.
205. J. Hermans, Jr., and H. A. Scheraga, *J. Am. Chem. Soc.*, **83,** 3293 (1961).
206. S. Beychok and J. Steinhardt, *J. Am. Chem. Soc.*, **81,** 5679 (1959).
207. J. Steinhardt, R. Ona, and S. Beychok, *Biochemistry*, **1,** 29 (1962).
208. C. F. Tanford and Y. Nozaki, *J. Biol. Chem.*, **241,** 2832 (1966).
208a. E. Antonini, J. Wyman, M. Brunori, C. Fronticelli, E. Bucci, and A. Rossi-Fanelli, *J. Biol. Chem.*, **240,** 1096 (1965).
209. E. Breslow and F. R. N. Gurd, *J. Biol. Chem.*, **237,** 371 (1962).
210. L. M. Riddiford and H. A. Scheraga, *Biochemistry*, **1,** 95 (1962).
211. S. N. Timasheff and J. A. Rupley, unpublished experiments quoted in Ref. *199*.
212. A. Hvidt and S. O. Nielson, *Advan. Protein Chem.*, **21,** 21 (1966).
213. A. Wishnia and M. Saunders, *J. Am. Chem. Soc.*, **84,** 4235 (1962).
214. M. Eigen, *Angew. Chem.*, **75,** 489 (1963).
215. E. M. Bradbury, W. C. Price, and Wilkinson, *J. Mol. Biol.*, **4,** 39 (1962).
216. M. Saunders, H. A. Jung, and W. L. Hamilton, *J. Am. Chem. Soc.*, **89,** 472 (1967).
217. G. Ehrlich and G. B. B. M. Sutherland, *J. Am. Chem. Soc.*, **76,** 5268 (1954).
218. C. C. McDonald and W. D. Phillips, *J. Am. Chem. Soc.*, **89,** 6332 (1967).
219. W. D. Phillips, personal communication.
220. J. H. Bradbury and H. A. Scheraga, *J. Am. Chem. Soc.*, **88,** 4240 (1966).
221. J. H. Bradbury and P. Wilairat, *Biochem. Biophys. Res. Commun.*, **29,** 84 (1967).
222. D. H. Meadows, J. L. Markley, J. S. Cohen, and O. Jardetzky, *Proc. Natl. Acad. Sci., U.S.*, **58,** 1307 (1967).
223. O. H. Griffith and H. M. McConnell, *Proc. Natl. Acad. Sci. U.S.*, **55,** 8 (1966).
224. J. C. A. Boeyens and H. M. McConnell, *Proc. Natl. Acad. Sci. U.S.*, **56,** 22 (1966).
225. L. J. Berliner and H. M. McConnell, *Proc. Natl. Acad. Sci. U.S.*, **55,** 708 (1966).
226. T. J. Stone, T. Buckman, P. L. Nordio, and H. M. McConnell, *Proc. Natl. Acad. Sci. U.S.*, **54,** 1010 (1965).
227. A. M. Crestfield, W. H. Stein, and S. Moore, *J. Biol. Chem.*, **238,** 2413 (1963).
228. A. M. Crestfield, W. H. Stein, and S. Moore, *J. Biol. Chem.*, **238,** 2421 (1963).
229. D. G. Hoare and D. E. Koshland, Jr., *J. Am. Chem. Soc.*, **88,** 2057 (1966).
230. T. Okuyama and T. Satake, *J. Biochem.* (*Tokyo*), **47,** 454 (1960).
231. K. Satake, T. Okuyama, M. Ohashi, and T. Shinoda, *J. Biochem.* (*Tokyo*), **47,** 654 (1960).
232. A. R. Goldfarb, *Biochemistry*, **5,** 2570 (1966).
233. A. R. Goldfarb, *Biochemistry*, **5,** 2574 (1966).
234. C. Tanford, *J. Am. Chem. Soc.*, **72,** 441 (1950).
235. P. D. Boyer, *The Enzymes* (P. D. Boyer, H. Lardy, and K. Myrbäck, eds.). Vol. I, Part II, Academic, New York (1959).
236. R. Cecil and J. R. McPhee, *Advan. Protein Chem.*, **14,** p. 255 (1959).
237. R. Benesch, *Sulfur in Proteins* (R. Benesch et al., ed.), Academic, New York (1959).
238. J. F. Riordan, W. D. C. Wacker, and B. L. Vallee, *Biochemistry*, **4,** 1758 (1965).
239. A. K. Balls and H. N. Wood, *J. Biol. Chem.*, **219,** 245 (1956).

240. G. E. Perlmann, *J. Biol. Chem.*, **241,** 153 (1966).
241. K. Kurihara, H. Horinishi, and K. Shibata, *Biochim. Biophys. Acta*, **74,** 678 (1963).
242. Y. Hachimori, K. Kurihara, H. Horinishi, A. Matsushima, and K. Shibata, *Biochim. Biophys. Acta*, **105,** 167 (1965).
243. A. Wormall, *J. Exptl. Med.*, **51,** 295 (1930).
244. M. Sokoalovsky, J. F. Riordan, and B. L. Vallee, *Biochemistry*, **5,** 3582 (1966).
245. J. F. Riordan, M. Sokolovsky, and B. L. Vallee, *Biochemistry*, **6,** 358 (1967).
246. L. K. Ramachandran, *Chem. Rev.*, **56,** 199 (1956).
247. F. W. Putnam in *The Proteins* (H. Neurath and K. Bailey, eds.), Vol. 1, Academic, New York, 1953, p. 893.
248. W. L. Hughes, Jr., and R. Straessle, *J. Am. Chem. Soc.*, **72,** 452 (1950).
249. H. Edelhoch, *J. Biol. Chem.*, **237,** 2778 (1962).
250. R. T. Simpson and B. L. Vallee, *Biochemistry*, **5,** 1760 (1966).
251. W. Ferdinand, W. H. Stein, and S. Moore, *J. Biol. Chem.*, **240,** 1150 (1965).
252. D. D. Ulmer, *Biochemistry*, **5,** 1886 (1966).
253. E. Stellwagen, *J. Biol. Chem.*, **242,** 602 (1967).
254. S. Pontremoli, E. Grazi, and A. Accorsi, *Biochemistry*, **5,** 3568 (1966).
255. I. Covelli and J. Wolff, *J. Biol. Chem.*, **242,** 881 (1967).
256. F. M. Robbins, L. G. Holmes, and M. J. Kronman, unpublished results.
257. F. M. Robbins and L. G. Holmes, unpublished results.
258. I. Covelli and J. Wolff, *Biochemistry*, **5,** 860 (1966).
259. J. Wolff and I. Covelli, *Biochemistry*, **5,** 867 (1966).
260. M. E. Friedman, H. A. Scheraga, and R. Goldberger, *Biochemistry*, **5,** 3770 (1966).
261. J. F. Riordan, W. D. C. Wacker, and B. L. Vallee, *Nature*, **208,** 1209 (1965).
262. R. F. Steiner, *Biochemistry*, **5,** 1964 (1966).
263. W. G. Gordon and J. Ziegler, *Arch. Biochem. Biophys.*, **57,** 80 (1965).
264. U. Brodbeck, W. L. Denton, N. Tanahashi, and K. E. Ebner, *J. Biol. Chem.*, **242,** 1391 (1967).
265. R. T. Simpson and B. L. Vallee, *Biochemistry*, **5,** 2531 (1966).
266. J. G. Cory and E. Frieden, *Biochemistry*, **6,** 116 (1967).
267. M. Aoyama, K. Kurihari, and K. Shibata, *Biochim. Biophys. Acta*, **107,** 257 (1965).
268. E. Margoliash and A. Schejter, *Advan. Protein Chem.*, **21,** 113 (1966).
269. S. M. Rosen and O. M. Rosen, *Biochemistry*, **6,** 2094 (1967).
270. E. L. Pugh and B. L. Horecker, *Arch. Biochem. Biophys.*, **122,** 196 (1967).
271. M. Sokolovsky and B. L. Vallee, *Biochemistry*, **6,** 700 (1967).
272. A. Previero, E. Scoffone, C. A. Benassi, and P. Pajetta, *Gazz. Chim. Ital.*, **93,** 849 (1963).
273. B. Witkop, *Advan. Protein Chem.*, **16,** 221 (1961).
274. N. M. Green and B. Witkop, *Trans. N.Y. Acad. Sci.*, **26,** 659 (1964).
275. T. F. Spande, N. M. Green, and B. Witkop, *Biochemistry*, **5,** 1926 (1966).
276. T. Takahashi, K. Hamaguchi, K. Hayashi, I. Imoto, and M. Funatsu, *J. Biochem. (Tokyo)*, **58,** 385 (1965).
277. T. Peters, *Compt. Rend. Trav. Lab. Carlsberg*, **31,** 227 (1959).
278. A. Patchornik, W. B. Lawson, E. Gross, and B. Witkop, *J. Am. Chem. Soc.*, **82,** 5923 (1960).
279. E. Stellwagen and S. Van Rooyan, *J. Biol. Chem.*, **242,** 4801 (1967).
280. D. E. Koshland, Jr., Y. D. Karkhanis, and H. G. Latham, *J. Am. Chem. Soc.*, **86,** 1448 (1964).
281. H. R. Horton and D. E. Koshland, Jr., *J. Am. Chem. Soc.*, **87,** 1126 (1965).
281a. J. R. Spies and D. C. Chambers, *Anal. Chem.*, **21,** 1294 (1949).

282. Y. Hachimori, H. Horinishi, K. Kurihara, and K. Shibata, *Biochim. Biophys. Acta*, **93,** 346 (1964).
283. B. R. Dasgupta, E. Rothstein, and D. A. Boroff, *Anal. Biochem.*, **11,** 555 (1965).
284. N. B. Oza and C. J. Martin, *Biochem. Biophys. Res. Commun.*, **16,** 195 (1964).
285. T. A. Bewley and C. H. Li, *Nature*, **206,** 624 (1965).
286. T. Viswanatha, W. B. Lawson, and B. Witkop, *Biochim. Biophys. Acta*, **40,** 216 (1960).
287. R. F. Steiner, *Arch. Biochem. Biophys.*, **115,** 257 (1966).
288. F. J. Hartdegan and J. A. Rupley, *Biochim. Biophys. Acta*, **92,** 625 (1964).
288a. A. Previero, M. A. Colletti-Previero, and P. Jolles, *J. Mol. Biol.*, **24,** 261 (1967).
289. F. J. Hartdegan and J. A. Rupley, *J. Am. Chem. Soc.*, **89,** 1743 (1967).
289a. R. E. Weber and C. Tanford, *J. Am. Chem. Soc.*, **81,** 3255 (1959).
290. S. J. Singer, *Advan. Protein Chem.*, **17,** 1 (1962).
291. H. Edelhoch and R. F. Steiner, *Biochim. Biophys. Acta*, **60,** 365 (1962).
291a. N. M. Green, *Biochem. J.*, **90,** 564 (1964).
292. H. F. Fisher and D. G. Cross, *Arch. Biochem. Biophys.*, **110,** 217 (1965).
293. H. Horinishi, Y. Hachimori, K. Kurihara, and K. Shibata, *Biochim. Biophys. Acta*, **86,** 477 (1964).
294. M. Sokolovsky and B. L. Vallee, *Biochemistry*, **5,** 3574 (1966).
295. L. Weil, W. G. Gordon, and A. R. Buchert, *Arch. Biochem. Biophys.*, **33,** 90 (1951).
296. G. Oster, J. S. Bellin, R. W. Kimball, and M. E. Schrader, *J. Am. Chem. Soc.*, **81,** 5095 (1959).
297. L. A. A. E. Sluyterman, *Biochim. Biophys. Acta*, **60,** 557 (1962).
298. J. M. Brake and F. Wold, *Biochim. Biophys. Acta*, **40,** 171 (1960).
299. W. W. Ray, Jr. and D. E. Koshland, Jr., *Brookhaven Symp. Biol.*, No. 13, 135 (1960).
300. G. R. Stark, W. H. Stein, and S. Moore, *J. Biol. Chem.*, **236,** 436 (1961).
301. W. L. Koltun, R. N. Dexter, R. E. Clark, and F. R. N. Gurd, *J. Am. Chem. Soc.*, **80,** 4188 (1958).
302. J. A. Duke, M. Bier, and F. F. Nord, *Arch. Biochem. Biophys.*, **40,** 424 (1952).
303. T. Inagami, *J. Biol. Chem.*, **240,** 3453 (1965).
304. E. Shaw, M. Mares-Guia, and W. Cohen, *Biochemistry*, **4,** 2219 (1965).
305. G. Schoellmann and E. Shaw, *Biochemistry*, **2,** 252 (1963).
306. V. Massey and B. S. Hartley, *Biochim. Biophys. Acta*, **21,** 361 (1956).
307. J. R. Whitaker and B. J. Jandorf, *J. Biol. Chem.*, **223,** 761 (1956).
308. L. Weil and T. S. Seibles, *Arch. Biochem. Biophys.*, **54,** 368 (1955).
309. S. Pale'us, *Acta Chem. Scand.*, **8,** 971 (1954).
310. H. Theorell and A. Åkesson, *J. Am. Chem. Soc.*, **63,** 1812 (1941).
311. E. Stellwagen, Biochem. Biophys. Res. Commun., **23,** 29 (1966).
312. M. Nakatani, *J. Biochem.* (*Tokyo*), **48,** 469 (1960).
313. L. J. Banaszak, P. A. Andrews, J. W. Burgner, E. H. Eylar, and F. R. N. Gurd, *J. Biol. Chem.*, **238,** 3307 (1963).
314. L. J. Banaszak and F. R. N. Gurd, *J. Biol. Chem.*, **239,** 1836 (1964).
315. J. H. Park, C. F. Agnello, and E. Mathew, *J. Biol. Chem.*, **241,** 769 (1966).
316. M. L. Bender, *Chem. Rev.*, **60,** 53 (1960).
317. J. Teipel and R. L. Hill, unpublished observations quoted in Ref. *208*.
318. H. Horinishi, K. Kurihara, and K. Shibata, *Arch. Biochem. Biophys.*, **111,** 520 (1965).
319. U. W. Kenkare and F. M. Richards, *J. Biol. Chem.*, **241,** 3197 (1966).
320. E. A. Barnard and W. D. Stein, *J. Mol. Biol.* **1,** 339 (1959).
321. W. D. Stein and E. A. Barnard, *J. Mol. Biol.*, **1,** 350 (1959).
322. V. Tomasek, E. S. Severin, and F. Sorm, *Biochem. Biophys. Res. Commun.*, **20,** 545 (1965).

323. E. B. Ong, E. Shaw, and G. Schoellmann, *J. Am. Chem. Soc.*, **86,** 1271 (1964).
324. E. B. Ong, E. Shaw, and G. Schoellmann, *J. Biol. Chem.*, **240,** 694 (1965).
325. G. Kartha, J. Bello, and D. Harker, *Nature*, **213,** 862 (1967).
326. J. Bello, personal communication.
327. G. G. Hammes and H. A. Scheraga, *Biochemistry*, **5,** 3690 (1966).
328. R. W. Woody, M. E. Friedman, and H. A. Scheraga, *Biochemistry*, **5,** 2034 (1966).
329. L.-K. Li, J. P. Riehm, and H. A. Scheraga, *Biochemistry*, **5,** 2043 (1966).
330. J. P. Riehm and H. A. Scheraga, *Biochemistry*, **5,** 99 (1966).
331. H. C. Watson and J. C. Kendrew, *Nature*, **190,** 663 (1961).
332. J. Hermans, Jr., and L. Wan Lu, *Arch. Biochem. Biophys.*, **122,** 331 (1967).
333. J. T. Edsall, P. J. Flory, J. C. Kendrew, A. M. Liquori, E. Nemethy, G. N. Ramachandran, and H. A. Scheraga, *J. Mol. Biol.*, **15,** 399 (1966).
334. E. Breslow, *J. Biol. Chem.*, **239,** 486 (1964).
335. M. F. Perutz, *J. Mol. Biol.*, **13,** 646 (1965).
336. H. C. Watson, J. C. Kendrew, and L. Stryer, *J. Mol. Biol.*, **8,** 166 (1964).
337. J. Jollès, J. Janreugui-Adell, and P. Jollès, *Compt. Rend.*, **258,** 3926 (1964).
338. J. A. Rupley, *Proc. Roy. Soc.*, **167B,** 416 (1967).
339. T. T. Herskovits and M. Laskowski, Jr., *J. Biol. Chem.*, **243,** 2123 (1968).
340. K. Hayashi, T. Shimoda, T. Imoto, and M. Funatsu, *J. Biochem* (*Tokyo*), **64,** 365 (1968).
341. R. Sakakibara and K. Hamaguchi, *J. Biochem.* (*Tokyo*), **64,** 613 (1968).
342. T. A. A. Dopheide and W. M. Jones. *J. Biol. Chem.*, **243,** 3906 (1968).
343a. C. J. Menendez, T. T. Herskovits, and M. Laskowski, Jr., *Biochemistry*, **8,** 5044 (1969).
343b. C. J. Menendez and T. T. Herskovits, *Biochemistry*, **8,** 5052 (1969).
344a. I. Putter, A. Barreto, J. L. Markley, and O. Jardetzky, *Proc. Natl. Acad. Sci.*, *U.S.*, **64,** 1396 (1969).
344b. J. L. Markley, I. Putter, and O. Jardetzky, *Science*, **161,** 1249 (1968).
344c. J. L. Markley, I. Putter, and O. Jardetzky, *Z. Nal. Chem.*, **243,** 367 (1968).
344d. G. C. K. Roberts and O. Jardetzky, *Advan. Prot. Chem.*, **24** (1969).
344e. D. H. Meadows, O. Jardetzky, R. M. Epand, H. Ruterjans, and H. Scheraga, *Proc. Natl. Acad. Sci.*, *U.S.* **60,** 766 (1968).
345. T. Spande, M. Wilchek, and B. Witkop, *J. Am. Chem. Soc.*, **90,** 3256 (1968).
346. P. Cuatrecasas, S. Fuchs, and C. B. Anfinsen, *J. Biol. Chem.*, **243,** 4787 (1968).
347. K. Morihara and K. Nagami, *J. Biochem.* (*Tokyo*), **65,** 321 (1969).

AUTHOR INDEX

Numbers in parentheses are reference numbers and indicate that an author's work is referred to although his name is not cited in the text. Numbers in italics give the page on which the complete reference is listed.

C

D

E

F

G

H

I

M

R

S

Y

Z

SUBJECT INDEX

R

S

T

U

V

X